Wetland
Crisis and Options

The Editor

Goutam Kumar Saha passed B.Sc. (Honours in Zoology) from Burdwan University securing first class first position in the year 1981. In M.Sc. examination too, he stood first amongst the successful candidates in Zoology and received University gold medal. He was awarded his Ph.D degree from Calcutta School of Tropical Medicine and The University of Burdwan. He has published 160 research papers in different journals of International repute and authored 7 books/technical monographs in his credit. He has actively participated in 44 National and 14 International seminars and symposia and extensively visited different countries like Germany, France, Netherlands, Belgium, Italy, China, Singapore, Thailand, Australia, Korea and Canada in connection with his research work. He has successfully carried out a good number of major research and development projects sponsored by national and international funding agencies. He started his professional carrier as a Lecturer in Zoology at Post Graduate Department of Zoology, Darjeeling Govt. College, Darjeeling and at present working as a Professor of Zoology, University of Calcutta. He is an honorary fellow of Indian College of Allergy and Applied Immunology and The Zoological Society, Kolkata. Besides his excellence in the academic field, he is a good sportsman and represented the University as its blue and participated several National games.

Wetland
Crisis and Options

— Editor —

Goutam Kumar Saha

Professor of Zoology
University of Calcutta

2016

Associated Publishing Company®

A Division of

Astral International Pvt. Ltd.

New Delhi – 110 002

Cataloging in Publication Data--DK
Courtesy: D.K. Agencies (P) Ltd. <docinfo@dkagencies.com>

Wetland : crisis and options / editor, Goutam Kumar Saha.
pages cm
Includes bibliographical references and index.

ISBN 978-93-86071-03-3 (International Edition)

1. Wetlands--India. 2. Wetland biodiversity--India. 3. Ecosystem services--India. 4. Wetland conservation--India. I. Saha, Goutam Kumar, editor.

QH541.5.M3W48 2016 DDC 577.680954 23

Published by : **Associated Publishing Company®**
A Division of
Astral International Pvt. Ltd.
– ISO 9001:2008 Certified Company –
4760-61/23, Ansari Road, Darya Ganj
New Delhi-110 002
Ph. 011-43549197, 23278134
E-mail: info@astralint.com
Website: www.astralint.com

Respectfully dedicated to the sweet memory of my beloved father

Late Sri Gopinath Saha

who inspired me in every step of my career

Foreword

Wetlands are among the important habitats for a variety of organisms and have a profound impact on mankind directly or indirectly. As the part of greater human desire for utilizing every possible resource, the wetlands too are not spared. The human interference knowingly or unknowingly in many instances have interrupted the natural cycles occurring in wetlands. Consequently, significant portion of the global biodiversity is at stake. Thus, precise knowledge on proper understanding, optimum harvest, systematic research, proper community outreach with regard to wetlands are the utmost need of the moment for sustainable development and conservation. In this context, the book entitled **"Wetland : Crisis and Options"** by **Prof. Goutam Kumar Saha,** a senior Professor, Department of Zoology, University of Calcutta perfectly fills the void and is expected to enrich our understanding on the wetlands and its importance in maintaining global biodiversity.

The book is robustly conceived and is well-organized. The initial chapters introduce the readers to the rich faunal wealth of Indian wetlands followed by the chapters describing a broad range of services provided by the wetlands. The final chapters critically elaborate the concerns and challenges existing in wetlands and the initiatives undertaken to combat those issues. As the book is written in simple language, the general readers are expected to gain better information on the dynamics and importance of wetlands. The elaborate research work incorporated in this book can be a feast for scholars around the globe and is believed to provide a window for understanding prospects and challenges in the Indian wetlands.

I would like to congratulate **Prof. Saha** for his extensive work in achieving the present form of the book. The hard work of all the contributing authors also deserves appreciations. I am hopeful that this book will receive a huge public attention and be successful in conveying the message pertaining to the roles of wetlands in sustainable resource development and global biodiversity conservation.

Prof. Sugata Marjit

Vice-Chancellor
University of Calcutta
Kolkata

Prelude

Wetland - an interface between terrestrial and aquatic ecosystems are the most precious life-sustaining water resources and play crucial role as ecosystem service provider. They provide food, filter water and offer a unique habitat for many different species. India has a rich variety of wetland habitats due to varied topography and eco-climatic regimes across the country and each of them is ecologically unique, with its own faunal wealth. Although wetlands occupy only two percent of the surface area of earth, yet they are the most productive ecosystems of the world, providing drinking water, fish, fodder, fuel, wildlife habitat, play vital role in global cycling of nutrients, a potential source of carbon sequestration, regulate groundwater replenishment and the quality and quantity of water table, stabilize and protect shoreline from the devastating effects of natural calamities such as hurricanes, cyclones and tsunamis. They also play a vital role in the sustenance of different life forms and overall a safe and resourceful reservoir of biodiversity and often harbors many endemic as well as endangered species. Wetlands offer a welcome pit-stop, offering protection and food before the migratory birds on their way to final destination. Wetlands often provide livelihood for numerous people living around them. Many species of wetland flora and fauna are often harvested by the local communities for personal and commercial use. Above all, significantly influences the religious, historical or archeological value to many cultural reformations around the world. Thus, wetlands contribute in no small way to our quality of life-indeed, to our very survival. Inspite of their multifaceted use and importance, they are still undervalued, naturally less attended. However, the concern is that over half of the wetlands in the world have been lost in the last century and the remaining wetlands have been degraded to varying degrees because of the adverse influence of human activities and continually subjected to various anthropogenic disturbances. Many wetlands have been drained and transformed by various anthropogenic activities, *viz.* unplanned urbanization, agricultural extension, industrial development, road construction, impoundment, over extraction of resources, etc. Wetlands are thus invaluable as well as a finite natural resource to man's varied activities. As a result of increased human demand for water, the wetland habitats are being subjected to increased levels of human disturbance around the world. Moreover, the value of wetland waters gets increased as ecosystems become more stressed and their goods and services are declined. As a matter of fact, conservation aspects of wetland ecosystem and its biodiversity deserve special attention. Considering their vital role as ecosystem service provider, the present compilation has been initiated to highlight different aspects of wetland ecosystem and to generate awareness among students, stakeholders and common people for their proper management as well as sustenance of this unique habitat. The present compilation is enriched with 27 lead articles contributed by the authorities in their respective field and is meant to serve as a resource material for both undergraduate and postgraduate students, researchers, academicians, ecologists and environmentalists, nature lovers as well as common people. The present compilation provides diversified information ranging from microscopic zooplankton to charismatic mega forms including water hyacinth, macrophytes, aberrant groups, aquatic insects, molluscs, both migratory and resident birds, mammals associated with wetland habitats. The book also highlights the role of wetlands as potential ecosystem service provider. At the same time discusses the problems and crisis in and around functioning of this unique ecosystem in greater details and suggest conservation measures. I express my deep sense of gratitude to the contributors for their whole hearted cooperation in this endeavour. I am extremely grateful to prof. Sagata Marjit, Vice-chancellor, University of Calcutta, Kolkata for providing the foreword of the book. I am indebted to my beloved teacher Professor Subrata Roy, Former Professor of Zoology and Dean, Faculty of Science, University of Burdwan for critically going through all the manuscripts and providing valuable suggestions. Special thanks are also due to Dr. Gautam Aditya, Ms. Swarnali Mukherjee and Dr. Avisek Basu for their editorial assistance. I am thankful to Mr. Prateek Mittal of Astral International Pvt. Ltd., New Delhi for timely publication of the volume.

Goutam Kumar Saha

Contents

List of Contributors

1. DR. ABHISEK BASU
   Assistant Professor, Department of Zoology, Victoria Institution (College), Kolkata
2. DR. AMBARISH MUKHERJEE
   Professor, Department of Botany, The University of Burdwan, Burdwan
3. MR. ANINDYA SUNDAR BHUNIA
   Research Scholar, Aquatic Toxicology Laboratory, Department of Zoology, University of Calcutta, Kolkata
4. DR. ANIRBAN SINHA
   Research Scholar, Ecology and Wildlife Biology Laboratory, Department of Zoology, Maulana Azad College, Kolkata
5. DR. ANIRUDDHA MUKHOPADHAYAY
   Professor, Department of Environmental Science, University of Calcutta, Kolkata
6. DR. APURBA RATAN GHOSH
   Professor, Department of Environmental Science, The University of Burdwan, Burdwan
7. DR. ASISH KUMAR GHOSH
   Former Director, Zoological Survey of India and Director, Centre for Environment and Development, Kolkata
8. DR. ASOK KANTI SANYAL
   Former Additional Director, Zoological Survey of India, Kolkata and Chairman, West Bengal Biodiversity Board, Kolkata

9. DR. AVIJIT MAJUMDAR

   Associate Professor, Department of Zoology, The University of Burdwan, Burdwan

10. MS. BARNALI SARKAR

    Research Scholar, Entomology and Wildlife Biology Research Laboratory, Department of Zoology, University of Calcutta, Kolkata

11. DR. DEBNATH PALIT

    Assistant Professor, Department of Botany, Durgapur Government College, Durgapur

12. MR. DIPENDRA SHARMA

    Research Scholar, Entomology and Wildlife Biology Research Laboratory, Department of Zoology, University of Calcutta, Kolkata

13. MR. DIBYENDU SAHA

    Research Scholar, Department of Zoology, The University of Burdwan, Burdwan

14. DR. GAUTAM ADITYA

    Associate Professor, Ecology Laboratory, Department of Zoology, University of Calcutta, Kolkata

15. DR. GOUTAM KUMAR SAHA

    Professor of Zoology, Entomology and Wildlife Biology Research Laboratory, Department of Zoology, University of Calcutta, Kolkata

16. SK. HABIBUR RAHAMAN

    Research Scholar, Department of Zoology, The University of Burdwan, Burdwan

17. MR. JOY CHAKRABORTY

    Research Scholar, Department of Zoology, The University of Burdwan, Burdwan

18. DR. K. A. SUBRAMANIAN

    Scientist D, Zoological Survey of India, Kolkata

19. DR. KAUSHIK DEUTI

    Herpetology Division, Zoological Survey of India, Kolkata

20. MS. LAK TSHEDEN THEENGH

    WWF-India, Khangchendzonga Landscape Programme, Gangtok, Sikkim

21. DR. MALAY KANTI DEV ROY

    Crustacea Section, Zoological Survey of India, Kolkata

22. DR. MITALI RAY

    Scientist, Aquatic Toxicology Laboratory, Department of Zoology, University of Calcutta, Kolkata

23. DR. MOUMIT ROY GOSWAMI

Assistant Professor, Department of Environmental science, Netaji Nagar College for Women, Kolkata

24. DR. MUNMUN CHAKRABORTY

Research Scholar, Aquatic Bioresource Research Laboratory, Department of Zoology, University of Calcutta, Kolkata

25. DR. NEPAL CHANDRA NANDI

Former Additional Director, Zoological Survey of India, Kolkata

26. DR. PRANABES SANYAL

School of Oceanographic Studies, Jadavpur University, Kolkata. Former Field Director, Sunderban Tiger Reserve

27. DR. PARTHA SARATHI GHOSE

WWF-India, Khangchendzonga Landscape Programme, Gangtok, Sikkim

28. DR. PARTHIBA BASU

Associate Professor, Ecology Research Unit, Department of Zoology, University of Calcutta, Kolkata

29. DR. PAULAMI MAITY

Assistant Professor, Department of Zoology, Lady Brabourne College, Kolkata

30. DR. PRANTIK HAZRA

Research Scholar, Ecology and Wildlife Biology Laboratory, Department of Zoology, Maulana Azad College, Kolkata

31. MS. PRIYADARSHINEE SHRESTHA

WWF-India, Khangchendzonga Landscape Programme, Gangtok, Sikkim

32. DR. RINA CHAKRABORTY

Former Joint Director, Zoological Survey of India, Kolkata

33. DR. SAMIR BANERJEE

Former Professor of Zoology, University of Calcutta, Kolkata

34. MR. SANTANU GUPTA

Research Scholar, P. G. Department of Conservation Biology, Durgapur Govt. College, Durgapur

35. MR. SANTANU MITRA

Crustacea Section, Zoological Survey of India, Kolkata

36. DR. SHREYA BRAHMA

Research Scholar, Entomology and Wildlife Biology Research Laboratory, Department of Zoology, University of Calcutta, Kolkata

37. DR. SAILESH CHATTERJEE

Professor, Department of Forest Entomology, Birsa Agricultural University, Ranchi, Jharkhand

38. DR. SAJAL RAY

Professor of Zoology, Aquatic Toxicology Laboratory, Department of Zoology, University of Calcutta, Kolkata

39. MS. SARMISTHA SAHA

Research Scholar, Ecology Research Unit, Department of Zoology, University of Calcutta, Kolkata

40. DR. SOUMYAJIT BANERJEE

Research Scholar, Entomology and Wildlife Biology Research Laboratory, Department of Zoology, University of Calcutta, Kolkata

41. DR. SUBHENDU MAZUMDAR

Assistant Professor, Shibpur Dinobundhoo Institution (College), Howrah

42. MR. SUBINOY MONDAL

Research Scholar, Department of Environmental Science, The University of Burdwan, Burdwan

43. DR. SUMIT HOMECHAUDHURI

Professor of Zoology, Aquatic Bioresouce Research Laboratory, Department of Zoology, University of Calcutta, Kolkata

44. MS. SRIMOYEE BASU

Research Scholar, Entomology and Wildlife Biology Research Laboratory, Department of Zoology, University of Calcutta, Kolkata

45. MR. TAPAN SAHA

Scientist, Institute for Environmental Studies and Wetland Management, Kolkata

46. DR. TARAK NATH KHAN

Associate Professor, Ecology and Wildlife Biology Laboratory, Department of Zoology, Maulana Azad College, Kolkata

Part I

Wetland: Faunal Wealth

Chapter 1

Indian Wetlands: Problems, Prospects and Potentials

Goutam Kumar Saha

Entomology and Wildlife Biology Research Laboratory, Department of Zoology, University of Calcutta, Kolkata

Introduction

Wetlands are the unique transitional condition or *ecotone* zone between dry land and water bodies and an interface between terrestrial and aquatic ecosystems, consisting chiefly of hydric soil saturated with water, either permanently or seasonally and support characteristic flora and fauna. According to the Ramsar Convention "wetlands are areas of submerged or water saturated lands, both natural or artificial, permanent or temporary, with water that is static or flowing, fresh or brackish, or salty including areas of marine water, the depth of which at low tide does not exceed six meters." This definition is widely used in the Indian context and as such the term wetland includes all inland water bodies even rice fields. Many wetlands show a seasonal appearance with free overlying water during the rainy season and often become almost marshy or dry during summer and the periodic fluctuations in water level influence the productivity and species composition of wetland community. There are various types of wetlands, like marshes, swamps, peat lands, flood plains, etc. and are mostly freshwater ecosystems. The natural wetlands of this country also include temporary or permanent ponds, lakes, floodplain marshes and swamps of freshwater nature. Nevertheless, there are some wetlands which have high salt concentrations, such as inland saline marshes, coastal salt marshes and mangrove swamps. Besides, the estuaries and coastal lagoons are also considered as brackish water wetlands. There are also numerous man-made wetlands formed in areas influenced by human activities. Numerous need based small reservoirs have been built up for the purpose of irrigation, water supply, electricity, fisheries and flood control, in the drier regions add to the countries wetland wealth. Substantial waterlogged areas with wetland vegetation formed by spillover from irrigation channels or reservoirs are the examples of such man-made wetlands. In India, there are about 1,550 large reservoirs, covering a total area of more than 14,500 sq km and 100,000 small and medium reservoirs that cover another 11,000 sq km. Innumerable fishponds and extensive paddy fields are also considered as wetlands, which are often modified from the earlier marshes. The Millennium Ecosystem Assessment (2005) study estimated that wetlands cover 7 per cent of the earth's surface and are amongst the most productive ecosystems on the earth, deliver 45 per cent of its natural productivity and provide many important services and goods to human society (Ghermandi *et al.*, 2010; ten Brink *et al.*, 2012). It recycles nutrients, purifies and provides drinking water, fodder and fuel, reduces flooding, recharges groundwater, facilitates aquaculture, buffers the shoreline against erosion, provides a habitat for wildlife and offers avenues for recreation. Globally, 1.5-3 billion people depend on wetlands as a source of drinking water as well as food and livelihood security. At the same time, wetlands are one of the most explored and threatened habitats of the world and are considered as wastelands, particularly in Indian context leading towards an unperceived ecological crisis. Since they are ecologically sensitive and adaptive systems, they are under continuous tremendous stress due to rapidly expanding human population, large scale changes in land use pattern, rapid urbanization, industrialization, agricultural intensification, burgeoning development projects and improper use of watersheds, resulting in a steady decline in their hydrological, economic and ecological functioning.

Wetlands of India

Wetlands exhibit enormous diversity according to their genesis, geographical location, water regime and chemistry, dominant species, soil and sediment characteristics. The diverse eco-climatic regimes across the country resulted in a variety of wetland systems ranging from high altitude cold desert wetlands to hot and humid wetlands in coastal zones with its diverse flora and fauna. First survey undertaken by the Department of Science and Technology in 1976, revealed a total of 1,193 wetlands in India, covering an area of 39045 sq km, of which 572 were natural, 542 were man-made, seven included both natural and man-made habitats and the remainder were unclassified. Of these wetlands, 938 were freshwater, 134 brackish water and 19 coastal in nature. Most of the wetlands (418) were used for irrigation purposes, 369 sites for fishing, 90 for fish culture, 161 for grazing, 30 for waste disposal and 19 for reed-gathering. However, there are many responsible for multiple uses, 138 sites are used both for fishing and irrigation. Subsequent second survey revealed the total area of wetlands (excluding rivers) in India to be 5,82,860 sq km (18.4 percent of the country's area), 70 percent of which comprises areas under paddy cultivation.

Characteristic Features of Wetland

Climate

The environment of the wetlands is moist with relatively higher amount of humidity and receives around 70 percent of the annual rainfalls during the southwest monsoon. Nevertheless, the temperature, rainfall and other climatic parameters largely vary depending on the location of the wetland. In addition, frequent variations are also noticed in the same wetland in different years. Dissolved oxygen (DO) content of the wetlands usually varies between 4-10 ppm.

Vegetation

Wetlands in India support more than 1500 aquatic and wetland species. Besides, more than 40 percent algae and over 12 percent of the Pteridophytes of India are found in the freshwater wetlands. Types of wetland plants are also remarkably diverse owing to the large scale variation in topography and huge climatic gradient across the country, as well as, due to local variation in rainfall. However, all of these wetland plants have broadly been grouped under following categories

1. Free-floating hydrophytes (*Eichhornia crassipes, Hygroryza aristata, Lemna perpusilla, Pistia stratiotes* etc.)
2. Suspended hydrophytes (*Ceratophyllum demersum, Eriocaulon setaceum, Hydrilla verticillata*, etc.)
3. Submerged-anchored hydrophytes (*Blyxa auberii, Cabomba caroliniana, Ottelia alismoides, Vallisneria* sp. etc.)
4. Anchored hydrophytes with floating leaves (*Aponogeton natans, Nelumbo nucifera, Sagittaria guayanensis* etc.)
5. Anchored hydrophytes with floating shoots (*Geissaspis cristata, Ipomoea aquatic, Ludwigia adscendens* etc.)
6. Emergent-anchored hydrophytes (*Acorus calamus, Aeschynomene aspera, Bacopa monnieri, Hydrocera triflora, Limnophila aromatic, Typha* spp. etc.).

Faunal Richness

It is estimated that at least 12280 faunal species (16 percent of the known faunal diversity) belonging to various taxa are found in Indian freshwater environment. Majority of indigenous fishes are largely wetland dependent - 131 species under 67 genera, 28 families and 10 orders of cartilaginous fishes (Kar *et al.*, 2000), whereas 2415 species of bony fishes belonging to 902 genera, 226 families and 30 orders (Prasad *et al.*, 2002) are reported from Indian region, with high degree (8.75 per cent) of endemism. Mangroves serve as critical hatching and nursery grounds and coral reefs are good food source for many tropical fish species. Frogs are the major amphibian species in most wetlands as they require both terrestrial and aquatic habitats - the tadpoles control algal populations and the adults forage on insects and serve as an indicator of wetland health. Saltwater or estuarine crocodiles (*Crocodylus porosus*) are found in the brackish waters along the coast, estuaries, and mangroves, where as marsh crocodiles or *muggers* (*Crocodylus palustris*) and gharials (*Gavialis gangeticus*) are found in some rivers. Various kinds of snakes, lizards and turtles are also fairly common in wetland habitats. Many species of small mammals along with some large herbivores like wild water buffalo, swamp deer, Manipur brow-antlered deer etc. live within and around wetlands. In addition to these vertebrate fauna, many species of invertebrates including various insects are found in wetlands, submerged in water or in the soils, as well as on the surface water of the wetlands. Besides, wetlands are some of the most diverse wildlife habitats on earth.

Inspite of available water and nutrients, flora and fauna of wetlands require to overcome the periods of both flood and drought conditions, often need to survive at reduced oxygen level in both the soil and water, as well as, contend with accumulated salts or other pollutants. At the same time, finding food, shelter and successful mating is also challenging. Therefore, wetland plants and animals develop a variety of physical and behavioral adaptations to deal with the particular set of challenges in these unique biomes.

Potentials

Wetlands are the most precious life-sustaining water resources, each of them is ecologically unique and provide a range of vital ecological services and goods and significantly contribute towards a healthy environment and are often designated 'kidneys of the landscape'. They provide food, filter water and offer a unique habitat for many different species. Although wetlands occupy only two percent of the surface area of earth, yet they are the most productive ecosystems of the world and are very important in terms of their ecological, economic, cultural, spiritual and recreational values. They play vital role in global cycling of sulfur, nitrogen, phosphorus and a potential source of carbon sequestration and also play a vital role in the sustenance of different life forms. Valuable functions of a wetland include recycling of nutrients, water purification, flood control, maintenance of stream flow and groundwater recharge. Besides, wetlands also serve in providing drinking water, fish, fodder, fuel, wildlife habitat, control rate of runoff in urban area, buffer shorelines against erosion and recreation to the society. Some of the important ecological services rendered by the natural wetlands are:

1. Multiple-use Water Services

Wetlands such as tanks, ponds, lakes, and reservoirs have long been providing multiple-use water services which include water for irrigation, domestic needs, fisheries, groundwater recharge; flood control and recreational uses.

2. Storage Reservoir and Flood Attenuation

Wetlands act as natural rainwater harvesting sites through collection of precious rainwater within it and distribute it as and when required. Wetlands in close vicinity of rivers also act as buffers to control flood and river flow. When the level of river rises, water flows into the wetlands and similarly when river water level decreases, water from these wetlands gushes into the river thus maintaining the average flow of the river. Wetlands also provide an important ecological service in storing flood waters. Wetland systems of floodplains are formed from major rivers downstream and these floodplains actually act as natural storage reservoirs. During flood, the depth and speed of excess water gets reduced as they spread over wide area of these floodplains and thus prevent sudden and damaging effects of flood downstream.

3. Groundwater Replenishment

Drinking water and irrigation of crops across the country largely depends on groundwater. In wetlands, water usually percolates slowly through porous sediments, layers of soils and overlying rocks into aquifers. Some other wetland systems are directly linked to groundwater. Thus wetlands play crucial role in groundwater replenishment and also in regulating the quality and quantity of water found beneath the ground. Wetlands not only act as groundwater recharging sites, sometimes they can even provide easy access to groundwater by converting it into surface water.

4. Water Purification

Wetlands improve water quality by acting as natural water purification systems, removing silt and absorbing nutrients and toxins. Purification of water is rendered by the vegetations in two ways- through nutrient retention and sediment traps. Many wetland flora absorb certain nutrients in some of their body parts and those nutrients are retained in the system until the plant dies or harvested. Besides, reed beds and submerged vegetations present in the wetlands act as physical barriers; slow down the water flow and trap nutrients in the sediments. Many species of algae and plants have remarkable capacity of accumulating heavy metals in their mucilage and leaves and are known as hyper-accumulators. Some hydrophytes also have the capacity to remove toxic substances and heavy metals. Water hyacinth (*Eichhornia crassipes*), duck weeds (*Lemna* spp.) and water ferns (*Azolla* spp.) store iron and copper commonly found in wastewater. Many fast-growing emergent wetland plants such as cattail (*Typha* spp.) and reed plants (*Phragmites* spp.) also aid in the role of heavy metal up-take. Animals such as the oyster can filter more than 200 liters (53 gallons) of water per day while grazing for food, removing nutrients, suspended sediments, and chemical contaminants in the process.

5. Shoreline Stabilization and Protection

The devastating effects of natural calamities such as hurricanes, cyclones and tsunamis may be minimized through wetlands. Tidal and inter tidal wetlands protect and stabilize coastal zones. Similarly coral reefs and mangroves also act as a protective barrier along the coastline of this country. Mangroves absorb the impact of sea storms and significantly decrease the near surface wind speed. Similarly, the coral reefs, tidal and inter-tidal wetlands effectively reduce the speed and height of tidal surges and storm waters.

6. Carbon Sequestration

Swamps, mangroves, peat lands, mires and marshes play an important role in carbon cycle. Though wetlands contribute about 40 per cent of the global methane (CH_4) emissions, they have the highest carbon (C) density among terrestrial ecosystems and relatively greater capacities to sequester additional carbon dioxide (CO_2) (Pant and Verma, 2010). Wetlands sequester carbon through high rates of organic matter inputs and reduced rates of decompositions as proposed by Pant *et al.* (2010). Wetland soils may contain as much as 200 times more carbon than its vegetation. However, drainage of large areas of wetlands and their subsequent cultivation at many places had made them a net source of CO_2. Restoration of wetlands can reverse them to a sink of atmospheric CO_2 (Lal, 2008). As estimated by IPCC (2000) carbon sequestration potential of restored wetlands (over 50 year period) is about 0.4 tonnes C/ha/year. In India, coastal wetlands are playing a major role in carbon sequestration. Overall, mangroves are able to sequester about 1.5 metric tone of carbon per hectare per year, and the upper layers of mangrove sediments have high carbon content, with conservative estimates indicating the levels of 10 per cent (Kathiresan and Thakur, 2008). However, mangroves were also found to be emitting methane (CH_4), one of the primary greenhouse gases, which was around 19 per cent of their carbon sequestration potential.

7. Reservoir of Biodiversity

Wetlands of India, estimated to be 58.2 million hectares, are important repositories of aquatic biodiversity and often harbors many endemic species. Some vertebrates and invertebrates depend on wetlands for their entire life cycle while others only associate with these areas during particular stages of their life. Several wetlands also serve as important nursery areas for many commercial fish stalks. Species richness and productivity are higher in many mudflats, salt marshes, mangroves and sea grass beds. Freshwater wetlands alone support 20 percent of the known range of biodiversity in this country (Deepa and Ramachandra, 1999). In India, lakes, rivers and other freshwater bodies support a large diversity of biota representing almost all taxonomic groups. Wetlands support specialized flora and fauna and serve as suitable habitat for fish and other aquatic fauna. In addition, wetlands are important feeding, breeding and nesting sites for a number of endangered species. It is a paradise of bird watchers and many birds find hostage in wetlands all around the world. Of the 310 wetland bird species in India, 51 (16 per cent) are Threatened of which 34 are Globally Threatened (four Critical, seven Endangered, 23 Vulnerable); 16 Near Threatened (NT) and one Data Deficient (DD). The total numbers of aquatic plant species exceed 1200 and they provide a valuable source of food, especially for waterfowl (Prasad *et al.*, 2002). The freshwater ecosystems of Western Ghats alone support 290 species of fishes, 77 species of molluscs, 171 species of odonates, 608 species of aquatic plants and 137 species of amphibians within its periphery of 136,800 km^2 area. Out of these, almost 53 per cent of fish, 36 per cent of mollusc, and 24 per cent of aquatic plant species are endemic to this region (Molur *et al.*, 2011). Moreover, abundance of invertebrates in the mud of the inter-tidal mudflats often serve as major foraging ground for many species of migratory avifauna, particularly waders.

8. A Refuge for Migrating Bird

When winter sets in across the northern hemisphere, it triggers the most fascinating mass movement of countless birds over vast distances. The wetlands serve as 'stop-over site' for many migratory birds, offering protection and food before the birds continue on to their final destination and act as refuge for various species of waterfowls.

9. Wetland Products

Wetlands often provide livelihood for numerous people living around them. Many species of wetland flora and fauna are often harvested by the local communities for personal and commercial use. Cultivation of paddy, jute and pisciculture are possibly three most important and traditional practices associated to the wetland ecosystems. Fisheries, in particular, provide great economic support to people and ensure livelihood of millions. Wetlands yield fuel wood for cooking, thatch for roofing, fibres for textiles and paper making, and timber for building. Some of the major supplementary vegetables include water spinach, helencha, alligator weed, vegetable fern, taro, marsh barbel, Asiatic pennywort, narenga, waterlily etc. Some medicinal plants of wetlands are brahmi (*Bacopa monnieria*), gotu kola (*Centella asiatica*), sweet flag or *vacha* (*Acorus calamus*), watercress (*Nasturtium officinale*), lady's tresses orchid (*Spiranthes* spp.). Medicines are extracted from their bark, leaves, and fruits. Wetland plants also provide tannins and dyes, used extensively in leather industry. Besides, *Hydrilla* spp., *Aponogiton* spp., *Vallisneria natans*, etc. are often used as aquarium plants. However, overexploitation of some of these commercially important products has emerged as one of the major problem for the sustainable use of wetlands.

10. Pollution Abatement

Wetlands act as a sink for contaminants in many agricultural and urban landscapes. From an economic perspective too, wetlands have been suggested as a low cost measure to reduce point and non-point pollution (Bystrom *et al.*, 2000). Wetlands are polluted through agricultural runoff and discharge of untreated sewage and other waste from urban areas. Under normal conditions, wetlands do retain pollutants from surface and sub-surface runoff from the catchment and prevent them from entering into streams and rivers (Verhoeven *et al.*, 2006). However, because of increased urbanization and land use changes, the nutrient loading in wetlands far exceed their capacity to retain pollutants and remove them through different ways and adversely affects the wetland water quality and its biodiversity. Moreover, the role of wetland plants in ameliorating heavy metal pollution both in a microcosm and natural condition is well established (Dhir *et al.*, 2009). For example, *Typha*, *Phragmites*, *Eichhornia*, *Azolla*, and *Lemna* are some of the identified potent wetland plants for heavy metal removal (Rai, 2008). Several workers claimed that constructed wetlands are considered to be a viable option for treatment of municipal waste water and in pollution abatement on experimental scale (Billore *et al.*, 1999; Kaur *et al.*, 2012).

11. Climate Change Mitigation

Wetlands perform two-fold functions in mitigating the climate change. On one hand various wetlands, particularly tropical mangroves and coastal salt marshes act as carbon sinks. It has been estimated that globally the wetlands store approximately 44.6 Tg of carbon every year. On the other hand, prevent and reduce the devastating effects of storms, tidal surge and sea-water rise through their ability to store and regulate water.

12. Avenues for Recreation

Wetlands everywhere provide important leisure facilities - canoeing and fishing, shell collecting and bird watching, swimming and snorkeling, hunting and sailing.

13. Cultural Value

Throughout history humans have gathered around wetlands and these areas have played a pivotal role in human development including their religious, cultural, historical or archeological values around the world.

Wetlands at Risk

Inspite of their multifaceted use, benefits and importance, they are still undervalued, naturally less attended and thus often designated as wasteland. The main concern is that over half of the wetlands in the world have been lost in the last century and the remaining wetlands have been degraded to varying degrees because of the adverse influence of human activities. As a result of increased human demand for water, the wetland habitats are being subjected to increased levels of human disturbance around the world. Freshwater wetlands are much utilized, depended upon and exploited ecosystems in terms of sustainability and well-being (Molur *et al.*, 2011). In Asia alone, about 5000 km^2 of wetland area are lost annually to agriculture, dam construction, and other uses (McAllister *et al.*, 2001). Further, dependence on water and other resources in this environment has initiated enormous pressures on the ecosystem, resulting loss in species diversity and populations (Molur *et al.*, 2011). As a result many wetland dependent species including 21 per cent of bird species; 37 per cent of mammal species; and 20 per cent of freshwater fish species are either extinct or globally threatened (MEA, 2005).

The major threats outlined include habitat destruction and encroachments through drainage and landfill for infrastructure development and rapid urbanization, indiscriminate agricultural expansion, over extraction of resources, specially fish, discharge of waste water and industrial effluents, harmful fertilizer and pesticide runoff, uncontrolled siltation, weed infestation and indiscriminate livestock grazing. Unfortunately, the abuse of wetlands reduces their ability to perform useful functions such as water retention and flood control, to supply services and, in many cases, valuable products. Besides, implementation of remedial measures incurs heavy financial and environmental costs. As a matter of fact, conservation aspects of wetland ecosystem and its biodiversity deserve special attention.

☆ **Urbanization and land use changes:** During last few decades a substantial increase in total population in India has taken place, with an average decadal growth rate of around 22 per cent and urban population, in particular has increased eightfold (Bassi and Kumar, 2012). This magnitude of growth exerted tremendous burden and pressure on wetlands and flood plain areas to meet the growing demand for water, food and shelter. Accordingly, total cultivated land in India has been increased from about 129 to 156 m ha and such an increase in area for both agricultural and non-agricultural use was at the cost of conversion of flood plain areas, primary forests, grasslands and associated freshwater ecosystems (Zhao *et al.*, 2006) and a large portions of so called wasteland have been reclaimed for agriculture purpose. Impact of urbanization is equally alarming on natural water bodies in the cities due to urban sprawl (Ramachandra and Kumar, 2008). Majority of the wetlands, located particularly in urban and peri-urban landscapes are being converted to industrial, agricultural and residential use. Ironically, the mangrove forests are being cleared to make room for agricultural land, human settlements and infrastructure (such as port, harbours) and industrial areas. More recently, clearing for tourist developments, shrimp aquaculture, and salt farms has given priority. Construction of dams and irrigation channels alter the water regimes of the rivers and other associated wetlands both quantitatively and qualitatively, reduce the amount of water reaching mangrove forests, increasing the salinity level of water, make them unfit to survive.

☆ **Over extraction of bioresources:** Over extraction of wetland resources and products, overfishing in particular has posed a serious threat to the sustenance of this resourceful habitats throughout the globe. The ecological balance of food chains and species composition of fish communities can also be altered. Overuse of mangrove trees as firewood, construction wood, wood chip and pulp production, charcoal production and animal fodder also threatens the future of the forests.

☆ **Destruction of coral reefs:** Coral reefs provide the first barrier against currents and strong waves. When they are destroyed, the stronger-than-normal waves and currents reaching the coast can undermine the fine sediment in which the mangroves grow. This can prevent seedlings from taking root and wash away nutrients essential for mangrove ecosystems.

☆ **Pollution:** Water in most of the rivers, lakes, streams and wetlands has been heavily degraded, mainly due to agricultural runoff of pesticides and fertilizers, and industrial and municipal wastewater discharges, resulting widespread eutrophication (Liu and Diamond, 2005). As a result of intensification of agricultural activities over the past few decades, fertilizer consumption in India has increased to a considerable extent and 10-15 per cent of the nutrients added to the soils through fertilizers eventually find their way to the surface water system (Indian Institute of Technology, Kanpur, 2011). High nutrient contents stimulate undesirable algal growth, leading to eutrophication of surface water bodies. Besides, runoff from agricultural fields is the major source of non-point pollution for the Indian rivers particularly flowing through Indo-Gangetic plains (Jain *et al.*, 2007), resulting contamination of drinking water which is more expensive to purify for drinking or other industrial uses and considerable reduction in fish production (Verhoeven *et al.*, 2006). Besides, untreated wastewater in most of the Indian urban centres also contributes significantly to pollution of water bodies to a considerable extent (Central Pollution Control Board, 2009).

☆ **Invasive species:** Alien invasive species have had severe impacts on local aquatic flora and fauna, and can upset the natural balance of an ecosystem.

☆ **Climate change:** Global climate change is expected to become an important indicator for loss and change in wetland ecosystem (UNESCO, 2007). This is specially important for Indian subcontinent where the mean atmospheric temperature and frequency of intense rainfall events has increased, while the number of rainy days and total amount of annual precipitation have decreased due to increase in the concentration of greenhouse gases such as CO_2, CH_4 and N_2O in the atmosphere (Bates *et al.*, 2008). Empirical evidences suggest that high altitude wetlands and coastal wetlands (including mangroves and coral reefs) are some of the most sensitive areas to climate change (Patel *et al.*, 2009). For instance, climate change induced rising level of glacial fed high altitude lakes, such as Tsomoriri in Ladakh, has submerged its islands inside the lake, used by endangered migratory birds like the black-necked crane and barheaded goose for their breeding purposes (Chandan *et al.*, 2008). In case of the coastal wetlands such as Indian part of Sunderban mangrove, rising sea surface temperature and sea level rise due to thermal expansion, could affect the fish distribution and lead to the destruction of significant portion of mangrove ecosystem and would diminish their critical role as natural buffers against tropical cyclones resulting in loss of lives and livelihoods (UNESCO, 2007; CSE, 2012). Mangrove forests are extremely sensitive to current rising sea levels caused by global warming and climate change as they require stable sea levels for long-term survival. The inland natural wetlands, especially those in arid and semi-arid regions, will be impacted through alteration in its hydrological regime due to changes in precipitation, runoff, temperature and evapo-transpiration (Patel *et al.*, 2009). Climate change induced rising temperature and declining rainfall pattern alter the freshwater inflows to wetland ecosystems (Bates *et al.*, 2008; Erwin, 2009) and can aggravate the problem of eutrophication, leading to algal blooms, fish kills, and dead zones in the surface water (Gopal *et al.*, 2010). As per estimates, India will lose about 84 per cent of coastal wetlands and 13 per cent of saline wetlands with climate change induced sea water rise of 1 m (Blankespoor *et al.*, 2012). As a result there will be adverse consequences on wetland species, especially those that cannot relocate to suitable habitats, as well as migratory species that rely on a variety of wetland types throughout their life cycle.

Conservation Initiatives and Policy Statement

Across the globe the wetlands are under continuous threat as these habitats are among the most heavily exploited, impacted and degraded of all ecological systems. Half of them have already been destroyed in the past 100 years alone. More specifically, our current over-use of fresh water resources and projected future increases pose serious threats not only to the continued maintenance and functioning of wetland ecosystems and their biological diversity, but to the essence of human well-being. In reality, management of wetlands has received inadequate attention in the national water sector agenda. In India, wetlands are undervalued and hardly drew attention in water resources management and development plans. Though India is a signatory to both Ramsar Convention on Wetlands and the Convention of Biological Diversity and the primary responsibility for the management of these ecologically sensitive ecosystems is in the hands of Government of India itself, there seem to be no clear cut regulatory framework for conservation of wetlands.

Surprisingly, till date there is no separate legal provision for wetland conservation in India. However, it is indirectly regulated by number of related legal instruments like Indian Fisheries Act 1857, Indian Forest Act 1927, Wildlife (Protection) Act 1972, Water (Prevention and Control of Pollution) Act 1974, Environmental (Protection) Act 1986, Wildlife (Protection) Amendment Act 1991, Biodiversity Act 2002 etc. Provisions under these acts range from protection of water quality and notification of ecologically sensitive areas towards conserving, maintaining, and augmenting the floral, faunal and avifaunal biodiversity of the country's aquatic bodies.

Until recently, the policy statement for wetland conservation in India was virtually non-existent and ground level action plan was initiated by the international commitments made under Ramsar Convention through identification of few selected wetlands and implementation of some basic management practices to restore their quality. During 1985-1986, National Wetland Conservation Programme (NWCP) was launched in close collaboration with concerned State Government, involving only designated Ramsar sites and stringent measures were undertaken to prevent further degradation and shrinkage of the identified water bodies due to encroachment, siltation, weed infestation, catchment erosion, agricultural run-off carrying pesticides and fertilizers, and wastewater discharge. Subsequently in 1993, National Lake Conservation Plan (NLCP) was initiated to focus on lakes particularly those located in urban and peri-urban areas which are subjected to anthropogenic pressures. Since 1995, National River Conservation Plan (NRCP) is operational, with an objective to improve the water quality of the major Indian rivers through the implementation of pollution abatement measures. Lately, the National Environmental Policy 2006 first of all recognized the importance of wetlands in providing numerous ecological services (MoEF, 2007) and realized the need of a legally enforceable regulatory mechanism for identified wetlands, to prevent their degradation and enhance their conservation (Dandekar *et al.*, 2011). Further, based on the directives of National Environment Policy, 2006 and recommendations made by National Forest Commission, Government of India notified the Wetlands (Conservation and Management) Rules, 2010 along with the formation of Central Wetlands Regulatory Authority (CWRA) and Expert Group on Wetlands (EGOW) to monitor management action plans (MoEF, 2012). In the recent past, National Water Policy, 2012 has been drafted with a view to recognize the need for conservation of river corridors and water bodies (including wetlands) in a scientifically planned manner. However, the efforts are too meager as only few selected wetlands like wetlands selected under Ramsar Convention, wetlands in ecologically sensitive and important areas, wetlands recognized as UNESCO World Heritage site; high altitude wetlands etc. are being considered under the management preview. Thus, a great majority of the wetlands continue to be ignored in the policy process. It is essential to identify other ecologically important wetlands, implement and execute management interventions in order to maintain their ecological integrity and wise use of these underutilized and underestimated resources.

Gaps and Priority Areas

Considering the great ecological importance of wetlands in maintaining the stability of the nature and as crucial for the very survival of human and their economic well-beings, more attention has to be given to the conservation and management of these precious habitats. Despite the implementation of some activities and measures for their conservation and proper management, there are still some gaps and constraints with special reference to policy statements and legal framework, floral and faunal inventory, diversity at species and genetic level, coordination among stakeholders and involvement of locals. To overcome all this fundamental attributes and to ensure proper and sustainable management of wetlands and its rich biodiversity, emphasis or priority should be given to the following strategies in an integrated manner, rather than addressed them individually.

- ✰ Priority should be given to prepare a detail and reliable inventory and scientific documentation of the wetland bio-resources as well as physico-chemical features related to the wetland eco-health including their present status and traditional uses and values.
- ✰ To formulate an integrated policy and plan at the national level considering the productivity of the biomass, carrying capacity of the habitat and need of the local people.
- ✰ To develop a national level comprehensive legislation for conservation, wise use and sustainable management of wetlands and its biodiversity.
- ✰ To formulate mechanism for the execution of primary objectives of international conventions and acts and regulations properly.
- ✰ To identify critical wetlands and suggest special management programme for their sustenance.
- ✰ To strengthen resource base, specially human resource and expertise in the existing institution and local organizations to render appropriate attention and service.
- ✰ To develop a database and information network for proper dissemination of knowledge.
- ✰ To promote environmental friendly activities like rational use of wetland resources, ecotourism, conservation of traditional knowledge and practices and encourage income-generating schemes, rehabilitation of degraded wetlands etc.
- ✰ To raise awareness among the local stakeholders about the plight of this degraded and neglected ecosystem.

Acknowledgements

The author is grateful to the University authorities for providing all possible help to this endeavor. Special thanks are also due to Dr. Gautam Aditya, Dr. Abhisek Basu, Ms. Swarnali Mukherjee for their help and cooperation during the preparation of manuscript.

References

Bassi N and Kumar MD. 2012. Addressing the civic challenges: perspective on institutional change for sustainable urban water management in India. *Environ. Urban. Asia,* 3 (1): 165-183.

Bates BC, Kundzewicz ZW, Wu S and Palutikof JP. (eds.). 2008. *Climate Change and Water*. [Technical Paper VI], Intergovernmental Panel on Climate Change, Geneva.

Billore SK, Singh N, Sharma, JK. *et al.*, 1999, Horizontal subsurface flow gravel bed constructed wetland with Phragmites karka in Central India. *Water Sci. Technol.*, 40 (3): 163-171.

Blankespoor B, Dasgupta S and Laplante B. 2012. *Sea level Rise and Coastal Wetlands: Impacts and Costs.* [Policy Research Working Paper 6277] The World Bank, Washington, DC.

Bystrom O, Andersson H and Gren I. 2000. Economic criteria for using wetlands as nitrogen sinks under uncertainty. *Ecol. Econ.*, 35 (1): 35-45.

Central Pollution Control Board (CPCB). 2009. *Status of Water Supply, Wastewater Generation and Treatment in Class-I Cities and Class-II Towns of India.* Central Pollution Control Board, Ministry of Environment and Forests, Government of India, New Delhi.

Centre for Science and Environment (CSE). 2012. *Living With Changing Climate: Impact, Vulnerability and Adaptation Challenges in Indian Sundarbans.* Centre for Science and Environment, New Delhi.

Chandan P, Chatterjee A and Gautam P. 2008. Management planning of Himalayan high altitude wetlands: a case study of Tsomoriri and Tsokar wetlands in Ladakh, India. In. *Proc. Taal 2007*, The 12th World Lake Conference (Sengupta M and Dalwani R. eds.), Jaipur, India.

Cowardene LM, Carter V, Golet FC and LaRoe ET. 1979. *Classification of wetlands and deepwater habitats in the United States.* U.S. Fish and Wildlife Services, Washington D.C. pp.103.

Dandekar P, Bhattacharya S and Thakkar H. 2011. Wetland (Conservation and Management) Rules 2010, Welcome, But a Lost Opportunity: This Cannot Help Protect the Wetlands. South Asia Network on Dams, Rivers and People, New Delhi.

Deepa RS and Ramachandra TV. 1999. Impact of urbanization in the interconnectivity of wetlands. *Proc. National Symposium on Remote Sensing Applications for Natural Resources: Retrospective and Perspective*, pp. XIX-XXI.

Dhir B, Sharmila P and Saradhi PP. 2009. Potential of aquatic macrophytes for removing contaminants from the environment. *Crit. Rev. Environ. Sci. Technol.*, 39 (9): 754-781.

Erwin KL. 2009. Wetlands and global climate change: the role of wetland restoration in a changing world. *Wetl. Ecol. Manage.*, 17 (1): 71-84.

Ghermandi A, van den Bergh JCJM, Brander LM. *et al.*, 2010. Values of natural and human-made wetlands: a meta-analysis. *Water Resour. Res.*, 46 (12): 1-12.

Gopal B, Shilpakar R and Sharma E. 2010. *Functions and Services of Wetlands in the Eastern Himalayas: Impacts of Climate Change* [Technical Report 3]. International Centre for Integrated Mountain Development, Kathmandu, Nepal.

Indian Institute of Technology (IIT). 2011. *Trends in Agriculture and Agricultural Practices in Ganga Basin: An Overview.* Indian Institute of Technology, Kanpur.

Intergovernmental Panel on Climate Change (IPCC). 2000. *Special Report on Land Use, Land-Use Change, and Forestry: Summary for Policy makers.* Geneva, Switzerland.

Jain CK, Singhal DC and Sharma MK. 2007. Estimating nutrient loadings using chemical mass balance approach. *Environ. Monit. Assess.* 134 (1–3): 385–396.

Kar D, Nagarathna AV and Ramachandra TV. 2000. Fish diversity and conservation aspects in an aquatic ecosystem in India. *Proc. 6th International Wetland Symposium,* INTECOL, Quebec, Canada.

Kathiresan K and Thakur S. 2008. *Mangroves for the Future: National Strategy and Action Plan, India.* Ministry of Environment and Forests, New Delhi.

Kaur R, Dhir G, Kumar P. *et al.,* 2012. Constructed wetland technology for treating municipal wastewaters. *ICAR News,* 18 (1): 8-9.

Lal R. 2008. Carbon sequestration. *Philos. Trans. Roy. Soc.,* B 363 (1): 815-830.

Liu JG and Diamond J. 2005. China's environment in a globalizing world. *Nature,* 435: 1179-1186.

McAllister DE, Craig JF, Davidson N. *et al.,* 2001. *Biodiversity Impacts of Large Dams.* International Union for Conservation of Nature and United Nations Environmental Programme, Gland and Nairobi.

Millennium Ecosystem Assessment (MEA), 2005. *Ecosystems and Human Well-being: Wetlands and Water Synthesis.* World Resources Institute, Washington, DC.

Ministry of Environment and Forests (MoEF), 2007. *Conservation of Wetlands in India: A Profile (Approach and Guidelines).* MoEF, Government of India, New Delhi.

Ministry of Environment and Forests (MoEF), 2012. *Annual Report 2011–2012.* MoEF, Government of India, New Delhi.

Molur S, Smith KG, Daniel BA and Darwall WRT. 2011. *The Status and Distribution of Freshwater Biodiversity in the Western Ghats, India.* International Union for Conservation of Nature, Cambridge and Gland.

Pant N and Verma RK. 2010. *Tanks in Eastern India: A Study in Exploration.* International Water Management Institute-TATA Water Policy Research Program and Centre for Development Studies, Hyderabad and Lucknow, India.

Patel JG, Murthy TVR, Singh TS and Panigrahy S. 2009. Analysis of the distribution pattern of wetlands in India in relation to climate change. In. *Proc. Workshop on Impact of Climate Change on Agriculture* (Panigrahy S, Ray S. and Parihar JS. eds.) Int. Soc. for Photogrammetry and Remote Sensing, Ahmedabad.

Prasad SN, Ramachandra TV, Ahalya N. *et al.,* 2002. Conservation of wetlands of India – a review. *Trop. Ecol.,* 43 (1): 173-186.

Rai PK. 2008. Heavy metal pollution in aquatic ecosystems and its phytoremediation using wetland plants: an eco-sustainable approach. *Int. J. Phytoremed.,* 10 (2): 133-160.

Ramachandra TV and Kumar U. 2008. Wetlands of greater Bangalore, India: automatic delineation through pattern classifiers. *Electr. Green J.*, 1 (26): 1-22.

ten Brink P. Badura T, Farmer A and Russi D. 2012. *The Economics of Ecosystem and Biodiversity for Water and Wetlands: A Briefing Note.* Institute for European Environmental Policy, London.

United Nations Educational, Scientific and Cultural Organization (UNESCO), 2007. *Case Studies on Climate Change and World Heritage.* UNESCO World Heritage Centre, France.

Verhoeven JTA, Arheimer B, Yin C and Hefting MM. 2006. Regional and global concerns over wetlands and water quality. *Trends Ecol. Evol.*, 21 (2): 96-103.

Zhao S, Peng C, Jiang H. *et al.*, 2006. Land use change in Asia and the ecological consequences. *Ecol. Res.* 21 (6): 890-896.

Chapter 2

Insect Diversity in Wetland Ecosystem of West Bengal

Asok Kanti Sanyal

West Bengal Biodiversity Board, Kolkata

Introduction

The association of man and wetlands is ancient. The distribution of man and its early non-human ancestors was governed by water sources, the wetlands. The civilization of Indus Valley, Nile delta, Tigris and Euphrates were established and flourished due to either flood plains, fertile lands and availability of man's basic necessities for survival- water for drinking and agriculture. The history says the first sign of civilization are definitely around the wetland areas. But in recent past the wetlands have not been given due importance and have been considered as wastelands and breeding grounds of disease carrying insects, difficulties and danger. Of late, the idea has changed, the wetlands are now recognised as prime natural resource and most valuable ecosystem. The amazingly high productivity of wetlands has been brought to the attention of scientists and researchers. Marshes and swamps having productivity higher than the bush, tropical rainforests are being looked upon as resources of food rather than as breeding grounds of disease carrying animals. Wetland is defined depending on its situation and the attitude of people who look at it. These have made the definition in more than fifty different ways to include a wide spectrum of habitats, of which the most useful one is of Cowardin *et al.* (1979), according to them *"wetlands are lands transitional between terrestrial and aquatic systems where the water table is usually at or near the surface or the land is covered by shallow water"*. The wetland which includes different habitats like inland, coastal and marine should have the following three essential attributes:

- ✰ Water must be present for at least seven successive days during the growing season.

- At least for some part of the year, the area supports hydrophytic vegetation.
- The substrate has hydric soils that are saturated or flooded for a long period to become anaerobic (lacking in oxygen) in their upper layers.

Wetlands of West Bengal

The state of West Bengal is endowed with different types of freshwater and brackish water wetland habitats including ponds, tanks, lakes, marshes, swamps, *bheries*, reservoirs, rivers and streams. The state has coastal wetlands comprising of inshore coastal waters as per the IUCN criteria. The freshwater wetlands in West Bengal are distributed over 3.35 lakh ha excluding rivers, canals and paddy fields. Maximum share of the total area is covered by tanks and ponds with 2.76 lakh ha, followed by *beels*, oxbows and derelict water bodies of 0.42 lakh ha and reservoirs of 0.17 lakh ha (Anonymous, 1996). The major river system in the state is Bhagirathi-Hugli. Besides there are other important freshwater rivers like Tista, Mahananda, Jaldhaka, Jalangi, Churni, Dwarakeswar, Ajay, Damodar, Rupnarayan, Haldi, etc. The total cultivable brackish water wetlands in West Bengal is 0.405 million ha and the freshwater rivers and wetlands have been estimated at 3,11,192.52 ha in West Bengal. The biological and physico-chemical characteristics of wetlands in the state greatly vary according to major landforms, *viz.*, mountains and hills, highland and plateau and plains. The wetlands with varied forms and characteristics support widely diverse faunal resources specifically of insects.

Wetland Insects – Ecological Adjustment

Wetland harbours very rich and diverse faunal groups representing almost all the phyla and other higher taxonomic groups. The large plants and microscopic algae that grow in wetlands provide foods for small animals specially insects. They also enjoy the shelter given by the floral communities. These small insects and other animals in turn become a food source for larger insects such as damselflies and dragonflies. The insects live in semi-terrestrial habitats by moving easily through the moist but airy layers of living stems and rotting vegetation just above the waterline. The insects like many other aquatic animals can move up or down when water level gradually rises and falls during wet or dry periods. Mostly the predator and scavenger insects in their adult form live on the water surface and others stay below it, but for breathing air they need to rise frequently. The larvae of many wetland insects are adapted to aquatic habitat with gills for extracting oxygen under water and some respire in low oxygen area by unique red blood pigments. The most peculiar habit of aquatic insect larvae is climbing the tall stem of marshland plants, to dry their wings in preparation for an aerial life. The standing water and sluggish seepages of wetland are usually well colonized by small, free swimming insects, like the mosquito larvae, or wrigglers, feed on microscopic plankton and decaying organic matter. The worm like larvae of midges and crane flies, burrowing mayfly nymphs and the larvae of caddies flies live in still water making nests either from hollow plant stems or twigs, or by gluing plant fragments or small stones on to a parchment living in still water. All these wetland insects together with animals and microscopic life forms play a major part in cleansing run-off-water and are important links in the food chain because they help to reprocess the finest particles of organic decay. They in turn become food for large predatory animals in wetlands. The insect abundance and biomass are distributed differently among vegetation zones within wetlands along the gradient. Biomass does not differ consistently along the trophic gradient but increased with increasing emergent vegetation cover. The wetland insects live permanently or temporarily in and around water bodies for food and shelter. They are grouped as aquatic, wetland dependent and wetland associated mainly for shelter, food, egg laying and development of larva and nymph.

Wetland Insects of West Bengal: A Retrospect

The first two reports of wetland insects from West Bengal were made by Gurney (1906, 1907) and Annandale (1907a, 1907b), subsequently followed by Kemp (1917a, 1917b), Hora (1933,1934), Nandi (1984), Nandi and Misra (1987), Das and Nandi (1999), Mandal and Nandi (1989). In the recent past, Mitra and Mitra (2009) recorded insect species from Sunderbans brackish water ponds and Matla river. Earlier, Brehm (1950) while studying freshwater fauna of India added few insect species in the list. Mukherjee (1975) again reported insect species from Sunderban wetlands. Khan (1979) studied primary productivity and trophic status of two water bodies of Kolkata and recorded several insect species. Ram *et al.* (1982) worked on Odonata fauna of Kolkata. Ghosh and Sen (1987) made a contribution on the ecological history of Kolkata wetlands. De *et al.* (1989) and De and Sengupta (1993) made attempts to study the wetland fauna of Kolkata which included a number of insect species specially coleopteran insects. Next year Ghosh and Chattopadhyay (1994) and later Roy and Nandi (2008) published reports including insect fauna from wetlands of East Kolkata. Ghosh and Chattopadhyay (1990, 1994) studied the biological resources of Brace Bridge wetlands and Santragachi jheel, Howrah and also reported insect species. Bhattacharya and Gupta (1991) recorded freshwater inhabiting insects of West Bengal. Nandi *et al.*, through a series of works (1993, 1999, 2001a, 2001b, 2004, 2005, 2007, 2013) and Roy and Nandi (2010) extensively studied wetland fauna with reports of insect species in the districts of West Bengal like 24 Parganas (both North and South), Howrah, Hooghly, Birbhum, Alipurduar, Jalpaiguri, Darjeeing, Purulia, Bankura, and flood plain areas. Chakrabarti and Saha (1993) studied physico-chemical and hydrobiological features of three water bodies in Darjeeling hills. Scientists of Zoological Survey of India like Das *et al.* (1993), Bal and Basu (1994a, 1994b), Biswas and Mukhopadhyay (1995), Biswas *et al.* (1995a, 1995b, 2008), Das (1998), Ghosh (1990, 1998), Prasad (1998) and Varshney (1998) recorded many insect species under the order Plecoptera, Hemiptera, Coleoptera, Odonata and Lepidoptera. Pahari *et al.* (1997, 1999) studied aquatic insects of east and west Midnapore districts. Mukherjee *et al.* (2000) and Mukherjee and Nandi (2006) studied ecology, biodiversity and management of Rabindra Sarovar Lake in Kolkata. Bhattcaharya (2000) studied insect fauna associated with water hyacinth in wetlands of West Bengal. Paria and Konar (2000) reported impact of environment on pond ecosystem in Jalpaiguri district. Khan and Ghosh (2001) studied diversity of aquatic insects in freshwater wetlands of south eastern West Bengal. Bhat *et al.* (2001) reported faunal components from Kistobazar Nala at Purulia. Khan (2002) made a study on the ecology and faunal diversity of two flood plain lakes. Alfred and Nandi (2002) published a review article on freshwater wetlands and their faunal resources. Jana *et al.* (2006) made a report on insect fauna in industrial and non-industrial areas in West Bengal. Pal and Nandi (2006) reported Phytofaunal community of two freshwater lakes. Saha *et al.* (2007) and Roy and Nandi (2008) recorded hemipteran insects from East Kolkata Wetlands. Ghosh and Nandi (2009) studied community based wetland management at Brace Bridge Nature Park, Kolkata. Jana *et al.* (2009) recorded aquatic insects from Midnapore town. Dutta and Bhattacharya (2010) studied on microinvertebrates in freshwater wetlands. Pramanik *et al.* (2010) reported animal species from Kulik Bird Sanctuary, Raigunj. Chandra (2011) made an exhaustive review of insect species recorded from the States and Union Territories in India. Sanyal *et al.* (2012) and Nandi *et al.* (2013) recorded the insect species so far described and recorded from West Bengal.

Wetland Insects of West Bengal: Taxonomic Account

The recent estimate made by Chandra (2011), suggests 63,760 species of insects distributed over 658 families under 27 orders and three classes are known from India. The most abundant orders are Coleoptera, Hemiptera, Diptera, Lepidoptera, Hymenoptera, Orthoptera and Thysanoptera. These 8 orders constitute the bulk (94 per cent) of insect fauna, while the remaining 21 orders are represented by only 6 per cent of species of the total number of insects species (5818) so far known from West Bengal. As many as 622 species of insects have been reported from wetlands of West Bengal. These belong to the orders Ephemeroptera, Odonata, Plecoptera, Hemiptera, Coleoptera and Diptera (Chandra, 2011; Nandi *et al.*, 2013).

Order Odonata

These insects are commonly known as 'dragonflies' (Anisoptera) and 'damselflies' (Zygoptera). The adults are flying insects, mate in air and females release eggs in water in a gelatinous strand or in aquatic plants. The nymphs live in water, carnivorous and voracious feeder. The adults are predacious insects. In India a total of 499 species belonging to 17 families are found. West Bengal has the representatives of 15 families with 93 genera and 180 species (Sanyal *et al.*, 2012).

Order Plecoptera

These insects are popularly known as 'stoneflies' and are generally found in and around high altitude hill streams. Some species may occur in ponds and pools. Females lay eggs on water surface or oviposit on some submerged substratum. Nymphs are aquatic and are found under stones in streams. The adults are weak fliers. Out of 130 Indian species, 21 species belonging to 10 genera and 4 families are represented in West Bengal (Sanyal *et al.*, 2012). All the species are so far known from wetlands in Darjeeling district.

Order Ephemeroptera

This ancient group of insects evolved about 290 million years ago, are commonly known as 'mayflies'. They are found near the unpolluted streams, ponds, rivers and lakes. After mating, females lay their eggs in water and the nymphs are aquatic. The adults are very short lived having lifespan of few hours to few weeks. 124 species under 46 genera and 12 families are so far known from wetland habitats of India (Chandra, 2011), out of which 17 species have been reported from West Bengal wetlands (Nandi *et al.*, 2013).

Order Hemiptera

Popularly known as 'water bugs' are adapted in aquatic, semi-aquatic and terrestrial habitats. These insects have distinctively piercing-sucking mouth parts. Some hemipteran insects swim upright (Corixidae-water boatman), many swim downwards (Notonectidae-back swimmer), some walk on water surfaces (Gerridae-water striders) or crawl in shallow water (Nepidae-water scorpion) with the help of hind legs. The semi aquatic bugs live on aquatic weeds. Of 966 species from West Bengal, 65 species are known from aquatic habitat (Nandi *et al.*, 2013).

Order Coleoptera

Coleopteran insects or beetles represent the largest group of organisms at the order level and are found in wetlands with abundant aquatic vegetation and detritus. India holds about 17,455 species and West Bengal shares 1,570 species under 524 genera and 25 families (Sanyal *et al.*, 2012). From aquatic habitats of West Bengal 140 species of Coleoptera have been recorded (Nandi *et al.*, 2013).

Order Diptera

These true flies are found almost in everywhere. Adult dipteran insects are terrestrial but larvae and some of the larvae of many of their families are aquatic. Aquatic forms are found in the silt and sand underlying water, on the surface of rocks, logs or vegetation, or in water columns. Larvae and pupae usually swim by the characteristic wriggling movements of the body. Of the 6,337 species reported from India, 681 species are known from West Bengal. The number of wetland dipteran species in West Bengal is not yet ascertained, but as many as 196 species of chironomids have been reported from West Bengal wetlands (Nandi *et al.*, 2013).

District-wise Diversity and Distribution

The analysis of number of wetland species of insects under different orders in West Bnegal shows that Diptera is represented by highest number of species (196) and the other five orders in respect of species richness are Odonata (183), Coleoptera (140), Hemiptera (65), Plecoptera (21) and Ephemeroptera (17) as shown in Figure 2.1.

The district wise distribution pattern of species of six insect orders in West Bengal wetlands shows that maximum number (273) of species are known from Darjeeling district followed by Kolkata (194), Burdwan (102), North 24-Parganas (97), Jalpaiguri (88), South 24- Parganas (81) and Murshidabad (61), in other districts species number is below 50 (Figure 2.2 and Table 2.1).

The distribution and abundance of wetland insect species under each order in different districts of West Bengal show that highest number of dipteran species is recorded from Burdwan (77). Likewise coleopteran species from Kolkata (79), hemipteran species from Kolkata (41) and odonata and ephimeropteran species from Darjeeling (99) and Jalpaiguri (13) districts respectively (Figure 2.3 and Table 2.1).

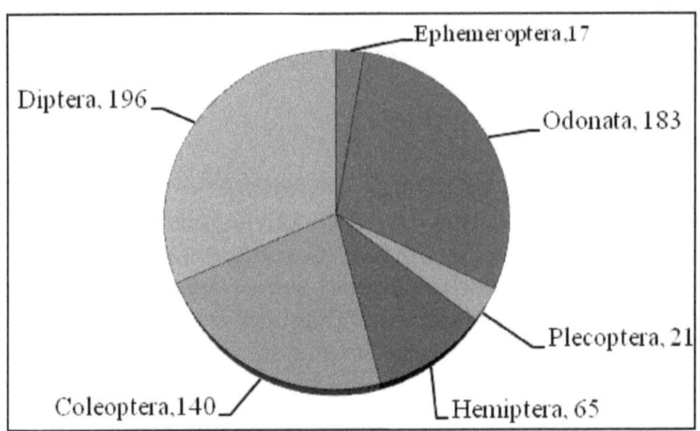

Figure 2.1: Species Diversity of Insect Groups in Wetlands of West Bengal.

Table 2.1: Distribution of Insect Species in Wetland Ecosystems of different Districts of West Bengal

Districts	*No. of Species under each Order*						*Total No. of Species*
	Ephemero-ptera	*Odonata*	*Plecoptera*	*Hemiptera*	*Coleoptera*	*Diptera*	
Darjeeling	7	99	21	18	54	74	273
Jalpaiguri	13	50	–	12	5	8	88
Coochbehar	–	12	–	1	–	4	17
Alipurduar	–	6	–	1	–	–	7
Maldah	2	24	–	1	–	3	30
Murshidabad	1	12	–	12	21	15	61
Hooghly	–	20	–	7	6	11	44
Burdwan	3	10	–	6	6	77	102
Bankura	3	23	–	8	–	9	43
Birbhum	3	18	–	7	5	11	44
Purulia	4	11	–	6	19	5	45
Purba Medinipur	4	16	–	3	12	8	43
Paschim Medinipur	4	16	–	5	12	9	46
Nadia	3	6	–	6	2	14	31
Howrah	–	31	–	4	3	11	49
North 24-Pgs.	6	35	–	25	13	18	97
South 24-Pgs.	5	36	–	22	9	9	81
Kolkata	10	43	–	41	79	21	194
Uttar Dinajpur	–	13	–	1	–	2	16
Dakshin Dinajpur	1	15	–	1	–	2	19

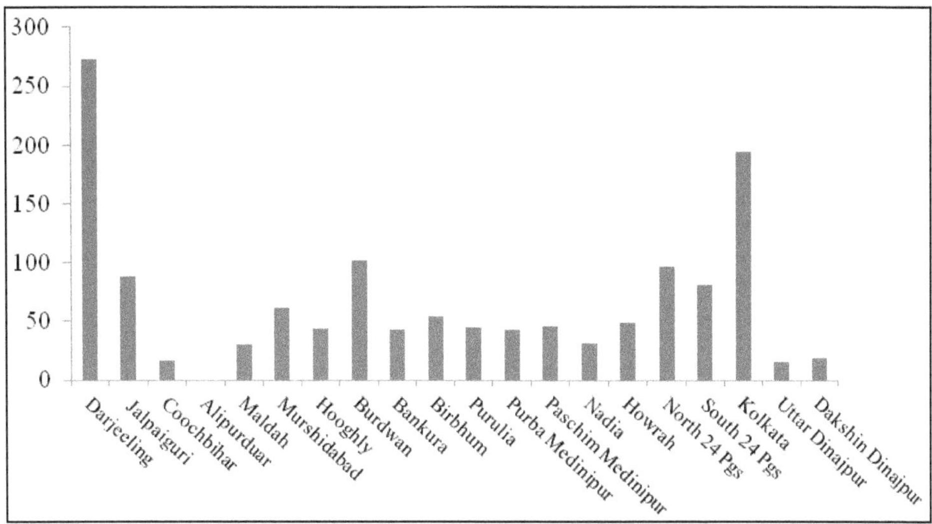

Figure 2.2: District-wise Distribution of Insect Groups in Wetlands of West Bengal.

As the plecopteran species are known from Darjeeling only, district wise species richness can not be ascertained.

Wetland Insects - Endemic Species of West Bengal

Nandi *et al.* (2013) reported 331 animal species which are described from West Bengal are endemic to India. Like-wise, out of 622 insect species described and recorded from West Bengal wetlands, maximum number of endemic species (134) has been reported from order Diptera. The order Coleoptera includes three endemic and Ephemeroptera and Odonata contain one endemic species each (Figure 2.4).

Wetland Insects at Risk

The real threat is the ecological impact of anthropogenic disturbances on wetland ecosystem. The major threats are changes in habitats due to abuse and pollution particularly in high fragile ecosystems such as freshwater and brackish water wetlands. The threats include encroachment, reclamation, habitat destruction, habitat alteration, siltation, shrinkage of wetlands, damming of rivers, urbanization, industrialization, timber cutting, over-exploitation, weed infestation, eutrophication and pollution. The adverse effect of all these threats caused pollution of streams, particularly through drainage and siltation and resulted profound changes in aquatic insect communities. The loss of native insect population was also effected due to indiscriminate conversion of natural habitat for agricultural and construction activities for various industrial/commercial purposes and due to introduction of exotic insect species for the control of pests and weeds in wetlands.

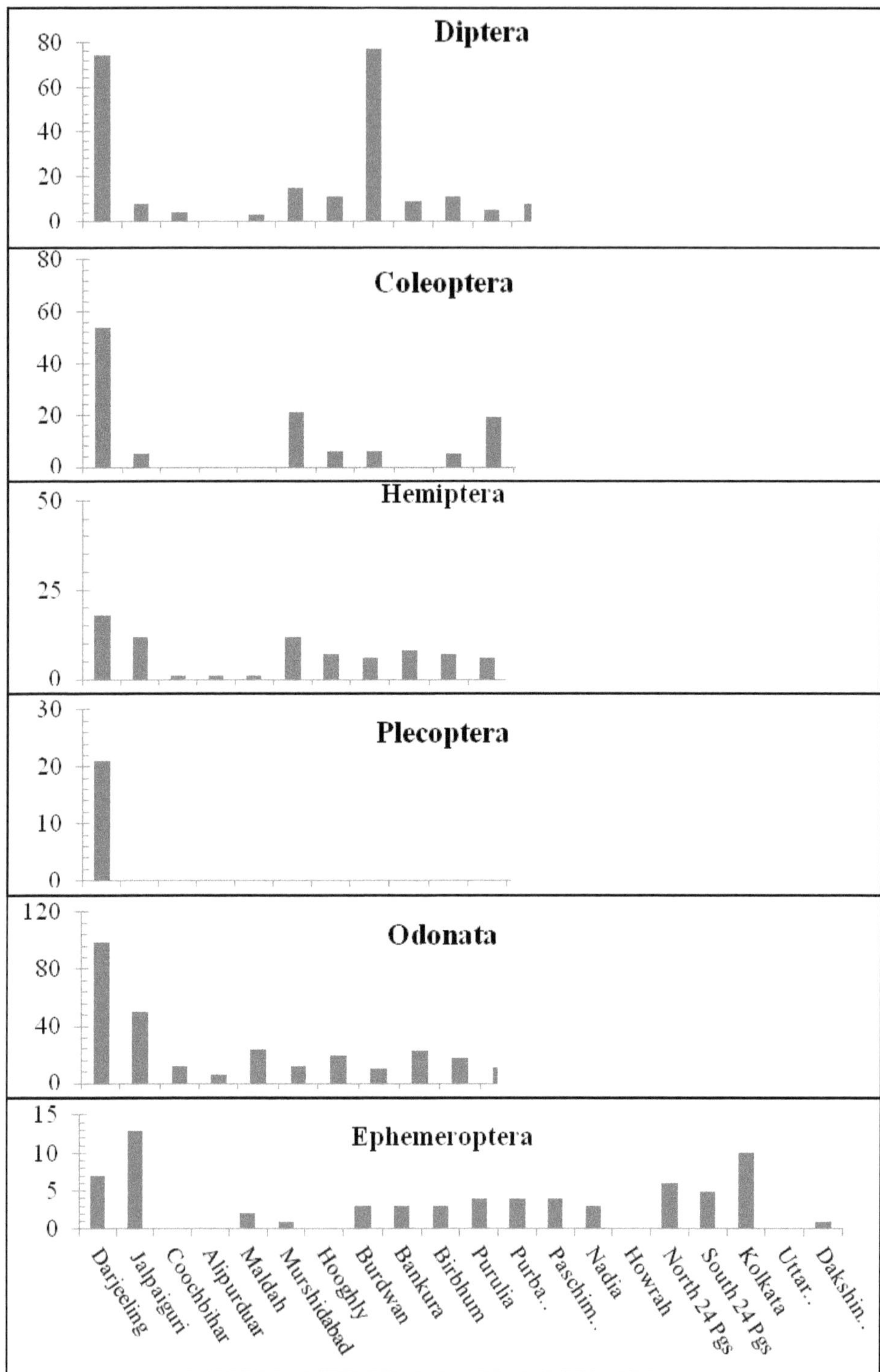

Figure 2.3: District-wise Distribution of Insect Orders in Wetlands of West Bengal.

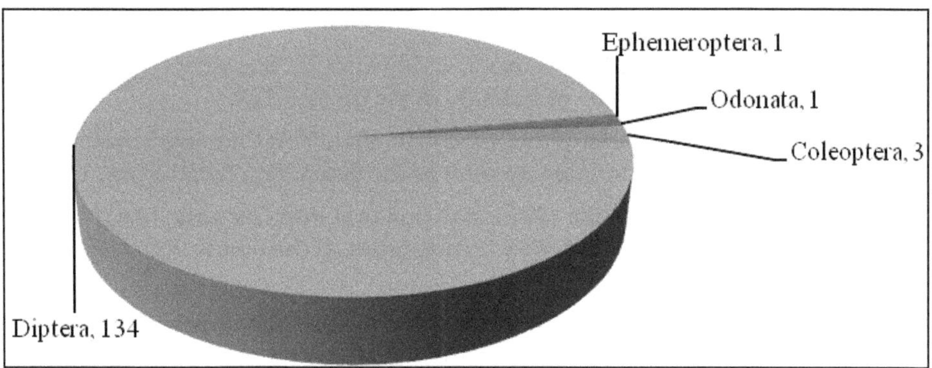

Figure 2.4: Number of Endemic Species of Insect in Wetlands of West Bengal.

Conservation Initiatives

The first and foremost measures necessary for conservation, restoration and management of wetland habitat are protection and preservation of the diversified features of wetland environment as well as bioresources as possible at various spatial levels in a heterogeneous landscape. This natural floral and faunal assemblage will afford opportunities for maximal growth of native and endemic insect species and help to suppress invasive alien species. The various measures for conservation of wetland ecosystem to minimize threats to wetland insects will be effective with creation of awareness among local communities and to involve them in conservation and management activities.

Conclusion

The insects are predominant biota in all continents including Antarctica. It is commonly recognised that insects constitute about 75-80 per cent of the total animal species on this planet. Their successful survival since, late Palaeozoic or early Mesozoic periods (200-250 million years ago) was due to their minute size, high fecundity, capacity for flight and dispersal, capability for living in all possible habitats, extremes of adaptability, presence of chitinous exoskeletons, etc. The state of West Bengal is rich in wetland insect population due to presence of wetlands in ecologically diverse zones like high altitude Himalayan regions, Gangetic plains and also saline and brackish water wetlands in mangrove areas in Sunderbans. These unique wetland habitats have enriched the insect population to a great extent. Unfortunately, in recent times indiscriminate anthropogenic activities are causing threats to wetlands and their faunal resources including insects. Thus the information on wetlands in the state and the status of insect fauna therein and their values need to be increasingly understood and communicated to the community, making them aware about necessity of conservation of wetlands. It is equally important that they should be aware of the wetland management plans taking into account the conservation of biodiversity and they are to be associated with the management strategies for their successful implementation.

References

Alfred JRB and Nandi NC. 2002. Wetlands: Freshwater. In. *Ecosystems of India*. ENVIS Centre, Zoological Survey of India, Kolkata. pp. 165-193.

Annandale N. 1907a. The fauna of the brackish ponds at Port Canning, Lower Bengal. Introduction and Preliminary account of the fauna. *Rec. Indian Mus.*, 1: 35-43.

Annandale N. 1907b. The fauna of brackish ponds at Port Canning, Lower Bengal. 6. Observations on Polyzoa, with further notes on the ponds. *Rec. Indian Mus.*, 1: 197-205.

Anonymous 1996. A long term multidisciplinary research approach and report on mangrove ecosystems of Sundarbans. Department of Marine Sciences, University of Calcutta. pp. 1-92.

Bal A and Basu RC. 1994a. Hemiptera: Belostomidae, Nepidae, Notonectidae and Pleidae. In. *Fauna of West Bengal, State Fauna Series 3,* Zoological Survey of India, Part-5: 511-534.

Bal A and Basu RC. 1994b. Hemiptera: Mesovellidae: Hydrometridae, Vellidae and Gerridae. In. *Fauna of West Bengal, State Fauna Series 3,* Zoological Survey of India, Part-5: 535-558.

Bhatt JP, Jatin A, Bhaskar A and Pandit MK. 2001. Pre-impoundment study of biotic communities of Kistobazar Nala in Purulia, West Bengal. *Curr. Sci.*, 81(10): 1332-1337.

Bhattacharya DK. 2000. Insect fauna associated with large water hyacinth in freshwater wetlands of West Bengal. *Diversity and Environment*. Proc. Nat. Seminar on Environ. Biol. (Aditya AK and Haldar P eds.). Daya Publishing House, New Delhi. pp. 145-148.

Bhattacharya DK and Gupta B. 1991. Freshwater wetland inhabiting insects of West Bengal. *Environ. and Ecol.*,9(4): 995-998.

Biswas BK, Hassan MA and Mukherjee A. 2008. Abundance and seasonal variation of benthic macroinvertebrates in an ox-bow lake in West Bengal, India. *Environ. and Ecol.*, 26(4B): 2112-2115.

Biswas S and Mukhopadhyay P. 1995. Insecta: Coleoptera: Hydrophilidae. In. *Fauna of West Bengal, State Fauna Series 3,* Zoological Survey of India, Part 6a: 143-168.

Biswas S, Mukhopadhyay P and Saha SK 1995a. Insecta: Coleoptera: Adephaga, Family Dytiscidae. In: *Fauna of West Bengal, State Fauna Series 3,* Zoological Survey of India, Part-6a: 77-120.

Biswas S, Mukhopadhyay P and Saha SK.1995b. Insecta: Coleoptera: Adephaga, Family Gyrinidae and Haliplidae. In: *Fauna of West Bengal, State Fauna Series 3,* Zoological Survey of India, Part-6a: 121-142.

Brehm V.1950. Contributions to the freshwater fauna of India. *Rec. Indian Mus.*, 48(1): 9-28.

Chakrabarty D and Saha GK.1993. The physico-chemical and hydrobiological features of three stagnant water sources in Darjeeling Hills. *Geobios,* 20(2): 61-65.

Chandra K. 2011. Insect fauna of States and Union Territories in India. *ENVIS Bulletin: Arthropods and their Conservation in India (Insects and Spiders),* 14(1): 189-218.

Cowardin LM, Carter V, Golet FC and LaRoe ET. 1979. Classification of wetlands and deepwater habitats of the United States. Fish and Wildlife Service FWS/OBS-79/31, pp.103.

Das AK and Nandi NC. 1999. Fauna of Sundarbans biosphere Reserve. *ENVIS Newsletter,* Zoological Survey of India, 5(1 and 2): 4-9.

Das BC. 1998. Plecoptera. In. *Faunal Diversity in India.* (Alfred *et al.,* eds.). ENVIS Centre, Zoological Survey of India. pp. 179-182.

Das BC, Chatterjee A and Dutta AL. 1993. Insecta: Plecoptera. In. *Fauna of West Bengal, State Fauna Series 3,* Zoological Survey of India, Part-4: 169-177.

De M and Sengupta T.1993. Beetles (Coleoptera: Insecta) of wetlands of Calcutta and surroundings. *Rec. zool. Surv. India,* 93: 103-138.

De M, Bhuinya S and Sengupta T. 1989. A preliminary account of major wetland fauna of Calcutta and its surroundings. *Encology,* 34(4): 5-11.

Dutta A and Bhattacharya DK. 2010. Diversity and seasonal abundance of macroinvertebrate in some freshwater wetlands of West Bengal. *Environ. and Ecol.,*28(3): ·1544-15497.

Ghosh AK. 1990. Biological resources of wetlands of East Calcutta. *Indian J. Landscape System and Ecol. Studies,* 13(1): 10-23.

Ghosh AK and Chattopadhyay S. 1990. Biological resources of Brace Bridge wetlands. *Indian J. Landscape System and Ecol. Studies,* 13(2): 189-207.

Ghosh AK and Chattopadhyay S. 1994. Biological resources of peri urban wetlands: Santragachi jheel, Howrah district, West Bengal. *Indian J. Landscape system and Ecol. Studies,* 17(1): 1-7.

Ghosh AK and Nandi NC. 2009. Community-based wetland management-A case study of Brace Bridge Nature Park (BBNP), Kolkata, India. In. *Sustaining the World's Wetlands: Setting Policy and resolving conflicts.* (Smardoninger R. eds.), Drodrecht, Heidelberg, London, NewYork, pp. 125-150.

Ghosh D and Sen S. 1987. Ecological history of Calcutta Wetlands Conservation. *Environmental Conservation,* 14(3): 219-226.

Ghosh LK. 1998. Hemiptera. In. *Faunal Diversity in India* (Alfred *et al.,* eds.). ENVIS Centre, Zoological Survey of India, Calcutta. pp. 233-241.

Gurney R.1906. On some freshwater entomostracans in the collection of Indian Museum, Calcutta. *J. Asiat. Soc. Bengal,* 2: 273-281.

Gurney R. 1907. Further notes on Indian Freshwater Entomostracans. *Rec. Indian Mus.,* 1: 21-33.

Hora SL. 1933. Animals in brackish water at Uttarbhag, Lower Bengal. *Curr.Sci.,* 1: 12-38.

Hora SL.1934. Brackish water animals of the Gangetic delta. *Curr.Sci.,* 2(11): 426-427.

Jana G, Misra KK and Bhattacharya T. 2006. Diversity of some insect fauna in industrial and non-industrial areas of West Bengal, India. *J. Insect Conservation,* 10: 249-260.

Jana S, Pahari PR, Dutta TK and Bhattacharya T. 2009. Diversity and community structure of aquatic insects in a pond in Midnapore town,West Bengal, India. *J. Environ.Biol.,* 30(2): 283-287.

Kemp SW. 1917a. Notes on the fauna of Matlah River in the Gangetic delta. *Rec. Indian Mus.,*13: 233-244.

Kemp SW 1917b. Notes on the fauna of the Matlah river in the Gangetic delta. *Rec. Indian Mus.,* 13(4): 233-241.

Khan RA.1979. Primary productivity and trophic status of two tropical water bodies of Calcutta, India. *Bull.zool.Surv.India,* 2: 129-138.

Khan RA. 2002. The ecology and faunal diversity of two flood plain lakes. *Rec. zool. Surv. India,* Occ.Paper No.195: 1-57.

Khan RA and Ghosh LK. 2001. Faunal diversity of aquatic insects in freshwater wetlands of south eastern West Bengal. *Rec. zool.Surv. India, Occ. Paper* No.194: 1-104.

Mandal AK and Nandi NC.1989. Fauna of Sundarban mangrove ecosystem, West Bengal, India. *Fauna of Conservation Areas,* Zoolgical Survey of India, Kolkata, No.3: 1-116.

Mitra A and Mitra B. 2009. *Pictorial handbook of dragon and damselflies (Odonata : Insecta) of mangroves of Sunsderbans.* Zoolgical Survey of India, Kolkata, pp. 1-56.

Mukherjee AK. 1975. The Sundarbans of India and its biota. *J. Bombay Nat. Hist. Soc.,* 72(2): 1-20.

Mukherji M and Nandi NC. 2006. Ecology, biodiversity and management of Rabindra Sarovar in Kolkata, West Bengal. In. *Ecology of Lakes and Reservoir* (Sakhare SV ed.). Daya Publishing House, New Delhi. pp. 36-53.

Mukherji M, Pal S and Nandi NC. 2000. Macroinvertebrate diversity of some urban wetlands in Calcutta. In. *Biodiversity and Environment* (Aditya AK and Halder P eds.). Daya Publishing House, New Delhi. pp. 136-144.

Nandi NC. 1984. Faunal resources of Sunderban: its prospectives and predicaments. *Environ. and Ecol.,* 2: 99-102.

Nandi NC, Das SR, Bhuinya S and Dasgupta JM.1993. Wetland Faunal Resources of West Bengal.1. North and South 24-Parganas Districts. *Rec. Zool. Surv. India,* Occ. Paper No. 150: 1-50.

Nandi NC and Misra A.1987. Bibliography of the Indian Sundarbans with special reference to fauna. *Rec. Zool. Surv. India,* Occ. Paper No. 97: 1-135.

Nandi NC, Das AK and Dey A. 2013. *Wetland faunal Diversity of West Bengal.* WBBB and NBI, Kolkata. pp. 1-256.

Nandi NC, Mukhopadhyay P, Ghosh SK and Das SK. 2004. Notes on aquatic entomofauna of Narathaly Lake of Buxa Tiger Reserve, West Bengal. *Rec.zool. Surv.India,* 102(1-2): 53-56.

Nandi NC, Venkataraman K, Das SR. *et al.,* 1999. Wetland Faunal Resources of West Bengal. 2. Some selected wetlands of Haora and Hughly Districts. *Rec. Zool. Surv. India,* 97(4): 103-143.

Nandi NC,Venkataraman K, Das SR and Bhuinya S. 2001a. Faunal diversity of wetlands in the Botanical Garden, Howrah, West Bengal. *Rec.zool. Surv. India,* 99(1-4): 111-129.

Nandi NC, Venkataraman K, Das SR. *et al.,* 2001b. Wetland Faunal Resources of West Bengal. 3.Birbhum District. *Rec. zool.Surv.India,* 99(1-4): 135-156

Nandi NC, Venkataraman K, Das SR. *et al.,* 2005. Wetland Faunal Resources of West Bengal.4. Darjiling and Jalpaiguri Districts. *Rec. Zool. Surv. India,* 104(1-2): 1-25.

Nandi NC, Venkataraman K, Das SR and Das SK 2007. Wetland Faunal Resources of West Bengal-5. Bankura and Puruliya districts. *Rec. Zool. Surv. India,* 107(2): 61-91.

Pahari PR, Dutta TK and Bhattacharya T.1997. Aquatic insects of Midnapore district-I (Insecta, Coleoptera, Dytiscidae). *Vidyasagar Univ.J. Biosciences,* 3: 45-51.

Pahari PR, Dutta TK and Bhattacharya T. 1999. Aquatic insects of Midnapore district-I (Coleoptera, Hydrophilidae). *Vidyasagar Univ. J. Biosciences,* 3: 45-51.

Pal S and Nandi NC. 2006. Phytofaunal community of two freshwater lakes of West Bengal, India. *Rec.zool.Surv.India,* Occ.Paper No. 248: 1-146.

Paria T and Konar SK. 2000. Environmental impact on pond ecosystem in Jalpaiguri district, West Bengal. *Environ.* and *Ecol.,*18(3): 624-629.

Pramanik AK, Santra KB and Manna CK. 2010. Abundance and diversity of plants and animals in the Kulik bird sanctuary, Raiganj, West Bengal, India. *J. Biodiversity,* 1(1): 13-17.

Prashad M. 1998. Odonata. In. *Faunal Diversity in India* (Alfred *et al.,* eds.). ENVIS Centre, Zoological Survey of India, Kolkata. pp. 171-122.

Ram R, Srivastava VD and Prasad M. 1982. Odonata (Insecta) fauna of Calcutta and surrounding. *Rec. zool. Surv. India,* 80: 169-196.

Roy M and Nandi NC. 2008. Diversity and distribution of aquatic entomofauna of East Calcutta Wetlands. *Environ. and Ecol.,* 26(4C): 2211-2214.

Roy M and Nandi NC. 2010. Faunal diversity of Indrabeel, Puruliya, West Bengal. *J. Environ. and Sociobiol.,* 7(1): 53-55.

Saha N, Aditya G, Bal A and Saha GK. 2007. Comparative study of functional response of common hemipteran bugs of East Calcutta Wetlands, India. *International Review of Hydrobiology,* 92: 242-257.

Sanyal AK, Alfred JRB, Venkataraman K. *et al.*, 2012. *Status of Biodiversity of West Bengal*. Zoological Survey of India, Kolkata. pp. 1-969.

Varshney RK. 1998. Insecta. In. *Faunal diversity in India* (Alfred *et al.*, eds.). ENVIS Centre, Zoological Survey of India, Kolkata. pp. 145-157.

Chapter 3

Brachyuran Crab Diversity, Distribution, Behaviour and Adaptation in Sundarban Delta

Malay Kanti Dev Roy[1] and Nepal Chandra Nandi[2]

[1]*Crustacea Section, Zoological Survey of India, Kolkata*
[2]*Social Environmental and Biological Association, Kolkata*

Introduction

Sundarban delta is the largest land-sea interface in the world, rich in habitat, plant and animal diversities, comprises of a total geographical area of about 10,000 sq km, of which, 6,000 sq km comes under Bangladesh and 4,110 sq km under Indian part. Indian Sundarban comprises of coastal wetlands, including mangroves, estuaries, brackish water impoundments and freshwater ponds. In the Indian part of Sundarban delta, mangrove forest area is about 2180 sq km, coast line of 64 km long and inshore and estuarine area of 2085 sq km, which includes estuarine rivers, tidal creeks, canals, sea-shore and mud-flats. The coastal waters and innumerable criss-crossed estuaries provide suitable habitats for highly rich and diverse faunal communities including brachyuran crabs (Mandal and Nandi, 1989; Chaudhuri and Choudhury, 1994) and anomuran crabs or hermit crabs (Reddy, 1995).

Historical Resume

The first report on brachyuran crab from Sundarban delta dates back to 1861 when A. Milne Edwards described a portunid crab, *Goniosoma rostratum* from the mouth of the River Ganges. Later, Henderson (1893) also recorded the same species from Sundarban along with a varunid crab, *Varuna litterata.* Subsequently, Alcock in a series of publications (1895, 1896, 1899 and 1900) recorded 28 species from Sundarban coast. In 1908, De Man reported the occurrence of four species of crabs which included one new species. Kemp (1917, 1919a, 1919b and 1923) added 10 more species including five new taxa from Hugli and Matla rivers. Chopra (1933a, 1933b, 1934 and 1935) recorded as many as 51 species and varieties belonging to 27 genera and 12 families to the crab fauna of West Bengal. In 1995, crabs of Hugli-Matla estuary were published by a number of scientists of Zoological Survey of India. These studies revealed the occurrence of 13 species of xanthid (Deb, 1995), 15 species of portunid (Bhadra, 1995), 24 species of grapsid (Ghosh, 1995) and 21 species of ocypodid crabs (Bairagi, 1995). Deb (1998) published a list of 127 species of crabs of 16 families from the state, major portion of which represented Sundarban delta, with a comprehensive account of 81 species including two new species. Dev Roy and Nandi (2008), Dev Roy (2010) and Nandi *et al.* (2013) contributed to the list of crustacean fauna from West Bengal including crabs.

Precisely, brachyuran crabs of Sundarban mangroves (including Tiger Reserve Area) has been investigated by a number of workers (Mandal and Misra, 1985; Saha, 1986; Chakraborty *et al.*, 1986; Chakraborty and Choudhury, 1992a, 1992b; Chaudhuri and Choudhury, 1994). Mandal and Nandi (1989) published an exhaustive list of brachyuran crabs comprising of 52 species under 26 genera and 11 families from Sundarban mangroves which was further elaborated by Dev Roy and Nandi (2008) that contained two new records from the Sundarban delta, namely, *Thalamita crenata* and *Macrophthalmus (Mareotis) tomentosus*. Brachyuran crabs of commercial importance from coastal waters of West Bengal were studied by some workers (Dutta, 1973; Dev Roy and Nandi, 1991, 2001, 2004; Nandi and Pramanik, 1994; Paul *et al.*, 1999). Taxonomic aspects of commercial crabs were contributed by Dev Roy and Nandi (1991) and Nandi and Pramanik (1994), while biometrical studies (Nandi and Dev Roy, 1996), biology of crabs (Pearse, 1932), burrow pattern and behaviour (Nandi and Dev Roy, 1989; Nandi *et al.*, 1998; Dubey *et al.*, 2013), feeding behavior (Paul and Nandi, 1999), population ecology (Chakraborty and Choudhury, 1992a), seasonal abundance (Paul *et al.*, 2012), zonation (Chakraborty and Choudhury, 1992b), bionomics (Hora, 1933), fishery (Hora, 1935; Nandi and Ghatak, 1985; Nandi *et al.*, 1993; Nandi and Pramanik, 1994; Dev Roy and Nandi, 2002, 2004; Pramanik and Nandi, 2004, 2011, 2012; Pramanik, 2014a, Pramanik, 2014b; Nandi and Pramanik, 2015), landing (Nandi and Pramanik, 1986) and export of mud crab were documented in the works of Nandi and Pramanik (1994),Dev Roy and Nandi (2012) and Dev Roy (2014). However, in the present communication our extensive and in-depth investigation since 1980's on crab diversity, distribution, zonation, behaviour and adaptive features of brachyuran crabs of Sundarban delta are summarized and highlighted.

Materials and Methods

Both extensive and intensive surveys on brachyuran crabs were conducted in different parts of Sundarban delta since 1980s. Collections made during these surveys along with materials present in National Zoological Collection of the Zoological Survey of India as well as literature studies constitute the basis for identification and enlisting of the brachyuran crab diversity occurring in Sundarban delta. Distribution, zonation, habit, habitat of the brachyuran species were investigated and analysed based mainly on field observations. Special efforts were made to observe, investigate and understand the burrowing habit, zonation pattern and adaptation in behaviour and morphological features in response to environment. Foraging activities and substratum characteristics were noted in the field, while anatomical peculiarities were mainly studied in the laboratory.

Results and Discussion

Species Diversity

Altogether 135 species of brachyuran crabs under 72 genera and 26 families have been collected and/or reported herein from estuarine and marine areas of Sundarban delta (Table 3.1). Among these, 69 species are marine and 70 estuarine which includes 16 overlapping species occurring in both costal and estuarine waters. The family-wise distribution of these species reveals highest diversity of brachyuran crabs in the families Portunidae (23 species) followed by Sesarmidae (13 species) and Leucosiidae, Varunidae and Macrophthalmidae each represented by 10 species. The families Dromiidae, Raninidae, Aethridae, Corystidae, Ethusidae and Scalopidiidae contain the least number of species being represented by one species each. Habit and habitats of these species are shown in Table 3.1.

Table 3.1: Diversity of Species, Habit, Habitat and Zonation of Crabs Recorded from Sundarban Delta

Sl.No.	*Family and Species*	*Habit*	*Habitat*
	Family **Dromiidae**		
1.	*Conchoecetes artificiosus* (Fabricius, 1798)	NS	CW
	Family **Raninidae**		
2.	*Raninoides personatus* Henderson, 1888	NS	CW
	Family **Aethridae**		
3.	*Drachiella morum* (Alcock, 1896)	NS	CW
	Family **Calappidae**		
4.	*Calappa lophos* (Herbst, 1790)	B	CW
5.	*Calappa pustulosa* Alcock, 1896	B	CW
	Family **Matutidae**		
6.	*Ashtoret lunaris* (Forskål, 1775)	S, B	EW/CW
7.	*Matuta planipes* Fabricius, 1798	S, B	EW/CW
8.	*Matuta victor* (Fabricius, 1781)	S, B	EW/CW
	Family **Corystidae**		
9.	*Jonas indica* (Chopra, 1935)	B	CW
	Family **Dorippidae**		
10.	*Dorippe quadridens* (Fabricius, 1793)	NS	CW
11.	*Dorippoides facchino* (Herbst, 1785)	NS	EW/CW
12.	*Neodorippe callida* (Fabricius, 1798)	NS	EW/CW

Contd...

Table 3.1–*Contd...*

Sl.No.	*Family and Species*	*Habit*	*Habitat*
	Family **Ethusidae**		
13.	*Ethusa indica* Alcock, 1894	NS	CW
	Family **Menippidae**		
14.	*Menippe rumphii* (Fabricius, 1798)	W	CW
15.	*Myomenippe hardwicki* (Gray, 1831)	W	EW/CW
	Family **Scalopidiidae**		
16.	*Scalopidia spinosipes* Stimpson, 1858	NS	CW
	Family **Iphiculidae**		
17.	*Iphiculus spongiosus* Adams and White, 1848	NS	CW
18.	*Pariphiculus mariannae* (Herklot, 1852)	NS	CW
	Family **Leucosiidae**		
19.	*Arcania erinaceus* (Fabricius, 1793)	NS	CW
20.	*Arcania septemspinosa* (Fabricius, 1787)	NS	CW
21.	*Ixa cylindrus* (Fabricius, 1787)	NS	CW
22.	*Ixa inermis* Leach, 1817	NS	CW
23.	*Myra elegans* Bell, 1855	NS	CW
24.	*Myra fugax* (Fabricius, 1798)	NS	CW
25.	*Philyra globus* (Fabricius, 1775)	NS	CW
26.	*Euclosia rotundifrons* Chopra, 1933	NS	CW
27.	*Leucosia craniolaris* (Linnaeus, 1758)	NS	CW
28.	*Seulocia rhomboidalis* (De Haan, 1841)	NS	CW
	Family **Epialtidae**		
29.	*Doclea armata* De Haan, 1839	W	CW
30.	*Doclea canalifera* Stimpson, 1857	W	CW
31.	*Doclea hybrida* (Fabricius, 1798)	W	CW
32.	*Doclea muricata* (Fabricius, 1787)	W	CW
33.	*Doclea ovis* (Fabricius, 1787)	W	CW
34.	*Doclea rissonii* Leach, 1815	W	CW
35.	*Hyastenus diacanthus* (De Haan, 1839)	NS	CW
36.	*Phalangipus longipes* (Linnaeus, 1758)	NS	CW
	Family **Hymenosomatidae**		
37.	*Hymenicoides carteri* Kemp, 1917	NS	EW/FW
38.	*Neorhynchoplax inachoides* (Alcock, 1900)	NS	EW
39.	*Neorhynchoplax nasalis* (Kemp, 1911)	NS	EW
40.	*Neorhynchoplax woodmasoni* (Alcock, 1900)	NS	EW
41.	*Trigonoplax unguiformis* (De Haan, 1839)	NS	EW

Contd...

Table 3.1–*Contd...*

Sl.No.	*Family and Species*	*Habit*	*Habitat*
	Family **Parthenopidae**		
42.	*Cryptopodia angulata* H. Milne Edwards and Lucas, 1841	NS	CW
43.	*Enoplolambrus pransor* (Herbst, 1796)	NS	CW
	Family **Galenidae**		
44.	*Galene bispinosa* (Herbst, 1783)	NS	CW
45.	*Halimede fragifer* De Haan, 1835	NS	CW
46.	*Halimede tyche* (Herbst, 1801)	NS	CW
47.	*Parapanope euagora* De Man, 1895	NS	CW
	Family **Pilumnidae**		
48.	*Benthopanope indica* Stimpson, 1858	NS	EW
49.	*Eurycarcinus bengalensis* Deb, 1998	B	EW
50.	*Eurycarcinus natalensis* (Krauss, 1843)	B	EW/CW
51.	*Eurycarcinus orientalis* A. Milne Edwards, 1867	B	EW/CW
52.	*Heteropanope glabra* Stimpson, 1858	NS	EW
53.	*Heteropanope neolaevis* Deb, 1998	NS	EW
54.	*Heteropilumnus ciliatus* (Stimpson, 1858)	NS	EW
55.	*Typhlocarcinus nudus* Stimpson, 1858	NS	EW
	Family **Portunidae**	NS	
56.	*Charybdis (Charybdis) affinis* Dana, 1852	S	CW
57.	*Charybdis (Charybdis) callianassa* (Herbst, 1801)	S	CW
58.	*Charybdis (Charybdis) feriata* (Linnaeus, 1758)	S	CW
59.	*Charybdis (Charybdis) helleri* (A. Milne Edwards, 1867)	S	CW
60.	*Charybdis (Charybdis) lucifera* (Fabricius, 1798)	S	CW
61.	*Charybdis (Charybdis) miles* (De Haan, 1852)	S	CW
62.	*Charybdis (Charybdis) orientalis* Dana, 1852	S	CW
63.	*Charybdis (Charybdis) rostrata* (A. Milne Edwards, 1861)	S	EW/CW
64.	*Charybdis (Charybdis) variegata* (Fabricius, 1798)	S	CW
65.	*Charybdis (Goniohellenus) truncata* (Fabricius, 1798)	S	CW
66.	*Charybdis (Goniohellenus) vadorum* Alcock, 1899	S	CW
67.	*Lissocarcinus arkati* Kemp, 1923	S	CW
68.	*Portunus (Achelous) granulatus* (H. Milne Edwards, 1834)	S	CW
69.	*Portunus (Lupocycloporus) gracilimanus* (Stimpson, 1858)	S	CW
70.	*Portunus (Monomia) gladiator* Fabricius, 1758	S	CW
71.	*Portunus (Portunus) pelagicus* (Linnaeus, 1758)	S	CW/EW
72.	*Portunus (Portunus) pubescens* (Dana, 1852)	S	CW
73.	*Portunus (Portunus) sanguinolentus* (Herbst, 1790)	S	CW
74.	*Portunus (Xiphonectes) hastatoides* Fabricius, 1798	S	CW

Contd...

Table 3.1–*Contd...*

Sl.No.	*Family and Species*	*Habit*	*Habitat*
75.	*Portunus (Xiphonectes) pulchricristatus* (Gordon, 1931)	S	CW
76.	*Scylla serrata* (Forskål, 1775)	S	BW/EW
77.	*Scylla tranquebarica* (Fabricius, 1798)	S	BW/EW
78.	*Thalamita crenata* (Latreille, 1829)	S	CW
	Family **Xanthidae**		
79.	*Liagore erythematica* Guinot, 1971	NS	CW
80.	*Orphanoxanthus microps* (Alcock and Anderson, 1894)	NS	CW
	Family **Grapsidae**		
81.	*Metopograpsus latifrons* (White, 1847)	R, CL	BW/EW/CW
82.	*Metopograpsus messor* (Forskål, 1775)	R, CL	BW/EW/CW
83.	*Pachygrapsus propinquus* De Man, 1888	NS	BW/EW
	Family **Sesarmidae**		
84.	*Clistocoeloma merguiense* De Man, 1888	B	BW/EW
85.	*Episesarma mederi* (H. Milne Edwards, 1853)	B	BW/EW
86.	*Metasesarma obesum* (Dana, 1851)	B	BW/EW
87.	*Muradium tetragonum* (Fabricius, 1798)	B	BW/EW
88.	*Neosarmatium smithi* (H. Milne Edwards, 1853)	B	BW/EW
89.	*Parasesarma pictum* (De Haan, 1835)	B	BW/EW
90.	*Parasesarma plicatum* (Latreille, 1803)	B, CL	BW/EW
91.	*Perisesarma bidens* (De Haan, 1835)	B,CL	BW/EW
92.	*Pseudosesarma edwardsi* (De Man, 1887)	B	BW/EW
93.	*Sesarmops impressum* (H. Milne Edwards, 1837)	B	BW/EW
94.	*Sesarmops intermedium* (De Haan, 1835)	B	BW/EW
95.	*Sesarmoides longipes* (Krauss, 1843)	B	BW/EW
96.	*Sesarmoides kraussi* (De Man, 1888)	B	BW/EW
	Family **Varunidae**	B	
97.	*Metaplax crenulata* (Gerstaecker, 1856)	B	BW/EW
98.	*Metaplax dentipes* (Heller, 1865)	B	BW/EW
99.	*Metaplax distincta* H. Milne Edwards, 1852	B	BW/EW
100.	*Metaplax indica* H. Milne Edwards, 1852	B	BW/EW
101.	*Metaplax intermedia* De Man, 1888	B	BW/EW
102.	*Parapyxidognathus deianira* De Man, 1888	NS	BW/EW
103.	*Ptychognathus dentatus* De Man, 1892	NS	BW/EW
104.	*Ptychognathus onyx* Alcock, 1900	NS	BW/EW
105.	*Pyxodognathus fluviatilis* Alcock, 1900	NS	BW/EW
106.	*Varuna litterata* (Fabricius,1798)	S, B	BW/EW/CW

Contd...

Table 3.1–*Contd...*

Sl.No.	*Family and Species*	*Habit*	*Habitat*
	Family **Dotillidae**		
107.	*Dotilla blanfordi* Alcock, 1900	B	BW/EW
108.	*Dotilla intermedia* De Man, 1888	B	BW/EW
109.	*Dotillopsis brevitarsis* (De Man, 1888)	B	BW/EW
110.	*Ilyoplax gangeticus* (Kemp, 1919)	B	BW/EW
111.	*Ilyoplax stapletoni* (De Man, 1908)	B	BW/EW
112.	*Scopimera globosa* De Haan, 1835	B	BW/EW
113.	*Scopimera investigatoris* Alcock, 1900	B	BW/EW
114.	*Scopimera proxima* Kemp, 1919	B	BW/EW
	Family **Macrophthalmidae**		
115.	*Macrophthalmus (Macrophthalmus) brevis* (Herbst, 1804)	NS	BW/EW
116.	*Macrophthalmus (Macrophthalmus) crassipes* H. Milne Edwards, 1834	NS	BW/EW
117.	*Macrophthalmus (Macrophthalmus) sulcatus* H. Milne Edwards, 1852	B	BW/EW
118.	*Macrophthalmus (Macrophthalmus) transversus* (Latreille, 1817)	B	BW/EW
119.	*Macrophthalmus (Mareotis) depressus* Rüppell, 1830	B	BW/EW
120.	*Macrophthalmus (Mareotis) pacificus* Dana, 1851	NS	BW/EW
121.	*Macrophthalmus (Mareotis) teschi* Kemp, 1919	NS	BW/EW
122.	*Macrophthalmus (Mareotis) tomentosus* Souleyet, 1841	B	BW/EW
123.	*Macrophthalmus (Paramareotis) erato* De Man, 1888	B	BW/EW
124.	*Venitus dentipes* (Lucas in Guérin-Méneville, 1836)	NS	BW/EW
	Family **Ocypodidae**		
125.	*Ocypode ceratophthalma* (Pallas, 1772)	B	BW/EW/CW
126.	*Ocypode macrocera* H. Milne Edwards, 1837	B	BW/EW/CW
127.	*Ocypode platytarsis* H. Milne Edwards, 1852	B	BW/EW/CW
128.	*Uca dussumieri* (H. Milne Edwards, 1852)	B	BW/EW
129.	*Uca lactea* (De Haan, 1835)	B	BW/EW
130.	*Uca rosea* (Tweedie, 1937)	B	BW/EW
131.	*Uca triangularis* (A. Milne Edwards, 1873)	B	BW/EW
132.	*Uca vocans* (Linnaeus, 1758)	B	BW/EW
	Family **Pinnotheridae**		
133.	*Nepinnotheres cardii* (Bürger, 1895)	C	CW
134.	*Pinnotheres boniensis* Stimpson, 1858	C	BW/EW
135.	*Pinnotheres mactricola* Alcock, 1900	C	CW

CW: Coastal Water (Marine); BW: Brackish Water; EW: Estuarine Water; FW: Fresh Water; CL: Climber; CO: Commensalic; B: Burrower; R: Runner; S: Swimmer; W: Walker; NS: Not Studied.

Distribution and Zonation

Inter-tidal crabs living in mangroves/deltaic region of Sundarbans have been found to fall broadly under following three groups on the basis of habitat preferences:

1. **Crabs always living in burrows**: The following crabs are included under this category: *Uca dussumieri, U. rosea, U. lactea, U. triangularis, Ocypode ceratophthalma, O. macrocera, O. platytarsis, Dotilla blanfordi, D. intermedia, Dotillopsis brevitarsis, Ilyoplax gangeticus, I. stapletoni, Scopimera globosa, S. investigatoris, S. proxima, Metaplax crenulata, M. dentipes, M. distincta, M. indica, M. intermedia, Episesarma mederi, Muradium tetragonum, Parasesarma pictum, P. plicatum, Perisesarma bidens, Sesarmops intermedium, Sesarmoides longipes, S. kraussi, Pseudosesarma edwardsii* and *Neosarmatium smithi.*
2. **Crabs living among crevices of mangrove trees, pneumatophores and logs**: In this category are included 6 species, such as, *Metopograpsus latifrons, M. messor, Parasesarma plicatum, Perisesarma bidens, Eurycarcinus orientalis* and *Ashtoret lunaris*. Of these, the pilumnid crab, *Eurycarcinus orientalis* was found in decaying mangrove logs at Gona Island and at Matla river near confluence of Bidya river. It was also noted in burrows at Haldibari and Chhotahardi. The calappid crab, *Ashtoret lunaris* was also often collected from dead stumps at Saimari Point and Bhangaduni Island.
3. **Crabs always associated with bivalves**: This category includes one species, namely, *Nepinnotheres cardii*. It is to be noted that during survey works, the present authors noted the occurrence of an unidentified crab species within the mantle cavity of the bivalve, *Anadara granosa* in mangrove habitats of Prentice Island.

Crabs are characteristic components of coastal fauna and have received much attention all over the world. Distribution of these animals can be correlated with the tolerance of organisms to extremes of physical conditions such as temperature, salinity and water loss (Powers and Bliss, 1983). In general, crabs inhabiting upper shore levels have a tendency to have higher and lower lethal temperature limits than their counterparts from the lower shore. The ability to withstand water loss and osmotic stress is more in species occupying at the upper shore compared to those from the lower tide levels (Edney, 1962; Macintosh, 1982, 1984).

Zonation

Like other shore animals in India and elsewhere, brachyuran crabs of Sundarban delta display distinct zonation pattern in response to their different degrees of adaptation. Several factors govern the distribution and zonation of crabs, such as, physical nature of substratum, tidal exposure and period of tidal inundation, rate of siltation, vegetation, salinity, temperature, food and predation (Teal, 1958; Miller, 1961; Kerwin, 1971). The zonation patterns of some selected brachyuran species of Sundarban delta are given in Table 3.2.

Table 3.2: Zonation Pattern of some Selected Brachyuran Crabs of Sundarban Delta

Intertidal zones	*Occurrence of Brachyuran Crabs in the Preferred Zone*
Low Tide Level	*Scylla serrata, S. tranquebarica, Metaplax crenulata, M. intermedia, Macrophthalmus (Mareotis) tomentosus, Uca dussumieri*
Middle Tide Level – Low Tide Level	*Ocypode ceratophthalma, Dotilla blanfordi*
Middle Tide Level – High Tide Level	*Ocypode macrocera*
High Tide Level – Extreme High Water Spring	*Uca rosea, U. triangularis, Metopograpsus latifrons, M. messor, Episesarma mederi, Muradium tetragonum, Perisesarma bidens, Parasesarma pictum, Sesarmops intermedium, Neosarmatium smithi*

Several studies on distribution and zonation of brachyuran crabs across different Indian shores (Joel *et al.*, 1986; Chakraborty and Choudhury, 1992b; Ravichandran *et al.*,2001; Dev Roy and Nandi, 2012) indicate variations in zonation patterns from shore to shore and even within the same shore as also revealed from the present study.

Behavioural Pattern

Brachyuran crabs of Sundarban delta exhibit interesting behaviours in relation to habit and habitat of the crabs (Figure 3.1). Broadly, the behaviour of brachyuran crabs of Sundarban delta may conveniently be categorized into four major groups based on their life styles as indicated by Schafer (1954) and Warner (1977) which include moving (*e.g.*, walking, running or climbing) swimming, burrowing, inconspicuousness and associative (*e.g.*, commensal, symbiotic or parasitic). Besides these, they display diverse ways of feeding and social behaviour. However, in general, crabs are primarily known ecologically as runners, climbers, swimmers and burrowers.

Runners

In the Sundarban delta, semi-terrestrial shore crabs, *viz.*, ocypodids and grapsids, are usually 'rapid runners'. These land crabs are always alert and ready to race off after prey or run for cover at the sight of predator. Such sights were common in Sagar, Frasergunj, Bakkhali and Sainmari beaches.

Climbers

Walking or running on vertical or near vertical surfaces, *viz.*, embankment structures, jetties, mangrove tree trunks, twigs, *etc.*, is a specialised activity of spider crabs (Warner, 1977). But mangrove forest inhabiting grapsid and sesarmid species are quite apt in climbing trees at the approach of flow tide.

Swimmers

Portunid crabs are well known swimmers, though a wide variety of marine and estuarine crabs of the genera, *viz.*, *Varuna*, *Ashtoret* and *Matuta* of Sundarban delta are comfortably adapted for swimming. In portunid crabs, last pair of legs

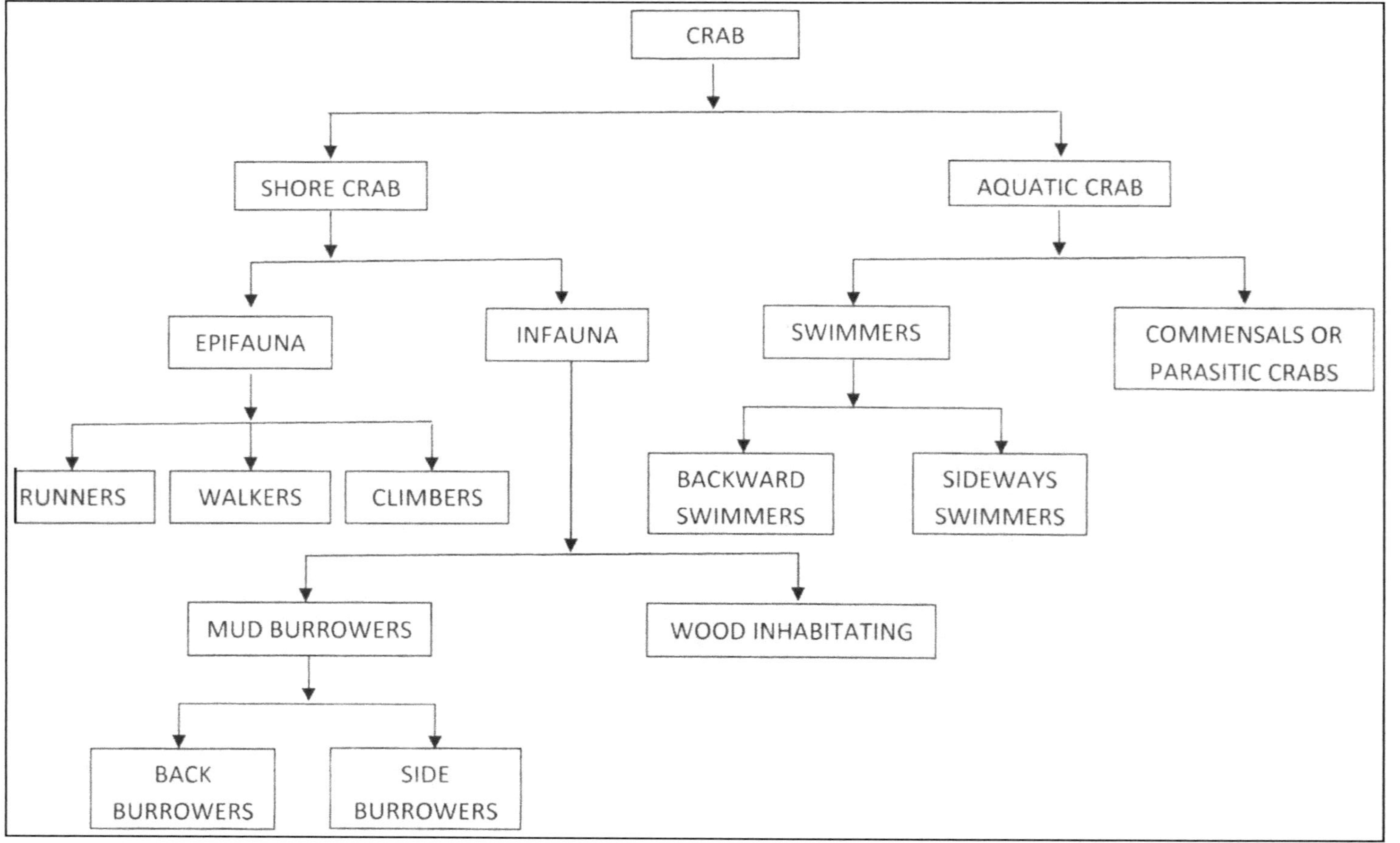

Figure 3.1: Schematic Representation of Crab Fauna of Sundarban Delta Based on Habit and Habitat.

is modified into paddles and are aligned in such a way to permit forward and backward strokes to produce a sculling action while swimming (Spirito, 1972).

Burrowers

Mangrove crabs of Sundarban delta are mostly burrowers, inhabiting self-constructed burrows in the intertidal mudflats. They belong to the families Ocypodidae, Dotillidae, Portunidae and Sesarmidae. The burrows protect the crabs from extremes of temperature and salinity stresses and also save them from predators and aggression from neighbouring crabs. The burrows are also important as an arena for courtship. Burrowing is most diverse and specialised in dotillids. Some members of the family Dotillidae have been found to make "igloo" in Sundarbans as in Chulkati and Ajmalmari-Kaikolmari. Mud crabs of Sundarban show side burrowing habit, digging simple or complex burrows depending on the nature of substratum and threats from man and wild animals (Nandi and Pramanik, 1994; Pramanik, 2014a). The depth of the burrows of *Scylla serrata* crabs in Sundarban region varied from 0.7-1.45 m, while the width of the burrow measured nearly twice the depth or thickness of the sheltering crab (Nandi and Dev Roy, 1989).

In addition to above life style oriented behaviours, brachyuran crabs of Sundarban mangroves show specialized feeding and social behaviours as follows:

Feeding Behaviour

Primarily brachyuran mangrove crabs are as usual opportunistic omnivores with a preference for animal food. Still, they have some tendencies of specialization towards particular types of diet deriving their distinction as carnivorous, hervivorous, algivorous, detritivorous, etc. The grapsid crabs of Sundarban as well as some sesarmid species (*viz., Sesarma edwardsi*), show preference for herbivorous/ detritivorous diet scooping vigorously from tidal flats (Paul and Nandi, 1999). The portunid species, on the other hand, depend on animal food, their mouth parts and appendages are specialized for catching prey species. Distinct specialization is shown by ocypodid crabs in response to their deposit feeding habit in sandy beaches of Sagar Island. The ocypodid crabs have developed spoon-tipped hairs in the second maxilliped as an adaptation to inhabit sandy-mud habitats (Miller, 1961).

Social Behaviour

Besides waving, threat and combat behaviour, brachyuran crabs show interesting courtship behaviour with morphological dimorphism, conspicuously in cheliped size. Such social behaviour along with sexual dimorphism allowing the recognition of sex and recognition anomalous behaviour (not retreating) which also allows the recognition of receptiveness (Warner, 1977).

Adaptive Features

Brachyuran crabs particularly mangrove crabs of Sundarban delta are morphologically, physiologically and behaviourally well adapted to their environment. Briefly, the morphological and behavioural adaptations of crabs of Sundarban delta are as follows:

Morphological and Physiological Adaptations

Morphological adaptations in brachyuran crabs of Sundarban delta, like other parts of the world, are in response to life style, habit, habitat preference and also in response to food, feeding behaviour and to some extent with courtship behaviour. In general, most of the brachyuran crabs have light but stiff chitinous exoskeleton with tips of their walking legs adapted for clinging, while swimming crabs have their last pair of legs flattened into paddles to exert propulsive force for swimming purpose. Burrowing crabs are of two categories such as back burrowers and side burrowers. Back burrowers have flattened digging legs which also help in swimming. However, spectacular adaptations are noticed in response to different feeding methods which include modifications in chelae (such as, strong crushing chela, fast cutting chela, serrated spoon-tipped chela, slender sharp-toothed chela, etc.) and in the modifications of the opposing surfaces of dorsal and lateral teeth of the gastric mill in response to different diet (Warner, 1977).

Majority of mangrove crabs of Sundarban delta as usual are capable of breathing in water. They extract oxygen in two ways: (i) some crabs (also called non-pumpers), such as,*Ocypode*, *Uca* and *Grapsus* can retain water in the gill chamber and circulate a stream of air through the water. *Ocypode* and *Uca* are additionally provided with special respiratory openings between the third and fourth walking legs to take up water from thin films (ii) others recirculate the branchial water over the body surface (Macnae, 1968). This method is characteristic of the family Sesarmidae (*Sesarma* sp.), Varunidae (*Metaplax* sp.), Grapsidae (*Metopograpsus* sp.) and macrophthalmids, such as, *Macrophthalmus* spp. (Macnae, 1968). Furthermore, it was also observed that *Episesarma mederi* can remain out of water for more than ten hours. Several modifications of gas exchange surfaces in crabs include: (i) gills are reduced in numbers (6 pairs in *Ocypode* in comparison to 9 pairs in marine forms as reported by Edney, 1960), (ii) gill size is reduced (Gray, 1957; Veerannan, 1974), (iii) gill volume is reduced in relation to body volume (Ayers, 1938) and (iv) gills are thickened and stiffened to avoid collapse. In order to compensate the gill surface area, in some members of the terrestrial and semi-terrestrial crabs the branchial chamber is partitioned into an upper vascularized chamber functioning as lung and a lower regular gill chamber. In *Ocypode ceratophthalma* and in some other ocypodid crabs, branchial covers develop lacunar system which are highly vascularized and often adorned with respiratory tufts (Edney, 1960; Bliss, 1968).

Behavioural Adaptation

Crabs of Sundarban delta display different types of behavioural mechanisms to combat desiccation and temperature, which include nocturnal activity and burrowing habit.

a. Nocturnal Habit

Some of the mangrove crabs of Sundarban delta, such as, *Ocypode ceratophthalma*, *Muradium tetragonum* and *Episesarma mederi*, are nocturnal or crepuscular in habit. They take advantage of lower temperature and increased moisture from moist substrates during evening and at night or dawn to combat desiccation.

b. Burrowing Habit

A good number of mangrove crabs of Sundarban delta are burrowers. The burrows frequently descend to water table, standing pools and a damp substratum.

It may be mentioned that crabs living on the lower shore do not experience the problems of desiccation as the habitat is wetted by every high tide of the fortnightly lunar cycle. Their activities are, however, affected by the daily incoming tides. Consequently, crabs inhabiting such areas feed intensively throughout the period of exposure as this zone remains under water during each high tide and also in the neap tide. But crabs inhabiting higher shore level and the supra-littoral zones are subjected to more prolong exposure. Therefore, crabs inhabiting the higher shore level visit their burrows periodically in order to cool and replenish the body water lost by exposure and evaporation. Macintosh (1984) observed the frequency of burrow visits by the fiddlers to be more as the shore become drier and hotter.

Summary

A total of 135 species of brachyuran crabs belonging to 72 genera and 26 families have been identified and reported herein from marine and estuarine areas of Sundarban delta. Of these, 69 species are marine and 70 species estuarine, with 16 overlapping species. The family Portunidae (23 species) represents the highest diversity, followed by Sesarmidae (13 species) and Leucosiidae, Varunidae and Macrophthalmidae by 10 species each. The families Dromiidae, Raninidae, Aethridae, Corystidae, Ethusidae and Scalopidiidae are represented by one species each. Habits of these crabs are categorized as burrower, runner, swimmer, climber and commensalic inhabiting other animals. They are found broadly in various aquatic habitats like coastal water, brackish water impoundment; estuarine river water and freshwater pond. Both morphological and behavioural adaptations of crabs of Sundarban delta are found in response to life style, habit, habitat, food, feeding, and courtship behaviour. They have light chitinous exoskeleton, walking legs adapted for clinging, last pair of legs flattened into paddles for swimming, etc. Burrowing crabs are both back burrowers and side burrowers. They construct complex burrows to escape from predation and have flattened digging legs. Nocturnal habits of the crabs are correlated to combat desiccation and temperature.

Acknowledgement

The senior author is thankful to Dr. K. Venkataraman, Director, Zoological Survey of India, Kolkata for the facilities provided.

References

Alcock A. (1895-1901). Carcinological Fauna of India (Nos. 1-6. The Brachyura, Oxyrhyncha, Oxystomata, Cyclometopa, Primigenia or Dromiacea and Catometopa or Grapsoidea). *J. Asiat. Soc. Bengal*, 64: 157-291; 65: 134-296; 68: 67-233; 68: 111-119, 123-126; 69: 279-456. Reprint edition in 1984. International Books and Periodical Supply Service, New Delhi.

Ayers JC. 1938. Relationship of habitat to oxygen consumption by certain estuarine crabs. *Ecology*, 19: 523-527.

Bairagi N. 1995. *Ocypodidae: Decapoda: Crustacea. Estuarine Ecosystem Series, Part 2: Hugli-Matla Estuary*. Zoological Survey of India, Calcutta. pp. 263-287.

Bhadra S. 1995. *Portunidae: Decapoda: Crustacea. Estuarine Ecosystem Series, Part 2: Hugli Matla Estuary*. Zoological Survey of India, Calcutta. pp. 249-262.

Bliss DE. 1968. Transition from water to land in decapods crustaceans. *Am. Zool.*, 8 : 355-392.

Chakraborty SK and Choudhury A. 1992a. Population ecology of fiddler crabs (*Uca* spp.) of the mangrove estuarine complex of Sundarbans, India. *Tropical Ecology*, 33: 78-88.

Chakraborty SK and Choudhury A. 1992b. Ecological studies on the zonation of brachyuran crabs in a virgin mangrove island of Sundarbans, India. *J. mar. biol. Ass. India.*, 34(1 and 2): 189-214.

Chakraborty SK, Choudhury A and Deb M. 1986. Decapoda Brachyura from Sundarban mangrove estuarine complex, India. *J. Beng. Nat. Hist. Soc.*, 5(1): 55-68.

Chaudhuri AB and Choudhuri A. 1994. Mangroves of Sundarbans, India, IUCN. 1: 1-247. Bangkok, Thailand.

Chopra B. 1933a. Notes on Crustacea Decapoda in the Indian Museum. 3. On the Decapod Crustacea collected by the Bengal Pilot Service off the mouth of the River Hughli: Dromiacea and Oxystomata. *Rec. Indian Mus.*, 35(1): 25-52.

Chopra B. 1933b. Further notes on Crustacea Decapoda in the Indian Museum. 4. On two new species of Oxystomus crabs from the Bay of Bengal. *Rec. Indian Mus.*, 35(1): 77-86.

Chopra B. 1934. Notes on Crustacea Decapoda in the Indian Museum.6. On a new Dromiid and a rare Oxystomus crab from the Sandheads off the mouth of the Hooghly River. *Rec. Indian Mus.*, 36(4): 477-481.

Chopra B. 1935. Notes on Crustacea Decapoda in the Indian Museum.8. On the Decapod crustacean collected by the Bengal Pilot Service off the mouth of the river Hugli. Brachygnatha (Oxyrhyncha and Brachyrhyncha). *Rec. Indian Mus.*, 37(4): 463-514.

De Man JG. 1908. The fauna of brackish ponds at Port Canning, Lower Bengal. Part 10. Decapod crustacean with an account of a small collection from brackish water near Calcutta and in the Dacca district, Eastern Bengal. *Rec. Indian Mus.*, 2(3): 211-231.

Deb M. 1995. *Crustacea: Xanthidae. Estuarine Ecosystem Series, Part 2: Hugli-Matla Estuary*, Zoological Survey of India, Kolkata. pp. 217-228.

Deb M. 1998. *Crustacea: Decapoda. State Fauna Series 3: Fauna of West Bengal, Part 10*, Zoological Survey of India, Kolkata. pp. 345- 403.

Dev Roy MK. 2010. Diversity and distribution of crustacean fauna in wetlands of West Bengal. *J. Environ. and Sociobiol*, 7(2): 147-187.

Dev Roy MK. 2014. Crustacean resources of Sundarban mangroves. *ENVIS Newsletter*, Zoological Survey of India, Kolkata, 20(1 and 2): 2-8.

Dev Roy MK and Nandi NC. 1991. Crabs of coastal West Bengal and Andaman Islands – their recognition and fishery information. *J. coastal agric. Res. Inst.*, 9(1-2): 69-75.

Dev Roy MK and Nandi NC. 2001. Crustacean biodiversity in mangrove ecosystems of Sundarbans and Bay Islands. *Bull. natn. Inst. Ecol.*, 11: 9-23.

Dev Roy, MK, and Nandi NC. 2002. Crab fishery resources of marine and estuarine ecosystems of Sundarban coast, West Bengal. *SDMRI Research Publication* No. 2: 111-119.

Dev Roy MK and Nandi NC. 2004. Crustacean fishery resources of coastal West Bengal and their conservation issues. *J. Environ. and Sociobiol.*, 1(1 and 2): 71-80.

Dev Roy MK and Nandi NC. 2008. Checklist and distribution of brachyuran crabs of West Bengal, India. *J. Environ. and Sociobiol.*, 5(2): 191-214.

Dev Roy MK and Nandi NC. 2012. *Brachyuran crabs (Crustacea). Fauna of Andaman and Nicobar Islands, State Fauna Series*, Zoological Survey of India, Kolkata, 19(1): 185-236.

Dubey SK, Chakraborty DC, Chakraborty S and Choudhury A. 2013. Burrow architecture of Red Ghost Crab *Ocypode macrocera* (H. Milne Edwards, 1852): a case study in Indian Sundarbans. *Explor. Anim. Med. Res.*, 3(2): 136-144.

Dutta SN. 1973. The edible crabs of deltaic West Bengal. *Seafood Export J.*, 5(12): 25-28.

Edney EB. 1960. Terrestrial adaptations. In. *The Physiology of Crustacea* (Waterman TH ed.), Academic Press, New York and London. pp. 367-393.

Edney EB. 1962. Some aspects of the temperature relations of fiddler crabs (*Uca* spp.). In. *Biometerology*, (Tromp SW ed.), Pergamon Press: Oxford, New York. pp. 79-85.

Ghosh SK. 1995. *Crustacea: Decapoda: Grapsidae. Estuarine Ecosystem Series, Part 2: Hugli Matla Estuary*, Zoological Survey of India. pp. 229-248.

Gray IE. 1957. A comparative study of the gill area of crabs. *Biol. Bull.*, 112: 32-42.

Henderson JR. 1893. A contribution to Indian Carcinology. *Trans. Linn. Soc. Lond. Zool.*, ser. 2, 5: 325-458.

Hora SL. 1933. A note on the bionomics of two estuarine crabs. *Proc. zool. Soc. Lond.*, 11: 881-884.

Hora SL. 1935. Crab fishing at Uttarbhag, Lower Bengal. *Curr. Sci.*, 3(11): 543-546.

Joel DR, Raj PJS and Raghavan R. 1986. Distribution and zonation of Shore crabs in the Pulicat Lake. *Proc. Indian Acad. Sci. (Anim. Sci.).*, 95(4): 437-446.

Kemp S. 1917.Notes on Crustacea Decapoda in the Indian Museum. X. Hymenosomatidae *Rec. Indian Mus.*, 13(5): 243-279.

Kemp S. 1919a. Notes on Crustacea Decapoda in the Indian Museum. XII. Scopimerinae. *Rec. Indian Mus.*, 16(5): 305-348.

Kemp S. 1919b. Notes on Crustacea Decapoda in the Indian Museum. XIII. The Indian species of *Macrophthalmus*. *Rec. Indian Mus.*, 16(5): 383-394.

Kemp S. 1923. Notes on Crustacea Decapoda in the Indian Museum. XVI. On two interesting crabs from the mouth of the river Hughly. *Rec. Indian Mus.*, 25(4): 405-409.

Kerwin JA. 1971. Distribution of the fiddler crab (*Uca minax*) in relation to marsh plants within a Virginia estuary. *Chesapeake Sci.*, 12(3): 180-183.

Macnae W. 1968. A general account of the fauna and flora of mangrove swamps and forests in the Indo-West Pacific Region. *Adv. Mar. Biol.*, 6: 73-270.

Macintosh DJ. 1982. Ecological comparisons of mangrove swamp and salt marsh fiddler crabs. *Proc. 1st International Wetland Conference,* National Institute of Ecology, Jaipur. pp. 243-257.

Macintosh DJ. 1984. Ecology and productivity of Malaysian mangrove crab populations (Decapoda: Brachura). *Proc. As. Symp. Mangr. Env. Res. and Manag.*, 1984: 354-377.

Mandal AK and Nandi NC. 1989. Fauna of Sundarban Mangrove Ecosystems, West Bengal, India. In. *Fauna of Conservation Areas*: 3. Zoological Survey of India, Kolkata, India, pp. 1-116.

Mandal AK and Misra A. 1985. Mud flats of Lower Bengal with special reference to macrobenthic fauna. In. *The Mangroves* (Bhosale LJ ed.). *Proc. Nat. Symp. Biol. Util. Cons. Mangroves*. pp. 425-431.

Miller DC. 1961. The feeding mechanisms of fiddler crabs, with ecological considerations of feeding adaptations. *Zoologica N. Y.*, 46: 89-100.

Milne Edwards A. 1861. Études Zoologiques sur les Crustacés recénts de la famille des Portuniens. *Archs. Mus. Natn. Hist. nat., Paris, sér.*, 10: 309-428.

Nandi NC, Das AK and Dey A. 2013. *Wetland Faunal Diversity of West Bengal.* WBBB and NBI, Kolkata, pp.1-256.

Nandi NC, Das SR, Bhuinya S and Dasgupta JM. 1993. Wetland faunal resources of West Bengal. 1. North and South 24-Parganas districts. *Rec. zool. Surv. India, Occ. Paper* No. 150: 1-50.

Nandi NC and Dev Roy MK. 1989. Burrowing activity and distribution of *Scylla serrata* (Forskål) from Hooghly and Matla estuaries, Sundarban, West Bengal. *J. Bombay nat. Hist. Soc.* 88(2): 167-171.

Nandi NC and Dev Roy MK. 1996. Biometrical studies on the Mud crab, *Scylla serrata* (Forskål) from Sundarban, West Bengal. *Seafood. Export J.*, 27(6): 17-22.

Nandi NC and Ghatak SS. 1985. Crabs of commercial importance from coastal West Bengal. *J. Indian Soc. Coastal agric. Res.*, 3(2): 131-135.

Nandi NC, Paul S and Dev Roy MK. 1998. Burrowing behavior of the Shore crab *Ocypode macrocera* H. Milne Edwards from Sundarban, West Bengal. *J. Bombay nat. Hist. Soc.*, 95(1): 140-141.

Nandi NC and Pramanik SK. 1986. A note on crab fishery and landing of *Scylla serrata* (Forskål) from Budhakhali, Sundarbans, West Bengal. *J. Indian Soc. Coastal agric. Res.*, 4(2): 151-153.

Nandi NC and Pramanik SK. 1994. *Crabs and Crab Fisheries of Sundarban.* Hindusthan Publishing Corporation (India), Delhi. pp. 1 -192.

Nandi NC and Pramanik SK. 2015. Wetland economics. 2. Crab production in brackishwater ponds of Sundarban, West Bengal. *Proc. Nat. Sem. on Aquaresources: Care and Concerns*, pp. 9.

Paul S and Nandi NC. 1999. Feeding behaviour of the grapsid crab *Sesarma edwardsi* De Man. *Rec. Zool. Surv. India*, 97(2): 265-267.

Paul S, Roy M and Dev Roy MK. 2012. Occurrence and seasonal abundance of sesarmiid crab, *Pseudosesarma edwardsii* (De Man) from Hugli-Matla estuary, West Bengal. *J. Environ. and Sociobiol.*, 9(2): 193-194.

Paul S, Sinha C, Dev Roy MK. *et al.*, 1999. Fish meal fishery resources and their conservation issues of Frasergunj - Bakkhali coast, West Bengal. *Proc. National Seminar on Integrated Coastal Zone Management.* pp. 40-41.

Pearse AS. 1932. Observations on the ecology of certain fishes and crustacean along the bank of the Matla River at Port Canning. *Rec. Indian Mus.*, 34(3): 289-298.

Powers LW and Bliss DE. 1983. Terrestrial adaptations. In. *The Biology of Crustacea* (Vernberg FJ and Vernberg HB eds.). Academic Press, New York. pp. 271-331.

Pramanik SK. 2014a. Sundarbaner Kankramara (In Bengali). Vivekananda Book Centre, Kolkata. pp. 1-185.

Pramanik SK. 2014b. *Sundarban: Jal-Jangal-Jiban.* Sahitya Prakash, Kolkata. pp. 1-175.

Pramanik SK and Nandi NC. 2004. *Dry Fish Production Profile of Indian Sundarban.* Classical Publishing Company, New Delhi. pp. 1-272.

Pramanik SK and Nandi NC. 2011. *Fisheries Sociology of Indian Sundarban.* Narendra Publishing House, New Delhi.

Pramanik SK and Nandi NC. 2012. Crab fattening (Chamber chas) – a promising enterprise in Indian part of Sundarban. *J. Environ. and Sociobiol.*, 9(1) : 78.

Ravichandran S, Soundarapandian P and Kannupandi T. 2001. Zonation and distribution of crabs in Pichavaram mangrove swamp, southeast coast of India. *Indian J. Fish.*, 48(2): 221-226.

Reddy KN. 1995. Hermit crabs (Crustacea: Decapoda). *Estuarine Ecosystem Series, Part 2: Hugli Matla Estuary,* Zoological Survey of India. pp. 199-215.

Saha D. 1986. Intertidal crabs of Sundarbans Tiger Reserve, West Bengal. *Environ. and Ecol,* 4(1): 177-178.

Schäfer W. 1954. Form und Funktion der Brachyuren-Schere. *Abh. senckenb. naturforsch. Ges.*, 489: 1-65.

Spirito CP. 1972. An analysis of the swimming behaviour of the portunid crab *Callinectes sapidus*. *Mar. Behav. Physiol.*, 1: 261-276.

Teal JM. 1958. Distribution of fiddler crabs in Georgia salt marshes. *Ecology*. 39(2): 185-193.

Veerannan KM. 1974. Respiratory metabolism of crabs from marine and estuarine habitats: an interspecific comparison. *Mar. Biol.,* 26: 35-43.

Warner GF. 1977. *The Biology of Crabs*. Elek Science, London. pp. 202.

Chapter 4

Chironomids Associated with Rice Agroecosystem of West Bengal: A Review

Sailesh Chattopadhyay[1] and Abhijit Mazumdar[2]

[1]Department of Forest Biology and Tree Improvement, Birsa Agricultural University, Ranchi

[2]Entomology Research Unit, Department of Zoology, University of Burdwan, Burdwan

Introduction

The areas under paddy cultivation in India is estimated to be 40.9 million hectares and these rice fields may serve as temporary wetlands and are considered to be important reservoirs of aquatic biological diversity (Bassia *et al.*, 2014), harboring many of the same species that breed in natural temporary ponds (Lawler, 2001). Thus, the rice agro-ecosystem plays a great role in the sustenance of regional biodiversity of various invertebrate and vertebrate groups. Rice fields, being temporary aquatic habitats with a generally predictable dry phase and a wet phase may scientifically be defined as an agronomically managed wetland ecosystem (Bambaradeniya and Amerasinghe, 2003) and several aquatic fauna dominate the wet phase (Roger and Kurihara, 1988). The rice fields may also be considered as modified marsh ecosystem and on account of agronomic practices make them less favourable for some organisms while temporarily more favourable for others.

Chironomidae is one of the cosmopolitan dipteran insects, representatives of which occur in all the zoogeographical regions including Antarctica (Ashe *et al.*, 1987) and are found in wide variety of aquatic habitats including lakes, ponds, marshes, swamps, river creeks, temporary pools, pitcher plants, bromeliads, tree-holes and leaf molds. Even man-made habitats such as sewage and water treated plants, fish pools, irrigation ditches and bird baths are also exploited by chironomids. Occasionally, they are found in marine, semi-terrestrial and terrestrial biotopes also. Individual species live in broad array of habitats from terrestrial to fully aquatic, however, a total of 339 genera and 4,147 species possess exclusively aquatic immature stages (Ferrington, 2008). Many species of chironomids undergo their immature development in habitats with rapidly change in suitability and generally follow two strategies for utilization of habitat – physiological or behavioral adaptations of larvae and repeated recolonization (Frouz *et al.*, 2003).

Besides, different microhabitats of the rice fields are occupied by one or more species of such larvae (Ree *et al.*, 1981, 1982; Ferrarese, 1993). Although they are often seen wriggling through open water, these larvae are hardly considered as typical nekton animals. Each chironomid taxon is thought to occupy a different niche in aquatic ecosystem responding individually to different chemical, physical and biological parameters.

Rice Field Chironomids: A Retrospect

Although rice fields have long been recognized as a pre-eminent habitat for invertebrate community, chironomids in particular, only few works related to this are available (Al-Shami *et al.*, 2010). Thienemann (1954) listed the species of midges from rice fields of southern Sumatra and west Java where the larvae were used as food of carp. The harmful activities of larvae of *Cricotopus trifasoiatus* to the rice seedlings in France were documented long back by Risbec (1952). Darby (1962) made a list of 36 species from the rice fields and its vicinities of California, USA and accounted 30 of them actually harbouring the paddies. A lot of species of chironomids was reported to injure seedling rice in Japan (Yokogi and Ueno, 1971; Ishihara, 1972). Hashimoto *et al.* (1981) gave a brief description on the adult morphology of 32 species of chironomids in the rice fields of Thailand. Sasa and Kikuchi (1986) made an account of 34 species from the rice fields of Japan and stated that at least 40 species of chironomids were found to be breeding in the rice paddy areas of Japan. According to Surakarn and Yano (1995), 166 species have been found to be associated with the rice paddies around the world so far. Ali (1995) described the rice field fauna from Malaysia and recorded the aquatic insects as food items for juvenile fish found in rice fields. From West Bengal, India, Chaudhuri and Chattopadhyay (1990) described 54 species of chironomidae occurring in the rice fields with a general account on their biology, emergence pattern, sex ratios and swarming and mating behavior. The chironomids inhabiting rice fields of India included 10 species previously reported from Thailand (Hashimoto *et al.*, 1981), 6 species from Japan (Sasa and Kikuchi, 1986) and 1 species from North America (Darby, 1962). Studies on various aspects of biology of the chironomids have been in progress since the late fifties in Europe, the then USSR and North America (Anderson *et al.*, 1965; Syrjamaki, 1965; Hilsenhoff, 1966; Oliver, 1971). Hafiz (1939) studied the chironomids from filter beds of water reservoir of Calcutta Corporation. After a long gap of nearly three decades, Chaudhuri *et al.* (1983), Chaudhuri and Ghosh (1986), Ali *et al.* (1987), Chattopadhyay *et al.* (1988), Chaudhuri *et al.* (2001) and Hazra *et al.* (2002) rejuvenated the study of biology of few species prevalent in India. The emergence of adults was related to the duration of larval period and was reported to occur throughout the year (Corbet, 1964). According to Palmen (1955) and Morgan and Waddell (1961), most species exhibit diel periodicity of emergence. Fischer and Rosin (1968) showed that dual control of temperature and light affected the diel emergence pattern in *Chironomus nuditersis* resulting peak emergence after the onset of dusk and a little lower in the dawn while, Ali (1980) observed a relation between the eclosion timing and the dual light changes. The emergence patterns of a good number of species were also presented by Danks and Oliver (1972a), Ali and Mulla (1979), Singh and Harrison (1982), Chaudhuri *et al.* (1983) and Chattopadhyay *et al.* (1988). There are several reports of sex ratio of chironomids deviating from the normal 1: 1. Darby (1962), Lindeberg (1971), Singh and Harrison (1982), Chaudhuri *et al.* (1983) and Chattopadhyay *et al.* (1988) reported the occurrence of unusual sex ratios of midges. The account on the population dynamics of larval chironomids appears meager. It was Darby (1962), who worked on the population dynamics of adult chironomids associated with Californian rice fields (USA) in relation to its physico-chemical factors. Carter (1976) made population study of chironomidae of Lough Neagh, the largest fresh water lake of British Isles. The ecological aspects of larval chironomids inhabiting lentic and lotic situations were exploited with benthic limnology and the relationships of the biota with the physico-chemical factors were reported by different workers (Lindegaard and Jonasson, 1979; Roy *et al.*, 1980; Parkin and Stahl, 1981; Ali and Baggs, 1982; Ali and Majori, 1984; Banaszak, 1984; Bass, 1986 and Lindegaard and Jonsson, 1987). Ree *et al.* (1981, 1982) included chironomid fauna in course of their study on population dynamics of aquatic invertebrates associated with the rice fields. Kikuchi *et al.* (1985) observed the seasonal prevalence of chironomid midges and mosquitoes by light traps set in a rice-paddy area in Tokushima (Japan).

Life Cycle Attributes

The studies of life cycles of chironomids inhabiting rice paddies are important as they present several interesting features. The five Oriental species are common in rice paddies and the life cycles subsequently studied in the laboratory are *Chironomus samoensis, Dicrotendipes pelochloris, Stictochironomus obscurum, Clinotanypus fuscosignatus* and *Procladius noctivagus*. The duration of each life stage was generally neglected in the earlier works. It could conveniently be determined only in the laboratory as done by Hilsenhoff (1966). The duration of life cycle was temperature-dependent and related to the overall length of the life cycle (Oliver, 1971). The time required for hatching of egg was recorded as 2.5 days at 24°C in *Chironomus atrells* (Anderson and Hitchcock, 1968), 4 days at 20°C in *C. zealandious* (Forsyth 1971) and 4-5 days at 18°C in some other species of chironomids. Similarly the duration was noted to be 3-4 days at 18°C in *Einfeldia synochrona* (Oliver, 1971) and 2-2.5 days and 3-3.5 days at 32 ± 2.5°C and 24.5 ± 3°C respectively in *Tanypus bilobatus* (Chaudhuri *et al.*, 1983). In the recent past, Chaudhuri and Ghosh (1986) reported that the duration of egg stage was 40-60 hours and 48-72 hours in *Kiefferulus barbatitarais* and 24-36 hours and 48-60 in *K. calligaster* at temperatures 32 ± 2.5°C and 24.5 ± 3°C respectively. Chattopadhyay *et al.* (1988) furnished the duration in case of *Polypedilum nubifer* to be 28-40 hours in summer (32 ± 2.5°C) and 42-72 hours in winter (24 ± 2.5°C). It may therefore be inferred that the hatching period varies with temperature, taking less time with the rise of temperature (in summer) and much with the fall of temperature (in winter). The larvae pass through four instars (Oliver, 1971; McCauley, 1974) to attain pupal stages. The duration of larval life stages of four rice field species in both summer and winter and one species only in winter have been studied by Chaudhuri and Chattopadhyay (1990). The larval duration of the members of chironominae required 16-29 days and 20-40 days in summer and winter respectively whereas, tanypodid member needed 45-50 days and 55-72 days respectively against the said temperature. Oliver (1971) stated that changes in the duration of larval period occurred over a wide range of latitude also. With increasing latitude or its climatic equivalent, the duration of larval period became longer and temperature happened to be one of the main factors responsible for this increase. According to Jonasson (1965) *Chironomus anthracinus* was recorded to have a two year life cycle, although some emerged after one year and growth was reduced both in summer (related to low oxygen concentration) and in winter (related to low temperature). The two year life cycle in other species was also reported by Hamilton (1965). Butler (1982) reported a seven year life cycle for two *Chironomus* species in arctic Alaskan tundra ponds. The processes involved in slowing down or cessation of growth and development had not received due attention. Developmental arrests with a cessation of feeding in both summer and winter were made known by different workers (Hilsenhoff, 1966; Jonasson and Kristiansen, 1967 and Armitage, 1970). From these studies, it become evident that temperature itself was not a factor which controlled larval life, rather it played a key role in determining the length of larval life and other factors, particularly the availability of food, were also important (Jonasson, 1965; Forsyth, 1971). As compared to the larval stage, the duration of the pupal stage was very brief ranging from a few hours to few days (Forsyth, 1971; Oliver, 1971; Chaudhuri *et al.*,1983; Chaudhuri and Ghosh, 1986; Chattopadhyay *et al.*, 1988) and was temperature dependent (Mundie, 1956; Hilsenhoff, 1966). The pupal life varied 21- 48 hours and 26-56 hours in chironominae and 46-66 and 51-75 hours in tanypodinae in summer and winter respectively.

The life span of adult chironomids lasted for a few days (Hein and Schmulbach, 1972). The emergence period was related to the duration of larval period (Chaudhuri *et al.*, 1983) and the adults were to fly immediately after emergence which seemed to be the same with the species reported by several researchers (Mundie, 1956; Hilsenhoff, 1966; Brundin, 1966; Forsyth, 1971 and Hein and Schmulbach, 1972). The rate of emergence of the species was highest during the middle of the total emergence period conforming the findings of Chaudhuri *et al.* (1983) and Chattopadhyay *et al.* (1988). According to Ali (1980), the periodicity was due to complex interaction of various environmental stimuli and endogenous rhythms. The phenomenon appeared to be typically regulated and maintained by an "internal clock" (the endogenous rhythms) which was said to be influenced by external stimuli or exogenous factors (Corbet, 1964). In the far north latitude, temperature was regarded as the most important factor determining diel emergence periodicities in the high Arctic chironomids (Oliver, 1968; Danks and Oliver, 1972b). Similar observations were reported in mosquitoes by Corbet (1964). In contrast, intensity of light appeared as the factor controlling the diel emergence of chironomids in the temperate zones. This study revealed that *Chironomus samoensis* emerged at dusk in summer establishing the assumption of Morgan and Waddell (1961) and a few did at dawn. *Dicrotendipes pelochloris* and *Stictochironomus obscurum* did not show a distinct periodicity in emergence, occurring mostly throughout the day, *i.e.* dawn to dusk and might be treated as a model as stated by Singh and Harrison (1982). The bimodal pattern of emergence of *Procladius noctivagus* and *Clinotanypus fuscosignatus* reported by Ali and Mulla (1979) and Ali (1980). However, a deviation was noticed in the peak of emergence related to the earlier onset of sunset in *Clinotanypus fuscosignatus* as stated before. Therefore, light intensity might probably be the inducing agent in the emergence of *D. pelochloris* and *S. obscurum*, thereby confirming the observations made by Oliver (1968). Since these two environmental clues were immediately linked together, their combined effects require further investigation.

There are several reports on sex-ratio deviating from the normal 1: 1. According to Palmen (1955), males emerged in less number than female which was later supported by Lindeberg (1971). In rice field chironomids higher percentage of females emerged in case of *Dicrotendipes pelochloris*, *Clinotanypus fuscosignatus* and *Procladius noctivagus* as observed by Singh and Harrison (1982) but reverse situation with a preponderance of males occurred in *Chironomus samoensis* and *Stictochironomus obscruun* (Hein and Schmulbach, 1972; Chaudhuri *et al.*, 1983 and Chattopadhyay *et al.*, 1991). A rhythm in emergence starting with an excess of males and ending with females predominance was recorded in most of the chironomids in question and similar observations was recorded by Miall and Hammond (1900) but no such rhythm was observed in *Stictochironomus obscurum*. Details of life stages of chironomid species recorded from rice field agroecosystem of West Bengal are depicted in Table 4.1.

Ricefield Chironomids: Economic Importance

The adults are commonly known as "blind mosquitoes" pose a variety of nuisance towards human and manmade structures within the flight and dispersal range of midges, water front residents, works and business (Ali, 1980) as well as people living in the rice paddies area too (Sasa and Kikuchi, 1986). Dense swarms of midges often prelude outdoor human activity since the adults may inhaled into the mouth, eyes and ears of an individual. Besides, the newly emerged adults release faecal deposits which stain stucco, paint and wall finish leading to great aesthetic and economic loss. They also soil automobiles, including the head-lights and windshields and thereby cause traffic hazards. At night, the adults are attracted to lights and swarm around indoor and outdoor fixture. The midges are often reported to clog air-conditioning filters. Accumulation of dead midges and the unsightly spider webs on ceiling in which adult midges are trapped cause additional difficulties requiring frequent washing and maintenance of home and other properties. The dead midges give a stench similar to rotting fish. At times, the dead adults accumulate on the roads in such quantities that they make the roads slippery, causing traffic hazards. Midges have also been reported to be a problem for plastic and paint industries where hordes of adults fly into the final products. In addition to cause such nuisance problems, the midges are associated with human allergic reactions such as asthma and rhinitis (Cranston, 1995). Earlier, Igarashi *et al.* (1985) and Sasa and Kikuchi (1986) demonstrated that rice field breeding midge species. *Polypedilum kyotoense* is

Table 4.1: List of Chironomid Species and Associated Life Stages (L: Larva, P: Pupa and A: Adult) Collected from the Rice Fields of Gangetic West Bengal

Genus: *Chironomus* Meigen	
C. circumdatus Kieffer	: A ♂
C. filitarsis Kieffer	: A ♂
C. javanus Kieffer	: A ♂
C. nudipes Kieffer	: A ♂
C. samoensis Edwards	: L, P, A ♂♀
C. striatipennis Kieffer	: A ♂
Genus: *Cryptochironomus* Kieffer	
C. fulvus (Johannsen)	: A ♂
C. judicius Chaudhuri and Chattopadhyay	: L, P, A ♂♀
C. rostratus (Kieffer)	: A ♂
C. subovatus Freeman	: A ♂
Genus: *Dicrotendipes* Kieffer	
D. pelochloris (Kieffer)	: L, P, A ♂♀
D. septemmaculatus (Becker)	: A ♂♀
Genus: *Endochironomus* Kieffer	
E. pekanus (Kieffer)	: A ♂
Genus: *Harnischia* Kieffer	
H. acuta (Goetghebuer)	: A ♂
H. incidata Townes	: A ♂
H. tenuitubercula Chaudhuri and Chattopadhyay	: A ♂
H. viridula (Linnaeus)	: A ♂
Genus: *Kiefferulus* Goetghebuer	
K. barbatitarsis (Kieffer)	: L, P, A ♂♀
K. calligaster (Kieffer)	: L, P, A ♂♀
Genus: *Microchironomus* Kieffer	
M. fuscitarsus (Guha and Chaudhuri)	: L, P, A ♂♀
M. tener (Kieffer)	: L, P, A ♂♀
Genus: *Paracladopelma* Harnisch	: A ♂
P. aratra Chaudhuri and Chattopadhyay	: A ♂
P. sacculifera Chaudhuri and Chattopadhyay	: A ♂
Genus: *Pentapedilum* Kieffer	
P. robusticeps Guha and Chaudhuri	: A ♂
P. uncinatum Goetghebuer	: A ♂

Contd...

Table 4.1–*Contd...*

Genus: *Polypedilum* Kieffer	
P. aegypgtium Kieffer	: A ♂
P. annulatipes (Kieffer)	: A ♂
P. ascium Chaudhuri *et al.*	: A ♂
P. chaudhurii Chaudhuri *et al.*	: A ♂
P. circulum Chaudhuri and Chattopadhyay	: A ♂
P. lineatum Chaudhuri *et al.*	: A ♂
P. nubifer (Skuse)	: L, P, A ♂♀
P. suturalis (Johannsen)	: A ♂
P. yapensis Tokunaga	: A ♂
Genus: *Stenochironomus* Kieffer	
S. hilaris (Walker)	: A ♂
S. longipalpis (Kieffer)	: A ♂
Genus: *Stictochironomus* Kieffer	
S. affinis (Johannsen)	: L, P, A ♂♀
S. obscurus (Guha and Chaudhuri)	: L, P, A ♂♀
Genus: *Xenochironomus* Kieffer	
X. laviventris (Kieffer)	: A ♂♀
Genus: *Cladotanytarsus* Kieffer	
C. conversus (Johannsen)	: A ♂
C. gloveri Ghosh and Chaudhuri	: A ♂
C. multispinulus Guha *et al.*	: A ♂
Genus: *Tanytarsus* v.d.Wulp	
T. bifurcus Freeman	: A ♂
T. commoni Glover	: A ♂
T. fuscimarginalis Chaudhuri *et al*	: A ♂
T. vinculus Chaudhuri *et al.*	: A ♂
Genus: *Clinotanypus* Kieffer	
C.fuscosignatus (Kieffer)	: L, P, A ♂♀
Genus: *Procladius* Skuse	
P.noctivagus (Kieffer)	: L, P, A ♂♀
Genus: *Tanypus* Meigen	
T. bilobatus (Kieffer)	: L, P, A ♂♀
T. grandis Chaudhuri *et al.*	: A ♂
T. lucidus Chaudhuri *et al.*	: A ♂
T. tenebrosus Chaudhuri *et al.*	: A ♂

considered to be an important allergen, causing bronchial asthma among people of that area. Mizukami *et al.* (priv. comm.) reported an interesting case study, where the most abundant species, *Tanytarsus oyamai* caused a severe asthmatic attack in a girl due to accidental inhalation of midges during jogging in the vicinity of rice paddies. It is also evident that the chironomid midges, *Chironomus circumdatus* and *Polypedilum nubifer* can elicit allergic reactions to susceptible individuals living in Kolkata and there is a potential risk mainly during their abundant emergence (Nandi *et al.*, 2014).

The larvae, known as "red or blood worms" have been reported to cause injury to rice crops and seedlings (Yokogi and Ueno, 1971; Ishihara, 1972; Yasumatsu *et al.*, 1979). The role of *Chironomus tepperi* as a pest of directly-sown rice crops in New South Wales, Australia has been well established (Stevens *et al.*, 2000), however, many other chironomid species occur within crops during the first few weeks after sowing. When *C. tepperi* was absent, suggesting the other chironomid species present in the fields is not routinely involved in causing significant crop damage (Stevens *et al.*, 2006). They are also treated as alternate prey for predators of rice pest (Hashimoto *et al.*, 1981).

Chironomids as Ecosystem Service Provider

The life stages of chironomids are also deemed to be useful indicator of water pollution (Wilson and McGill, 1979). The indicator communities are useful in pin-pointing localized pollution with a particular area of water pollution (Saether, 1979). At the same time, the worms are deemed to play as indicators of radioactive pollution (Curry, 1960). The juvenile stages are known to form an important dietary component of fresh water fishes (Kajak and Warda, 1968; Legner *et al.*, 1975). Larvae of genus *Ablabesmyia* has been indicated to serve as symbionts of fresh water mussels (Roback *et al.*, 1979). Like larvae, the pupae of midges are now used in the biological classification of rivers important for fisheries management, monitoring and control of pollution (Wilson, 1980). The community structure of chironomids in the rice agroecosystem of Penang, Malaysia followed the dynamic changes of the field including agronomic practices, patterns of water availability, and phases of rice plant growth (Al-Shami *et al.*, 2010). The zooplankton of rice fields is composed largely of benthic species associated with vegetation comprising primarily of chironomidae and oligochaetes (Kurihara, 1989). According to Fernando (1993), the diverse fish fauna feed on rice pests, chironomid larvae and ephydrids. The rice fields also provide spawning sites for both white and black fishes. Some of the smaller cyprinds, many silurids, ophiocephalids and anabantids inhabit the rice field and contiguous irrigation channel, ditches in Asia. In India, there is a long tradition of culturing fish in rice fields which is a variant of integrated fish culture popularly known as paddy-cum-fish culture or rice-fish systems. Chakraborty and Bhowmick (1985) recorded benthic organism count of 133 to 10,044 unit/m^2 brackish water paddies. Das *et al.* (2000) described the ecology of the DWR fields and advocated that through improved stocking the entire food chain may be utilized for increased food production. Varghese *et al.* (1973) studied the natural food items of *Heteropneustes fossilis, Clarias batrachas* and observed that chironomid larva constituted more that 16 per cent of its dietary component. Similarly, Thakur and Das (1985) found that the most important food item of fishes was larvae of chironomid. The paddy fields are also rich in dipteran larvae, especially chironomids, which efficiently make use of the minerals and nutrients, especially nitrogenous materials unused, and the fish like carp which feed on the chironomid, make use of an otherwise untapped energy source in the natural cycle. Some algae, which make use of the nutrients and sunlight available, are also very useful in fish culture system, for fishes like *Tilapia mossambica* and *Trichogaster pectoralis* feed on these algae. The excrements of the fish are also efficiently recycled, part going to the rice plant itself and part going to phytoplankton (algae) which become fish food and is thus fully recycled. Although *C. carpio* showed an overwhelming reliance on detritus aggregate throughout the season, there was a distinct but small dietary prey component which consisted mainly of chironomids and corixids.

Production of chironomid larvae in culturing media of various organic wastes was demonstrated by Jana and Pal (1990). Chironomid burrowing activity resulted in a substantial rise of soluble orthophosphate (7.4 to 46.6 per cent) in all treatments. Phosphorus input through bioturbation was significantly higher ($P < 0.05$) than that of chironomid excrement or natural solubilization of rock phosphate. Besides, Sokolova *et al.* (1992) studied the biology of *Chironomus piger* and emphatically revealed the role of the species in the self purification of a river. Bhattacharyay *et al.* (2005) recorded antennal deformities of chironomid larvae and their use in biomonitoring of the heavy metal pollutants in the river Damodar of West Bengal. Faria *et al.* (2007) used *Chironomus riparius* larvae to assess effects of pesticides from rice fields in adjacent freshwater ecosystems. Similarly, the deformities of *Chironomus* larvae as indicator of pollution stress in rice fields of Hooghly District, West Bengal (Saha and Mazumdar, 2013).

Control Measures

Owing to variety of breeding habits, the control of chironomids has become somewhat difficult and majority of their control studies are focused on chemical control only (Ali, 1996). The existing midge control procedures by physical and cultural methods such as rotational flooding and drying partial areas of breeding or habitat depth increases (Ali and Mulla, 1976) or substrate removal by mechanical means (Ali *et al.*, 1976) are not applicable or economically viable for the permanent and vast midge sources of different countries. Biological control through the predators has been explored and a number of parasites and pathogens in midges have recently been reported. The pathogens of chironomids include viruses (Huger *et al.*, 1970; Harkrider and Hall, 1979), rickettsiae (Federici *et al.*, 1976) and fungi (Weiser, 1976), but none of them were developed for midge control purposes. Even the known protozoan parasites of chironomids like microsporida (Hunter, 1968) and the ciliophores (Corliss, 1960) were not used purposefully for midge control. Some works were carried out on the biological control of the pest through use of predators like fishes particularly employing sunfishes, mosquito fishes, common carps, catfishes, pupfishes, young bass etc., macroinvertebrates such as leeches and dytiscid beetles (Legner *et al.*, 1975). According to Legner *et al.* (1975), *Dugesia dorotocephala* was mostly studied as invertebrate predator of chironomid larvae and it offered a good potential for biological control of midges because of its occurrence in the same habitat and also due to its easy mass rearing (Tasi and Legner, 1977). More studies on cultural and biological control techniques with minimum dependence on chemicals are needed for the integrated control of nuisance chironomids (Tabaru, 1987). The ecological manipulation of the midge environment with the use of natural enemies may also provide a solution to this nuisance problem. Anthropogenic activities including pollution-related decimation of aquatic benthic communities might allow introduction of invasive chironomids particularly eutrophic lakes and wastewater treatment areas and may accumulate contaminants in high concentrations (Failla *et al.*, 2015).

Summary

Chironomids belonging to 53 species of 18 genera under 2 subfamilies, chironominae and tanypodinae were explored from the rice fields of Gangetic West Bengal. The list of midge species inhabiting rice fields of India included 10 species previously reported from Thailand, 6 species from Japan and 1 species from North America. Of these species, morphological details of 27 species including life stages of 6 species have been documented in the light of modern terminologies and usages. The duration of entire life cycle (egg to adult) appeared to very approximately 18-38 days and 24-50 days in chironominae and 46-60 days and 60-83 days in tanypodinae in summer (32 ± 2.5°C) and winter 24.5 ± 2.5°C) respectively. The total emergence was long in winter than summer. Most species emerged at dusk and a few did at dawn. Some showed distinct bimodal pattern of emergence while others exhibited amodal throughout the day. The larval population was noted to exhibit a decline with the growth of rice plants and the complex of ecological factors played an important role in this phenomenon directly or indirectly. Besides, the chironomid community is an important and integral component of the food web connectance present in the rice agroecosystem. It is considered as an important dietary component of fish juveniles particularly the catfishes, tadpoles etc propagating in rice ecosystem. The larval activity plays an important role in bioturbation and self purification of water. On human health perspective the adults emerging from paddy fields cause allergic reactions to individuals residing nearby. In some rice growing areas, the larvae seem to damage the rice seedling. Recent studies suggest that the deformity occurring in chironomid larvae may be an effective early warning tool for monitoring the health of wetlands. Future studies should focus on the different ecological role of the chironomid community in rice agroecosystem and link to the ecological services.

Acknowledgements

The authors are thankful to the Head, Department of Zoology, Burdwan University and Head, Department of Forest Biology and Tree Improvement, Birsa Agricultural University, Ranchi for providing the infrastructure facilities. We are grateful to Dr. P.K.Chaudhuri, Retired Professor, Department of Zoology, Burdwan University, an eminent Chironomidologist for his valuable suggestions and comments.

References

Ali A and Baggs RD. 1982. Seasonal changes of chironomid populations in a shallow natural lake and in a manmade water cooling reservoir in Central Florida. *Mosquito News*, 42 (1): 76-85.

Ali A and Majori G. 1984. A short term investigation of chironomid midge (Diptera: Chironomidae) problem in saltwater lakes of Orbetello, Grosseto, Italy. *Mosquito News*, 44(1): 17-21.

Ali A and Mulla MS. 1976. Chironomid larval density at various depths in the Southern California water percolation reservoir. *Environmental Entomology*, 5(6): 1071-1074.

Ali A and Mulla MS. 1979. Diel periodicity of eclosion of adult chironomd midges in a residential recreational lake. *Mosquito news*, 39: 360-364.

Ali A. 1996. A concise review of Chironomid midges (Diptera: Chironomidae) as pests and their management. *Journal of Vector Ecology*, 21(2): 105-121.

Ali A, Chaudhuri PK and Guha DK. 1987. Description of *Stictochironomus affinis* (Johannsen) (Diptera: Chironomidae), with notes on its behavior. *Florida Entomologist*, 70(2): 259-267.

Ali A, Mulla MS and Pelsue FM. 1976. Removal of substrate for the control of chironomid midges in concrete – lined flood control channels. *Environmental Entomology*, 5: 755-758.

Ali A. 1980. Diel adult eclosion periodicity of nuisance chironomid midges of Central Florida. *Environmental Entomology*, 9 (4): 365-370.

Ali A. 1995. Nuisance, economic impact and possibilities for control. In. *The Chironomidae: Biology and Ecology of Non-Biting Midges* (Armitage PD, Cranston PS and Pinder LVC. Eds.). Chapman and Hall, London, UK. pp. 339-364.

Al-Shami SA, Salmah Md RC, Hassan AA and Azizah MNS. 2010. Temporal distribution of larval Chironomidae (Diptera) in experimental rice fields in Penang, Malaysia. *Journal of Asia-Pacific Entomology*, 13: 17-22.

Anderson JF and Hitchcock SW. 1968. Biology of *Chironomus atrella* in a tidal cove. *Ann. Entomol. Soc. Am.*, 61: 1597-1603.

Anderson LD, Bay EC and Mulla MS. 1965. Aquatic midge investigations in Southern California. *Proc. Calif. Mosq. Contr. Assoc.*, 33: 32-33.

Armitage PD. 1970. The Tanytarsini (Diptera, Chironomidae) of a shallow woodland lake in South Finland, with special reference to the effect of winter conditions on the larvae. *Ann. Zool. Fenn.*, 7: 313-322.

Ashe P, Murray DA and Reiss F. 1987. The Zoogeographical distribution of Chironomidae (Insecta : Diptera). *Annls Limnol.*, 23 (1): 27- 60.

Bambaradeniya CNB and Amerasinghe FP. 2003. *Biodiversity associated with the rice field agroecosystem in Asian countries: a brief review.* Working paper 3. Colombo, Sri Lanka: International Water Management Institute (IWMI).

Banaszak J. 1984. Density and biomass of benthic Chironomidae (Diptera) in a lake and Melioration channel, situated in agricultural landscape. *Pol. Arch. Hydrobiol.*, 31 (4): 353-363.

Bass. D. 1986. Larval chironomidae (Diptera) of Big Thicket streams. *Hydrobiologia*, 135 : 271-285.

Bassia N, Dinesh KM, Sharma A and Pardha-Saradhia P. 2014. Status of wetlands in India: A review of extent, ecosystem benefits, threats and management strategies. *Journal of Hydrology: Regional Studies*, 2: 1-19.

Bhattacharyay G, Sadhu AK, Mazumdar A and Chaudhuri PK. 2005. Antennal deformities of chironomid larvae (Diptera: Chironomidae) and their use in biomonitoring of the heavy metal pollutants in the river Damodar of West Bengal, India. *Environmental Monitoring and Assessment*, 108: 67-84.

Brundin L. 1966. Transantarctic relationships and their significance, evidenced by chironomid midges. With a monograph of the subfamilies Podonominae and Aphroteniiae and the austral Heptagyiae. *K. Svenska Vetensk Acad. Handl.*, 11 : 1-472.

Butler MG.1982. A 7 year life cycle for two *Chironomus* species in arctic Alaskan tundra ponds (Diptera : Chironomidae). *Canadian Journal Zoology*, 60 (1): 58-70.

Carter CE. 1976. A population study of the Chironomidae (Diptera) of Lough Neagh. *Oikos.*, 27 : 346-354.

Chakraborty RK and Bhowmick ML. 1985. Ecology and productivity of brackish water paddy cum fish culture on river Hooghly. *Journal of Indian Society of Coastal Agricultural Research*, 3(2): 137-148.

Chattopadhyay S, Dutta T and Chaudhuri PK. 1988. Morphology and biology of *Polypedilum nubifer* (Skuse) from India (Diptera: Chironomidae). *Journal Bengal natural History Society*, 7(2): 29-41.

Chaudhuri PK and Chattopadhyay S. 1990. Chironomids of the rice paddy areas of West Bengal, India (Diptera: Chironomidae). *Tijdschrift vor Entmologie*, 133(2): 149-195.

Chaudhuri PK and Ghosh M 1986. Two Indian species of *Kiefferulus* Goet. (Diptera: Chironomidae). *Systematic Entomology*, 11: 277-292.

Chaudhuri PK, Hazra N and Alfred JRB. 2001. Check-list of Chironomid midges (Diptera) of India. *Oriental Insects*, 35: 335-390.

Chaudhuri PK, Nandi SR and Ghosh M. 1983. Postembryonic stages of *Tanypus bilobatus* (Kieffer) n. comb. (Diptera, Chironomidae) with a note on its biology. *Arch. Hydrobiol.*, 97 (1): 122-133.

Corbet PS. 1964. Temporal patterns of emergence in aquatic insects. *Canadian Entomologist*, 96 : 264-279.

Corliss JC. 1960. *Tetrahymena chironomi* sp. Nov., a ciliate from midge larvae and the current status of facultative parasitism in the genus *Tetrahymena*. *Parasitology*, 50: 111-153.

Cranston PS. 1995. Medical significance. In. *The Chironomidae: Biology and Ecology of Non-Biting Midges* (Armitage PD, Cranston PS and Pinder LVC eds.), Chapman and Hall, London, UK. pp. 31-61.

Curry LL. 1960. Midges larvae as indicators of radioactive pollution. *Proc. Ind. Waste Conf. Purdue Univ.,* 106: 269-280.

Danks HV and Oliver DR. 1972a. Seasonal emergence of some high arctic Chironomidae (Diptera). *The Canadian Entomologist,* 104 : 661-686.

Danks HV and Oliver DR. 1972b. Diel periodioities of emergence of some high arotic chironomidae (Diptera). *The Canadian Entomologist,* 104 : 903-916.

Darby RE. 1962. Midges associated with California rice fields, with special reference to their ecology (Diptera: Chironomidae). *Hilgardia,* 32 : 1-206.

Das DN, Roy B and Mukhopadhyay PK. 2000. Survey on farmers practices of fish farming in deep water rice environments in West Bengal, India. *Aquaculture,* 1(2): 125-139.

Failla AJ, Vasquez AA, Fujimoto M and Ram JL. 2015. The ecological, economic and public health impacts of nuisance chironomids and their potential as aquatic invaders. *Aquatic Invasions,* 10 (1): 1-15.

Faria MS, Nogueira AJA and Soares AMVM. 2007. The use of *Chironomus riparius* larvae to assess effects of pesticides from rice fields in adjacent freshwater ecosystems. *Ecotoxicology Environmental Safety,* 67: 218-226.

Federici BA, Kramer WL and Mulla MS. 1976. A disease in *Chironomus frommeri* caused by a Rickettsiella – like organism. *Proc. Calif. Mosq. Contr. Assoc.,* 44: 123.

Fernando CH. 1993. Rice field ecology and fish culture – an overview. *Hydrobiologia,* 259: 91-113.

Ferrarese U. 1993. Chironomids of Italian Rice Fields. *Netherlands Journal of Aquatic Ecology,* 26: 341-46.

Ferrington LC Jr. 2008. Global diversity of non-biting midges (Chironomidae: Insecta-Diptera) in freshwater. *Hydrobiologia,* 595: 447-455.

Fischer J and Rosin S. 1968. Einfluss von Licht and Temperatur auf die Schlupf – Aktivitat von *Chironomus nuditarsis. Revue Suisse Zool.,* 75: 538-549.

Forsyth DJ. 1971. Some New Zealand Chironomidae (Diptera). *J.R.Sec. New Zealand.,* 1(2): 113-144.

Frouz J, Matena J and Ali A. 2003. Survival strategies of chironomids (Diptera: Chironomidae) living in temporary habitats: a review. *European Journal of Entomology,* 100: 459-465.

Hafiz HA.1939. Observations on the bionomics of the midges *Chironomus* (*Limnochironomus*) *tenuiforceps* (Kieffer) occurring on the filter beds of the Calcutta Corporation at Palta, near Calcutta (Chironomidae: Diptera). *Rec. Indian Mus.,* 41: 225-231.

Hamilton AL. 1965. An analysis of a freshwater benthic community with special reference to the Chironomidae. Ph.D Thesis, Dept. of Zoology, Univ. of British Columbia. pp. 216.

Harkrider JR and Hall IM. 1979. The effect of an entomopovirus on larval population of an undescribed midge species in the *Chironomus decorus* complex under laboratory conditions. *Environmental Entomology,* 8: 631-635.

Hashimoto H, Wongsiri T, Wongsiri N. *et al.,* 1981. Chironomidae from rice fields of Thailand with descriptions of 7 new species. *Technical Bulletin Entomology and Zoology Division,* Department of Agriculture, Bangkok, Thailand, 7: 1-47.

Hazra N, Saha GK, and Chaudhuri PK. 2002. Records of orthoclad species from the Darjeeling–Sikkim Himalayas of India (Diptera: Chironomidae), with notes on their ecology. *Hydrobiologia,* 474 (1-3): 41-55.

Hein J and Schmulbach JC. 1972. Biology of the pupal and imago stages of *Chironomus pallidivittatus* (Chironomidae: Diptera). *Proc. S.D. Acad. Sci.,* 51: 80-88.

Hilsenhoff WL. 1966. The biology of *Chironomus plumosus* (Diptera: Chironomidae) in Lake Winnebago, Wisconsin. *Annual Entomological Society of America,* 59: 465-473.

Huger AM, Kreig A, Emschermann P and Gotz P. 1970. Further studies on *Polypoxvirus chironomi,* an insect virus of the pox group isolated from the midge *Chironomus luridus. Journal Invertebrate Pathology,* 15: 253-261.

Igarashi T, Saeki Y, Okada T. *et al.,* 1985. Two cases of bronchial asthma induced by chironomid midges. *Chiryogaku,* 14: 122-126.

Ishihara T. 1972. *An introduction to the study of agricultural insects of Japan.* Yokendo, Tokyo. pp. 310.

Jana BB and Pal G. 1990. Production of chironomid larvae in culturing media of various organic wastes. *Limnologica,* 21(1): 281-285.

Jonasson PM and Kristiansen J. 1967. Primary and secondary production in lake Esrom. Growth of *Chironomus anthracinus* in relation to seasonal cycle of phytoplankton and dissolved oxygen. *Int. Revue ges. Hydrobiol. Hydrogr.,* 52 : 163-217.

Jonasson PM. 1965. Factors determining population size of *Chironomus anthracinus* in Lake Esrom. *Mitt. Int. Ver. Limnol.,* 13 : 139-162.

Kajak Z and Warda J. 1968. Feeding of benthic non-predatory chironomidae in lakes. *Ann. Zool. Fenn.,* 5: 57-64.

Kikuchi M, Kikuchi T, Okubo S and Sasa M. 1985. Observations on the seasonal prevalence of chironomid midges and mosquitoes by light traps set in a rice paddy area in Tokushima. *Japan J. Sanit. Zool.,* 36: 333-342.

Kurihara Y. 1989. Ecology of some rice fields in Japan as exemplified by some benthic fauna with some notes on management. *Hydrobiologia,* 74: 507-548.

Lawler SP. 2001. Rice fields as temporary wetlands: A review. *Israel Journal of Zoology,* 47: 513-528.

Legner EF, Medved RA and Hauser WJ. 1975. Predation by the desert pupfish, *Cyprinodon macularius* on *Culex* mosquitoes and benthic chironomid midges. *Entomophaga*, 20: 23-30.

Lindeberg B. 1971. Parthenogenetic strains and unbalanced sex rations in Tanytarsini (Diptera, Chironomidae). *Annales Zoologici Fennici*, 8: 310-317.

Lindegaard C and Jonasson PM. 1979. Abundance, population dynamics and production of zoobenthos in Lake Myvatn, Iceland. *Oikos*, 32: 202-227.

Lindegaard C and Jonsson E. 1987. Abundance, population dynamics and high production of chironomidae (Diptera) in Hjarback Fjord, Denmark, during a period of eutrophication. *Entomologica Scandenavia Supplement*, 29: 293-302.

McCauley VJE. 1974. Instar differentiation in larval Chironomidae (Diptera). *Canadian Entomologist*, 106: 179-200.

Miall LC and Hammond AR. 1900. The structure and life history of the Harlequin fly (*Chironomus*). Clarendon Press, Oxford. pp.196 pp.

Morgan NC and Waddell AB. 1961. Diurnal variation in the emergence of some aquatic insects. *Transactions of the Royal Entomological Society* of *London*, 113: 123-137.

Mundie JH. 1956. The biology of flies associated with water supply. *Instn. Publ. Hlth Engrs J.*, 55: 178-193.

Nandi S, Aditya G, Chowdhury I. *et al.*, 2014. Chironomid midges as allergens: evidence from two species from West Bengal, Kolkata, India. *Indian Journal of Medical Research* 139: 921-926.

Oliver DR. 1968. Adaptations to arctic Chironomidae. *Ann. Zool. Fenn.*, 5: 111-118.

Oliver DR. 1971. Life history of the Chironomidae. *Annual Review of Entomology*, 16: 211-230.

Palmen E.1955. Diel periodicity of pupal emergence in natural population of some chironomids (Diptera). Ann. *Zool. Soc.vanamo*, 17 : 1-30.

Parkin RB and Stahl JB. 1981. Chironomidae (Diptera) of Baldwin lake, Illinois, a cooling reservoir. *Hydrobiologia*, 76: 119-128.

Ree HI and Kim HS. 1981. Studies on Chironomidae (Diptera) in Korea I. Taxonomical Study on Adults of Chironomidae. *Proc. Coll. Natur. Sci., SNU.*, 6(1) : 123-226.

Ree HI, Hong HK, Shim JC and Lee JS. 1981. A study on seasonal prevalence of population of the mosquito larvae and other aquatic invertebrates in rice fields in Korea. *Korean Journal of Zoology*,24 (3): 151-161.

Ree HI, Shim JC, Kim CL and Lee WJ. 1982 *Studies on population dynamics of vector mosquito larvae and other aquatic animals including predators breeding in rice fields in 1982*. Report, NIH, Korea, 19: 137-149.

Risbec J. 1952. Les insects nuisibles au ris dans le nidi de la France. Les Risiculteurs de France. *Bulletin 18, Etude techmique* 51: 14-19.

Roback SS, Bereza DL and Vidrine MF. 1979. Description of an *Ablabesmyia* (Diptera: Chironomidae: Tanypodinae) Symbiont of Unionid fresh – water mussels (Mollusca: Bivalia: Unionacea), with notes on its biology and Zoogeography. *Transaction American Entomological Society*, 105: 577-620.

Roger PA and Kurihara Y. 1988. Floodwater biology of tropical wetland rice fields. In. *Proceedings of the First International Symposium on paddy soil fertility*. Chiang Mai, Thailand, University of Chiang Mai. pp. 275-300.

Sadler WO. 1935 Biology of the midge *Chironomus tentans* Fabrisus, and methods for its propagation. *Cornell Univ. agric. Ex. Stn. Mem.*, 173: 1-25.

Sæther OA. 1979. Chironomid communities as water quality indicators. *Holarctic Ecology*, 2: 65-74.

Saha D and Mazumdar A. 2013. Deformities of *Chironomus* sp. larvae (Diptera: Chironomidae) as indicator of pollution stress in rice fields of Hooghly District, West Bengal. *Journal of Today's Biological Sciences: Research and Review*, 2 (2): 44-54.

Sasa M and Kikuchi M. 1986. Notes on the chironomid midges of the subfamilies Chironominae and Orthocladiinae collected by light traps in a rice paddy area in Tokushima (Diptera, Chironomidae). *Jap. J. Sanit. Zool.*, 37 (1): 17-39.

Singh MP and Harrison, AD. 1982. Diel periodicities of emergence of midges (Diptera: Chironomidae) from a wooded stream on the Niagara Escarpment, Ontario. *Aquatic Insects*, 4(1): 29-37.

Sokolova N Yu, Paliy AV and Izvekova BI. 1992. Biology of *Chironomus piger* Str. (Diptera: Chironomidae) and its role in the self-purification of a river. *Aquatic Ecology*, 26(2): 509-512.

Stevens MM, Fox KM, Warren GN. *et al.*, 2000. An image analysis technique for assessing resistance in rice cultivars to root feeding chironomid midge larvae (Diptera: Chironomidae). *Field Crops Research*, 66: 25-36.

Stevens MM, Helliwell S and Cranston PS. 2006. Larval chironomid communities (Diptera: Chironomidae) associated with establishing rice crops in southern New South Wales, Australia. *Hydrobiologia*, 556: 317-325.

Surakarn R and Yano K. 1995. Chironomidae (Diptera) recorded from paddy fields of the world: a review. *Makunagi/Acta Dipterologica*, 18: 1-20.

Syrjamaki J. 1965. Laboratory studies on the swarming behaviour of *Chironomus* Strenzkei Fittkau in litt (Dipt., Chironomidae). I. Mechanism of swarming and mating. *Ann. Zool. Fenn.*, 2: 145-152.

Tabaru Y, Moriya K and Ali A. 1987. Nuisance midges (Diptera: Chironomidae) and their control in Japan. *Journal American Mosquito Control Association*, 3(1): 45-49.

Tasi SC and Legner EF. 1977 Expenential growth in oulture of the planarian mosquito predator, Dugesia dorotocephals (Woodworth). *Mosquito News*, 37: 474-478.

Thakur NK and Das P. 1985. Synopsis of biological data on Singhee *Heteropneustes fossilis* Bloch, 1794. *Bulletin No.39. Central Inland Fisheries Institute*, Barrackpore, India.

Thienemann A. 1954. Chironomus Leben, Verbreitung und Wirtschaftliche Bedeutung dor Chironomiden. *Binnengewasser*, 20 : 834.

Varghese TJ, Devaraj KV and Satyanarayana Rao GP. 1973. Food of juveniles of the catfish, *Clarias batrachus* (Linn.). *Journal of the Inland Fisheries Society of India*, 5: 78-81.

Weiser J 1976 The intermediary host for the fungus *Coelomomyces chironomi*. *J. Invert Pathol.*, 28 : 273-274.

Wilson RS and McGill JD. 1979. The use of chironomid pupal exuviae for biological surveillance of water quality. *Dept. Envir. Lond. Tech. Mem.*, 18.

Wilson RS. 1980. Classifying rivers using chironomid pupal exuviae. In. *Chironomidae (Ecology, Systematics, Cytology and Physiology*, (Murray DA. ed.). Pergamon Press, Oxford and New York. pp. 209-216.

Yasumatsu K, Hashimoto H and Chang YD. 1979. Chironomid fauna of Korea and their role in the rice agroecosystem. *IRRN*,4(4): 17-18.

Yokogi K and Ueno R. 1971. *Wasabi (Japanese horseradish, Wasabia japonica)*. Nohson-Gyoson Bunka Kyouksi, Tokyo. pp. 136.

Chapter 5

Less Known Invertebrate Community in Wetland Ecosystems of West Bengal

Santanu Mitra

Crustacea Section, Zoological Survey of India, Kolkata

Introduction

Wetlands of West Bengal harbour some interesting animals of different little known and less attended phyla, such as Bryozoa, Porifera, Cnidaria, Sipuncula, Echiura, Echinodermata and Brachiopoda. Although most of these phyla hold comparatively few species which thrive in pond ecosystems or restricted to the shallow brackish-water or strictly in marine water of intertidal zone, these animals have shown a variety of adaptive features and play an important role in respective ecosystems.

Phylum: Bryozoa

The bryozoa or ectoprocta are small benthic, sessile, aquatic invertebrates growing as colonies of connected zooids on submerged substrates. They feed on suspended organic particles which they capture by the whorls of ciliated tentacles (lophophore). Of the estimated 8000 extant species of bryozoans (Ryland, 2005), only a small number are found in freshwater habitats. Till now 94 bryozoan species are found in freshwater belonging to 24 genera and 10 families (Massard and Geimer, 2008). Most of these species belong to the exclusively freshwater inhabiting class Phylactolaemata and reproduce asexually by means of statoblasts (Buoyant floatoblasts and fixed sessoblasts), which are very important in identification of species. Fresh water bryozoan colony is adhering to the surface of any substratum inside the water bodies *i.e.* aquatic weeds, logs, stones, bricks or any other artificial substratum in the ponds, lakes, water reservoirs, streams, and rivers.

Literature on the freshwater bryozoa of India reveals that very little attention has been paid to this group. Annandale (1911) dealt with this group in Indian sub continent. Whereas Rao (1962, 1972) and Kulsreshtha (1976) worked out on freshwater bryozoa in western and middle part of India. At present only 17 species of freshwater bryozoa are recorded from India of which 9 species are reported from West Bengal (Samanta, 1998). Interestingly, all the nine species which are found in West Bengal were reported from different freshwater bodies in and around the city of Kolkata (Annandale, 1911). The Only endemic bryozoans species of West Bengal *Victorella bengalensis* (Class: Gymnolaemata, Order: Ctenostomata, Family: Paludicellidae) was reported from Sunderban and are well adapted in brackish water environment, and it's the only species that does not belong to freshwater. *Hislopia lacustris* and *H. moniliformis* belongs to family Hislopidae, are widely distributed in the ponds and lakes of southern West Bengal, adapted in lotic ecosystems of ponds and are grow on the aquatic plants and any submerged articles in ponds and lakes. Besides, a total of six species of bryozoans belongs to class Phylactolaemata (Order : Plumatellida; family : plumatellidae) are purely adapted in freshwater condition. Among those, genus *Plumatella* is dominated in species level diversity, represented by four species, *Plumatella diffusa, P. emerginata, P. fruticosa,* and *P. javanica.* All these species are represented in southern as well as northern West Bengal, while *P. fruticosa* are only species reported from cold water of Darjeeling district. The other two species *Hyalinella punctata* and *Stolella indica* are well adapted in warm water of southern and western West Bengal.

Phylum: Porifera

Phylum Porifera is a group of aquatic invertebrates, meaning pore bearers and most of them live in marine environment but some are well adapted in freshwater conditions too, while very few are reported from the brackish water of estuarine environment. Due to poorly organized body structure and loss of locomotory activity, these groups have a tendency for high morphological plasticity which finally leads to pronounced polymorphism. The diversity and population of these animals are strikingly depending upon the water quality and seasonal changes.

Till date, there is no report on the presence of any marine sponge from West Bengal coast, though after the pioneering work of Annandale (1907a, 1907b, 1907c, 1909, 1911, 1915) on the freshwater sponges of India including those of West Bengal, Soota and Pattanayak (1982) recorded seven species from West Bengal. Later, Soota (1991) and Pattanayak (1998) described 16 species under nine genera from West Bengal. It may be interesting to note that so far nine species (*Spongilla (Euspongilla) alba, S. (Eunapius) carteri, S. (Eunapius) crassussima, S. (Eunapius) fragilis* sub sp. *decipiens, S. (Eunapius) fragilis* sub sp. *calcuttana, Trochospongilla latouchiana, T. phillottiana, Ephydatia meyeni* and *E. crateriformis* under five genera and one family are recorded from a single medium sized pond inside the Indian Museum premises, Kolkata; so far this is the only pond which shows such remarkable freshwater sponge diversity as reported by Pattanayak (1998). So far 31 species belonging to 18 genera of freshwater sponges have been recorded from India, 51.6 per cent are found in West Bengal, particularly the wetlands in and around Kolkata are most important regarding the diversity of the sponges, So far as the local status of availability of

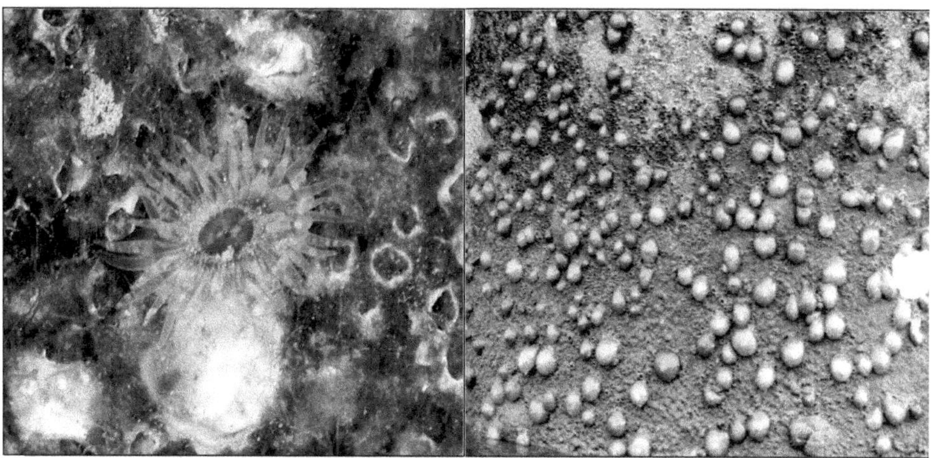

1. *Diadumene schilleriana* with open tentacle in High tide

2. A colony of *Diadumene schilleriana* with closing their tentacle in low tide

3. *Paracondylactis sinensis* opening its tentacles at low tide at sandy habitat

4. *Edwardsia jonesii*

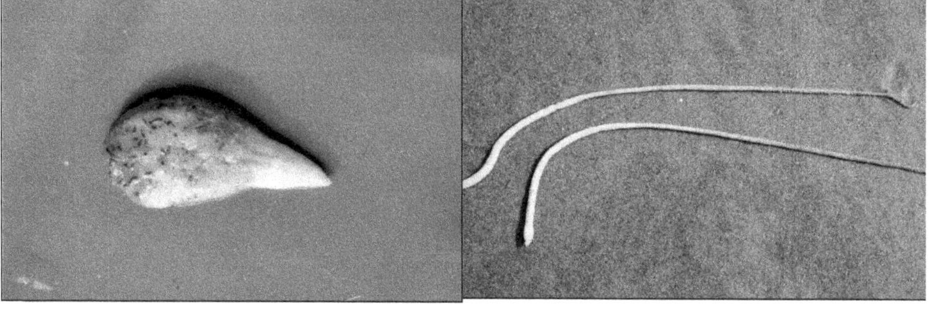

5. *Cavernularia* sp.

6. *Virgularia* sp.

Figure 5.1: Some Cnidarians from wetland ecosystems of West Bengal.

the freshwater sponges in West Bengal is concerned, only three species are found abundant whereas 13 species are rare and interestingly three species are endemic to our country. Taxonomically all these freshwater sponges are belong to Family Spogillidae of the Order Haplosclerida and Class Demospongiae. *Spongilla lacustris* is generally found in very clear lotic water bodies. It has an irregular base and surface extended into long tubular lobes and like many other freshwater sponges, this species harbor varying number of zoochlorellae and as a result they show a different colour morph, from light grey to bright green (Penney and Racek, 1968). As light is one of the most important factor for photosynthesis by the symbiotic algae present inside the sponge body, these sponges prefer to grow on the low depth or peripheral water mass of a pond or lacustrine ecosystem. In contrast, *S. alba* has a bulky growth of very large encrustations with a more or less smooth surface, this species has been reported from slightly saline water of North and South 24 Parganas. *Eunapius carteri* has a wide distributional range, from southern districts to Malda and Jalpaiguri districts of West Bengal. In comparison to *E. calcattanus,* the surface of later species is quite smooth, as they mainly prefer lotic water bodies. Interestingly, *Corvospongilla lapidosa* an endemic freshwater sponge of India is reported from the district of North 24 Parganas.

Phylum: Cnidaria

Phylum Cnidaria is a large taxonomically diversified group and consist of several organisms with diversified forms and shapes, commonly known as sea anemones, sea whips, sea fans and sea pansies, hydras, corals and variety of jellyfishes. They are linked together by their simple anatomical design, possession of nematocysts and their carnivorous feeding habits. Globally there are about 10,000 species of Cnidarians, belong to five classes Hydrozoa, Scyphozoa, Cubozoa, Anthozoa and Polypodizoa. Diversity of Cnidarian fauna is quite good in India as per global context (8.5 per cent), comprised of 842 species; In West Bengal, this group is represented so far by 47 species only, of which remarkably 12 species (26 per cent) are considered as endemic to this country.

Cnidarian fauna from the wetlands of West Bengal were studied by several authors, as this group is very diverse. Jellyfishes were extensively studied by Annandale (1907b, 1915), Ritche (1915), Rao (1931), Kramp (1958), Haldar and Choudhury (1995), Goswami (1992); whereas sea anemones were studied by Stoliczka (1869), Misra (1975, 1976), Rao and Misra (1980, 1983, 1986), Misra and Soota (1981), Paruleker (1990), Goswami (1992), Bairagi (1998). There is only a single report of the cnidarian species as meiofauna, from West Bengal, *Halmmohydra sagarensis*, which was collected from the interstitial sand of intertidal sea beaches of Sagar Island. Due to its immense ecological role, and a valuable source of potential biomedical compounds, more taxonomical work should be needed to explore the species level diversity of this diversified animal group.

Among the cnidarians, class Hydrozoa, is represented by 22 species, all of them are pelagic except the only benthic one *Obelia spinulosa* (Order: Leptothecata; Family: Campanulariidae) generally called as moss animal and often encrust thickly on the gastropodan shell in Sunderban and sandy coastal intertidal locality of West Bengal.

The extreme pelagic species like, *Porpita porpita, Velella velella* and *Physalia physalis* belong to family Porpitidae (Order Chondrophora) and family Physalidae (Order Siphonophora) are very rare in coastal waters due to its distribution and availability are exclusively depended upon the water currents, wind patterns etc. *Hydra vulgaris* (Family: Hydridae; Order: Actinulia) are well adapted in different lotic system, even flourished in different ponds of Kolkata. Leaf litter, semi terrestrial water plants are needed for their survival in pond ecosystem. *Diphyes bojani, Octophialucium indicum, Aequorea pensilis* and *Tamoya gargantua* are repoted from the coastal waters of sandheads, and are not a regular visitor of the coastal waters. Representatives of family Bougainvillidae (*Bimeria vestia*) and family Corynidae (*Dicyclocoryne filamentata*) are reported to be well adapted in brakish water pond ecosystem. *Blackfordia virginica* (Family: Eucopidae; Order: Leptothecata) are reported from brackish water wetlands of Saltlake, its a marine species and seems to well adapted to brackish water envirronment. *Eirene ceylonensis* and *E. menoni* belong to family Eirenidae are adapted in low saline part of the riverine ecosystem and are reported from Matla river of Canning and from Ballykahl at Howrah district; *Moerisia gangetica* and *M. gemata* (Family: Moeriicidae; Order: Limnomeducae) are freshwater medusae, well adapted for lower saline part of lower riches of Gangetic estuary and later one was also reported from brakishwater pond at Canning; *Limnocnida indica* of family Limnocnididae are reported only from freshwater part of Ganges. A total seven species of scyphozoan, commonly called as 'jellyfishes' are purely pelagic ones and a occasional visitors to coastal waters and also enters the lower reaches of Gangetic estuary for spawning. One of the most interesting group of cindarians are the 'sea anemones' belong to class Anthozoa of the sub class Hexacorallia and order Actiniaria. All the 12 representative species of this group are well adapted in muddy or sandy habitat of estuarine or marine coastal wetland of this state. *Edwardsia jonesii* and *E. tinctrix* belong to family Edwardsiidae are two species of a very small anemones adapted in muddy intertidal zones of narrow creeks of lower reaches of Gangetic estuary and also in the small estuaries like Jalda, Digha, Rasulpur and Junput of coastal Medinipur district. Worm like body of *Edwardsia jonesii* is covered by a cuticle of rusty red colour which help them to attract the prey animals; *Pelocoetes exul* and *Phytocoetes gangeticus* belong to the family Heliactiidae are soft mud burrowing forms in the intertidal zone, found at mudflat just beside the narrow creeks of the middle tidal area, the former species are very interesting as they posses tentacles with many branches. *Diadumene schilleriana,* the only representative species of family Diadumenidae in the wetlands of West Bengal. This species has a small stalk and they need a hard substrata to keep them attach and occupied any hard substratum like concrete jetty, boulders, rocks, wooden poles submerged in estuarine and coastal intertidal area in Sunderban and other small estuaries of West Bengal coast and also prefer to attach on trunks and pneumatophore of mangroves plants. At Digha-Shankarpur coast *Diadumene schilleriana* flourish on the concrete embankment of upper and supra-littoral coastal zone, thus they may be able to protect their tentacles and other organs by contracting them in the body cavity to avoid sun-dry for a prolonged period of low tide condition. This species often found on the surface of living animals like hermit crabs, horse-shoe crab and sometimes also on plastics bags which are occasionally found as floating in waters.

Family Actinidae has only single species from the state West Bengal *Paracondylactis sinensis*. This species with a good population are distributed in the lower-intertidal zone of sandy zone of coastal area and sandy-mud substrata of estuarine belt. This species has a large soft cylindrical body with 96 long tentacles and are adapted to live in sand or mud by digging hole and place themselves in the holes fully, only the tentacles surrounding the mouth hole are visible on the surface. When disturbed they may able to move their body in the holes totally, by quick discharging the water and in this way they protect themselves from the enemies as well as from the direct sunlight during low tide condition. Order Pennatulacea belongs to sub-class Octocorallia, of class Anthozoa are represented with three families and eight species, among them *Virgularia* sp. is very common and distributed along the coastal zone of this state in muddy or sandy substrata of lower intertidal zone. They have long worm like stalk to attach themselves in substrata, and a hard and thin stick like skeleton on which the zooids are attchaed, zooids feed on the zooplankton from the incurent waters, but in low tide condition they are also capable to hide their body inside the sand or mud where they attach to the substrata.

Phylum: Echiura

Phylum Echiura includes soft bodied, almost defenseless sedentary creatures, well adapted to live in burrows and are commonly called as spoon worms for the special spoon like appearance of their contracted proboscis. A total of 33 species under nine genera are recorded from India, of which three species belonging to single genus and single family are reported from the estuarine wetlands of West Bengal (Haldar, 1998). These three species are *Anelassorhynchus branchiorhynchus*, *A. dendrorhynchus* and *A. microrhynchus* that belongs to family Thalassematidae of order Echiuroinea and Class Echiurida. *A. branchiorhynchus,* an endemic Indian species, lives in hard mud intertidal area in the estuarine zone of Gangetic mouth and also in the smaller estuarine area like Jalda, Digha mohana, Rasulpur etc. of coastal belt of West Bengal. They make a U shaped burrow, commensally associated with polychaetes, isopods, amphipods and enjoy all the advantages of host's activity and performance. The other two species (*A. dendrorhynchus* and *A. microrhynchus*) also enjoy almost the same habitat and they are adapted in lower salinity and prefer soft muddy substrata just beside the creeks. In Digha mohana, the present author observed the later species on the mud bank of Champa canal about 1.5 km far from the sea mouth.

Phylum: Sipuncula

Sipunculans, are commonly called as pea-nut worms, due to the shape of their contracted trunk and 37 species belonging to ten genera of this phylum are reported from India. In West Bengal only three species of three different genera and two families are recorded (Haldar,1998), of which only a single species *Phascolosoma arcuatum* was reported from the intertidal wetland area of Sunderbans, and later the present author recorded it from the small estuarine zone of the coastal West Bengal. This species remarkably adapted to live in semi-terrestrial habitat, and are found in the intertidal zone from mean high water spring tide to mean low water spring tide levels, being exposed to a marked degree of varying salinity from 5 per cent to 25 per cent in the mangrove area of West Bengal. They also possess an adaptive coloration to mimic the mud and may be able to contract and expand their body to a considerable extent.

Phylum: Echinodermata

Echinoderms being exclusively marine inhabitants, comprise of a very few species that can tolerate a low salinity condition. In West Bengal this group is represented by 22 species belonging to 19 genera, 15 families and four classes (Sastry, 1998). In class Asteroidea, only two species *Astropecten indicus* and *A. euryacanthus* visit the coastal intertidal waters and are adapted to crawl on the sandy bottom of the coastal littoral area and often take shelter in to the sand when at danger. Among the Ophiuroids, *Ophiactis modesta* is the most common inhabitant of the intertidal mud and often take shelter inside rotten woods of mangroves. The only Echinoidis *Temnopleurus toreumaticus* are reported from the coastal shallow as well as intertidal water. Class Holothuroidea is represented by three species from wetlands of coastal shallow waters. *Acuadina molpadioides* of the family Caudinidae are known to occur in mid-littoral waters of Digha-Junput coastal belt. They have sausage shaped body, with 15 small tentacles. The present author observed this species to form a breeding ground in low-mid littoral area of Shankarpur–Jalda coastal area. Another holothuroid *viz., Thorsonia investigatoris,* once a common resident of intertidal area of Digha beach, now has become very rare. They make a V shaped appearance by burrowing the sand and putting the whole body into the sand; only open the two end of its slender body on the upper side of the sand. *Protenkyra simisilis* reported from intertidal mud of Sunderban is a worm like slender holothuroid that enjoys a mangrove habitat.

Phylum: Brachiopoda

Phylum Brachiopoda represented by a single species from the coastal wetland of West Bengal, *Lingula anatina* of family Lingulidae, is commonly known as lamp shell. It is flourished from Cambrian times to the present and come down to the ages with little changes and often referred to as *Living fossil*. Soota and Reddy (1976) reported it from intertidal zone of Digha mohana-Shankarpur coastal area. Present author also witnessed a population of this species in the same area. In Digha-mohana, the habitat of *Lingula anatina* occurs in either side of the tidal creeks, substrata are generally soft muddy area, however, sometimes decomposed black soil and sand mixed mud were preferred for living and juvenile-bed was found in fine soft mud only. The Lingulid beds are only 2-8 meter in width. Both the banks of the creeks covered by patchy mangroves and mudflats are highly exposed during low tide, while during high tide the areas are inundated by sea water at a depth of 0.5 to 1.2 meter. As these animals are filter feeder they enjoy the mass of zooplankton of estuarine area despite the high level of turbidity of the estuarine water.

Acknowledgement

The author wishes to express his deep sense of gratitude to Dr. K. Venkataraman, Director, Zoological Survey of India, Kolkata, for providing infrastructure facilities.

Thanks are also due to Dr. Kousik Deuti of Zoological Survey of India for his constant encouragement.

References

Annandale N. 1907a. Notes on the freshwater Fauna of India. IX. Description of new Freshwater sponges from Calcutta, with a record of two known species from the Himalayas and a list of the Indian forms. *J. Proc. Asiat. Soc. Beng.*, 3 : 15-26.

Annandale N.1907b. The fauna of brackish ponds at Port Canning, Lower Bengal. *Rec. Indian. Mus.*, 1(1): 47-74.

Annandale, N. 1907c. Notes on freshwater sponges. *Rec. Indian Mus.*, 1: 387-392.

Annandale N.1909. Preliminary note on a new genus of *Phylacto laematous* Polyzoa, *Rec. Indian. Mus.*, iii: 279.

Annandale N. 1911. Freshwater sponges, Hydroids and Polyzoa. *The Fauna of British India*. Taylor and Francis, London. pp. 1-261

Annandale N. 1915. Fauna of Chilka lake. The Coelenterate of the Lake, with an account of the Actiniaria of brackish waters in the Gangetic delta. *Mem. India. Mus.*, 5 : 65-114.

Bairagi N. 1998. Cnidaria: Sea Anemones. In. *Fauna of West Bengal, State Fauna Series, Zoological Survey of India, 3.* (Part 11): 29-44.

Goswami Bharati B.C. 1992. Marine Fauna of Digha Coast of West Bengal, India. *J. mar. boil. Ass. India*, 34(1 and 2): 115-137.

Haldar B.P. 1998. Echiura and Sipancula. *Satae fauna series-3:* In. *Fauna of West Bengal., State Fauna Series, Zoological Survey of India, 3.* (Part 10): 1-16.

Haldar B.P. and Choudhury A. 1995. Medusae: Cnidaria. *Hugli Matla Estuary, Estuarine Ecosystem Series,* Part 2: 9-30.

Kramp P.L. 1958. Hydromedusae in the Indian Museum. *Rec. Indian. Mus.*, 53: 339-376.

Massard JA and Geimer G. 2008. Global diversity of bryozoans (Bryozoa or Ectoprocta) in freshwater: an update. *Bull. Soc. Nat. luxemb.*, 109: 139-148.

Misra A. 1975. A note on the collection and narcotization *Paracodylactis* sp. from Sagar Island. *Newsl. Zool. Surv. India.*, 1(3): 46-47.

Misra A. 1976. On the distribution of *Edwardsia jonesii* Seshya and Cuttress on the coast of India. *Newsl. Zool. Surv. India.*, 2(4): 161-162.

Misra A and Soota TD. 1981. On the occurence of the Sea-anemone *Phytocoeteopsis ramunnii* Panikkar in a tidal creek of Sagar Island, India. *Bull. Zool. Surv. India.*, 4(2): 151-153.

Parulekar AH. 1990. Actiniarian Sea Anemone fauna of India. In. *Marine biofueling and power plants*. Proceeding of marine biodeterioration with reference to power plant cooling systems, IGCAR, Kalpakkam, pp. 218-228.

Pattanayak JG. 1998. *Freshwater Sponges*. In. *Fauna of West Bengal. Zoological Survey of India. State fauna series 3.* Part 11: 1-27.

Penney JT and Racek AA. 1968. Comprehensive revision of worldwide collection of freshwater sponges (Porifera : Spongillidae). *Bull. US. natn. Mus.* No. 272. pp. 184.

Rao HS. 1931. Notes on Scyphomedusae in the Indian Museum. *Rec. Indian. Mus.,* 33: 25-62.

Rao CG and Misra A. 1980. On a new species of *Halammohydra* (Actinulida: Hydrozoa) from Sagar Island, India. *Bull. Zool. Surv. India,* 3(1 and 2) : 113-114.

Rao CG and Misra A. 1983. Studies on the meiofauna of Sagar Island. *Proc. Indian Acad. Sci. (Anim. Sc.),* 92(1) : 73-85.

Rao CG and Misra A. 1986. The Meiofauna and Macrofauna of Digha beach, West Bengal, India. *Rec. zool. Surv. India,* 83(3 and 4): 31-49.

Rao KS. 1972. Studies on the freshwater Bryozoa. III. The Bryozoa of the Narmada system. *Proc. Second. Inter. Conf. Bryozoa.* IBA. Durham.

Rao KS. 1976. Studies on fresh water Bryozoa. IV. The Bryozoa of Rajasthan, India. *Rec. Zool. Surv. Ind.,* 69: 329-345.

Rao KS. and Kulsreshtha KS. 1962. Studies on freshwater Bryozoa. I. The Bryozoa of Vindhyan region. *J. Univ. Sougor.,* 2B: 50-64.

Ryland J. 2005. Bryozoa: an introductory overview. *Denisia,* 16: 9-20.

Ritche J. 1915. The Hydrids of the Indian museum II. *Annulella gemmata,* a new and remarkable brackish water hydroid. *Rec. Indian. Mus.,* 11: 541-568.

Samanta TK. 1998. Fresh Water Bryozoa. In. *Fauna of West Bengal, Zoological Survey of India: State Fauna Series* 3.part 10: 445-461.

Sastry DRK. 1998. Echinodermata. In. *Fauna of West Bengal, Zoological Survey of India: State Fauna Series* 3: part 10: 463-489.

Soota TD. 1991. Freshwater Sponges of India. *Rec. zool. Surv. India,* Occ. Paper No. 138 : 116.

Soota TD and Pattanayak JG. 1982. On some freshwater sponges from the unnamed collection of the Zoological Survey of India. *Rec. zool. Surv. India,* 80 : 215-229.

Soota TD and Reddy KN. 1976. On the distribution and habitat of the brachiopod Lingula in India. *Newsletter. Zool. Surv. India,* Kolkata 2(6): 235-237.

Stoliczka F. 1869. On the anatomy of *Sagarita schilleriana* in brackish water at Port Canning. *J. Asiat. Soc. Bengal.,* 38(2): 28-63.

Chapter 6

Indigenous Ornamental Fish Resources of West Bengal

Abhisek Basu*

Department of Zoology,
University of Calcutta, Kolkata

Introduction

Ornamental fishes can be defined as attractive colourful fishes of peaceful nature, that are kept as pets in confined spaces of an aquarium or a garden pool with the purpose of enjoying their beauty for fun and fancy (Dey, 1996). Since ornamental fishes are usually kept in glass aquarium, these are also popularly known as aquarium fishes. Ornamental fishes are the most popular pets in the world (Singh, 2005) and aquarium keeping has emerged as the second most popular hobby in recent years next to photography (Chapman, 1997). It offers a feast to our eyes and relaxation to the mind, especially when we feel tired or depressed. Mills (1990) viewed aquarium fishes as visually exciting objects. They may have unique shapes, colouration, body forms and movements. Ornamental fishes are also called 'living jewels' for their beautiful colours and playful behaviour. They are typically small sized, colourful and most often bizarre shaped in appearance (Dey, 1996). However, these fishes need not necessarily be always colourful. In fact, certain fish species loved by aquarist are quite ugly, in such cases the peculiar appearance is a source of attraction for the aquarium lovers and naturalists (Dey, 1996).

Ornamental Fish: Global Scenario

With the inspiring popularity of aquarium keeping in households in many parts of the world ornamental fish has become an important part in international

---

* *Present Address*: Department of Zoology, Victoria Institution (College), Kolkata.

trade and has become a global industry (Tlusty *et al.*, 2013). The world trade of ornamental fish is valued at about USD 9.0 billion USD (FAO, 2004) to 15-30 billion USD (Penning *et al.*, 2009) with an annual growth rate of 6 percent. The USA is the largest market for importing ornamental fishes valued at USD 60.0 million USD annually, followed by Japan (32.9 million USD) and Germany (21.0 million USD). Singapore is the top exporter. In the international trade of aquarium fish, the freshwater fish species represent about 90 percent in terms of value, against 10 percent for marine species (Lem, 2001) and involves ~5300 freshwater and 1802 marine fish (Hensen *et al.*, 2010; Rhyne *et al.*, 2012).

Ornamental Fish: Indian Scenario

Indian waters possess a rich diversity of ornamental fishes with over 200 varieties of indigenous species (Raghavan *et al.*, 2007, 2009, 2011). Indian domestic trade in ornamental fish is flourishing. The present domestic demand is higher than the supply. By virtue of possessing vast and varied aquatic and icthyo-faunal resources and favourite climate, the country has great potential to increase the present level of export to about USD 30 million every year (Vinci, 1998; Swain and Jena, 2002). The climate of India is almost similar to that of the other countries in eastern Asia and several varieties of Indian freshwater ornamental fishes are well known in the international market. While huge exploitation of few of these fish species (Raghavan *et al.*, 2011) raises concern, many others are yet to be harvested optimally to sustain ornamental fish trade as a viable livelihood. This is particularly relevant considering the extent of diverse fish species that qualify as ornamental fish in Indian as well as international markets. The later holds more promise since newer varieties of ornamental fishes have greater demand in overseas market and attract higher prices than classical ones. To promote ornamental fish trade and monitor the dynamic nature of price and profit, an assessment of the fish species diversity of a particular geographical region is necessary.

Diversity of Indigenous Ornamental Fishes in West Bengal

Fish diversity study enables highlighting the species richness and existing population trends and size structure. Exploitation of the selected fish species can be further carried out following the surveillance of the fish population and the corresponding market value. In Indian context, assessment of fish diversity with emphasis for their potential in ornamental fish trade has largely been carried out using the Western Ghats as a focal area (Bhat, 2003, 2004; Ananda *et al.*, 2007; Sreekantha *et al.*, 2007; Raghavan *et al.*, 2008). In all these instances the effort was given to highlight the richness of the fish species and their population trends and size structure. In Indian context, an approximate 806 fish species inhabits freshwater (Talwar and Jhingran, 1991), of which, 266 species (recorded and reported) are available in the North East, belonging to 114 genera under 38 families and 10 orders. However, recent studies suggest a rich aquatic biodiversity with 2319 fish species of which 838 (Lakra and Sarkar, 2010) to 930 (Kar *et al.*, 2006) inhabiting freshwaters of India. This is approximately 33.13 percent of total Indian freshwater fishes and as many as 52 indigenous ornamental fish species occurring in the North East reportedly have overseas demand (Bhattacharjya *et al.*, 2003).

Notable earlier studies on the freshwater fishes of India include the works of, Hamilton (1822); Shaw and Shebbeare (1937); Hora (1921a, 1921b, 1930, 1937, 1939,1940, 1943, 1951, 1953); Misra (1959); Menon (1999); Jayaram (1981, 1999); Sen (1982,1985); Talwar and Jhingran (1991); Dey and Kar (1989a, 1989b, 1989c, 1990). Some studies on the fish diversity of North-East India include Nath and Dey (1997, 2000); Kar and Dey (2000,2002); Kar and Barbhuiya (2000, 2004); Kar *et al.* (2002 a, 2002b, 2002c, 2006); Kar (1984, 1990, 2003, 2005a, 2005b); Jhingran (1975); Ponniah and Gopalakrishnan (2000); Ponniah and Sarkar (2000); Johal (2005); Jayaram (2006, 2010); Vishwanath *et al.*(2007); Rema Devi and Indra (2010) and Acharjee and Barat (2013). Despite all these, the knowledge of the fish fauna of tropical Asia is still in its exploratory phase particularly in India where survey work is still incomplete (Le´ve^que *et al.*, 2008) and a recent validated checklist on the freshwater fishes of India having ornamental value is presently lacking, and hence there is considerable uncertainty on the total number of species that occur in the country (Raghavan *et al.*, 2013).

On the other hand, the availability of several significant publications, and more importantly as a result of the collaborative work of IUCN and their partners through the global freshwater biodiversity assessments (Allen *et al.*, 2010; Molur *et al.*, 2011), are greatly contributing to our knowledge on freshwater biodiversity. However, inspite of tremendous significance in determining productivity and calculating species diversity, not many studies have been done on the fish population dynamics, ichthyo-diversity and conservation of fishes in the lentic ecosystems in North Eastern India in particular as compared to elsewhere in India (Jhingran and Tripathi, 1969) and the world (Janzen, 1981).

In case of West Bengal, ichthyiofaunal records of Terai Duars (Northern part of West Bengal), are largely restricted to few fragmented studies in recent years (Mukherjee and Sarkar, 2005; Sarkar and Pal, 2008; Mukherjee *et al.*, 2011), which mostly substantiate the earlier observation (Shaw and Shebbeare, 1937; Hora and Gupta, 1940) on species records only.

The study of fish diversity in the northern part of West Bengal dates back in 1937 (Shaw and Shebbeare, 1937), which continued with some other works by Hora and Gupta (1940), Menon (1962), Sen (1992), Sehgal (1999), Barat *et al.* (2005), Mukherjee and Sarkar (2005), Sarkar and Pal (2008), Chakraborty and Bhattacharjee (2008), Mukherjee *et al.* (2011) and Patra *et al.* (2011). But these works mostly fail to emphasize particularly on the indigenous ornamental fish fauna of this particular region. In case of southern part of the state the scenario is more or less the same except some fragmentary works by Ghosh *et al.*(2002); Panigrahi *et al.* (2009); Basu *et al.* (2012); Saha and Patra (2013) and Paul and Chanda (2014). An up-to-date checklist of ornamental fish species of West Bengal is provided in Table 6.1.

Probable Threats to Indigenous Ornamental Fishes

See the Figure 6.1.

Table 6.1: District-wise List of Indigenous Ornamental Fish available in West Bengal

Name of the Fish	*Family*	*Distribution*
Amblypharyngodon mola (Hamilton, 1822)	Cyprinidae	DAR, COACH, JAL, UD, DD, MAL, MUR, BIR, NAD, BUR, BANK, PUR, HOOG, HOW, MED, NP, SP.
Amphipnous cuchia (Hamilton, 1822)	Synbranchidae	COACH, JAL, UD, DD, MAL, MUR, BIR, NAD BUR BANK, PUR, HOOG, HOW, MED, NP, SP.
Anabas testudineus (Bloch, 1792)	Anabantidae	DAR, COACH, JAL, UD DD, MAL, MUR, BIR, NAD BUR BANK, PUR, HOOG, HOW, MED, NP, SP
Anguilla bengalensis bengalensis (Gray, 1831)	Anguillidae	NP, NAD SP, HOW, HOOG, BUR
Ailia coila (Hamilton, 1822)	Schilbeidae	MAL, NAD, BUR, HOOG, HOW, NP, SP
Aplocheilus panchax (Hamilton, 1822)	Aplocheilidae	DAR, COACH, JAL, UD, DD, MAL, MUR, BIR, NAD, BUR, BANK, PUR, HOOG, HOW, MED, NP, SP
Barilius vagra (Hamilton, 1822)	Cyprinidae	COACH, JAL
Barilius shacra (Hamilton, 1822)	Cyprinidae	COACH, JAL
Badis badis (Hamilton, 1822)	Nandidae	DAR, COACH, JAL, UD, DD MAL, MUR, BIR, NAD BUR BANK, PUR, HOOG, HOW MED, NP SP
Brachydanio rerio (Hamilton, 1822)	Cyprinidae	DAR, COACH, JAL, UD, DD, MAL, MUR, BIR, NAD, BUR, HOOG, HOW, MED, NP, SP
Bagarius bagarius (Hamilton, 1822)	Sisoridae	DAR, COACH, JAL, UD, MAL, MUR, NAD, BUR, HOOG, HOW, NP, SP
Botia dario (Hamilton, 1822)	Cobitidae	DAR, JAL
Boleophthalmus boddarti (Pallas, 1770)	Gobiidae	NP, SP, MED
Chanda ranga (Hamilton, 1822)	Ambassidae	DAR, COACH, JAL, UD, DD, MAL, MUR, BIR, NAD, BUR, BANK, HOOG, HOW, MED, NP, SP
Chanda nama (Hamilton, 1822)	Ambassidae	DAR, COACH, JAL, UD, DD, MAL, MUR, BIR, NAD, BUR, BANK, HOOG, HOW, MED, NP, SP
Channa punctata (Bloch, 1793)	Channidae	DAR, COACH JAL, UD, DD, MAL, MUR, BIR, NAD, BUR, BANK, PUR, MED, NP, HOOG, HOW, SP

Contd...

Table 6.1–*Contd...*

Name of the Fish	*Family*	*Distribution*
Channa striata (Bloch, 1793)	Channidae	DAR, COACH, JAL, UD, DD, MAL, MUR, BIR, NAD, BUR, BANK, PUR, HOOG, HOW, MED, NP, SP
Channa marulius (Hamilton, 1822)	Channidae	UD, DD, MAL, MUR, BIR, BUR, NAD BANK, HOOG, HOW, MED, NP, SP
Channa gachua (Hamilton, 1822)	Channidae	DAR, COACH, JAL, UD, DD, MAL, MUR, NAD, BUR, HOOG, MED, NP, SP
Chelonodon patoca (Hamilton, 1822)	Tetraodontidae	NAD, BUR, PUR, HOOG, HOW, MED, NP, SP
Colisa fasciata (Bloch and Schneider, 1801)	Osphronemidae	DAR, COACH, JAL, UD, DD, MAL, MUR, BIR, NAD, BUR, BANK, PUR, HOOG, HOW, MED, NP, SP
Colisa lalia (Hamilton, 1822)	Osphronemidae	DAR, COACH, JAL, UD, DD, MAL, MUR, BIR, NAD, BUR, BANK, PUR, HOOG, HOW, MED, NP, SP
Colisa chuna (Hamilton, 1822)	Osphronemidae	COACH, JAL, UD, DD, MAL, MUR, BIR, NAD, BUR, BANK, PUR, HOOG, HOW, MED, NP, SP.
Chela laubuca (Hamilton, 1822)	Cyprinidae	DAR, COACH, JAL, UD, DD, MAL, MUR, BIR, NAD, BANK, HOW, HOOG MED, BUR, NP, SP
Chagunius chagunio (Hamilton, 1822)	Cyprinidae	DAR, COACH, JAL
Chaca chaca (Hamilton, 1822)	Chacidae	DAR, COACH, JAL NAD, BUR
Danio devario (Hamilton, 1822)	Cyprinidae	DAR, COACH, JAL, UD, DD, MAL, MUR, BIR, NAD, BUR, HOOG, HOW, MED, NP, SP
Esomus danricus (Hamilton, 1822)	Cyprinidae	DAR, COACH, JAL, UD, DD, MAL, MUR, BIR, NAD BUR, BANK, PUR, HOOG, HOW, MED, NP, SP
Glossogobius giuris (Hamilton, 1822)	Gobiidae	DAR, COACH, JAL, UD, DD, MAL, MUR, BIR, NAD BUR, BANK, PUR, NP, SP, HOOG, HOW, MED
Gagata cenia (Hamilton, 1822)	Sisoridae	DAR, COACH, JAL, UD, DD, MAL, BUR
Garra annandalei (Hora, 1921)	Cyprinidae	DAR, JAL
Glyptothorax telchitta (Hamilton, 1822)	Sisoridae	DAR, COACH, JAL, MAL, BUR
Hara hara (Hamilton, 1822)	Sisoridae	DAR, COACH, JAL, BUR, NAD

Contd...

Table 6.1–*Contd...*

Name of the Fish	*Family*	*Distribution*
Labeo bata (Hamilton, 1822)	Cyprinidae	DAR, COACH, JAL, UD, DD, MAL, MUR, BIR, BANK, BUR, PUR, HOOG, NAD, KOL, NP, SP, HOW, NAD, MED
Labeo calbasu (Hamilton, 1822)	Cyprinidae	DAR, COACH, JAL, UD, DD, MAL, MUR, BIR, BANK, BUR, PUR, HOOG, NAD, KOL, NP, SP, HOW, NAD, MED
Lepidocephalichthys guntea (Hamilton, 1822)	Cobitidae	DAR, COACH, JAL, UD, DD, MAL, MUR, BIR, NAD, BANK, PUR, HOOG, BUR HOW MED NP SP
Lepidocephalus thermalis (Valenciennes, 1846)	Cobitidae	DAR, COACH, JAL
Mystus bleekeri (Day, 1877)	Bagridae	UD, DD MAL, BIR, NAD, BUR, PUR, HOW, HOOG, MED, NP, SP
Mystus cavassius (Hamilton, 1822)	Bagridae	NAD, BUR, HOOG, MED, HOW, NP, SP
Mystus aor (Hamilton, 1822)	Bagridae	COACH, JAL, UD, DD MAL, MUR, BIR, NAD, BUR, BANK, PUR, HOW, HOOG, MED, NP, SP
Mystus gulio (Hamilton, 1822)	Bagridae	NAD, BUR, BANK, PUR, HOOG, HOW MED, NP, SP
Mystus tengara (Hamilton, 1822)	Bagridae	NAD, BUR, HOOG, HOW, MED, NP, SP
Mystus vittatus (Bloch, 1794)	Bagridae	UD, DD MAL, BIR, NAD, BUR, PUR, HOW, HOOG, MED, NP, SP
Moringua raitaborua (Hamilton, 1822)	Moringuidae	NP, SP
Mastacembelus armatus (Lacepède, 1800)	Mastacembelidae	DAR, COACH, JAL, UD, DD, MAL, MUR, BIR, NAD, BUR, BANK, PUR, HOOG, HOW, MED, NP, SP
Macrognathus pancalus (Hamilton, 1822)	Mastacembelidae	DAR, COACH, JAL, UD, DD, MAL, MUR, BIR, NAD, BUR, HOW, MED, BANK, PUR, HOOG, NP, SP
Macrognathus aral (Bloch and Schneider, 1801)	Mastacembelidae	DAR, COACH, JAL, UD, DD, MAL, MUR, BIR, NAD, BUR, BANK, PUR, HOOG, HOW, MED, NP, SP
Muraenesox cinereus (Forsskål, 1775)	Muraenesocidae	NP, SP
Nandus nandus (Hamilton, 1822)	Nandidae	UD, DD, MAL, MUR, BIR, NAD, BUR, BANK, PUR, HOOG, HOW, MED, NP, SP

Contd...

Table 6.1–*Contd...*

Name of the Fish	*Family*	*Distribution*
Notopterus notopterus (Pallas, 1769)	Notopteridae	DAR, COACH, JAL, UD, DD, MAL, MUR, BIR, NAD, BUR, BANK, PUR, HOOG, HOW, MED, NP, SP
Notopterus chitala (Hamilton, 1822)	Notopteridae	MAL, MUR, BIR, NAD, BUR, HOOG, HOW, MED, NP, SP
Nemacheilus beavani (Günther, 1868)	Balitoridae	DAR, COACH, JAL
Ompok pabo (Hamilton, 1822)	Siluridae	UD, DD, MAL, MUR, BIR, NAD, BUR, BANK, PUR, HOOG, HOW, MED, NP, SP
Ompok bimaculatus(Bloch, 1794)	Siluridae	UD, DD, MAL, MUR, BIR, NAD, BUR, BANK, PUR, HOOG, HOW, MED, NP, SP
Pungasius pungasius (Hamilton, 1822)	Pangasidae	UD, DD, MAL, MUR, BIR, BANK, BUR, PUR, HOOG, NP, SP, HOW, NAD, MED
Pterengraulis atherinoides (Linnaeus, 1766)	Engraulidae	COACH, UD, DD, BIR, BUR, PUR, HOOG, NP, SP, HOW, MED
Puntius gelius (Hamilton, 1822)	Cyprinidae	COACH, UD, DD, MAL, MUR, BIR, BANK, BUR, PUR, HOOG, NP, SP, HOW, NAD, MED
Puntius ticto (Hamilton, 1822)	Cyprinidae	DAR, COACH, JAL, UD, DD, MAL, MUR, BIR, BANK, BUR, PUR, HOOG, NP, SP, HOW, NAD, MED
Puntius sophore (Hamilton, 1822)	Cyprinidae	DAR, COACH, JAL, UD, DD, MAL, MUR, BIR, BANK, BUR, PUR, HOOG, NP, SP, HOW, NAD, MED
Puntius phutunio (Hamilton, 1822)	Cyprinidae	JAL, UD, DD, MAL, MUR, BIR, NAD, BUR, BANK, PUR, HOOG, HOW, MED, NP, SP
Puntius terio (Hamilton, 1822)	Cyprinidae	UD, DD, MAL, MUR, BIR, NAD, BUR, BANK, PUR, HOW, HOOG, MED, NP, SP
Puntius conchonius (Hamilton, 1822)	Cyprinidae	DAR, COACH, JAL, UD, DD, MAL, MUR, BIR, NAD, BUR, BANK, PUR, HOOG, HOW, MED, NP, SP
Puntius sarana sarana (Hamilton, 1822)	Cyprinidae	DAR, COACH, JAL, MAL, BIR, UD, DD, MUR, NAD, BUR, BANK, PUR, HOOG, HOW, MED, NP, SP
Scatophagus argus (Linnaeus, 1766)	Scatophagidae	NAD, BUR, BANK, HOOG, HOW, MED, NP, SP

Contd...

Table 6.1–*Contd...*

Name of the Fish	*Family*	*Distribution*
Somileptes gongota (Hamilton, 1822)	Cobitidae	DAR, COACH, JAL
Salmostoma bacalia (Hamilton, 1822)	Cyprinidae	DAR, COACH, JAL, UD, DD, MAL, MUR, BIR, NAD BUR BANK, PUR, HOOG, HOW, MED, NP, SP
Stigmatogobius sadanundio (Hamilton, 1822)	Gobiidae	MED, NP, SP
Terapon jarbua (Forsskål, 1775)	Terapontidae	HOOG, MED, NP, SP
Tetraodon cutcutia (Hamilton, 1822)	Tetraodontidae	UD, DD, MAL, MUR, BIR, NAD, BUR, BANK, HOOG, HOW, MED, NP, SP
Tetraodon fluviatilis (Hamilton, 1822)	Tetraodontidae	NAD, BUR, HOOG, MED, NP, SP
Tetraodon inermis (Hamilton, 1822)	Tetraodontidae	MED, NP, SP
Wallago attu (Bloch and Schneider, 1801)	Siluridae	MAL, MUR, BIR, NAD, BUR, HOOG, HOW, MED, NP, SP
Xenentodon cancila (Hamilton, 1822)	Belonidae	DAR, COACH, JAL, UD, DD, MAL, MUR, BIR, NAD, BUR, BANK, PUR, HOOG, HOW, MED, NP, SP

Source: Basu *et al.* (2012) and Saha and Patra (2013).

DAR: Darjeeling; JAL: Jalpaiguri; UD: Uttar Dinajpur; DD: Dakshin Dinajpur; MAL: Maldah; NAD: Nadia; MUR: Mursidabad; COACH: Coachbehar; BIR: Birbhum, BUR: Burdwan; BANK: Bankura; PUR: Purulia; HOOG: Hoogley;HOW: Howrah; MED: Medinipur; NP: North 24 Parganas; SP: South 24 Parganas; KOL: Kolkata.

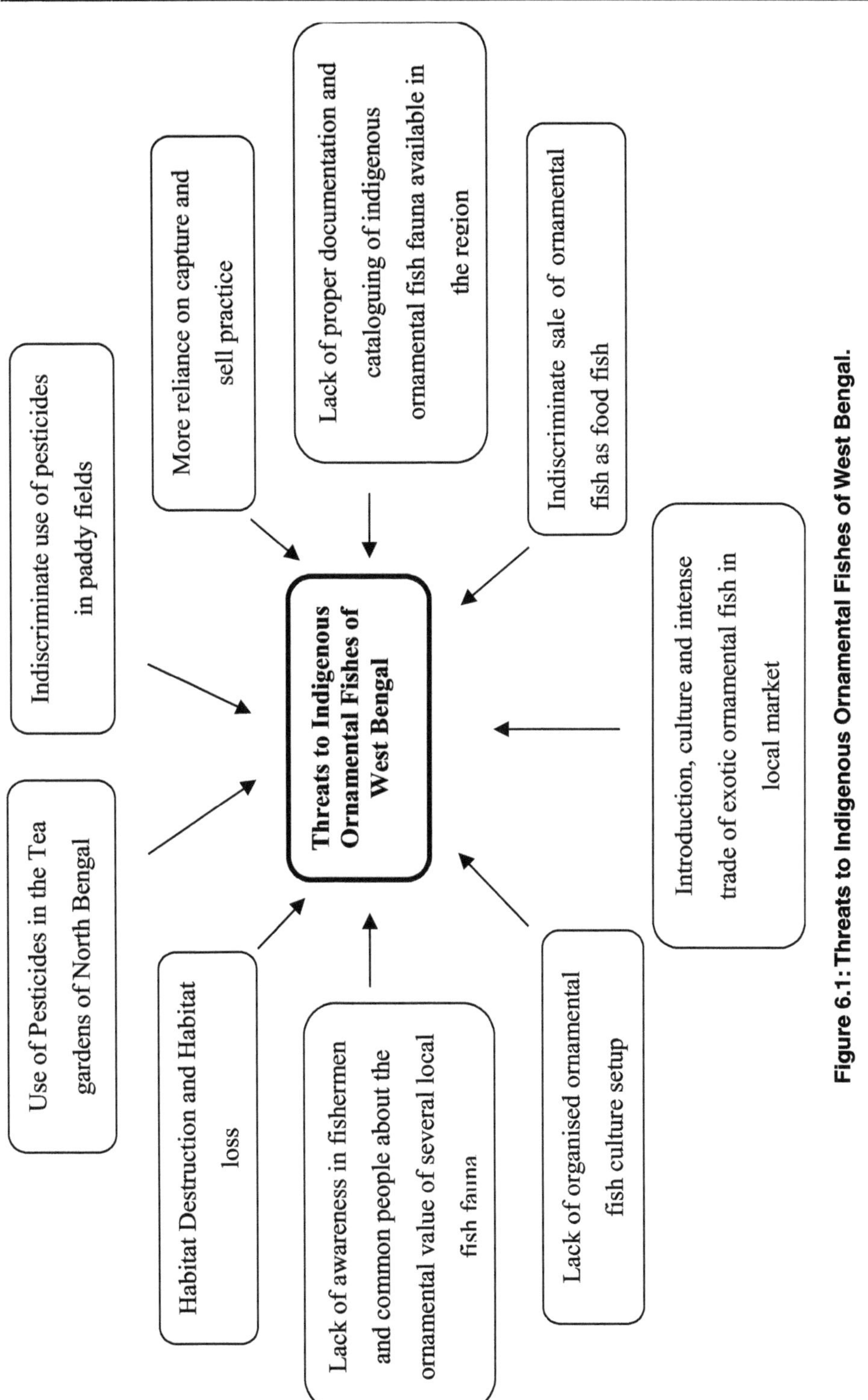

Figure 6.1: Threats to Indigenous Ornamental Fishes of West Bengal.

Some Commercially Important Indigenous Ornamental Fishes of West Bengal

Amblypharyngodon mola

Notopterus notopterus

Badis badis

Danio rerio

Colisa fasciata

Colisa lalia

Chanda nama

Chanda ranga

Some Commercially Important Indigenous Ornamental Fishes of West Bengal

Esomus danricus

Channa marulius

Channa striata

Nandus nandus

Lepidocephalichthys guntea

Mystus vittatus

Macrognathus pancalus

Conclusion

Freshwater fishes comprise one of the most threatened vertebrate groups with close to one-third of the species in danger of extinction (Dudgeon, 2012) and several species are already extinct (Baillie and Butcher, 2012). Lack of proper information and knowledge (Cooke *et al.*, 2012), and the uneven distribution of conservation attention and investment (Darwall *et al.*, 2011) on freshwater ecosystems and their resources might be the reasons behind this vulnerability. Today, the fish diversity cataloguing and its appropriate management is a great challenge. The precise number of existing fish species remains to be determined and experts feel that the final total may be considerably higher (Le´ve^que *et al.*, 2008). Reports say that aspects of biodiversity and conservation of the icthyo-fauna in the Asian region, in comparison to that of Africa, Europe and North America have been relatively less documented (Nguyen and De Silva, 2006). However, last few decades the freshwater fish diversity of India has been updated by several authors. As a result the dynamic pattern of exploitation and the inclination in the fish species exploitation are highlighted to enable appropriate conservation effort. In parity with these views and considering ornamental fish trade of West Bengal as a potential livelihood source, more detailed and long-term studies on the indigenous ornamental fish species diversity is essential. An overview of the existing literature on the fish fauna in different lotic and lentic ecosystems suggests that North Bengal is a hotspot of fish resources with prospect for exploitation in ornamental fish trade. The Terai and Duars region of the North Bengal (northern Part of West Bengal) is very much similar geographically and meteorologically both with the northeastern part of India and is also known for its rich biodiversity (Jha *et al.*, 2004). It is linked to the borders of the Darjeeling-Sikkim Himalaya biodiversity hotspot, thereby bearing significance as a bio-geographical region with diverse flora and fauna.

Wetlands are productive, ecologically sensitive and adaptive ecosystems and provide many important services to human society as well as support a rich resource of biodiversity which certainly includes fishes. The water quality is one of the major aspects of wetland ecosystem and it also affects the fauna and flora associated with it. From past few years the negative economic, social, and environmental consequences of declining water quality in wetlands has become an issue of concern for India, especially for West Bengal. In West Bengal, the deteriorating water quality has become very alarming in the case of small water bodies such as lakes, bills and ponds. These water resources are one of the major habitats and breeding grounds for many ornamental ichthyofauna in West Bengal, especially in Terai-Duars region (northern part of west Bengal). Apart from this, these water sources have also been utilised for several economic purposes such as for fisheries. Various anthropogenic activities such as construction of dams, over exploitation of wetlands for social and economic purposes, use of pesticides in the surrounding crop-fields and in tea gardens (in Terai-Duars region) has mainly caused the depletion of water quality of wetlands in West Bengal causing a potent threat to its biodiversity especially to the vulnerable fish species. Apart from these, growing urbanisation along with its associated problems of potable water supply, sanitation, water pollution, flooding, and depletion of groundwater are also diminishing the wetlands and its associated fish fauna. So it is the call of the time to conserve these wetlands of West Bengal, which could easily be used as the breeding ground for small indigenous fish species (SIS) of ornamental value and will certainly contribute to the enrichment of biodiversity of this ornamental ichthyofauna, and again it might help in the commercial exploration of these living jewels.

In West Bengal as a whole, a range of river fans, annual streams, lakes and wetlands characteristically constitute the freshwater bodies providing diverse habitats for fish species. Sustainable utilization of fish resources of these freshwater bodies involving local people can be a probable option for expansion of trade in indigenous ornamental fishes occurring in the region, which will generate enormous employment opportunities for the people ensuring earning of valuable foreign exchange for the country as well as will ensure the conservation issue.

Summary

West Bengal is sanctified with a rich source of indigenous ornamental fishes. More than 70 species of indigenous ichthyo-fauna of ornamental value have been reported from different parts of the state. Restricted information on the species richness and abundance in the region is a hindrance towards framing strategies for exploitation of species for ornamental fish trade. A detailed survey of the state for several natural indigenous ornamental fish fauna is thus justified. The results are also expected to provide an updated documentation of indigenous fish species of West Bengal to revive exploitation of conservation and sustainable use of freshwater fish resources.

Acknowledgement

I gratefully acknowledge Prof. (Retd.) Samir Banerjee, Department of Zoology, University of Calcutta and Dr. G. Aditya, Department of Zoology, University of Calcutta, for their valuable suggestions, constructive guidance and encouragement. I am also deeply acknowledged to the Head of the Department of Zoology, University of Calcutta, for providing me with all the infrastructural facilities.

References

Acharjee ML and Barat S. 2013. Ichthyofaunal Diversity of Teesta River in Darjeeling Himalaya of West Bengal, India. *Asian J. Exp. Biol. Sci.*, 41: 112-122.

Allen DJ, Molur S and Daniel BA. 2010. The Status and Distribution of Freshwater Biodiversity in the Eastern Himalaya. IUCN, Cambridge, UK and Gland, Switzerland; Zoo Outreach Organization, Coimbatore, India.

Ananda MO, Krishnaswamy J, Kumar A and Balif A. 2007. Sustaining biodiversity conservation in human-modified landscapes in the Western Ghats: Remnant forests matter. *Biological Conservation*, 143(10): 2363-2374.

Baillie JEM and Butcher ER. 2012. *Priceless or worthless? The world's most threatened species*. Zoolical Society of London, UK. pp. 123.

Barat S, Jha P and Lepcha RF. 2005. Bionomics and Cultural prospects of Katli, *Neolissocheilus hexagonolepis* (Mc Clelland) in Darjeeling district of West Bengal.

In. *Coldwater Fisheries Research and Development in North East Region of India* (Tyagi B, Shyam Sunder and Mohan M. eds.). NRCCWF, Bhimtal, Haldwani. pp. 66-69.

Basu A, Dutta D and Banerjee S. 2012. Indigenous ornamental fishes of West Bengal. *Recent Research in Science and Technology*, 4(11): 12-21.

Bhat A. 2003. Diversity and composition of freshwater fishes in river systems of Central Western Ghats. India. *Env. Biol. of Fishes*, 68(1): 25-38.

Bhat A. 2004. Patterns in the distribution of freshwater fishes in rivers of Central Western Ghats, India and their associations with environmental gradients. *Hydrobiol.*, 529(1-3): 83-97.

Bhattacharya BK, Sugunan VV and Choudhury M. 2003. Ichthyofaunistic resources of Assam with a note on their sustainable utilization. In. *Integration of Fish-Biodiversity Conservation and Development of fisheries in North-Eastern Region through Community participation.* Proc. Nat. Workshop, December, 12-13, 2001, NBFGR, Lucknow.

Chakraborty T and Bhattacharjee S. 2008. The Ichthyofaunal Diversity in the Freshwater Rivers of South Dinajpur District of West Bengal, India. *Jr. Bom. Nat. Hist. Soc.*, 105(3): 292-298.

Chapman, PSJ. 1997. *Ecology-Principle and Application.* Cambridge Universitry Press, Cambridge, UK. pp. 25.

Cooke SJ, Paukert C and Hogan Z. 2012. Endangered river fish: factors hindering conservation and restoration. *End. Sp. Res.*, 17: 179-191.

Darwall WRT, Smith KG, Allen DJ. *et al.* (eds.). 2011. *The diversity of life in African freshwaters: under water, under threat.* An analysis of the status and distribution of freshwater species throughout mainland Africa. IUCN, Gland, Cambridge.

Dey SC and Kar D. 1989a. Aquatic macrophytes of Lake Sone in Assam. *Environ. and Ecol.*, 7(1): 253-254.

Dey SC and Kar D. 1989b. An account of *Hilsa ilisha* (Hamilton) of Lake Sone in the Karimganj district of Assam. *Bang. J. of Zool.*, 17(1): 69-73.

Dey SC and Kar D. 1989c. Fishermen of Lake Sone in Assam: Their socio-economic status. *Sci. and Cult.*, 55: 395-398.

Dey SC and Kar D. 1990. Fish yield trend in Sone, a tectonic lake of Assam. Matsya., 15-16: 39-43.

Dey VK. 1996. *Ornamental fishes and Handbook of Aqua farming.* The Marine Products Export Development Authority, Cochin.

Dudgeon D. 2012. Threats to freshwater biodiversity globally and in the Indo-Burma Biodiversity Hotspot In. *The status and distribution of freshwater biodiversity in Indo-Burma.* (Allen DJ, Smith KG, Darwall WRT eds.). IUCN, Cambridge, UK and Gland, Switzerland. pp. 1-28.

Ghosh A, Mahapatra BK and Datta NC. 2002. Studies on Native Ornamental Fish of West Bengal with a Note on their Conservation. *Environ and Ecol.*, 20(4): 787-793.

Hamilton B. 1822. *Account of the Fishes found in the River Ganges and its tributaries.* Edinburgh, UK. pp. 405.

Hensen RR, Ploeg A and Fossa SA. 2010. *Standard Names for Freshwater Fishes in the Ornamental Aquatic Industry.* OFI Educational Publication 5. Ornamental Fish International, The Netherlands. pp. 146.

Hora SL and Gupta JC. 1940. On a Collection of Fish from Kalimpong Duars and Siliguri Terai, Northern Bengal. *Jr. Royl. Asi. Soc. Beng. Sc.,* 6(8): 77-83.

Hora SL. 1921a. Indian Cyprinoid fishes belonging to the Genus *Garra* with notes on the related species from other countries. *Rec. Indian. Mus.,* 22(633): 687.

Hora SL. 1921b. Fish and Fisheries of Manipur with some observations on those of the Naga Hills. *Rec. Indian. Mus.,* 22.

Hora SL. 1930. Ecology, Bionomics and Evolution of torrential fauna with special reference to the organs of attachment. *Phil. Tran. Royl. Soc. Lond.,* 218: 171-282.

Hora SL. 1937. Geographical distribution of Indian Freshwater Fishes and its bearing on the probable land connections between India and the adjacent countries. *Curr. Sc.,* 7: 351-356.

Hora SL. 1939. The Game Fishes of India, VIII: The Mahseers or the large-scaled barbs of India, I: The Putitor Mahseeer, *Barbus (Tor) putitora* (Hamilton). *J. Bom. Nat. His. Soc.,* 41: 272-285.

Hora SL. 1940. Dams and the problem of migratory fishes. *Curr. Sc.,* 9: 406-407.

Hora SL. 1943. The Game Fishes of India, XVI: The Mahseers or the large-scaled barbs of India, 9: Further observations on the Mahseers from Deccan. *Jr. Bom. Nat. Hist. Soc.,* 44: 1- 8.

Hora SL. 1951. Fish Geography of India. *Jr. Zool. Soc. India,* 3(1): 183-187.

Hora SL. 1953. Fish distribution and Central Asian Orography. *Curr. Sc.,* 22(4): 93-94.

Janzen DH. 1981. The peak in north American ichneumonid species richness lies between 38 and 420N. *Ecology,* 62: 532-537.

Jayaram KC. 1981. *The Freshwater Fishes of India, Pakistan, Bangladesh, Burma, Sri Lanka: A Handbook.* Zoological Survey of India, Kolkata. pp. 475.

Jayaram KC. 1999. *The Freshwater Fishes of the Indian Region.* Narendra Publ. House, Delhi. pp. 551.

Jayaram KC. 2006. *The Catfishes of India.* Narendra Publ. House, New Delhi. pp. 383.

Jayaram KC. 2010. *The freshwater fishes of the Indian region* (2nd edn.) Narendra Publ. House, New Delhi. pp. 625.

Jha P, Mandal A and Barat S. 2004. Mahananda Reservoir, W.B.: Its Ichthyofauna, Fishery and Socio- Economic Profile of Fish Production. *Fishing Chimes.* 24(6): 14-17.

Jhingran VG. 1975. *Fish and fisheries of India* (1st edn.). Hindustan Publ. Corporation, Delhi. pp. 160.

Jhingran VG and SD Tripathy. 1969. A Review of the measures adopted for the development of fisheries in reservoirs in India. *Proc. Sem. Ecol. Freshwater Res.* CIFRI. pp. 637-651.

Johal M. 2005. Recent innovations in the determinations in age determination using hard parts in Indian freshwater fishes. In. *New horizons in animal sciences* (Sobti R and Sharma V. eds.). Visual Publ. Company, Punjab. pp. 91–98.

Kar D. 1984. Limnology and Fisheries of Lake Sone in the Cachar District of Assam (India). Ph.D Thesis, University of Gauhati, Assam. pp. 201.

Kar D. 1990. Limnology and Fisheries of Lake Sone in the Cachar district of Assam (India). *Matsya,* 15(16): 209-213.

Kar D. 2003. Fishes of Barak drainage, Mizoram and Tripura. In. *Environment, Pollution and Management* (Kumar A, Bohra C and Singh LK. eds.). APH Publ. Corporation, New Delhi. pp. 604.

Kar D. 2005a. Fish Genetic Resources and Habitat Diversity of the Barak Drainage. In. *Aquatic Ecosystems, Conservation, Restoration and Management* (Ramachandra TV, Ahalya N and Rajsekara Murthy C eds.). Capital Publ. Company, Bangalore. pp. 68-76.

Kar D. 2005b. Fish fauna of river Barak, of Mizoram and of Tripura with a note on conservation. *Jr. Freshwater Biol.,* 16.

Kar D and Barbhuiya MH. 2000. Length-weight Relationship and Condition Factor in *Gudusia chapra* (Hamilton-Buchanan) and *Botia dario* (Hamilton-Buchanan) from Chatla Haor (flood plain wetland) in Cachar district of Assam. *Environ. and Ecol.,* 18(1): 227-229.

Kar D and Barbhuiya MH. 2004. Abundance and diversity of zooplankton in Chatla Haor, a floodplain wetland in Cachar district of Assam. *Environ. and Ecol.,* 22(1): 247-248.

Kar D and Dey SC. 2000. Yield and conservation of Indian major carps of Lake Sone in Assam. *Environ. and Ecol.,* 18(4): 1036-1038.

Kar D and Dey SC. 2002. On the occurrence of advanced fry of *Hilsa* (Tenuolosa) *ilisha* (Hamilton-Buchanan) in Chatla Haor seasonal wetland of Assam. *Proc. Zool. Soc. Calcutta,* 55(2): 15-19.

Kar D, Laskar BA and Nath D. 2002a. *Tor* sp. (Mahseer fish) in river Mat in Mizoram. *Aquaculture.,* 3(2): 229-234.

Kar D, Laskar BA, Mandal M. *et al.,* 2002b. Fish genetic diversity and Habitat parameters in Barak drainage, Mizoram and Tripura. *Indian Jr. Environ. and Ecoplan.,* 6(3): 473-480.

Kar D, Laskar BA, Nath D *et al.,* 2002c. *Tor progenius* (Mc Clelland) under threat in river Jatinga, Assam. *Sci. and Cult.,* 68(7-8): 211.

Kar D, Nagarathna AV, Ramachandra TV and Dey SC. 2006. Fish diversity and conservation aspects in an aquatic ecosystem in northeastern India. *Zoos' Print J.,* 21(7): 2308-2315.

Lakra WS and Sarkar UK. 2010. NBFGR-Marching ahead in cataloguing and conserving fish genetic resources of India. *Fishing Chimes*, 30: 102-107.

Lem A. 2001. Marine Ornamentals: Collection, Culture and Conservation. Proc. International Seminar on Marine Ornamentals. 6-8th June, 2001, Lake Buena Vista, USA.

Legendre P and Legendre L. 1998. Numerical Ecology (2nd ed.). Elsevier Science BV, Amsterdam. pp. 853.

Le´ve^que C, Oberdorff T, Paugy D. *et al.*, 2008. Global diversity of fish (Pisces) in freshwater. *Hydrobiol.*, 595: 545-567.

Menon AGK. 1962. A distributional list of fishes of the Himalayas. *Jr. Zool. Soc. India*, 14 (2): 23-32.

Menon AGK. 1999. *Checklist: Freshwater Fishes of India*. Occasional Paper No. 175, Zoological Survey of India, Kolkata. pp. 366.

Mills M. 1990. *An overview of ornamental fishes- Fresh and Marine water* (Tomey WA ed.), International Seminar on Ornamental Fishes. 25th July, 1990. pp.36.

Misra KS. 1959. An aid to the identification of commercial fishes of India and Pakistan. *Rec. Indian. Mus.*, 57(1-4): 1-320.

Molur S, Smith KG, Daniel BA and Darwall WRT. 2011. *The status and distribution of freshwater biodiversity in the Western Ghats, India*. IUCN, Cambridge and Zoo Outreach Organisation, Coimbatore.

Mukherjee M and Sarkar G. 2005. *Endangered Fishes of West Bengal, with a special reference to North Bengal on Environmental Degradation and Biodiversity*. Daya Publ. House, New Delhi. pp. 279.

Mukherjee M, Lepcha RF and Chakraborty C. 2011. *Fish and Fisheries of Himalayan and Terai Region of West Bengal with Ornamental touch*. Daya Publ. House. Delhi. pp. 342.

Nath P and Dey SC. 2000. Conservation of Fish Germplasm Resources of Arunachal Pradesh, *In. Fish Biodiversity of Northeastern India* (Ponniah AG and Sarkar UK eds.). NATP Publication No. 2, NBFGR, Lucknow. pp.49-67.

Nguyen TTT and De Silva S. 2006. Freshwater fin fish biodiversity and conservation: an Asian perspective. *Biodiversity Conservation*, 15(11): 3543-3568.

Panigrahi AK, Dutta S and Ghosh I. 2009. Selective study on the availability of indigenous fish species having ornamental value in some districts of West Bengal. *Sust. Aquacult.*, 14(4): 13-15.

Patra AK, Sengupta S and Datta T. 2011. Physico-chemical properties and Ichthyofauna Diversity in Karala river, A tributary of Teesta river at Jalpaiguri district of West Bengal, India. *Int. Jr. Appl. Biol. Pharmaceut. Technol.*, 2 (3): 47-58.

Paul B and Chanda A. 2014. Indigenous Ornamental Fish Faunal Diversity in Paschim Medinipur, West Bengal, India. *Int. Res. J. Biol. Sci.*, 3(6): 94-100.

Penning M, Reid G McG, Koldewey H. *et al.* (eds.). 2009. *Turning the Tide: A Global Aquarium Strategy for Conservation and Sustainability*. World Association of Zoos and Aquariums, Bern, Switzerland.

Ponniah AG and Gopalakrishnan A. 2000. *Endemic fish diversity of the Western Ghats*. NBFGR-NATP Publication, NBFGR, Lucknow, India. pp. 347.

Ponniah AG and Sarkar UK. 2000. *Fish biodiversity of North-east India*. NBFGR-NATP Publ., NBFGR, Lucknow, India. pp. 228.

Raghavan R, Prasad G, Ali A and Sujarittanonta L. 2007. 'Boom and bust fishery' in a biodiversity hotspot - is the Western Ghats (South India) losing its most celebrated ornamental fish, *Puntius denisonii* (Day)? *Curr. Sci.*, 92: 1671-1672.

Raghavan R, Prasad G, Ali A and Pereira B. 2008. Fish fauna of Chalakudy River, part of Western Ghats biodiversity hotspot, Kerala, India: patterns of distribution, threats and conservation needs., *Biodiversity and Conservation*, 17(13): 3119-3131.

Raghavan R, Prasad G, Ali A. *et al.*, 2009. Damsel in distress – the tale of Miss Kerala, *Puntius denisonii* (Day) an endemic and endangered cyprinid of Western Ghats biodiversity hotspot, India. *Aquat. Conserv. Mar. Freshwater Ecosyst.*, 19: 67-74.

Raghavan R, Ali A, Dahanukar N and Rosser A. 2011. Is the Deccan Mahseer, *Tor khudree* (Sykes 1839) (Pisces: Cyprinidae) fishery in the Western Ghats Hotspot sustainable? A participatory approach to stock assessment. *Fish. Res.*, 110: 29-38.

Raghavan R, Philip S, Dahanukar N and Ali A. 2013. Freshwater biodiversity of India: a response to Sarkar *et al.*, *Rev. Fish. Biol.*, 23: 547–554.

Rema Devi K and Indra TJ. 2010. *Checklist of the native freshwater fishes of India*. Zoological survey of India,Kolkata. pp. 24.

Rhyne AL, Tlusty MF, Schofield PJ. *et al.*, 2012. Revealing the appetite of the marine aquarium fish trade: the volume and biodiversity of fish imported into the United States. PLoS ONE, 7(5): e35808.

Saha MK and Patra BC. 2013. Resource potentiality of indigenous ornamental fishes in West Bengal. *Int. J Curr. Res.*, 5 (5): 1232-1238.

Sarkar T and Pal J. 2008. Studies on the Diversity of Fish in Different Reservoir and Rivers of Terai Region. *North Bengal University J. Anim. Sci.*, 2(2): 83-88.

Sehgal KL. 1999. Coldwater fish and fisheries in the Himalayas: rivers and streams. In. *Fish and fisheries at higher altitude-Asia* (Petr T ed.). FAO Fisheries Technical Paper No. 385: Rome, FAO. pp. 41-63.

Sen N. 1982. Studies on the Systematics, Distribution and Ecology of the Ichthyofauna of Meghalaya and their bearing on the Fish and Fisheries of the State. Ph.D Thesis, University of Gauhati, Assam. pp. 576.

Sen TK. 1985. The Fish Fauna of Assam and the Neighbouring Northeastern States of India. *Rec. Zool. Surv. India*, Occasional Paper No. 64: 1-216.

Sen TK. 1992. Freshwater fish. In. *Fauna of West Bengal State fauna series 3*. Zoological Survey of India, Calcutta.

Shaw GE and Shebbeare EO. 1937. The fishes of Northern Bengal. *Jr. Ryol. Asi. Soc. Beng. Sc.,* 3: 1-128.

Singh T. 2005. Emerging trends in world ornamental fish trade. *Info. fish Int.,* 24(3): 15-18.

Sreekantha Subash Chandran MD, Mesta D K. *et al.,* 2007. Fish diversity in relation to landscape and vegetation in central Western Ghats, India. *Curr. Sci.,* 92(11): 1592-1603.

Swain SK and Jena JK. 2002. Ornamental fish farming: A new dimension in aquaculture entrepreneurship. In. *Ornamental Fish Breeding and Culture for Entrepreneurship Development* (Swain SK ed.). CIFA, Bhubaneshwar. pp.1-6.

Talwar PK and Jhingran AG. 1991. *Inland Fishes of India and adjacent countries,* Vol. 1 and 2. Oxford and IBH Publ. Co. Pvt. Ltd. New Delhi.

Tlusty MF, Rhyne AL, Kaufman L. *et al.,* 2013. Opportunities for public aquariums to increase the sustainability of the aquatic animal trade. *Zoo Biol.,* 32: 1-12.

Vinci GK. 1998. Introduction to aquarium fishes. *Fishing Chimes,* 2(3): 25.

Vishwanath W, Lakra WS and Sarkar UK. 2007. *Fishes of North East India.* NBFGR, Lucknow, India. pp. 264.

Chapter 7

Fish Diversity of Riverine Systems along Eastern Himalayan Region

Munmun Chakraborty and Sumit Homechaudhuri

Aquatic Bioresource Research Laboratory, Department of Zoology, University of Calcutta, Kolkata

Introduction

India is one of the mega biodiversity countries in the world and occupies the ninth position in terms of freshwater mega biodiversity. The Eastern Himalayan Biodiversity Hotspot region and its foothills are very rich in both floral and faunal diversity including fish. A total of 218 species of fishes has been listed in whole Himalayas (Acharjee and Barat, 2013). Especially the region being the home of many large torrential rivers, fish diversity is also very rich and taxonomically interesting (Abell *et al.*, 2008). As such, the northern districts of the Indian state of West Bengal, specially the districts of Darjeeling and Jalpaiguri, lying within the Eastern Himalayan biodiversity hotspot range, hold a greater faunistic importance and are unique from the zoogeographical point of view. The fish fauna represents mostly the Chinese, Malayan and Indian elements of fishes of the Oriental realm. There is a lack of baseline information on freshwater fish species distribution and their ecological requirements throughout the Eastern Himalayas. It has been found that 31.3 per cent of the 1,073 freshwater species of fishes, molluscs, dragonflies and damselflies currently known in the Eastern Himalaya region, are assessed as Data Deficient, emphasizing the urgent need for new research in the region (Allen *et al.*, 2010). Biodiversity within inland water ecosystems in the Eastern Himalayan region is both highly diverse and of great regional importance to livelihoods and economies. Therefore, study of ichthyofaunal diversity and land usage practices assumes a great significance in this region considering conservation priority, endemecity and monitoring environmental health (Figure 7.1).

Like other Himalayan rivers, the Teesta river and its tributaries provide a fair ecological niche for many indigenous, and a few exotic fish species. This river forms the lifeline of the states of Sikkim and West Bengal, providing a source of livelihood to the locals. However, at present, the region with rich biodiversity and high level of endemism is under immediate threat of species extinction and habitat destruction mainly due to tremendous pressure from demotechnic growth and natural environmental changes. Scientific documentation of the Ichthyofaunal diversity of the river Teesta drainage basin is poor and there is scanty documentation on its stretch within West Bengal.

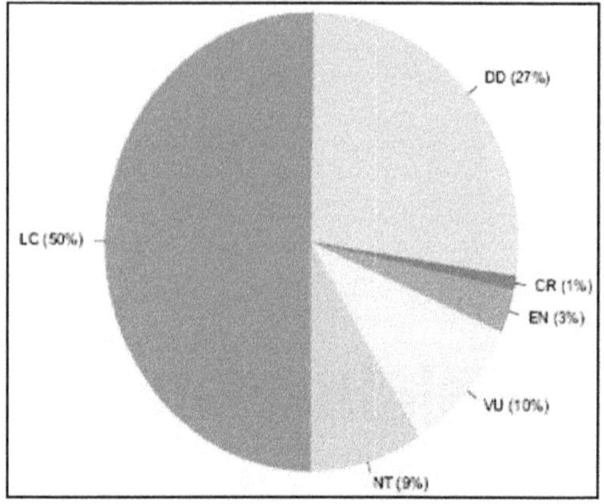

Figure 7.1: The Proportion (as per cent) of Freshwater Fish Species in each Red List Category in the Eastern Himalayan Assessment Region.

Fish Diversity and Ecological Drivers for Assemblage Patterns

The biodiversity and conservation of fish in the aquatic ecosystem have attracted the attention of various workers. In India several studies on fish biodiversity have been carried out in both freshwater as well as marine ecosystems. Many illustrations are available that substantiate the highly diverse nature of the aquatic ecosystems in India. Works on freshwater ecosystems indicated substantial biodiversity (Kumar, 2000; Martin *et al.*, 2000; Bhat, 2003; Dahanukar *et al.*, 2004; Kar *et al.*, 2006; Sarkar *et al.*, 2008; Negi and Rajput, 2012). River conservation and management activities in most countries including India suffer from inadequate knowledge of the constituent biota. Therefore, research is being pursued continually to develop conservation planning to protect freshwater biodiversity (Shahnawaz *et al.*, 2010; Lakra *et al.*, 2010; Thirumala *et al.*, 2011; Sarkar *et al.*, 2012; Tamboli and Jha, 2012; Kumar and Pandey, 2013).

Northeast India is considered as one of the hot spots of freshwater fish biodiversity in the world (Kottelat and Whitten, 1996). Scanning of literature shows many diverse works on fish diversity, conservation status and biogeography from this region (Ghosh and Lipton, 1982; Sen, 1985; Nath and Dey, 1989; Sarkar and Ponniah, 2000; Nath *et al.*, 2008). Faunistically, these areas are very rich especially the piscine populations are numerous in variety and are taxonomically interesting. This may be due to the presence of numerous streams and major rivers rushing to the plains through these regions. The chief rivers are Mahananda and Teesta with many tributaries like Sevoke, Atrai, Jaldhaka, Karala, Karotoyar, etc. Darjeeling district, excluding a major portion of Siliguri subdivision, lies on the Himalayan hill where as Jalpaiguri district mostly represents the plains with a belt of forest; the portion lying west of Teesta river is known as the "Terai" and the portion lying east of the river is known as the "Duars". This forest belt in the past was very dense and was full of swamps. The duars criss-crossed with mountain rivers, mostly support swamps of the sub-Himalayan region. During the initial stages, the most comprehensive account of the fish fauna of North Bengal was published by Shaw and Shebbeare (1937), who listed 131 species in their treatise. Hora and Gupta (1941) reporting on a small collection from Kalimpong Duars and Siliguri Terai added two more species to this list. Following the above pioneer work, Jayaram and Singh (1977) carried out a survey of North Bengal rivers and listed 96 species, of which 17 species were not reported by previous workers. Goswami *et al.* (2012) carried out fish diversity study of Northeast India, inclusive of the Himalayan and Indo Burma biodiversity hotspots zones and documented a checklist on their taxonomic status, economic importance, geographical distribution, present status and prevailing threats. They listed 422 fish species from Northeast India, belonging to 133 genera and 38 families. The maximum diversity is observed in the family Cyprinidae, which is represented by 154 species. The findings also provide 48 endangered, 69 near threatened, 103 vulnerable, 153 least concerned, 23 data deficient and 26 not evaluated species from the list. Allen *et al.* (2010) work out on the IUCN status of the freshwater biodiversity in the Eastern Himalaya. Bhatt *et al.* (2012) while studying diversity and distribution patterns of species along elevational gradients (50-3800 m) and understanding drivers behind these patterns, collated taxonomic and distribution data of fish species from 16 freshwater Himalayan rivers. Their studies indicate that various environmental factors influence distribution of fish species richness differently and the relationship varies in magnitude. Generalized linear model identified water discharge, basin area, temperature and phytoplankton as the most important factors influencing species richness.

In the recent past, Acharjee and Barat (2013) worked on the ichthyofaunal diversity of Teesta river only in Darjeeling Himalayan region of West Bengal, during March 2007 to February 2009, with special reference to their distribution, abundance and diversity through space and time and reported that the river stretch sustained about 65 rheophilic, cold water fish species belonging to 39 genera and 10 families, having ornamental, food and sport value. Fishes of the family Cyprinidae were found to be dominant followed by Sisoridae and Balitoridae.

Fish Guild Structure along Longitudinally Determined Ecological Zonation of River Teesta: A Case Study

Study Design

River Teesta, originating from north Sikkim and carving out verdant Himalayan temperate and tropical river valleys, traverses the Indian states of Sikkim and West Bengal and finally descends to Brahmaputra in Bangladesh. The total length of the river is 309 km, draining an area of 12,540 km^2. The present study area includes the course of the river Teesta in West Bengal divided into ecological zones based on elevation gradient and habitat types. The river stretch was divided in four zones *viz.* the upper stretch (Rishi khola and Rungpo) where elevations is higher with low temperatures; middle stretch (Teesta Bazaar) with comparatively low elevation; at Sevoke the river hits the plains; lastly the river plains (Gojoldoba, Domohoni and Haldibari). Along the longitudinal stretch of the river in West Bengal covering a distance of 142 km, each site was sampled at regular intervals (bi-annually with pre-monsoon and post-monsoon visits) when flow conditions were most stable and similar among sites (Table 7.1 and Figure 7.2). Local habitat attributes were recorded to find out any association with the variation in fish assemblages. Habitat variable for each sampling sites at each sampling operations were recorded in the field. Stream width was measured along three transects regularly spaced across the stream channel. Water depth, current velocity and temperatures were measured at the mid-point of each transect.

Fish sampling was carried out from December 2010 to March 2013 in every alternate six months at seven sampling areas (approximately 20-30 km apart) under the four zones covering the longitudinal gradient of river Teesta at Darjeeling and Jalpaiguri districts in West Bengal. After an initial pilot survey of the entire riverine stretch, these seven areas have been chosen based on different habitat patches, high fishing activity, accessibility and availability of local fish markets nearby (for gathering secondary data). Each sampling area was further divided into 4 sampling sites (approx. 1-2 km apart) totalling to 28 sites altogether. It was observed that 4 sampling sites per area were sufficient enough to represent the fish assemblage of the respective area. All the important freshwater aquatic microhabitats (riffles, pools, cascade, falls, etc.) were sampled using gill nets, cast nets, drag nets, and hooks and lines of varying dimensions. A sample reach of 50 m were fished for 2 hours at every sites using above mentioned fish nets as well as electro-fishing method using a single anode electro-fisher (300V, 3-4A, DC) operated by the same person. Captured fish specimens were counted and fixed in 10 per cent formalin solution and, after 48 hours, transferred to 70 per cent ethyl alcohol solution. Fishes were identified to the lowest taxonomic level using available literature (Shaw and Shebbeare, 1937; Menon, 1987; Day, 1889; Talwar and Jhingran, 1991 and Jayaram, 2006, 2010). All fish specimens have been deposited in the fish collection repertoire at the Zoological Survey of India, Kolkata. The status of the species on the IUCN Red List of Threatened Species has also been incorporated. The divisions of the zones is based on Aarts and Nienhuis (2003) and Aquatic Ecological System (AES) classification and also adds some later subdivisions based on the present occurrence of the zones.

Observation

A total of 16,703 number of fish specimens were collected overall, which includes 92 species belonging to 50 genera and 19 families from the longitudinal stretch of river Teesta in West Bengal (Table 7.2). Overall the fish species with highest abundances were *Barilius bendelisis, Puntius sophore, Schistura corica, Lepidocephalichthys guntea*. Fish diversity exhibits maximum value in the lower reaches of the river *viz.* Gojoldoba and Domohoni, mostly domimated by Cypriniformes (*Aspido pariamorar, Barilius bendelisis, Devario devario, Puntius sophore, Esomus danricus, Lepidocephalichthys guntea*) and Siluriformes (*Mystus bleekeri, Bagariu syarrelli, Glyptothorax telchitta, G. striatus, G. indicus, G. cavia*) fishes. Diversity in the middle regions *viz.* Teesta bazaar and Sevoke is limited and specialized (fish groups of mainly *Barilius* spp., *Schistura* spp. and *Garra* spp. dominate in this stretch) and lowest in further upper stretches *viz.* Rishi Khola and Rungpo. These groups of fishes are highly habitat specific and survive only in clear stream waters with

Table 7.1: Longitudinal Zonation Concepts with Additional Hydrological and Ecological Characteristics

Zones	Sampling Areas	Elevation	Temperature (°C)		Water Velociy (m/sec)	Stream Width (ft)	Stream Depth (ft)	Habitat guilds	Preferred Spawning Habitat
			AT	WT					
High-mid altitude zone	Rishi Khola	Moderate to high (1093-1000 ft) elevation water-sheds dominated by side slopes with gentle slopes and steep slopes.	21-23.7	18.5-21	2.1-2.9	1	1	Primary forest; hilly terrain.	Lithophils/ lithopelagophils
	Rungpo (26°10'21.94" N; 88°31'46.73" E)		21.1-24.5	19.5-21	1.9-2.2	15	31-32.5	Secondary forest	
Mid altitude zone	Teesta Bazaar (26°00'03.97" N; 88°26'31.80" E)	Moderate (628 ft) elevation watersheds dominated by side slopes and gentle slopes.	24-27	20.7-24.9	1.6-1.9	20	23-25	Secondary forest; ongoing construction work of Teesta Barrage project.	Lithoplegophils/ Speleophils
Low altitude-plain zone	Sevoke (26°53'25.37" N; 88°28'22.97" E)	Moderate to low elevation (500 ft) watersheds dominated by gentle slopes with substantial areas of flats and side-slopes; river hits the plain at this site.	18.5-25.2	15.4-18.5	1.3-1.6	7.4-21	24-25	Secondary forest	Lithoplegophils/ Psammophils
River plains	Gojoldoba (26°45'08.55" N; 88°35'05.04" E)	Low elevation (380-187 ft) dominated by flats, pastured land and urban inhabitation.	30.1-35.5	28.2-31	0.9-1.4	40	25-27	Secondary forest; Urban area; presence of Teesta Barrage	Phytophils/ Phytolothophils
	Domohoni (26°33'47.11" N; 88°45'39.34" E)		34-35.7	30.1-31	0.45-0.9	38	20-22	Agriculture land; Urban area	
	Haldibari (26°20'52.00" N; 88°54'16.76" E)		33.7-37.2	29.9-31.1	0.45-0.55	35	20-21	Agriculture land; Urban area	Phytophils/ polyphils

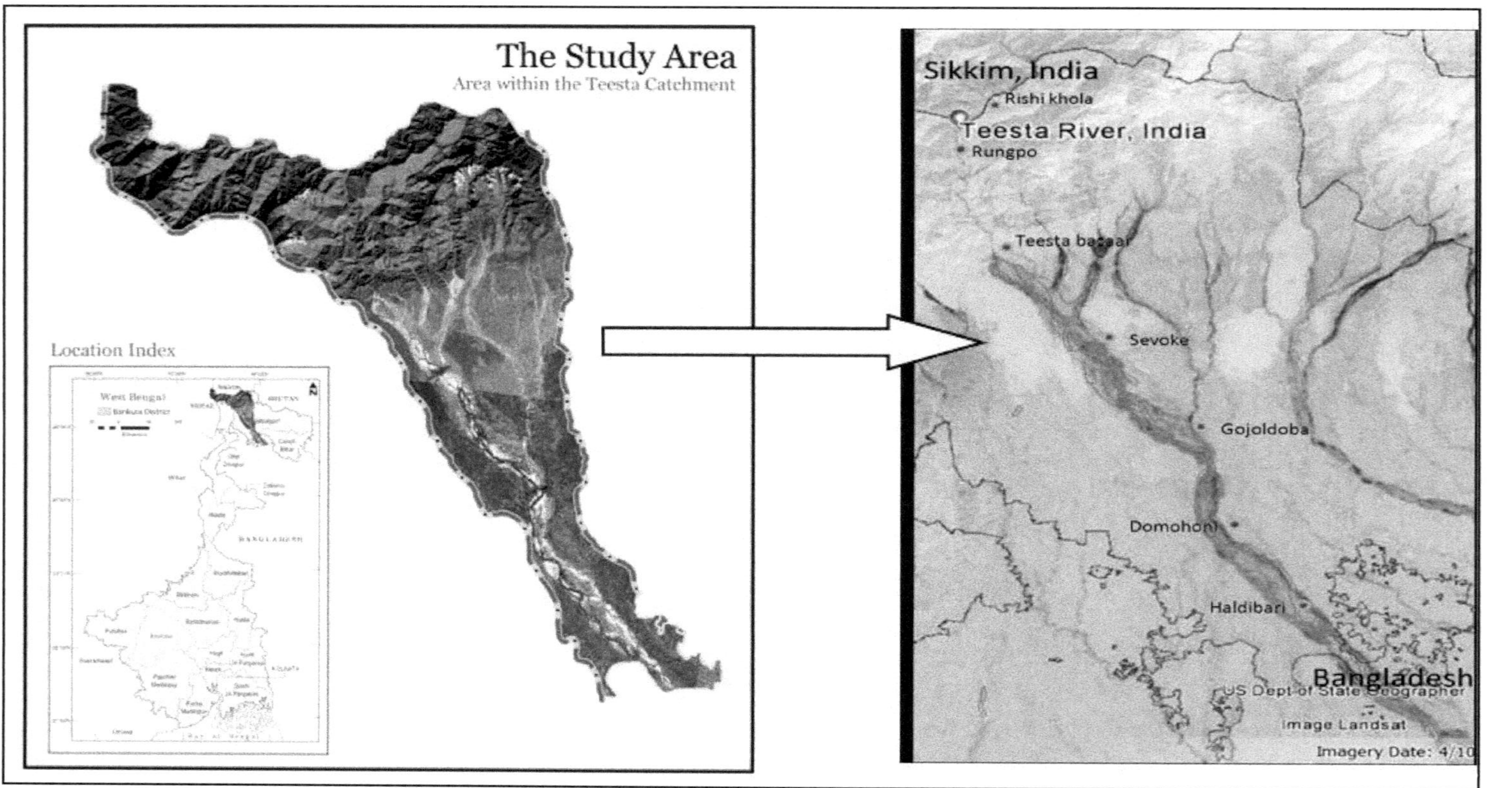

Figure 7.2: Distribution of River Teesta in West Bengal.

Table 7.2: Ichthyofaunal Diversity of the River Teesta in West Bengal, India

Order	*Family*	*Species*	*RK*	*RP*	*TB*	*S*	*G*	*D*	*H*	*RG*	*FPG*	*IUCN Status*
Cypriniformes	Cyprinidae	*Amblypharyngodon mola* (Hamilton, 1822)	–	–	–	–	+	+	–	PL	EU	LC
Cypriniformes	Cyprinidae	*Aspidoparia morar* (Hamilton, 1822)	–	–	–	–	+	–	+	PL	EU	LC
Cypriniformes	Cyrpinidae	*Aspidoparia jaya* (Hamilton, 1822)	–	–	–	–	–	–	+	PL	EU	LC
Cypriniformes	Cyprinidae	*Bangana dero* (Hamilton, 1822)	–	–	–	–	+	–	–	PP	LH	LC
Cypriniformes	Cyprinidae	*Barilius barna* (Hamilton 1822)	–	–	+	–	+	–	–	LI	RH	LC
Cypriniformes	Cyprinidae	*Barilius barila* (Hamilton, 1822)	–	+	+	+	+	–	+	LI	RH	LC
Cypriniformes	Cyprinidae	*Barilius bendelisis* (Hamilton, 1807)	–	–	+	+	+	–	–	LI	RH	LC
Cypriniformes	Cyprinidae	*Barilius shacra* (Hamilton 1822)	–	–	+	–	–	–	–	LI	RH	LC
Cypriniformes	Cyprinidae	*Barilius tileo* (Hamilton, 1822)	–	–	–	–	–	+	–	LI	RH	LC
Cypriniformes	Cyprinidae	*Barilius vagra* (Hamilton, 1822)	–	–	+	–	+	+	+	LI	RH	LC
Cypriniformes	Cyprinidae	*Crossocheilus latius latius* (Hamilton, 1822)	–	–	+	–	+	–	–	PL	RH	LC
Cypriniformes	Cyprinidae	*Danio dangila* (Hamilton, 1822)	–	–	+	–	–	–	–	PL	RH	LC
Cypriniformes	Cyprinidae	*Danio rerio* (Hamilton, 1822)	+	–	+	–	–	–	–	SP	RH	LC
Cypriniformes	Cyprinidae	*Devario aequipinnatus* (McClelland, 1839)	+	–	–	–	–	–	–	PL	RH	LC
Cypriniformes	Cyprinidae	*Devario devario* (Hamilton 1822)	–	–	–	–	+	+	–	PL	RH	VU
Cypriniformes	Cyprinidae	*Devario acuticephala* (Hora, 1921)	–	–	–	–	+	–	–	PL	EU	LC
Cypriniformes	Cyprinidae	*Esomus danricus* (Hamilton 1822)	–	–	–	–	+	+	–	PL	EU	LC
Cypriniformes	Cyprinidae	*Garra annandalei* (Hora, 1921)	–	–	+	–	+	–	–	LI	RH	LC
Cypriniformes	Cyprinidae	*Garra gotyla gotyla* (Gray, 1830)	–	–	+	+	–	–	–	LI	RH	LC
Cypriniformes	Cyprinidae	*Garra lamta* (Hamilton, 1822)	–	–	+	+	+	–	–	LI	RH	LC
Cypriniformes	Cyprinidae	*Labeo pangusia* (Hamilton 1822)	–	–	–	–	+	–	–	PH	LH	NT
Cypriniformes	Cyprinidae	*Labeo angra* (Hamilton, 1822)	–	–	–	–	–	+	+	PH	LH	LC

Contd...

Table 7.2–*Contd...*

Order	*Family*	*Species*	*RK*	*RP*	*TB*	*S*	*G*	*D*	*H*	*RG*	*FPG*	*IUCN Status*
Cypriniformes	Cyprinidae	*Neolissochilus hexagonolepis* (McClelland, 1839)	–	+	+	–	+	–	+	PS	RH	NT
Cypriniformes	Cyprinidae	*Neolissochilus hexastichus* (McClelland 1839)	–	–	+	–	–	–	–	PS	RH	NT
Cypriniformes	Cyprinidae	*Osteobrama cotiocotio* (Hamilton, 1822)	–	–	–	–	–	–	+	PL	LH	LC
Cypriniformes	Cyprinidae	*Psilorhynchus balitora* (Hamilton, 1822)	+	–	–	–	–	–	–	LT	RH	LC
Cypriniformes	Cyprinidae	*Psilorhynchus sucatio* (Hamilton 1822)	+	–	–	–	+	–	–	LT	RH	LC
Cypriniformes	Cyprinidae	*Puntius conchonius* (Hamilton, 1822)	–	–	–	–	+	–	–	PH/PL	EU	LC
Cypriniformes	Cyprinidae	*Pethia phutunio* (Hamilton, 1822)	–	–	–	–	+	–	–	PH/PL	EU	LC
Cypriniformes	Cyprinidae	*Puntius sarana* (Hamilton, 1822)	–	–	–	–	+	–	–	PH/PL	LH	LC
Cypriniformes	Cyprinidae	*Puntius sophore* (Hamilton 1822)	–	–	–	–	+	+	–	PH/PL	EU	LC
Cypriniformes	Cyprinidae	*Puntius terio* (Hamilton, 1822)	+	–	–	–	+	+	–	PH/PL	EU	LC
Cypriniformes	Cyprinidae	*Pethia ticto* (Hamilton, 1822)	–	–	–	–	+	+	–	PH/PL	EU	LC
Cypriniformes	Cyprinidae	*Raiamas bola* (Hamilton, 1822)	–	–	–	–	–	+	–	PP	EU	LC
Cypriniformes	Cyprinidae	*Rasbora rasbora* (Hamilton 1822)	–	–	–	–	+	–	–	PP	EU	LC
Cypriniformes	Cyprinidae	*Salmophasia bacaila* (Hamilton, 1822)	–	–	–	–	+	–	–	PP	LH	LC
Cypriniformes	Cyprinidae	*Salmophasia phulo* (Hamilton 1822)	–	–	–	–	+	–	–	PP	LH	LC
Cypriniformes	Cyprinidae	*Schizothorax richardsonii* (Gray 1832)	–	+	+	–	–	–	–	LT	RH	VU
Cypriniformes	Cyprinidae	*Tor tor* (Hamilton 1822)	–	+	–	–	–	–	–	LT	RH	NT
Cypriniformes	Nemacheilidae	*Acanthocobitis botia* (Hamilton, 1822)	–	–	+	–	+	–	–	LT	RH	LC
Cypriniformes	Nemacheilidae	*Aborichthys elongatus*Hora, 1921	–	–	–	+	–	–	–	LT/LI	RH	LC
Cypriniformes	Nemacheilidae	*Schistura corica* (Hamilton, 1822)	–	–	+	+	+	–	–	LT	RH	NT
Cypriniformes	Nemacheilidae	*Schistura devdevi* Hora, 1935	–	–	+	–	–	–	–	LT	RH	LC
Cypriniformes	Nemacheilidae	*Schistura multifasciata* (Day, 1878)	–	–	+	–	–	–	–	LT	RH	LC

Contd...

Table 7.2–*Contd...*

Order	*Family*	*Species*	*RK*	*RP*	*TB*	*S*	*G*	*D*	*H*	*RG*	*FPG*	*IUCN Status*
Cypriniformes	Nemacheilidae	*Physoschistura elongata*Sen and Nalbant, 1982	–	–	–	+	–	–	–	LT	RH	LC
Cypriniformes	Nemacheilidae	*Schistura savona* (Hamilton, 1822)	+	–	+	–	+	–	–	LT	RH	LC
Cypriniformes	Nemacheilidae	*Schistura scaturigina*McClelland, 1839	–	–	+	–	+	–	–	LT	RH	LC
Cypriniformes	Nemacheilidae	*Schistura beavani* (Günther, 1868)	–	–	+	–	–	–	–	LT	RH	VU
Cypriniformes	Nemacheilidae	*Schistura sikmaiensis* (Hora, 1921)	–	–	+	–	–	–	–	LT	RH	LC
Cypriniformes	Cobitidae	*Botia lohachata*Chaudhuri, 1912	–	–	+	–	+	–	–	PS	RH	LC
Cypriniformes	Cobitidae	*Botia rostrata* Günther, 1868	–	–	–	–	+	–	–	PS	RH	VU
Cypriniformes	Cobitidae	*Canthophrys gongota* (Hamilton, 1822)	–	–	–	–	+	–	–	PL	RH	LC
Cypriniformes	Cobitidae	*Lepidocephalichthys annandalei* (Chaudhuri, 1912)	–	–	–	–	–	+	–	PS	RH	LC
Cypriniformes	Cobitidae	*Lepidocephalichthys berdmorei* (Blyth, 1860)	–	–	–	–	+	–	–	PS	RH	LC
Cypriniformes	Cobitidae	*Lepidocephalichthys guntea* (Hamilton, 1822)	–	–	+	–	+	+	–	PS	EU	LC
Siluriformes	Amblycipitidae	*Amblyceps mangois* (Hamilton, 1822)	+	+	–	+	+	–	–	PL	EU	LC
Siluriformes	Bagridae	*Batasio tengana* (Hamilton, 1822)	–	–	–	–	+	–	–	PL	EU	LC
Siluriformes	Bagridae	*Mystus bleekeri* (Day 1877)	–	–	–	–	+	+	+	PL	EU	LC
Siluriformes	Bagridae	*Mystus tengara* (Hamilton, 1822)	–	–	–	–	+	–	–	PL	EU	LC
Siluriformes	Bagridae	*Mystus vittatus* (Bloch, 1794)	–	–	–	–	+	+	–	LT	RH	LC
Siluriformes	Chacidae	*Chaca chaca* (Hamilton 1822)	–	–	–	–	+	–	–	PP	EU	LC
Siluriformes	Eresthistidae	*Hara horai* Misra 1976	–	–	–	–	+	–	–	LI	RH	LC
Siluriformes	Eresthistidae	*Pseudolaguvia ribeiroi* (Hora 1921)	–	–	–	–	+	–	–	LI/PL	RH	LC
Siluriformes	Eresthistidae	*Pseudolaguvia foveolata* (Ng, 2005)	–	–	–	–	+	–	–	LI/PL	RH	DD
Siluriformes	Heterpneustidae	*Heteropneustes fossilis* (Bloch, 1794)	–	–	–	–	–	–	+	PP	EU	LC

Contd...

Table 7.2–*Contd...*

Order	*Family*	*Species*	*RK*	*RP*	*TB*	*S*	*G*	*D*	*H*	*RG*	*FPG*	*IUCN Status*
Siluriformes	Olyridae	*Olyra kempi* (Chaudhuri, 1912)	–	+	–	+	+	–	–	LT	RH	LC
Siluriformes	Olyridae	*Olyra longicaudata* (McClelland,1842)	–	–	–	–	+	–	–	LT	RH	LC
Siluriformes	Siluridae	*Ompok pabda* (Hamilton, 1822)	–	–	–	–	+	–	–	PP	RH	NT
Siluriformes	Sisoridae	*Bagarius yarrelli* (Sykes 1839)	–	–	–	–	+	–	–	PL	RH	LC
Siluriformes	Sisoridae	*Glyptothorax indicus* (Talwar, 1991)	–	–	–	–	+	–	–	LT/LI	RH	LC
Siluriformes	Sisoridae	*Glyptothorax telchitta* (Hamilton 1822)	–	–	–	–	+	–	–	LT/LI	RH	LC
Siluriformes	Sisoridae	*Glyptothorax cavia* (Hamilton, 1822)	–	–	–	–	+	–	–	LT/LI	RH	DD
Siluriformes	Sisoridae	*Glyptothorax conirostris* (Steindachner, 1867)	–	–	–	–	+	–	–	LT/LI	RH	NT
Siluriformes	Sisoridae	*Glyptothorax striatus* (McClelland, 1842)	–	–	–	–	+	–	–	LT/LI	RH	LC
Siluriformes	Sisoridae	*Gogangra viridescens* (Hamilton, 1822)	–	–	–	–	+	–	–	PL	RH	LC
Siluriformes	Sisoridae	*Pseudecheneis sulcata* (McClelland, 1842)	–	–	+	–	–	–	–	LT	RH	LC
Perciformes	Badidae	*Badis badis* (Hamilton, 1822)	–	+	–	+	+	–	–	PP	EU	LC
Perciformes	Channidae	*Channa gachua* (Hamilton, 1822)	–	–	–	–	+	–	–	PP	EU	LC
Perciformes	Channidae	*Channa marulius* (Hamilton, 1822)	+	–	–	–	+	–	–	PP	EU	LC
Perciformes	Channidae	*Channa punctata* (Bloch, 1793)	–	–	–	–	+	+	–	PP	EU	LC
Perciformes	Channidae	*Channa stewartii* (Playfair, 1867)	–	–	–	–	+	–	–	PS	RH	LC
Perciformes	Gobidae	*Glossogobius giuris* (Hamilton 1822)	–	–	–	–	+	–	–	PP	EU	LC
Perciformes	Osphronemidae	*Trichogaster fasciata* (Bloch and Schneider, 1801)	–	–	–	–	+	+	–	PH	EU	LC
Perciformes	Osphronemidae	*Trichogaster lalius* (Hamilton, 1822)	–	–	–	–	+	+	–	PH	EU	LC
Perciformes	Ambassidae	*Chanda nama* Hamilton, 1822	–	–	–	–	+	–	–	PP	EU	NT
Perciformes	Ambassidae	*Parambassis lala* (Hamilton, 1822)	–	–	–	–	+	+	–	PP	EU	LC
Perciformes	Ambassidae	*Parambassis ranga* (Hamilton, 1822)	–	–	–	–	–	+	–	PP	EU	NT

Contd...

Table 7.2–*Contd...*

Order	*Family*	*Species*	*RK*	*RP*	*TB*	*S*	*G*	*D*	*H*	*RG*	*FPG*	*IUCN Status*
Synbranchiformes	Mastacembelidae	*Macrognathus aral* (Bloch and Schneider, 1801)	–	–	–	–	–	+	–	PS	RH	LC
Synbranchiformes	Mastacembelidae	*Macrognathus pancalus* Hamilton 1822.	+	–	–	–	+	–	–	PH	EU	LC
Synbranchiformes	Mastacembelidae	*Mastacembelus armatus* (Lacepède, 1800)	–	–	–	–	+	–	–	PL	RH	LC
Synbranchiformes	Synbranchidae	*Monopterus hodgarti* (Chaudhuri, 1913)	–	–	–	–	+	–	–	PL	RH	LC
Beloniformes	Belonidae	*Xenentodon cancila* (Hamilton, 1822)	–	–	–	–	+	+	–	PH	EU	LC

RK: Rishi Khola; RP: Rungpo; TB: Teesta Bazaar; S: Sevoke; G: Gojoldoba; D: Domohoni; H: Haldibari; RG: Reproductive guild.

+: Present; –: Absent.

adequate water current, low temperature and with rocky substrate. Whereas, species abundance and richness again decreases in river plains *viz.* Haldibari (Table 7.3). This is attributed to limitations induced by shifting and homogenous substratum and high turbidity as after this point river Teesta enters the Bramhaputra river drainage. The species richness per zone increases downstream (Gojoldoba and Domohoni) but decreases further downstream (Haldibari). Five freshwater fish orders have been deduced with Cypriniformes being the most dominant followed by Siluriformes and Perciformes (Figure 7.3).

Table 7.3: Diversity Indices of the Fish Community at Different Regions of River Teesta, West Bengal

Sites	*Total Species (S)*	*Species Richness (d)*	*Pielou's Evenness (J')*
Rishi Khola	9	3.154	0.9307
Rungpo	7	1.78	0.9662
TeestaBazzar	22	4.543	0.9802
Sevoke	8	1.98	0.9819
Gojoldoba	65	11.56	0.966
Domohoni	20	4.556	0.9641
Haldibari	7	1.675	0.9677

Conclusion

To be effective, conservation efforts should be based on a solid conceptual foundation and a holistic understanding of natural river ecosystems. Such background knowledge is necessary to reestablish environmental gradients, to reconnect interactive pathways, and to reconstitute some resemblance of the natural dynamics responsible for high levels of biodiversity. The challenge for the future lies in protecting the ecological integrity and biodiversity of aquatic systems in the face of increasing pressures on our freshwater resources. This will require integrating sound scientific principles with management perspectives for future sustainability.

Landscape changes lead to habitat alterations, degradation, fragmentation and consequent depletion or loss of many species. The freshwater ecosystem is to be considered as an integral part of the general landscape (watershed/basin/catchment) and significant modifications of their catchments can have detrimental impacts on the native fish fauna. Therefore, a landscape-based approach for biodiversity assessment and management assumes great significance. However, landscape-based approach is yet to gain attention in the conservation or management of the biodiversity in general and freshwater fish diversity of Indian sub-continent in particular. Further, the development of a suitable protocol for measuring the index of biotic integrity would facilitate ecological and biological monitoring at lower costs than chemical monitoring. Therefore, the present study aimed at evaluating fish assemblages so that the riverine quality of the lower course of river Teesta in West Bengal can be assessed through Index of Biotic Integrity (IBI) assessment based on

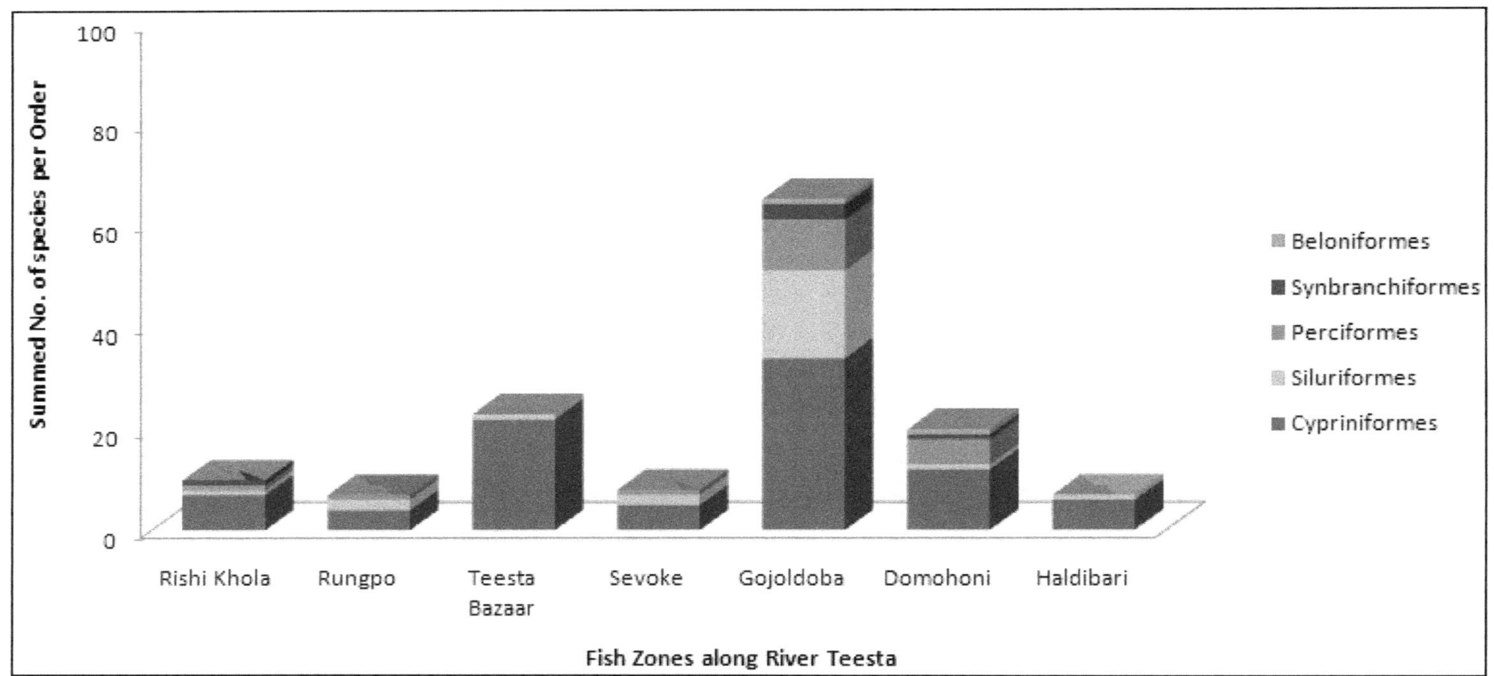

Figure 7.3: Taxonomic Composition of Fish Zones of River Teesta in West Bengal.

which future measures for the conservation and sustainable use of aquatic resources of the region can be conducted.

Summary

India is one of the mega biodiversity countries in the world and occupies the ninth position in terms of freshwater mega biodiversity. The Eastern Himalayan Biodiversity Hotspot region and its foothills are very rich in both floral and faunal diversity; 218 species of fishes has been listed in whole Himalaya. The Eastern Himalaya drained by the Brahmaputra has greater diversity of cold water fish than western Himalayan drainage. Especially the region being the home of many large torrential rivers, fish diversity is also very rich. Fish populations inhabiting these areas are numerous in variety and taxonomically interesting. As such, the northern districts of the Indian state of West Bengal, specially the districts of Darjeeling and Jalpaiguri, lying within the Eastern Himalayan Biodiversity Hotspot range, hold a great faunistic importance. The regions have a pronounced seasonal climate and lies north of the tropical belt. Moreover, being the integral part of the Eastern Himalaya, the region is regarded as freshwater biodiversity hotspot. The ichthyodiversity of this region is unique from the zoogeographical point of view. The fish fauna represents mostly the Chinese, Malayan and Indian elements of fishes of the Oriental realm. The chief rivers are Mahananda and Teesta, with many tributaries such as Murti, Atrai, Jaldhaka, Karala, and Karotoyar. There is a lack of baseline information on freshwater fish species distributions and their ecological requirements throughout the Eastern Himalaya. Therefore, study of fish biodiversity and land usage assumes a great significance in this region considering conservation priority, endemecity and monitoring environmental health.

A landscape-based approach is yet to gain attention in the conservation or management of the biodiversity in general and freshwater fish diversity of Indian sub-continent in particular. In this context, the present case-study area and evaluation have been planned. The course of the river Teesta within West Bengal was divided into ecological zones based on elevation gradient and habitat types *viz.* the upper stretch (Rishi khola and Rungpo) where elevations is higher with low temperatures; middle stretch (Teesta Bazaar) with low elevation; at Sevoke the river hits the plains; lastly the river plains (Gojoldoba, Domohoni and Haldibari). Analysis of ecological fish guilds was carried out to enhance the fish zonation pattern of river Teesta, and as a tool for assessing riverine health and ecological integrity of the river. Ecological characteristics of the fish species were used in this data set to analyze the longitudinal zonation concepts and to draw ecological patterns that were inferred from this analysis. The present data seem to confirm this prediction, in that maximum taxonomic richness (expressed as the number of taxonomic orders or as the number of species) and maximum number of feeding guilds is encountered in river plains zone *viz.*Gojoldoba. Flow preference and reproductive ecology of fishes are closely linked. Hence, analysis of ecological fish guilds not only enhanced the fish zonation pattern of river Teesta, but may also serve as tools for assessing riverine health and ecological integrity of the river in near future.

References

Aarts BGW and Nienhuis PH. 2003. Fish zonations and guilds as the basis for assessment of ecological integrity of large rivers. *Hydrobiologia*, 500: 157-178.

Abell R, Thieme ML and Revenga C. 2008. Freshwater Ecoregions of the World: A New Map of Biogeographic Units for Freshwater Biodiversity Conservation. *Bioscience*, 58(5): 403-414.

Acharjee ML and Barat S. 2013. Ichthyofaunal Diversity of Teesta River in Darjeeling Himalaya of West Bengal, India. *Asian Journal of Experimental Biological Sciences*, 4 (1): 112-122.

Allen DJ, Molur S and Daniel BA. 2010. The Status and Distribution of Freshwater Biodiversity in the eastern Himalaya. IUCN Publications Services, Gland, Switzerland.

Bhat A. 2003. Diversity and composition of freshwater fishes in river systems of Central Western Ghats, India. *Environmental Biology of Fishes*, 68: 25-38.

Bhatt JP, Manish K and Pandit MK. 2012. Elevational Gradients in Fish Diversity in the Himalaya: Water Discharge Is the Key Driver of Distribution Patterns. *PLoS ONE*, 7(9): e46237.

Dahanukar N, Raut R and Bhat A. 2004. Distribution, endemism and threat status of freshwater fishes in the Western Ghats of India. *Journal of Biogeograph.*, 31: 123-136.

Day F. 1889. *The fauna of British India: Fishes*. Taylor and Francis, London. 1: i–xx + 1-548.

Ghosh SK. and Lipton AP. 1982. Ichthyofauna of the N.E.H. Region with special reference to their economic importance. ICAR Spl. Bulletin No. B (ICAR Research Complex, Shillong). pp. 119-126.

Goswami UC, Basistha SK, Bora D. *et al.*, 2012. Fish diversity of North East India, inclusive of the Himalayan and Indo Burma biodiversity hotspots zones: A checklist on their taxonomic status, economic importance, geographical distribution, present status and prevailing threats. *International Journal of Biodiversity and Conservation*, 4(15): 592-613.

Hora SL and Gupta JC. 1941. On a collection of fish from Kalimpong, Duars and Siliguri Terai, North Bengal. *Journal of Asiatic Society of Bengal*, 47: 183-202.

Jayaram KC. 2006. *Catfishes of India*. Narendra Publishing House, Delhi. pp. 383.

Jayaram KC. 2010. *The Freshwater Fishes of the Indian Region* (2nd ed.). Narendra Publishing House, Delhi. pp. 616.

Jayaram KC and Singh KP. 1977. On a collection of fish from North Bengal. *Records of the Zoological Survey of India*, 72: 243-275.

Kar D, Nagarathna AV, Ramachandra TV and Dey SC. 2006. Fish Diversity and Conservation aspects in an aquatic ecosystem in Northeastern India. *Zoos' Print Journal*, 21(7): 2308-2315.

Kottelat M and Whitten T. 1996. Freshwater Biodiversity in Asia with special reference to Fish: World Bank Tech. Paper No. 343. The World Bank, Washington, DC. pp. 17-22.

Kumar AB. 2000. Exotic fishes and freshwater fish diversity. *Zoos' Print Journal*, 15 (11): 363-367.

Kumar J and Pandey AK. 2013. Present status of ichthyofaunal diversity and impact of exotics in Uttar Pradesh,

Journal of Experimental Zoology India, 16 (2): 429-434.

Lakra WS, Sarkar UK, Kumar RS. *et al.*, 2010. Fish diversity, habitat ecology and their conservation and management issues of a tropical River in Ganga basin, India. *Environmentalist*, 30: 306-319.

Martin P, Haniffa MA and Arunachalam M. 2000. Abundance and diversity of macroinvertebrates and fish in the Tamiraparani river, South India. *Hydrobiologia*, 430: 59-75.

Menon AGK. 1987. Fauna of India and the Adjacent Countries: Pisces, Vol. IV; Teleostei - Cobitoidae, Part I, Homalopteridae, Zoological Survey of India, Kolkata. pp. 259.

Nath P and Dey SC. 1989. *Fish and Fisheries of North East India*. Vol. I, pp.1-143.

Nath P, Dey SC, Dam D and Das DN. 2008. Fishery of Arunachal Pradesh (India): A potential hub in North-East Region. In. *Fish Genetic Resources* (Lakra WS, Singh AK and Mahanta PC. eds.). Narendra Publishing House, Delhi. pp. 73-82.

Negi RK and Rajput V. 2012. Fish Diversity in Two Lakes of Kumaon Himalaya, Uttarakhand, India. *Research Journal of Biology*, 2 (5): 157-161.

Sarkar UK and Ponniah AG. 2000. Evaluation of North East Indian Fishes for their Potential as Cultivable, Sport and Ornamental Fishes along with their Conservation and Endemics Status. In. *Fish Biodiversity of North East India* (Ponniah AG and Sarkar UK. Eds.). NBFGR, NATP Publ., 2: 11-30.

Sarkar UK, Pathak AK, Sinha RK. *et al.*, 2012. Freshwater fish biodiversity in the River Ganga (India): changing pattern, threats and conservation perspectives. *Review in Fish Biology and Fisheries*, 22: 251-272.

Sarkar UK, Pathak AK and Lakra WS. 2008. Conservation of freshwater fish resources of India: new approaches, assessment and challenges. *Biodiversity Conservation*, 17: 2495-2511.

Sen TK. 1985. The fish fauna of Assam and the neighbouring North Eastern States of India. *Records of Zoological Survey of India*, Occasional paper No. 64. pp. 1-216.

Shahnawaz A, Venkateshwarlu M, Somashekar DS and Santosh K. 2010. Fish diversity with relation to water quality of Bhadra River of Western Ghats (INDIA). *Environmental Monitoring and Assessment*, 161: 83-91.

Shaw GE and Shebbeare EO. 1937. The Fishes of North Bengal. *Journal of Royal Asiatic Society of Bengal, Science*, 3(1): 1-137.

Talwar PK and Jhingran AG. 1991. Inland fishes of India and adjacent countries (Vol.1 and 2). Oxford and IBH Publishing Company, New Delhi. pp. 1158.

Tamboli RK and Jha YN. 2012. Status of Cat Fish Diversity of River Kelo and Mand in Raigarh District, CG, India. *Journal of Biological Sciences,* 1(1): 71-73.

Thirumala S, Kiran BR and Kantaraj GS. 2011. Fish diversity in relation to physico-chemical characteristics of Bhadra reservoir of Karnataka, India. *Advances in Applied Science Research,* 2 (5): 34-47.

Chapter 8

Amphibians in Lentic Wetlands of West Bengal: Diversity, Behaviour and Adaptations

Kaushik Deuti

Herpetology Division, Zoological Survey of India, Kolkata

Introduction

The state of West Bengal is unique in the sense that it houses almost all the habitats and ecosystems found in different parts of India. The state extends from the alpine and sub-alpine areas in the Eastern Himalaya of the Darjeeling hills in the north, down to the moist evergreen rain forests and moist deciduous forests of the Duars areas at the Himalayan foothills to the Gangetic plains in central and lower Bengal to the dry deciduous forests of western Bengal and the mangrove forests on the sea coast of the Bay of Bengal. Thus the state also has an abundance of wetlands, *viz.*, mountain lakes in the Darjeeling hills; lakes, ponds, canals, ditches, jheels, beels, ox-bow lakes, estuaries etc in the north, central and southern Bengal, which support about 30 species of amphibians. This includes one species of salamander, 3 species of toads of the family Bufonidae, 1 species of mountain toad of the family Megohryidae, 5 species of burrowing frogs of the family Microhylidae, 14 species of aquatic frogs of the family Dicroglossidae, 2 species of true frogs of the family Ranidae and 4 species of tree frogs of the family Rhacophoridae.

Wetland Amphibians: A Review

A. Family Bufonidae (Toads)

The toads of the family Bufonidae in the fresh water lentic wetlands of West Bengal include 3 species. The common and scientific name of the species, their behaviour and adaptation are given below briefly:

1. Himalayan Toad (*Duttaphrynus himalayanus*)

The Himalayan toad is a very large toad (length upto 140 mm) and is commonly found all over the Darjeeling hills of northern West Bengal. It is found in all the villages, towns, tea gardens and forests in Darjeeling district and breeds in mountain lakes, ponds (locally called *'pokhries'*), small rain-water pools and ditches. It lays large number of eggs in very long egg strings which may contain upto 40000 eggs. These long egg strings are very often seen in the shallow Himalayan pools like a necklace of black beads throughout the monsoon season from March to October.

2. Common Indian Toad (*Duttaphrynus melanostictus*)

The common Indian toad is found in all the districts of West Bengal and may grow upto 120 mm in total length. They have cornified black bony ridges on the head and several black-tipped spiny warts on the rough skin. They also have large bean-shaped parotoid glands. They hide under stones, piles of bricks, in moist crevices of tree-trunks during the daytime but emerge at night to feed on the insects which are abundant in the rainy season. The adults are seen throughout the year, even in the winter months if the temperature does not fall severely. They do not seem to hibernate. They breed from June to September in all available freshwater wetlands in the plains and even in those that are highly polluted. They are prolific breeders with a single female laying upto 20000 eggs in long strings. The filling up of their breeding places for human habitation and rapid industrialization and dumping of toxic industrial effluents have decimated their numbers in recent times.

3. Marbled Toad (*Duttaphrynus stomaticus*)

The marbled toad is smaller (80-110 mm) than the common Indian toad and has a smoother skin with irregular flattened warts. There are no bony ridges on the head and the parotoid glands are flat and elliptical but not bean-shaped. The skin is patterned or marbled. They are more agile than the common Indian toads and burrow in sandy or wet soil. They prefer semi-arid areas but are found near wetlands in the breeding season. They aestivate during summer under earth at depths of almost a meter. They breed in shallow water bodies, during the monsoon. Eggs are laid in two parallel translucent strings which are pale yellowish-green in colour. A total of 9000-11000 eggs are laid, which take 24 hours to hatch. The tadpoles are small, not exceeding 12 mm in length. The species mainly occurs in southern West Bengal but is also known from Malda in central Bengal.

B. Family: Megophryidae

The mountain toads of the family Megophryidae in the fresh water lentic wetlands of West Bengal include only one species as the rest of the toads in this family found in West Bengal occur in lotic (running water) of mountain streams. The common and scientific name of the species, its behavior and adaptation to living in high-altitude wetlands is given below briefly.

1. Sikkimese Mountain Toad (*Scutiger sikkimensis*)

These high-altitude toads are small in size and are found above 3000 meters altitude in the Darjeeling hills of northern West Bengal. They are adapted to

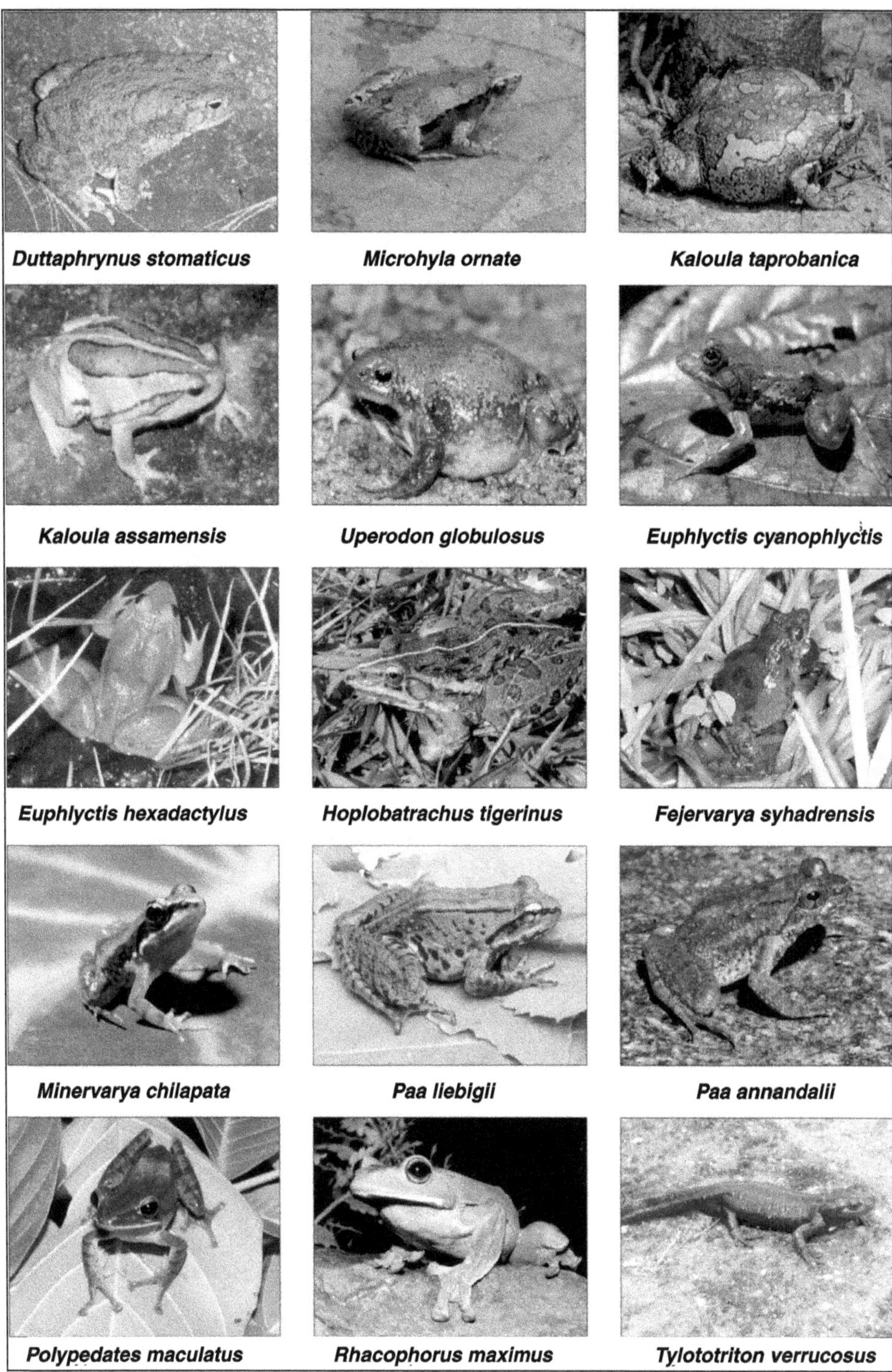

Figure 8.1: Some Common Amphibians of Wetland Ecosystems of West Bengal.

withstand very low air temperatures (upto -15°C) and have very rough skin which helps them to squeeze into the crevices of rocks found at such altitudes. They breed in high-altitude mountain lakes and the breeding season starts from April and may extend up to August before those areas turn too cold. However, their fecundity is less and the breeding rate is slower than their plain counterparts. They are found in the Singhalila National Park area of Darjeeling district, mainly seen between Tonglu, Gauribans and Sandakphu peak.

C. Family Microhylidae (Narrow-mouthed frogs)

The narrow-mouthed frogs of the family Microhylidae in the freshwater lentic wetlands of West Bengal include 5 species.

1. Ornate Narrow-mouthed Frog (*Microhyla ornata*)

This is a very small sized (20-25 mm) frog with pointed head but small bulging eyes. The tympanum as in all microhylids is in distinct. The tips of the fingers are flattened and two small but distinct oval pedal tubercles are present. The species is very common and active even during daytime in the rainy season. Normally found in the grasses and bushes on moist soil and under dry fallen bamboo leaves on the banks of ponds and ditches and even around small rain pools. Their numbers and calls become prominent after a heavy shower during the monsoon. They are very agile and in comparison to their size can cover a considerable distance by leaping. Despite their small size, the males have a loud and distinct call. Almost 500 eggs are laid covered by a transparent gelatinous envelope in separate flat slimy mats. Common in all the districts of West Bengal.

2. Indian Painted Frog (*Kaloula taprobanica*)

Medium-sized (25-50 mm) toad-like stout and colourful burrowing frogs, with a rounded snout. Tympanum is absent. Fingers bear short well developed truncate discs and toes are with dilated triangular discs. Inner pedal tubercle is large and shovel shaped. They are nocturnal and burrowing and emerge from their burrows during the monsoon, sometimes even during the day and are often seen to climb up into holes of tree-trunks and moist boundary walls of village houses. They have a deep guttural call and breed in dirty ponds and ditches often near cow sheds or in drains of houses in villages and towns. During pairing they float passively in the water of these wetlands. Found in the plains of northern and southern Bengal.

3. Assamese Painted Frog (*Kaloula assamensis*)

Known only from the Duars region of northern Bengal, this species occurs in the forested tracts where they seem to breed in temporary wetlands which develop during the monsoons inside the deeply vegetated areas. They are similar in size and looks to the Indian painted frog but have a yellowish mid-dorsal line. Not much is known about their breeding habits except that like all microhylids, they breed in the early part of the rainy season.

4. Sticky Frog (*Kalophrynus interlineatus*)

Known only from inside deep forests in the Duars region of northern Bengal, this species secretes a very sticky white fluid on being handled. Similar in size to the painted frogs, they seem to breed in the early part of the rainy season, in temporary wetlands and rain water pools which develop during the monsoons inside the forests. Not much is known about the breeding habits of this species. Recently been reported from Chilapata Reserve Forest in West Bengal after its first report in India from Orang National Park in Assam.

5. Balloon Frog (*Uperodon globulosus*)

This medium-sized (40-60 mm) burrowing frog is rather globular in shape with small beady eyes. Head is small and narrow, the snout is rounded. Tongue is oval and tympanum is absent. Tips of the fingers are rounded but do not bear discs. The hind legs are short. Two well developed, large and shovel-shaped (inner and outer) pedal tubercles are present. They are nocturnal and fossorial preferring loose sandy soil and are excellent burrowers using the strong pedal tubercles on the hind legs to quickly burrow and disappear underground. While burrowing the soil is dislodged by the sideways movement of the legs till the frog subsides into the ground with its eyes disappearing last and leaving no trace of its presence. They often dig into termite mounds and feed on the eggs and adults of termites. They breed in ponds, puddles and temporary rain water pools. A sticky substance secreted by both the partners, help them to remain tightly together during pairing. At the end of the pairing, the mating partners take several dips in water and struggle to free themselves. Found in both northern and southern West Bengal.

D. Family Dicroglossidae (Aquatic Frogs)

The aquatic frogs of the family Dicroglossidae have fully webbed feet and dorso-lateral eyes helping them to see above water while keeping their bodies submerged. The aquatic frogs of the family Dicroglossidae in the fresh water lentic wetlands of West Bengal include 14 species.

1. Skittering Frog (*Euphlyctis cyanophlyctis*)

Medium-sized (40-70 mm) pond frogs with fully webbed toes and pointed tips of fingers and toes. The eyes and nostrils project above the water surface when they float. Commonest species in the wetlands and is found in all types of water bodies. They float passively in water but when disturbed, skip smoothly over the water for some distance and then again float on the surface. Often they dive to escape pursuers. They can also tolerate a certain degree of salinity and can thrive in dirty drains and ditches contaminated with organic pollutants. They spend most of their life in water or in the vicinity of wetlands but may migrate to other water bodies in summer when the temporary pools dry up. They are highly vocal during the monsoon. The breeding season starts with the first rains and lasts till the end of the monsoon. Eggs are laid inside a frothy mass in standing water. Feed on small insects both during day and night. Seen in all the districts of West Bengal.

2. Green Pond Frog (*Euphlyctis hexadactylus*)

A giant leaf-green frog (80-140 mm) which is completely aquatic. Tips of the fingers and toes are pointed. Fingers are without webbing but the toes are fully webbed. There is a dermal fringe on the outer toes. An elongated digit-like inner metatarsal tubercle is present. A skin fold runs from behind the eye to the shoulder. They are found floating in old ponds and fishing *bheries* where dense green aquatic vegetation like *Azolla, Eichhornia, Wolffia, Pistia* and *Lemna* grow abundantly. The colour of their body merges with the floating vegetation so that they can easily camouflage and protect themselves from aerial as well as aquatic predators. They feed mainly on weeds but also on insects. Breeding season extends from June to October. Females lay about 2000 eggs in the wetlands. Found mainly in the coastal districts of West Bengal.

3. Indian Bull Frog (*Hoplobatrachus tigerinus*)

Large frogs (100-160 mm) with smooth skin and longitudinal glandular folds on the back. A skin fold runs from behind the eye to the shoulder. Snout is somewhat pointed. Tympanum is large, nearly equal to the eye diameter. Toes are fully webbed but the web does not reach the tip of the third toe. Yellowish or olive-green, with darker leopard-like spots.A yellowish-white median stripe present.They are semi-aquatic, good swimmers and hide among grasses, bushes at the edge of ponds, ditches and canals. They leap headlong into water and during the dry season they hide inside burrows or under heaps of leaves where they aestivate. They are solitary creatures, aggregating only in the breeding season. Males croak in chorus on the bank of the wetlands. They are carnivorous and cannibalism is also distinct. All kinds of small and large creatures are in their diet including other frogs, snakes, lizards, small birds and small mammals. The species is found in all the districts of West Bengal.

4. Jerdon's Bull Frog (*Hoplobatrachus crassus*)

Quite large frogs (60-100 mm) with flabby-shaped body and comparatively shorter legs than the Indian bull frog. Toes are entirely webbed with two segments of the fourth toe free. A highly developed, strongly compressed, shovel-shaped inner pedal turbercle is present. Skin on the back and limbs are highly granular with several interrupted warts. Greyish with darker stripes and patches on the back with 6-14 glandular longitudinal folds. Belly is however smooth. Burrowing and nocturnal, they spend the day hiding inside mud crevices on elevated banks of ponds, ditches and canals. They may also be found during the breeding season in temporary pools of water.

5. Syhadra Cricket Frog (*Fejervarya syhadrensis*)

Small-sized frogs (30-40 mm) with pointed snout, projecting beyond the mouth. Tips of the fingers and toes are swollen. Toes are half-webbed with three segments of the fourth toe free. A prominent skin fold runs from behind the eye to the shoulder. Some short and interrupted longitudinal glandular folds are present on the back. A narrow yellowish median streak is often present. The species is quite common in agricultural lands, gardens and forests. They are found in inundated paddy fields in large numbers and on the banks of ponds and ditches. During monsoon nights, the males call in chorus somewhat like the chirping of a cricket. Breed in pools of rain water accumulated in shallow depressions in paddy fields. Found in all the districts of West Bengal.

6. Odisha Cricket Frog (*Fejervarya odishaensis*)

Medium-sized frogs (30-50 mm) with rounded snout, projecting beyond the mouth. Tips of the fingers and toes are swollen. Toes are three-fourth webbed with two segments of the fourth toe free. A prominent skin fold runs from behind the eye to the shoulder. Many interrupted longitudinal glandular folds are present on the back. Lips and limbs are barred. A yellowish or whitish narrow or wide median streak is often present. The species is quite common in towns, agricultural lands and forests. They are found at the edges of inundated paddy fields and on the banks of ponds and ditches. If disturbed, they jump into water but return to the banks soon as they are not good swimmers. During monsoon nights, the males produce a harsh call. Breed in pools of rain water accumulated in shallow depressions on the ground. Found in the southern districts of West Bengal.

7. Terai Cricket Frog (*Fejervarya teraiensis*)

Medium-sized frogs (30-50 mm) with pointed snout, projecting beyond the mouth. Tips of the fingers and toes are slightly swollen. Toes are three-fourth webbed with two segments of the fourth toe free. A prominent skin fold runs from behind the eye to the shoulder. Some short and interrupted longitudinal glandular folds are present on the back. Both upper and lower lips and limbs are barred. A wide yellowish median streak is present. The species is quite common in tea gardens, villages and forests. They are also found on the banks of ponds and ditches. They are not good swimmers. During monsoon nights, the males produce a insect-like call. Breed in pools of rain water accumulated in shallow depressions on the forest floor and tea gardens. Found in the northern districts of West Bengal.

8. Nepalese Cricket Frog (*Fejervarya nepalensis*)

Small to medium-sized frogs (20-45 mm). Almost looks like terai cricket frog except the yellowish median streak is narrower. Behaviors are also same including breeding habit. The species is quite common in villages and forests. They are not good swimmers. The species is distinguished by its specific call. Found in the northern districts of West Bengal.

9. Pierre's Cricket Frog (*Fejervarya pierrei*)

Small to medium-sized frogs (20-40 mm) with pointed snout, projecting beyond the mouth. Many longitudinal glandular folds are present on the back. Both upper and lower lips and limbs are barred. A narrow whitish median streak is present. The species is not common and seen mainly in forests and show similar breeding habits. Found in the northern districts of West Bengal.

10. Chilapata Rain-pool Frog (*Minervarya chilapata*)

This recently described species is very small (20-28 mm) and mainly restricted to in and around Chilapata Reserve Forest in Jalpaiguri district of northern West Bengal. It has a pointed snout, projecting beyond the mouth. At both ends of the jaws are distinct and round rictal glands near to the joint with the forelimb. The limbs are slightly cross-barred. There is a narrow yellowish median streak. The species is mainly restricted to the deep forest where they are seen on the forested roads and moist forest floor with ample canopy cover. They breed for a short period during the monsoons in the small rain pools that form on the forest floor. Nothing is known about their breeding behaviour and life history.

11. Liebig's Bull Frog (*Paa liebigii*)

This large-sized frog (70-130 mm) has tuberculated skin and two dorso-lateral glandular folds on the back on either side. Snout is somewhat rounded. Tympanum is distinct, nearly half the eye diameter. Toes are partly webbed with two segments of the fourth toe free of webbing. Yellowish or olive-green, with darker black rounded spots. They are semi-aquatic, found near shallow pools and hill streams in the Eastern Himalaya. They are carnivorous as well as insectivorous. They feed on small creatures found in the hills. The species is found only in the wetlands of Darjeeling district of West Bengal.

12. Annandale's Frog (*Paa annandalii*)

This medium-sized frog (40-60 mm) also has tuberculated skin with very few glandular folds on the back. Tympanum is distinct, nearly one-third the eye diameter. Reddish or maroon, with yellowish specks. They are semi-aquatic, found near shallow pools and on forest floor near hill streams in the Eastern Himalaya. They are carnivorous and mainly feed on various insects. The species is found only in the wetlands of Darjeeling district of West Bengal.

13. Blanford's Frog (*Paa blanfordi*)

This medium-sized frog (40-50 mm) has tuberculated skin with no glandular folds on the back. Snout is rounded. Tympanum is less distinct, nearly one-third the eye diameter. Blackish brown with yellowish patches and stripes. They are semi-aquatic, found near shallow pools and on forest floor in the Eastern Himalaya. They mainly feed on various insects. The species is found only in the wetlands of Darjeeling district of West Bengal.

14. Burrowing Frog (*Sphaerotheca breviceps*)

This stout bodied, medium-sized burrowing frog (28-52 mm) has a short and rounded snout but a distinct tympanum, which is about half the diameter of the eye. Fingers are without webs and the tips of the fingers are swollen. Hind limbs are very short. Toes are slightly webbed. A highly developed, large, shovel-shaped inner pedal tubercle is present. Skin is smooth on the back but granular on the belly and underside of the thighs. Yellowish brown or grayish with darker spots or markings. The upper lip is with dark vertical bars and the limbs are with irregular dark cross bars. Nocturnal and subterranean in habit, they use their large pedal tubercles to burrow in loose earth, in search of food and shelter. At the onset of the rainy season they come out from their burrows and croak near pools of water. They enter into water only to breed and their eggs are laid in a clutch floating on the water surface. Feed mainly on underground dwelling insects. Found mainly in the southern and central districts of West Bengal.

E. Family Ranidae (True Frogs)

The true frogs of this family are slender-bodied with long muscular legs and partly webbed feet. Many are quite colourful and all breed in fresh water wetlands. Only two species are known from the freshwater lentic wetlands of West Bengal.

1. Reed Frog (*Hylarana tytleri*)

This small frog (30-40 mm) is slender and torpedo-shaped. Snout is pointed and projects beyond the mouth. Tympanum is very distinct, flesh-coloured and nearly equal to eye diameter. Tips of the slender fingers and toes are dilated into small discs. Toes are partly webbed. Leaf green or yellowish above with two golden or yellow lines running along the back from the eyelid to almost the vent. They are generally found among thick floating aquatic vegetation, on lily pads and among reed beds. They also conceal themselves among wooden logs which are partly submerged in water. They are quite agile and can leap well. Breeds in stagnant ponds or slow running canals filled with dense aquatic vegetation. Found in many freshwater wetlands of West Bengal.

2. Long-tongued Frog (*Hylarana leptoglossa*)

This medium-sized frog (30-50 mm) is slender and elongated. Snout is pointed and projected beyond the mouth. Tympanum is very distinct. Toes are partly webbed. Brownish in colour with the flanks more grayish. They are generally found on the leaf litter of the forest floor but breed in the ponds, lakes, ditches, canals etc. They have a typical duck-like call and are quite agile and can leap well. Common in the Duars area of northern West Bengal.

F. Family Rhacophoridae (Tree Frogs)

The tree frogs are well adapted for an arboreal life with the tips of their fingers and toes dilated into large suction discs. They mostly deposit their eggs inside foam-nests constructed by them that hang from the branches of trees and bushes over a pond or pool of water below, into which their tadpoles fall and develop. Represented in the freshwater wetlands of West Bengal by four species.

1. Common Indian Tree Frog (*Polypedates maculatus*)

These medium-sized (35-75 mm) tree frogs are slim, narrow-waisted with slender elongated limbs and goggling eyes. Snout is somewhat pointed. Tympanum is distinct, almost equal to eye diameter. Fingers are with rudimentary webs. Toes are almost half-webbed. Tips of fingers and toes are dilated into flattened, horse-shoe shaped adhesive discs. It has smooth skin and granular belly. Brown or grayish-yellow in colour, the belly is white. A dark brown marking extends from the nostril on both sides of the head, covering the eye right upto the middle of the flanks. Limbs are cross-barred and the undersides of the thighs are marked by dark yellow or purplish networks. Nocturnal and arboreal, they hide among vegetation during the day and call after sunset especially after showers during the rainy season. They lay 300-700 creamy white eggs inside globular foam-nests hanging a few inches to a few meters above the water surface of ponds and ditches into which their tadpoles fall and develop. The species is very common and occur in all the districts of West Bengal.

2. Terai Tree Frog (*Polypedates teraiensis*)

These large-sized (55-85 mm) tree frogs are slender, narrow-waisted with elongated limbs and large black goggling eyes. The snout is pointed. Tympanum is distinct, almost equal to eye diameter. Fingers are with rudimentary webs. Toes are almost half-webbed. Tips of fingers and toes are dilated into large, flattened, horse-shoe shaped adhesive discs. It has smooth skin and granular belly and underside of thighs. Brownish or yellowish in colour, the belly is white. A dark brown marking extends from the nostril on both sides of the head, covering the eye right upto the middle of the flanks. There are 4-6 longitudinal lines on the back. Limbs are cross-barred. Nocturnal and arboreal, they hide among vegetation during the day and call after sunset during the rainy season. They lay 400-900 creamy white eggs inside large globular foam-nests hanging a few feet above the water surface of ponds and ditches into which their tadpoles fall and develop. The species mainly occur in the northern districts of West Bengal.

3. Large Green Tree Frog (*Rhacophorus maximus*)

This large species (50-90 mm) of tree frog has a flabby skin with very large circular adhesive discs on toes with which they can cling onto branches of trees in forests. They have a large head with a pointed snout, large goggled eyes, narrow waists and elongated hind limbs. The fingers are partly webbed but the toes are fully webbed. Dark green in colour with white underparts; they can camouflage easily among leaves and branches of trees in moist evergreen forests, where they dwell. They congregate in the mountain lakes and pools to breed on the banks and lay foam-nests in water. Often many females lay their eggs in a single aggregated foam-nest, in a kind of communal foam-nesting. The species is found only in the wetlands of Darjeeling district of West Bengal.

4. Annandale's Tree Frog (*Chirixalus simus*)

This small tree frog (16-28 mm) is known from both north and south Bengal plains and may occur on trees, bushes, shrubs or grasslands in marshy areas of the state. Brown in colour with a number of longitudinal stripes on the back, it has no webbing between the fingers but much webbing on the toes. It lays 120-230 small white eggs in smaller foam-nests hanging from twigs or branches of trees over pools of water below. Multiple male amplexus with a single female has been observed and the small tadpoles take about a month to develop and metamorphose into tiny froglets (8-9 mm).

F. Family Salamandridae (Salamanders)

The salamanders of this family are from an ancient stock which is known to exist in Europe from the Jurassic era. Although common in Europe and Central Asia, only one species of this family occur in India.

1. Himalayan Salamander (*Tylototriton verrucosus*)

This is the only species of salamander (Order: Urodela) known to occur in India. It is a unique and rare tailed amphibian which is mostly known to inhabit the cool mountain lakes and temporary as well as permanent pools in the hills of Darjeeling district in northern West Bengal between the altitudes of 1330-2220 meters. It is a keystone species of the lentic freshwater wetland habitats in the Eastern Himalaya. The adults come down to feed and breed in the mountain lakes and pools from April-July, where they deposit their eggs on the emergent vegetation. The larvae are gilled and possess four legs with which they creep on the floor of these pools and feed on zooplankton and small insects. They metamorphose into juveniles by September and come out of the water. The adults are carnivorous feeding on earthworms, insects and woodlice. They hibernate during winter in the rock crevices and steep hillsides in the tea gardens and come down to breed the next year at the same very few water bodies in the Darjeeling hills. Thus, these wetlands are absolutely vital to the survival of the Himalayan Salamander.

Conclusion

The wetlands of West Bengal are the habitat of about 30 species of amphibians of the Order Anura (frogs and toads) and Urodela (salamanders) with distinct distribution pattern. The salamanders prefer a cold climate and are therefore restricted to the mountain lakes and shallow pools between 1300-2200 meters altitude. Out of the four species of toads, two (*Duttaphrynus melanostictus* and *D. stomaticus*) are found in the plain lands while *D. himalayanus* in the hills at lower elevations. The Sikkimese toad (*Scutiger sikkimensis*) is restricted to the higher elevations in the Darjeeling hills. Of the five species of burrowing narrow-mouthed frogs of the family Microhylidae, *Microhyla ornata, Kaloula taprobanica* and *Uperodon globulosus* are found in both the wetlands of north and south Bengal, while *Kaloula assamensis* and *Kalophrynus interlineatus* are restricted to the wetlands in the Duars area of northern parts of West Bengal. Out of the 14 species of aquatic frogs of the family Dicroglossidae, *Euphlyctis cyanophlyctis, Sphaerotheca breviceps, Hoplobatrachus tigerinus, H. crassus, Fejervarya syhadrensis, F. Odishaensis, F. teraiensis* are found in the wetlands of both northern and southern West Bengal while *Euphlyctis hexadactylus* is found only in the wetlands near the coastal districts. *Fejervarya nepalensis, F. pierrei, Minervarya chilapata* are restricted in the Duars area of the state while *Paa liebegii, P. annandali* and *P. blanfordi*are found in the wetlands of the Darjeeling Himalayas. Of the two species of true frogs of the family Ranidae, *Hylarana tytleri* is found in the wetlands of both northern and southern West Bengal while *H. leptoglossa* is confined only in the wetlands of northern West Bengal. Of the four species of tree frogs of the family Rhacophoridae, *Polypedates maculatus*is found all over West Bengal but *P. teraiensis* mainly restricted in northern West Bengal. *Rhacophorus maximus* occurs only in the wetlands of Darjeeling district while *Chirixalus simus* is found all over the state.

Summary

The wetlands of West Bengal encompassing lakes, *jheels, beels,* ponds, ditches, canals, ox-bow lakes, marshlands, swamplands and brackish-water marshes are the habitat of about 30 species of amphibians of the Order Anura (frogs and toads) and Urodela (salamanders). These include one species of salamander, 3 species of toads of the family Bufonidae, 1 species of mountain toad of the family Megophryidae, 5 species of burrowing frogs of the family Microhylidae, 14 species of aquatic frogs of the family Dicroglossidae, 2 species of true frogs of the family Ranidae and 4 species of tree frogs of the family Rhacophoridae. Even if all these 30 species do not live exclusively in these wetlands throughout the year, all of them depend on these wetlands to breed and continue their life cycle as their larvae (tadpoles) are all aquatic. Therefore, these wetlands are vital to the continued survival of these amphibians.

Acknowledgements

I thank my organization, the Zoological Survey of India for granting me necessary survey projects to study the amphibians in the different wetlands of the state of West Bengal and necessary laboratory facilities to identify them. I am grateful to my research guides Prof. Sushil Kumar Dutta and Dr. Indraneil Das for teaching me about amphibians.

References

Deuti K. 1996. The Himalayan Salamander. *Cobra*, 23: 35-37.

Deuti K. 1996. Ethological studies on the amphibians recorded from Bethuadahari Wildlife Sanctuary, West Bengal. *Zoosprint*, 11(5): 4-5.

Deuti K. 2001. Breeding ecology of Annandale's tree frog *Chirixalus simus* (Anura: Rhacophoridae) near Kolkata, West Bengal. *J. Bombay. Nat. Hist. Soc.*, 98(3): 341-346.

Deuti K and Goswami Bharati BC. 1995. *Amphibians of West Bengal Plains*. WWF – India, Kolkata.pp. 53.

Deuti K, Biswas S, Ahmed MF and Dutta SK. 2000. Rediscovery of *Chirixalus simus* Annandale, 1915 (Anura: Rhacophoridae) from Assam and West Bengal, eastern India. *Hamadryad*, 25(2): 215-217.

Deuti K and Hegde VD. 2007. *Handbook on Himalayan Salamander*. Nature Books India, New Delhi. pp. 43.

Deuti K and Sethy PGS. 2013. Catalogue of salamander and newt (Amphibia: Urodela/Caudata) specimens in the collection of the Zoological Survey of India. *J. Bombay. Nat. Hist. Soc.*, 110(1): 78-82.

Hegde VD and Deuti K. 2007. *Status survey on the Himalayan Salamander*. Zoological Survey of India, Kolkata, pp. 36.

Hegde VD and Deuti K. 2007. Distribution sites of the Himalayan Salamander (*Tylototriton verrucosus* Anderson) in the Darjeeling hills of northern West Bengal, India. *Cobra*, 1(2): 29-36.

Ohler A, Deuti K, Grosjean S. *et al.*, 2009. Small-sized dicroglossids from India, with the description of a new species from West Bengal, India. *Zootaxa*, 2209: 43-56.

Paul S, Biswas MC and Deuti K. 2007. First record of the Assam Painted Frog, *Kaloula assamensis* Das *et al.*, 2004 from West Bengal.*Cobra*, 1(3): 15-16.

Paul S, Biswas MC and Deuti K. 2007. *Kalophrynus orangensis* (Orang Sticky Frog) a new record for West Bengal. *Herpetological Review*, 38(1): 97-98.

Pratihar S and Deuti K. 2011. Distribution and threats of amphibian species in west Midnapore district, West Bengal, India. *Frogleg*, 17: 2-8.

Sarkar AK, Ray S and Biswas ML. 1992. Amphibia in State Fauna of West Bengal. Zoological Survey of India. *State Fauna Series*, 3(2): 67-100.

Chapter 9

Reptiles in Wetlands of West Bengal: Diversity, Behaviour and Adaptations

Kaushik Deuti

Herpetology Division, Zoological Survey of India, Kolkata

Introduction

Reptiles are generally adapted for a life on land as they breathe by lungs. However, many reptiles especially the turtles and crocodiles live almost all their life in the water. Some aquatic snakes and lizards can also swim and live in water during a considerable period of their life. The state of West Bengal abounds with different types of freshwater, riverine, brackish water and saline wetlands, stretching from the Eastern Himalaya of the Darjeeling hills in the north, down to the moist evergreen rain forests and moist deciduous forests of the Duars areas at the Himalayan foothills to the Gangetic plains in central and lower Bengal to the dry deciduous forests of western Bengal and the mangrove forests on the sea coast of the Bay of Bengal. Thus the state has number and variety of wetland habitats like mountain lakes in the hills; lakes, ponds, canals, ditches, *jheels, beels* etc in the north, central and southern regions of the state, the rivers with their tributaries and distributaries and the brackish water and saline marshes of the Sunderbans which support about 46 species of reptiles. This includes 2 species of crocodiles, 19 species of turtles, 3 species of lizards and 22 species of snakes.

Faunal Assemblages: A Review

A. Order: Crocodylia

Family: Crocodylidae (Crocodiles)

There are 25 species of crocodiles in the world of which the Estuarine Crocodile is the largest.

1. Estuarine Crocodile (*Crocodylus porosus*)

Also known as the salt-water crocodile, this is the largest species of crocodile and the largest living reptile in the world. It is found in the mangrove forested areas and estuaries of the Indian Sunderbans in the South 24 Parganas district of West Bengal and are responsible for a number of human casualties over the centuries. It is carnivorous and survives by preying on fish, crabs, shrimps, turtles, lizards, snakes, waterbirds and deer, wild boar, rhesus macaques and even tigers. It has a heavy snout with a pair of ridges running from the eyes to the center of the snout. Behind the eyes, it has four post-occipital scutes which are distinguishing features of this species. During the monsoon, the female gathers leaf litter and moist vegetation with its laterally compressed tail to construct a mound-nest in which she lays 60-80 large white eggs. She carefully guards her mound-nest often by sitting over it for long hours and days. After about 45 days, the hatchlings come out and when she hears their squeak, she digs open the nest, gathers the babies gently in her heavy jaws and carries them to the water. Even after releasing them into the water, she guards them for days. Inspite of all her best efforts, some egg predators like water monitor lizards will dig open her nest from the bottom and steal her precious eggs. Apart from the Sunderbans in West Bengal, they inhabit the Bhitarkanika mangroves in Odisha and Andaman and Nicobar Islands in India. Although their population has been declined drastically in the Sunderbans, they are still seen in good numbers in some parts of this Biosphere Reserve.

Family: Gavialiidae (Gharials)

This family is represented by a single living species although several fossil species indicate that the family was once widespread.

2. Gharial (*Gavialis gangeticus*)

The Gharial is a long-snouted specialized fish-eating crocodilian, found in some of the larger rivers of northern India. It is mainly restricted to the Ganga, Brahmaputra and Mahanadi rivers besides in Nepal and Bangladesh. It has a long, slender snout with more than a hundred sharp, interlocking teeth. The adult males have a distinctive knob at the tip of the snout which may help in producing sound but most probably is for sexual display. This resembles an earthen pot which is called 'ghara' in Hindi language and hence the common English name Gharial. The skin is olive tan with dark blotches or bands on the dorsum. The belly is pale yellowish. It is a specialized fish-eater but also a scavenger. It nests on sand banks in the middle of rivers, laying up to 60 eggs which hatch after about two months. Hatchlings are 17-18 cm in length. In West Bengal there are stray records of this crocodilian species from the Hughly river between Nabadwip and Murshidabad. A small population seems to still survive in the Hooghly district where there are some sand bars left for their breeding.

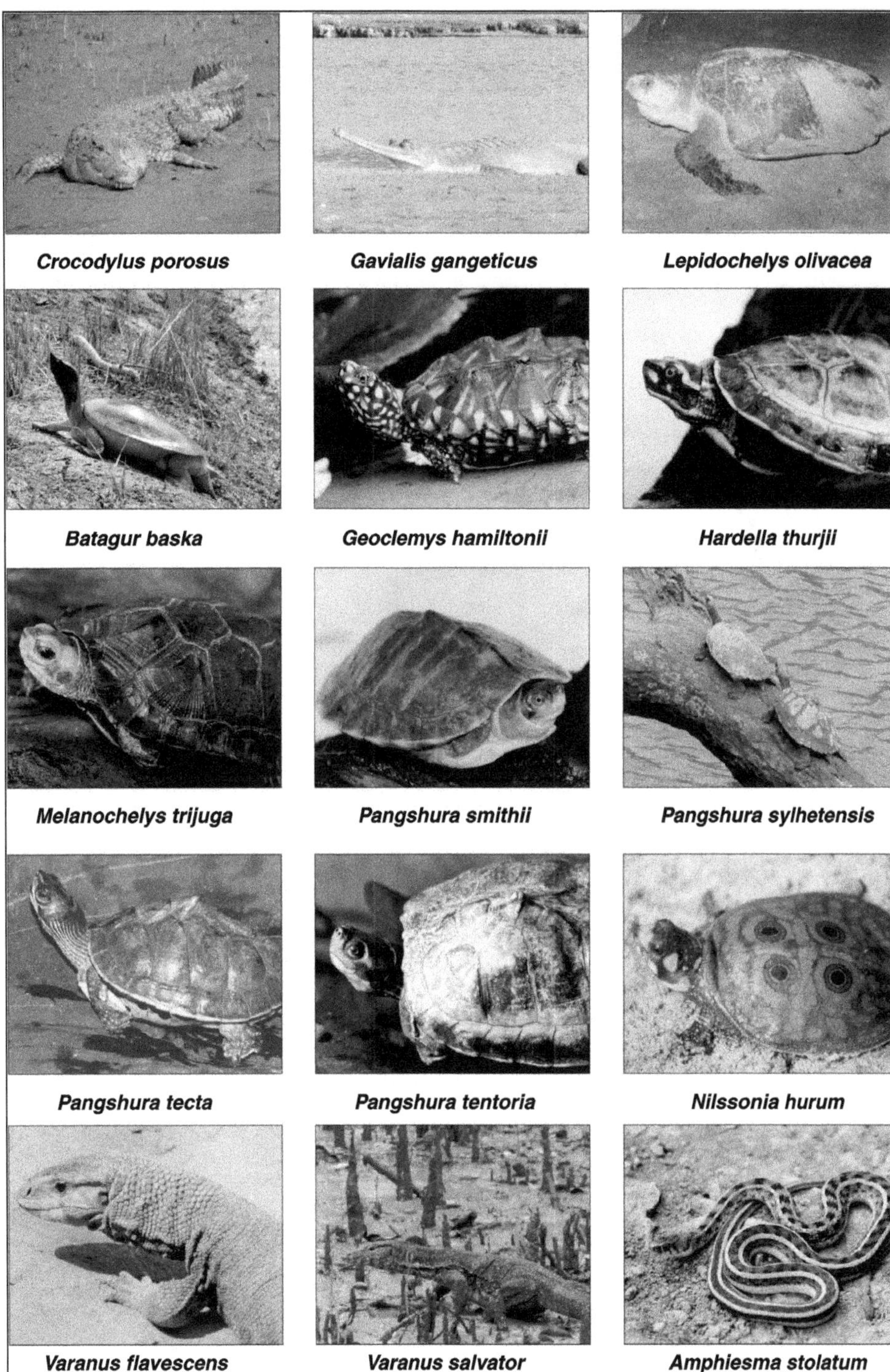

Figure 9.1: Some Common Reptilian Fauna Associated with Wetland Ecosystems of West Bengal.

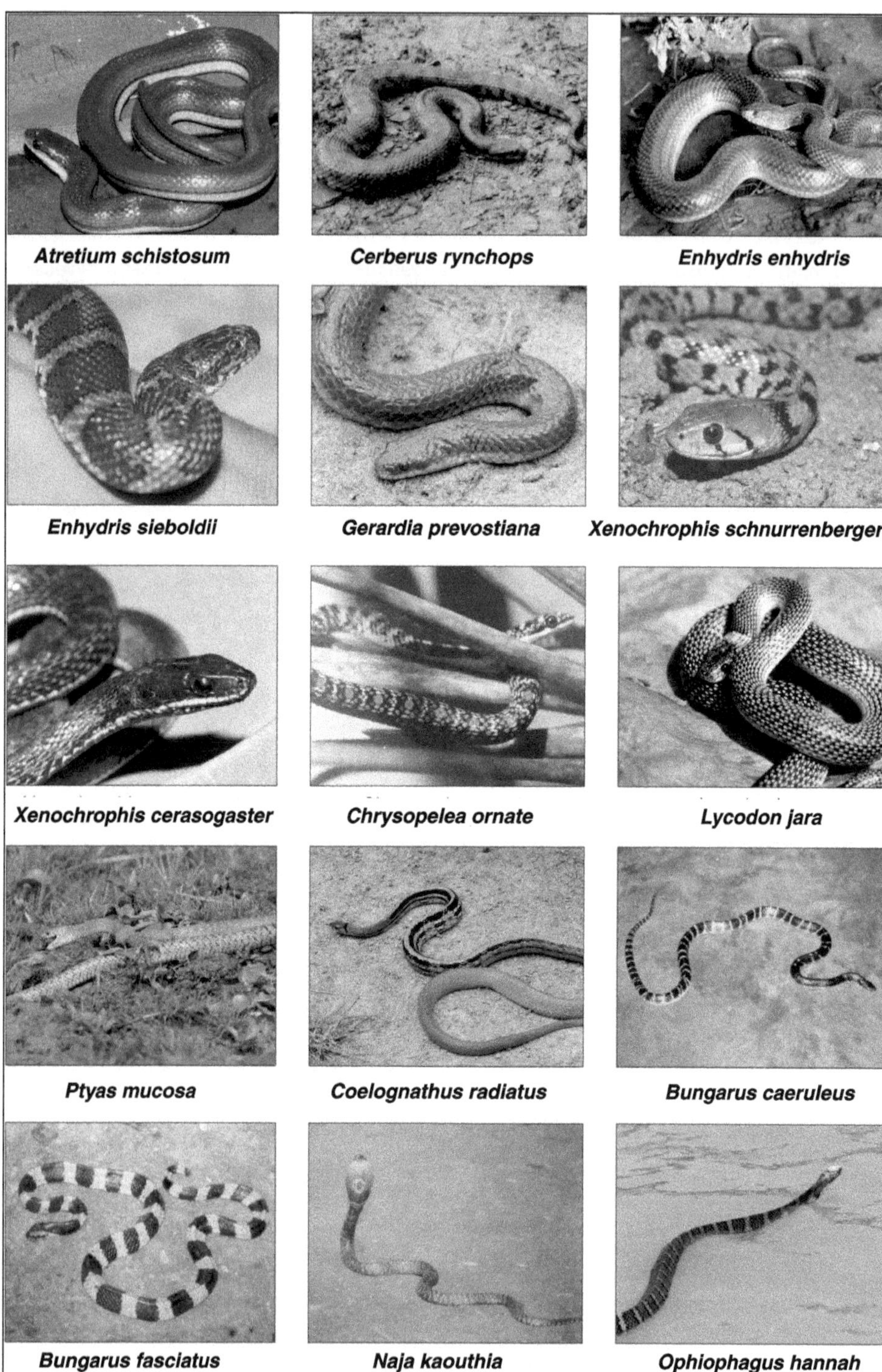

Figure 9.2: Some Common Reptilian Fauna Associated with Wetland Ecosystems of West Bengal.

B. Order: Chelonii

Family: Cheloniidae

1. Hawksbill Sea Turtle (*Eretmochelys imbricata*)

A large sea turtle with the upper jaw relatively narrow, elongated and forward projecting. The carapace is heart-shaped, with four pairs of imbricate (overlapping) costal scutes and two pairs of prefrontal scales. Carapace is olive-brown, juveniles with darker blotches. Associated with bays, estuaries and lagoons and there are isolated records from the Sunderbans of both West Bengal and Bangladesh. Feeds on sponges, algae, corals, jellyfishes and shellfishes.

2. Olive Ridley Sea Turtle (*Lepidochelys olivacea*)

The smallest and commonest of the sea turtles. Carapace is broad and heart-shaped with posterior marginals serrated; 5-9 pairs of juxtaposed costal scutes. Shell smooth and carapace olive-green or grayish-olive, plastron greenish-yellow. Use its front legs as flippers to swim. Hind legs are used for nesting. Some of the largest nesting aggregations called 'arribadas' occur in Odisha with several thousand turtles congregating to nest. A few hundred nest on the West Bengal coast on the beaches of Kalash, Mechua and Shaimari at the extreme southern part of the Sunderbans adjoining the Bay of Bengal. This turtle also enters into the mangrove forested areas of the Sunderbans to feed on fish, crabs, jellyfish, lobsters, shrimps, oysters and echinoderms.

Family: Geoemydidae/Bataguridae

3. River Terrapin (*Batagur baska*)

A large turtle from the mouths of the rivers, particularly the flooded mangrove forests of the Sunderbans. Carapace domed, plastron long. Head small with an upturned snout, forehead covered with small scales. Four claws on each forelimb. Carapace olive-grey or brown. Head similar coloured but lighter on the sides, plastron unpatterned yellow. Fruits of *Sonneratia* are an important food item along with leaves, fruits and stems of other mangrove plants. Most feeding occurs at high tide when such vegetation are exposed and fruits from low-hanging branches become more accessible from the water. Molluscs, crabs and fishes are also consumed. In the Sunderbans, they nest on the sea coast at Kedo, Mechua and Nagbarachar. A captive breeding center has been made at Sajnekhali in the Sunderbans for rearing this species.

4. Three-striped Roofed Turtle (*Batagur dhongoka*)

A large river turtle. Carapace is elevated and oval. Vertebral keel terminates in a knob on the third vertebra. Plastron narrow, truncated anteriorly and notched posteriorly. Snout slightly upturned. Upper jaw with a weak notch. Skin of head smooth anteriorly and divided into small irregular scales posteriorly. Carapace brownish grey with dark stripes on vertebral and marginal edges. Plastron yellow. Head and neck grayish cream with a cream or yellow stripe across face. Males are omnivorous, feeding on aquatic plants and molluscs while females are vegetarian.

The species basks on sand banks, rocks and projecting wooden stumps. Clutches of 20-35 eggs are laid which hatch in 56-90 days. Found in Ganga and Chambal rivers from Uttar Pradesh, Madhya Pradesh, Bihar, Jharkhand and Hugli river in West Bengal (southern Gangetic plains).

5. Painted Roofed Turtle (*Batagur kachuga*)

A beautiful river turtle with moderately domed and oval carapace. Vertebral keel with a knob and is most prominent on second vertebra. Plastron narrow, truncated anteriorly and notched posteriorly. Snout slightly upturned. Upper jaw strongly serrated. Carapace brownish-olive in males, dark brown or black in females. Plastron of both sexes are cream or yellow in colour. Basks on sand banks, rocks and logs and feeds on aquatic plants. A shy and non-aggressive species.

6. Leaf Turtle (*Cyclemys gemeli*)

Shell leaf-like in shape and colour to camouflage with leaf litter. Shell oval, depressed, bearing three keels. Enlarged scales on forehead. Plastron with a hinge in adult turtles. Carapace and plastron are brown with dark radiating lines. Omnivorous, ingesting figs as well as invertebrates. The turtle uses the back of its forelimbs to throw pieces of food into its mouth. Basks at the edge of water, diving in when disturbed; bottom walker than true swimmer. Sleeps on land at night. Nests are dug in the ground in which 2-4 elongated hard-shelled eggs are laid, hatch about 75 days later. Inhabits foothills and plains and more common in small rivers, streams and ponds. Found in Duars area of northern West Bengal, Northeast India, Bangladesh and southeastern Asia.

7. Spotted Pond Turtle (*Geoclemys hamiltonii*)

Turtle with an enlarged head for cracking open molluscan shells. Carapace with three interrupted keels. Head massive with a short snout. Shell black with yellow streaks and wedge-shaped marks. Head black with yellow spots. Neck grey with cream spots. Feeds on molluscs, aquatic plants, fruits and leaves. Inhabits ox-bow lakes and other standing water bodies. More than a single clutch can be laid in a year, each comprising 13-24 brittle hard-shelled eggs which hatch after 23-76 days. Found in the northern plains of India, in the Ganga and Brahmaputra drainages and in the wetlands in the southern part of West Bengal.

8. Crowned River Turtle (*Hardella thurjii*)

Males of this turtle are a third of the size of the females. Carapace fairly depressed with interrupted vertebral keel, marginals unserrated in adults. Plastron with a narrow posterior lobe, notched terminally and truncate anteriorly. Head large, snout projecting beyond the lower jaw. Inhabits slow-moving rivers, swamps, marshes, estuaries and large water bodies of southern West Bengal. Females may lay 30-100 eggs in a season from September to January and the clutch size varies between 8-13 eggs. The eggs may hatch after 275 days. Found in the Gangetic flood plains of northern India and Brahmaputra plains of Assam and Bangladesh, also in lower Bengal.

9. Black Turtle (*Melanochelys trijuga indopeninsularis*)

Carapace elongated, fairly elevated in adults. Nuchal small and triangular. Plastron truncated anteriorly and notched posteriorly. Head moderate with a short snout. Upper jaw notched. Toes fully webbed. Carapace brown. Males are larger than the females and possess longer tails that extend out of the carapace rim. They have concave plastrons. Inhabits standing waters with aquatic vegetation, although it may also be found in rivers. This turtle exhibits a nocturnal-crepuscular cycle, individuals are seen foraging after dark at the edge of the water bodies. Food includes freshwater prawns, snails, insect larvae, grass, water hyacinth, fruits and fishes. Widespread in peninsular and North-east India including northern and central West Bengal.

10. Indian Eyed Turtle (*Morenia petersi*)

A domed turtle of slow-moving and stagnant water bodies. Carapace smooth with a low vertebral keel. Head small with a pointed snout. Carapace green, olive or grey-black; vertebrals and costals with a green or yellow border. Head olive with three yellow stripes on face. Inhabits slow-moving waters of the Terai tracts at the Himalayan foothills. It occurs in standing water bodies such as weed-choked nullahs and ox-bow lakes as well as rivers. Basks on sand bars or on sand slabs along eroded river banks, dropping into water when threatened. Nests are constructed in December-January on loamy soil with sparse vegetation. Clutch of 6-10 eggs hatch in April-May. Found in isolated localities from northern India, Bihar, North Bengal and Assam and also Nepal and Bangladesh.

11. Brown Roofed Turtle (*Pangshura smithii*)

A low-domed, pale roofed turtle. Carapace elliptical and depressed. Medial keel weak, raised at posterior of vertebral scute. Plastron truncated anteriorly, notched posteriorly. Snout projecting beyond lower jaw. Upper jaw serrated and with a weak notch. Carapace light green in males and brown in females. Males are smaller than females and have relatively longer tails that are thicker at the base. Restricted to medium-sized rivers and their vegetation choked backwaters. Both aquatic plants and fish are consumed. Clutch consists of 7-9 elongated, brittle and hard-shelled eggs. Widespread from Punjab to West Bengal and adjacent areas like Nepal, Bangladesh and Pakistan.

12. Assam Roofed Turtle (*Pangshura sylhetensis*)

A small turtle with a highly elevated shell giving it a spike-like appearance. Carapace oval and serrated posteriorly and with 13 marginals. Snout slightly projecting, upper jaw weakly hooked. Carapace olive-brown with a pale brown vertebral keel. Red stripes on face, neck with light stripes. Plastron with large black blotches. Inhabits slow flowing hill-rivulets, ox-bow lakes and low-lying marshes in the plains as well as in the evergreen forests. Carnivorous and crepuscular. Feeds on small fishes. Clutches of 5 eggs are laid. Restricted to northern West Bengal, Assam, Meghalaya and Bangladesh.

13. Indian Roofed Turtle (*Pangshura tecta*)

A small, brightly coloured roofed turtle from ponds and other standing water bodies. Carapace elevated, oval with distinct vertebral keel on third vertebral scute which is spike like. Plastron truncated anteriorly, notched posteriorly. Head small with a projecting snout. Upper jaw serrated. Carapace brownish olive with light brown, red or orange stripe along first three vertebral scutes. Marginals with a narrow yellow border. Males are smaller than females and have relatively longer tails that are thicker at the base. Basks communally and several turtles are usually seen basking on logs in standing water bodies. Distributed in northern India, besides Pakistan, Nepal and Bangladesh. Often seen in ponds, jheels, beels and other water bodies in West Bengal.

14. Indian Tent Turtle (*Pangshura tentoria*)

A small roofed turtle, widespread in northern and central India. Carapace elevated, oval with a distinct vertebral keel on third vertebra. Plastron truncated anteriorly, notched posteriorly. Head small with projecting snout. Upper jaw serrated and unnotched with a single V-shaped ridge. Skin at the back of forehead with irregular scales. Digits entirely webbed. Males are smaller with comparatively longer and thicker tails. Colouration variable, shell olive-brown dorsally. Belly unpatterned or with a single black blotch on each scute. Found in northern and peninsular India as well as Pakistan, Nepal and Bangladesh. Primarily a riverine species, it is found in both small and large rivers. Basking turtles can be seen in the Ganga and Hughly rivers on rocks and tree trunks. Males and juveniles are more carnivorous than the females.

Family: Trionychidae

15. Indian Softshell Turtle (*Nilssonia gangetica*)

A large aggressive softshell turtle. Carapace low and oval. Five callosities in plastron. Snout slightly downturned. Carapace coloured grey-green or grey-black, usually with darker black reticulations. Males possess comparatively longer and thicker tails than females, the vent near the tail tip. Eye-like markings on dorsum of juveniles bordered with yellow. Inhabits rivers, ponds, lakes and reservoirs, sometimes burying itself in the mud at the bottom. Frequently basks on sandbars on the rivers. Males are known to become territorial during the breeding season, displaying aggression towards other males. Omnivorous, consuming water plants, various insects and invertebrates (specially molluscs) and also small vertebrates (fishes, frogs, flapshell turtles) and even scavenging on carrion. Known to grab waterfowl from beneath the surface of the water. Eggs spherical, brittle, hard-shelled and usually 13-35 eggs are laid. Distributed in the central and southern districts of West Bengal, northern India including the drainages of the Ganga and Mahanadi among the larger river systems, Nepal, Pakistan and Bangladesh.

16. Indian Peacock Softshell Turtle (*Nilssonia hurum*)

A beautiful softshell turtle marked with bright colours on the head. Shell with longitudinal rows of tubercles on posterior of carapace. Carapace low and oval. Five large plastral callosities. Head large with snout strongly downturned. Carapace olive with yellow rim. Plastron light grey. Forehead with black reticulations and large orange or yellow patches, across snout and on sides. Head and limbs olive. Juveniles with 4-6 dark-rimmed, yellow bordered ocelli (eye-like markings). Inhabits rivers, lakes, ponds and reservoirs. Snails, fish and larvae of mosquito are usually consumed. Nests between August-December, in clayey-sandy soil. Eggs spherical, hard-shelled. Found in the larger river systems of northern and eastern India including the central and southern districts of West Bengal as well as in Nepal and Bangladesh.

17. Narrow-headed Softshell Turtle (*Chitra indica*)

A bizarre-looking softshell turtle with a very narrow elongated head and its small eyes located very close to the snout near the nostrils. A short proboscis present. Shell oval and depressed. Dorsum dull olive or bluish grey with a pattern of dark reticulations. Carapace pattern continuing up to neck and outer surface of fore limbs. A V-shaped mark commences on nape and extends to carapace. Juveniles sometimes have four eye-like markings on the carapace or numerous black elongated spots. Inhabits sandy sections of medium or large rivers presumably because they are less turbid, making visibility better for prey capture. It does not bite but disables its victims with blows from the head. Small fishes are sucked head-first while the larger ones are swallowed gradually. It also ambushes fish underwater by burying itself in sand. Can also consume molluscs. Exudes an unpleasant smell when handled. Nests are constructed between August-September, in sandy or sandy-loam soils. Eggs brittle-shelled and spherical, clutches containing 65-190 eggs. Incubation period is 40-70 days. Found in Peninsular India but more widespread in rivers of northern India including central and southern West Bengal besides in Pakistan, Nepal and Bangladesh.

18. Indian Flapshell Turtle (*Lissemys punctata*)

A medium-sized domed softshell turtle with plastral flaps. Shell oval and low-domed with hinged anterior lobe. Cream or pale yellow plastron with a pair of flaps under which the hind limbs can be retracted. Tail short, scarcely extending out of the rim of the carapace margin in adult males and even smaller in the females. Carapace olive-green or olive-brown or grey-green with yellow spots or markings. Inhabits ponds, lakes, canals, swamps, inundated rice fields. Scavenges on animal corpses, also consumes fishes, frogs, tadpoles, crustaceans, molluscs and other invertebrates and various soft aquatic plants. Lays eggs between October-November, eggs nearly spherical, clutch varies between 5-15. Incubation period is about 9 months. Widespread in northern and peninsular India including water bodies of West Bengal and also found in Pakistan, Nepal, Sri Lanka, Bangladesh and northern Myanmar.

19. Asian Giant Softshell Turtle (*Pelochelys cantorii*)

The largest freshwater turtle in the Indian region, inhabiting coastal areas, can grow up to one meter in length and weigh upto 250 kgs. Found in the Sunderbans area of West Bengal although with no specific locality. Shell is low and depressed, elongated in young but oval in adults. Juveniles showing numerous tubercles on carapace and a low vertebral keel. Snout short, flat and rounded and with an extremely short proboscis. Interorbital space wider than greatest diameter of eye. Small flaps of skin on the gular region. Carapace olive or brown, spotted or streaked with lighter or darker shades with a lighter outer edge. Diet comprises of fish, shrimps, crabs, molluscs and aquatic plants. Between 20-28 eggs are laid at a time. The species migrates from the inland freshwater rivers to nest on the sea beaches. Recorded from both coasts of India and also Bangladesh, Myanmar, Thailand, Malaysia, Indonesia and Borneo.

C. Order: Squamata

Sub-Order: Sauria

Family: Varanidae (Monitor Lizards)

Monitor lizards are the largest of the living lizards. They are distinguished by their long and flattened body, long neck and long tail. The head is triangular, snout is elongated, eyes are with well-developed movable eyelids, teeth are recurved, tongue is elongated and forked similar to that of snakes and is flicked in and out to taste the chemical nature of the environs. Limbs are well-developed and the fingers and toes are armed with strong claws.

1. Bengal Land Monitor Lizard (*Varanus bengalensis*)

This is a large-sized monitor lizard with a large, flat and triangular head and an elongated snout, commonly seen in the countryside in North Bengal. Nostril is nearer to the eye. Body is cylindrical. Limbs are well-developed with strong claws. Tail is flattened on the sides. Adults are olive-grey or brownish above with sparse black spots. They do not wait for prey to come within striking distance but are swift and active predators of small mammals, birds, reptiles, frogs, fishes and invertebrates. Some are scavengers too. They are good swimmers and can remain submerged for a considerable time. When alarmed tries to stay still and escape notice. A cornered monitor will demonstrate by rising on its fore legs, hissing and lashing with its tail. It seeks its prey by both smell and sight. Fish and crabs caught in water are brought to the shore and eaten.

2. Yellow Monitor Lizard (*Varanus flavescens*)

A comparatively smaller golden-yellow monitor lizard with a short and convex head. Nostril is closer to the tip of the snout. Eyes are large. Tongue is long and forked. Tympanum is a large hole at the back of the head. Body is cylindrical and the head is strongly compressed. Brightly coloured with reddish brown blotches or bands on a yellowish-brown background. Associated with the low-lying wetlands of the Gangetic plains of West Bengal and in the swamplands. Quickly hides in burrows when approached. Incessantly searches for prey near the water bodies. Nest in self-excavated burrows during the monsoons between August-October. Usually lays 14-30 eggs.

3. Water Monitor Lizard (*Varanus salvator*)

The water monitor lizard is frequently seen in the Sunderbans and the extensive wetlands of southern West Bengal. It is abundant in the South 24 Parganas, Howrah and Hooghly districts. It is the largest monitor lizard in India with an elongated and flattened snout. Eyes are large. Nostrils are round and nearer to the tip of the snout than to the eye. Body is robust and cylindrical. Limbs thick with strongly developed claws. Tail strongly compressed with a crest above. Adults have yellow spots or rings in transverse rows on a grayish-brown back. They are semi-aquatic, diurnal and solitary. They are seen in both fresh and brackish waters. Inhabits a variety of wetlands from mangrove swamps, marshes, estuaries, rivers, canals even village ponds. Water monitors can swim up to two miles from one island to another.

Sub-Order: Serpentes

Family: Colubridae

1. Buff-striped Keelback Snake (*Amphiesma stolatum*)

A dainty striped non-venomous snake of the lowland plains growing upto 80 cm. Body slender with olive-grey or greenish-grey dorsum on which there are two bright longitudinal stripes. Eyes large, pupil round, 2-3 black stripes from eye to upper jaw. Two buff or orange stripes run down dorso-laterally as well. Underside white. Inhabits grasslands and bushy areas in the vicinity of lakes, ponds and canals as well as inundated paddy fields, plantations and backyard gardens. Active in the morning and at dusk when they hunt for frogs, insects, scorpions, fish and lizards. Docile but inflate fore body to display bluish-white inter scale colour when threatened or afraid. Clutches of 3-15 eggs are laid which hatch 36-62 days later. Two clutches may be laid in a year. Widespread throughout India in addition to Pakistan, Nepal, Sri Lanka, Bangladesh, Myanmar, China and southeast Asia. Commonly seen near all the wetlands of West Bengal.

2. Olive Keelback Water Snake (*Atretium schistosum*)

A small to medium sized snake (upto 1 m) from moist areas close to water bodies. Snout short, nostrils slit-like placed on top of snout. Scales distinctly keeled, lacking apical pits. Dorsum olive-brown or greenish-grey, unpatterned or with two series of small black spots, occasionally a dark lateral streak. Upper lip, outer row of scales and belly yellow. Fond of damp places near water and is also known from brackish coastal areas and near inundated agricultural fields. Diurnal, it is found near water bodies and other moist areas and is a good swimmer as well as adapted to climb low bushes. Diet comprises frogs, tadpoles, fish, prawns and crabs. Clutches of 10-32 eggs are laid between December to April and hatchlings measure 16-18 cm. Known from the Indian peninsula including West Bengal in addition to Nepal, Bangladesh and Sri Lanka.

3. Dog-faced Water Snake (*Cerberus rynchops*)

A common water snake from mangrove swamps and river mouths, with a projecting upper jaw, giving it a dog-like appearance. Head long and distinct from neck, eyes beady with rounded pupils. Scales distinctly keeled. It grows upto 1.25

meters. Dorsum dark-grey with faint dark blotches and a dark line across eyes. Belly yellowish cream with dark grey areas. Abundant in mangrove mudflats and inundated paddy fields in the Sunderbans, where they hide in crab holes, emerging at night. Diet comprises fish, such as mudskippers and gobies, as well as crabs and frogs. Widespread along the east coast of India including West Bengal, the Andaman and Nicobar Islands, its range extending upto Australia.

4. Common Smooth Water Snake (*Enhydris enhydris*)

A common water snake growing upto 80 cm, from southern and southeastern Asia. Head rather small, body medium-sized, stout and cylindrical. Head somewhat depressed and only slightly distinct from neck. Snout rounded, nostrils situated on upper surface of head, eyes small, pupils vertical. Dorsal scales smooth, lacking apical pits. Dorsum grayish-brown or olive-green, with a dark vertebral and two light lateral stripes from upper surface of head to tail. Belly yellowish cream, with brown spots that form three dark lines. Inhabits freshwater and sometimes also brackish water, including slow-moving rivers, marshes, ponds, lakes, wet paddy fields and water bodies with dense aquatic vegetation. Diurnal and crepuscular. Docile, rarely bites. Diet comprises fish, frogs, tadpoles and sometimes lizards. Mating occurs in December, being ovoviviparous produces several clutches a year, between January and June, each comprising 4-20 young. Widespread in eastern India including West Bengal, besides Nepal, Bangladesh, southern China and southeast Asia.

5. Siebold's Water Snake (*Enhydris sieboldii*)

Siebold's water snake is found in the freshwater and brackish water areas of the Sunderbans including a number of water bodies with aquatic vegetation. It is also known from eastern India including West Bengal, besides Bangladesh and southeast Asia.

6. White-bellied Mangrove Snake (*Fordonia leucobalia*)

A medium-sized (upto 1 m) extremely variable water snake in terms of colouration. Found in mangrove areas of the Sunderbans and other portions of the river close to the sea. Head short, wide, scarcely wider than neck. Head scales large and distinct. Loreal absent. Lower jaw short and scales smooth. Variable in colouration and pattern, may be dark grey or brown, with light spots or light grey, yellow or orange with dark spots. Belly pale cream, sometimes with small dark spots. Lips yellowish cream. Inhabits tidal rivers. Crabs form the main component of the diet, although small fish are also consumed. Ovoviviparous, producing 10-15 young at a time, hatchlings measuring 18 cm. The range of this coastal species extends from the east coast of India to Bangladesh, Myanmar and southeast Asia upto New Guinea and Australia.

7. Glossy Marsh Snake (*Gerardia prevostiana*)

A small snake (growing upto 53 cm) from the mangrove swamps of the Sunderbans. Head is distinct from neck, eye small with vertically elliptical pupil. Body relatively long, scales smooth, tail short. Dorsum unpatterned grey or brown.

Lips and lower scales of dorsum cream. Belly brownish cream with median dark streaks. Inhabits coastal areas, including mangrove swamps and is nocturnal. Soft-shelled crabs form the main diet although fish and shrimps are also eaten. Occurs on both coasts of India including West Bengal, besides Sri Lanka as well as isolated localities in southeast Asia, including Myanmar, Thailand and the Malay Peninsula.

8. Checkered Keelback Water Snake (*Xenochrophis piscator*)

The most common water snake in the wetlands of West Bengal (growing upto 1.75 m) is known to be most aggressive. When threatened, may display a narrow hood and hiss. Bite readily if caught or stepped upon. Body is cylindrical, eye is with rounded pupil. Dorsal scales strongly keeled. Dorsum olive-green or yellowish-grey with black spots arranged in a checkered pattern. Head brown with a black stripe from eye to upper lip and from postoculars to edge of mouth. Nape has an inverted V-marking. Underside usually white, sometimes may have dark edges between two ventrals. This water snake of the plains is abundant in the waterways, such as flooded paddy fields, drains, ponds, lakes, marshes and rivers. Active both by day and night. Fish and frogs form the typical diet although mice and small birds may be taken. Clutches of 30-70 eggs are laid. Females guard the eggs and incubation period is 37-90 days. Hatchlings measure 11 cm. Distributed from Pakistan to southern China and Thailand and are common throughout the Indian wetlands as well as those of West Bengal.

9. Bar-necked Keelback Water Snake (*Xenochrophis schnurrenbergeri*)

A related species of keelback water snake with olive-brown dorsum with 5-6 rows of black box pattern. A slight dark crossbar present on nape. Venter is white. It grows to about 80 cm and is also common in the wetlands of West Bengal as well as those of Nepal and Bangladesh.

10. Painted Keelback Water Snake (*Xenochrophis cerasogaster*)

A small brightly coloured keelback water snake found in the wetlands of West Bengal. Dorsum is reddish-brown which fades away at the posterior part of the body. Lip is yellow. Two mid-dorsal light brown or yellow lines run along the body axis. Venter is marbled with reddish black having white spots. It grows upto 50 cm. Found in the wetlands of Nepal, northern India and Bangladesh.

11. Ornate Flying Snake (*Chrysopelea ornata*)

This medium-sized (upto 1 m) slender colourful snake can be often seen among the stems of emergent aquatic plants in some wetlands of southern West Bengal. The head is depressed, dorsal scales smooth or feebly keeled with apical pits, vertebral scales not enlarged, ventrals with prominent scales pronounced keels laterally. Tail is long. Dorsum greenish-yellow or pale green with orange or red spots between dark cross bands. Head is black dorsally with yellow and black cross bars. Belly pale green with series of black lateral spots. Diurnal and arboreal in nature and is found on trees in forests and plantations. It feeds on lizards, frogs, small bats and rodents. Lays 6-20 eggs between May-June and incubation takes 65-80 days. Found throughout northern and north-eastern India besides Sri Lanka, Nepal, Bangladesh, Myanmar, China and southeastern Asia.

12. Common Wolf Snake (*Lycodon aulicus*)

A common house-dwelling small and slender snake (about 80 cm) also seen around the wetlands. Head is flat and distinct from neck. Snout is projecting beyond the lower jaw. Scales are smooth. Dorsum is reddish-brown or grayish-brown with 12-19 white cross bars, more marked on the anterior body. A broad white collar on neck. Eyes small and black. Belly cream or yellowish white. Inhabits low-lying plain lands, parks, gardens, human habitations, villages and marshlands. Nocturnal, emerging after dark to feed on geckos, small lizards, frogs, mice etc. Usually retreat under rocks, logs and debris. A good climber and can climb walls. Bite aggressively when handled and discharge a foul smell. Lays 3-11 eggs between February-July and more than one clutch may be produced in a year. Hatchlings emerge in September-October and measure 14-18 cm. Distributed all over India, in addition to Sri Lanka, Nepal, Bangladesh and Myanmar.

13. Yellow-speckled Wolf Snake (*Lycodon jara*)

A small, slender finely spotted wolf snake growing upto only 55 cm. A fairly long head with a somewhat flattened snout, not projecting beyond lower jaw. Dorsum brown or purple or olive-green, finely stippled throughout with yellow spots spread uniformly on each scale. Upper lip and lower surface unpatterned white with a white collar in juveniles. Inhabits forests, open areas and agricultural areas with bushes and scattered trees often near wetlands. Diet comprises frogs, lizards and small mammals. Oviparous. Known from West Bengal, north-eastern India and Nepal and Bangladesh.

14. Rat Snake (*Ptyas mucosa*)

A large rat-eating snake growing upto 2.7 meters. Head narrow. Eyes large with rounded pupil. Scales smooth. Dorsum yellowish brown, olivaceous brown to grayish black. Posterior of body with dark bands or reticulate pattern. Belly grayish white or yellow with caudal scales having black edges. Inhabits many forest types, plantations and abundant in agricultural fields, scrublands, grasslands and mangroves and near wetlands. Agile and fast moving. Forages both during day and night. Diet comprises rats, frogs, lizards, turtles, other snakes, birds and bats. They produce a variety of sounds from a hiss to a low growl. Combat dances are frequently reported, when two males are partially entwined, with the forebodies raised. The female lays 5-18 eggs and guards her eggs. Incubation period is 60 days and hatchlings measure 36-47 cm. Widespread in India including West Bengal and distributed from the middle east through Pakistan, Nepal, Bangladesh, Sri Lanka east to southern China and southeast Asia.

15. Copper-headed Trinket Snake (*Coelognathus radiatus*)

Another rat eating snake, it is more common in grasslands, forest edges, agricultural fields and wetlands. Body rather slender, snout relatively long. Scales keeled. Dorsum grayish-brown or yellowish-brown with four black stripes along the front of the body. A cream stripe runs along the upper two wide stripes. The lower stripes are narrower and generally broken up. Posterior part of body without any stripes. Underside light coloured. Head copper brown with three black radiating lines from the eyes. When threatened, it rears its head and part of its body, throwing the neck and body into a coil in a position of attack. Active during the cool hours of the morning and again at dusk. Its diet comprises birds and rats. 5-12 eggs are laid which hatch after 70-90 days and four clutches may be produced in a year. Widespread in the Indian region, the range extending to southern China and southeast Asia.

16. Indian Trinket Snake (*Coelognathus helena*)

A medium-sized relatively slender snake with smooth scales on the sides. Dorsum brown or olive with transverse bands and blotches on the sides. Neck with two dark narrow lines or a dark-edged white collar. A line radiates from behind the eyes to the angle of the jaw. Venter white. Found in the forests and plantations, especially in the vicinity of water bodies. When threatened, it also rears its head and part of its body, throwing the neck and body into a coil in a position of attack. Terrestrial and crepuscular in habit. Its diet comprises rats, lizards and birds. Usually 3-12 eggs are laid and more than a single clutch may be laid in a year. Widespread in the Indian region and also Sri Lanka, Nepal and Bangladesh.

17. Banded Racer (*Argyrogena fasciolatus*)

A slender and swift snake sometimes seen near the wetlands. Snout is strongly projecting with smooth dorsal scales. Adults are brown, sometimes with a blackish-red tint, anteriorly with narrow-white, brown and black cross bars. Posterior part of body with indistinct dark cross bars or spots. Belly unpatterned yellow or cream. Inhabits forests and dense bushes in parks and gardens near ponds and canals, concealing itself in rat holes and under rock piles. Diurnal and terrestrial, its diet comprises rats, mice, shrews, bats, frogs and lizards. Clutches of 2-7 eggs are laid in January which hatch in July. Known from Indian subcontinent including West Bengal.

18. Banded Kukri Snake (*Oligodon arnensis*)

Apparently looks like the venomous krait, although a completely harmless small snake. Body is stout and cylindrical. Snout is short and blunt. Scales are smooth. Dorsum is smooth usually with red or purple markings, lighter on flanks with 32-40 black cross bars that break up on flanks into streaks. Head with three dark arrow-shaped marks. Belly cream with indistinct lateral spots. A ground dwelling snake, inhabiting the leaf litter of forest floor and crevices of rocks and tree holes in disturbed habitats even near wetlands. Both diurnal and nocturnal although mostly active during cool rainy nights. Diet comprises rats, mice, lizards, frogs and reptile eggs. Generally 4-9 eggs are laid. Known from Indian subcontinent including West Bengal besides Pakistan, Nepal and Bangladesh.

Family: Elapidae

1. Common Krait (*Bungarus caeruleus*)

A medium-sized (1.75 m) terrestrial snake which is sometimes seen near wetlands. It is highly venomous and dangerous to humans, responsible for a large number of mortalities. The head is indistinct from neck with small eyes and the body is cylindrical. Scales are shiny; the vertebral scales are enlarged and hexagonal. Dorsum is steely-blue, black or dark brown with narrow white bands across the body. Belly is unpatterned cream. Widespread in the plains of West Bengal, in thinly wooded forests, secondary forests, plantations, wetlands, agricultural fields and near human habitation in the villages. Hunts at night, it hides under debris and vegetation at other times. During the day it is non-aggressive, flattening the body and attempting to hide its head under its body. At night, it may bite without provocation. The fangs are rather short, the venom highly toxic causing respiratory failure. Snakes are almost exclusive item in the diet although mice, frogs and lizards are also taken. Mating takes place between February and March. Lays 6-15 eggs that are guarded by the females and hatchlings measure 27-30 cm. Widespread in Pakistan, India, Nepal, Sri Lanka and Bangladesh.

2. Banded Krait (*Bungarus fasciatus*)

A dark and light banded snake (2.25 m). Body is triangular in cross-section with a short and stumpy tail. Head is slightly broader than neck. Dorsum with alternate black and yellow bands that are approximately equal in size. Top of the head with a V-shaped marking and small black eyes. Inhabits lowlands, swampy and marshy areas, drains and road culverts in the vicinity of villages. Although timid and non-aggressive by day, it is a highly venomous snake and is known to become bold by night. It eats water snakes, some land snakes, fish, frogs and skinks. Lays 4-14 eggs in April which hatch after an incubation period of 60 days, hatchlings measuring 25-30 cm. Widespread in eastern and north-eastern India extending upto southeast Asia.

3. Monocellate Cobra (*Naja kaouthia*)

A dangerously venomous snake that is responsible for a large number of human mortalities annually and can be easily identified by the monocle-like marking on the expanded hood. Head is distinct from body. Dorsum is dark brown, grayish brown or blackish brown, the hood marking typically a light circle with a dark center. Ventrally grayish or mottled. Throat region on ventral side has black bands. Inhabits wetter parts of the region and is known from agricultural fields and plantations as well as marshy and swampy areas, occasionally entering areas of human habitation. Typically crepuscular and nocturnal, eating rats, mice, frogs, lizards, fish and other snakes. When provoked, raise fore body, expand the hood, hiss loud and strike repeatedly. Clutches of 15-30 eggs are laid between January and March, with the female guarding the eggs during the incubation period, which is about 50 days. Distributed along northern and north-eastern India, the range includes Nepal and Bangladesh and extends to eastern China and southeast Asia.

4. King Cobra (*Ophiophagus hannah*)

The largest venomous snake with a great diversity of coloration, scalation and body proportions. Can grow upto 5.5 meters. Head is distinct from neck, covered with large black-edged shields. A pair of large occipital scales. Dorsum dark brown, olive brown or grey-black with narrow pale yellow or white bands in the young which mostly persist in adults. Ventrally shiny and mottled. Terrestrial as well as aquatic. Good swimmer and often seen swimming in small rivers in North Bengal and tidal creeks in the Sunderbans. They feed on other snakes and occasionally on monitor lizards and rodents. Females construct a mound nest by collecting fallen leaves in which eggs are deposited and guarded. About 20-45 eggs are laid which take 60-63 days to hatch. Hatchlings are 30-42 cm and immediately start hunting on their own. The adults are extremely agile, displaying a large and elongated hood when threatened. Attacks on humans are known. The enormous venom glands of the king cobra contain enough venom to kill an elephant. Known from the Himalayan foothills, south-western and north-eastern India, east to Indo-China and southeast Asia. Known throughout West Bengal but more in the Duars area of North Bengal and from the Sunderbans in the south.

Conclusion

The wetlands of West Bengal encompassing lakes, jheels, beels, ponds, ditches, canals, ox-bow lakes, marshlands, swamplands and brackish-water estuaries are the habitat of about 46 species of reptiles of the Order Crocodylia (crocodiles), Chelonia (turtles) and Squamata (lizards and snakes). The crocodiles prefer Hooghly riverine (gharial) and estuarine (estuarine crocodile) habitat of the Sunderbans. Of the 19 species of turtles, two are basically marine (Hawksbill and Olive Ridley of the Family Chelonidae), the rest are found in the rivers, ponds, lakes, jheels, beels, ox-bow lakes and marshy areas throughout the state as well as in the mangrove swamplands of the Sunderbans (12 species of Family Bataguridae and 5 species of Family Trionychidae).

The three species of monitor lizards (Sub-Order: Sauria and Family: Varanidae) can be seen in the wetlands of both north and south Bengal with the water monitor lizard being especially common and a major predator in the mangrove swamps of the Sunderbans.

The 18 species of aquatic snakes of the family Colubridae, all live near the wetlands and can swim well. Some are even found in the brackishwater of the Sunderbans like the glossy marsh snake (*Gerardia prevostiana*) and the white-bellied water snake (*Fordonia leucobalia*). The four species of venomous snakes of the family Elapidae are also seen in the wetlands with the common krait, monocellate cobra and the king cobra being more common in the Sunderbans.

Summary

The wetlands of West Bengal showed enormous variations including lakes, jheels, beels, ponds, ditches, canals, ox-bow lakes, marshlands, swamplands and brackish water marshes and support a wide variety of reptilian fauna. About 46 species of reptiles of the Order Crocodylia (crocodiles), Chelonia (turtles) and

Squamata (lizards and snakes) are reported from wetlands ecosystems of West Bengal. These include 2 species of crocodiles (family Crocodylidae); 19 species of turtles (2 species of family Chelonidae, 12 species of family Bataguridae and 5 species of family Trionychidae); 3 species of monitor lizards (family Varanidae) and 22 species of aquatic snakes (18 species of family Colubridae and 4 species of family Elapidae). Most of these 46 species do live in or near these wetlands throughout the year and depend on these wetlands to continue their life cycle. Therefore, these wetlands are vital to the continued survival of these aquatic reptiles.

Acknowledgements

I thank my organization, the Zoological Survey of India for granting me necessary survey projects to study the reptiles in the different districts of the state of West Bengal and necessary laboratory facilities to identify them. I am grateful to Dr. Indraneil Das, Mr. Romulus Whitaker, Mr. Dipak Mitra, Dr. Saibal Sengupta, Mr. Jayaditya Purkayastha and Mr. Anirban Chaudhuri for teaching me much about reptiles. I am specially thankful to Mr. Anirban Chaudhuri for some photos of reptiles of West Bengal.

References

Ahmed S and Dasgupta G. 1992. Reptilia in State Fauna of West Bengal. *Zool Surv Ind. State Fauna Series*. 3(2): 1-65.

Das I. 1995. *Turtles and Tortoises of India*. Oxford University Press. pp. 180.

Das I. 2002. *Snakes and other reptiles of India*. New Holland Publishers, UK. pp. 144.

Deuti K. 2009. Distribution and Status of the Endangered River Terrapin *Batagur baska* (Gray) in the Indian Sunderbans. In. *Freshwater Turtles and Tortoises of India*. Envis Bulletin, Wildlife Institute of India. 12(1): 53-56.

Deuti K. 2013. *Lizards of West Bengal*. Citadel Publishers, Kolkata, India. pp. 56.

Firoz Ahmed M, Das A and Dutta SK. 2009. *Amphibians and Reptiles of Northeast India*. Aaranyak, Guwahati, India. pp. 170.

Purkayastha J. 2013. *Reptiles of Assam*. Eastern Book House, Guwahati, India. pp. 146.

Chapter 10

Mammals of Indian Wetlands with Special Reference to West Bengal

Rina Chakraborty

Zoological Survey of India, Kolkata

Introduction

The term wetlands in India include a wide range of inland waters from high mountain lakes to rivers, swamps, bogs, ponds, brackish water lakes, estuaries and even the shallow coastal waters. In a nutshell, it could be defined as any water body whether stagnant or flowing at least for a major period of the year and transitional areas between dry terrestrial and permanent aquatic ecosystem. In India's diverse geographic and physiographic condition the mammalian species are adopted to live in every ecosystem present. The wetland mammals may be categorized as essentially wetland species or aquatic, wetland dependent and wetland associates. Wetland mammals reside permanently or temporarily in and around wetland for food, shelter and foraging. Splendid enough, a mammalian species which is wetland dependent or wetland associate in a particular bio-geographic area is not at all belongs to the same category in other areas too. Such as the Tiger, *Panthera tigris* is not a wetland species but the same live in quite ease in the deltaic wetland of the Sundarbans where it could be treated as wetland dependent species. The animal is so adjusted there that it can swim kilometers to cross a river for moving one island to another and feeds regularly on aquatic animals like fish, crab, turtle, monitor lizard etc., even when its regular prey like ungulates are available. Thus, categorization of wetland mammals depends mostly on its habitat types except essentially wetland species (aquatic). In the present review a brief account of wetland mammals with their distributional range in Indian limit and conservation status have been enumerated.

Essentially Wetland Mammals (Aquatic)

The essentially wetland mammalian species belongs to the orders Sirenia and

Cetacea. Altogether 15 species are enlisted of which 1 belongs to order Sirenia and the rest under 5 families of the order Cetacea as shown in Table 10.1.

Order: Sirenia

Family: Dugongidae

Species 1: *Dugong dugon* (Muller); Common Name: Dugong.

Only species under the order Sirenia, available in the Indian waters is the Dugong. It is an inhabitant of Gulf of Kutchch, Gulf of Mannar, Malabar coast and coasts of Andaman and Nicobar Islands in Indian waters. Now-a-days it is so scarce that only occasionally the species could be seen in the coastal waters of Indian mainland (Agrawal and Alfred, 1999).

Conservation Status: IWPA- Schedule I; CITES- Appendix I; IUCN- VU

Order: Cetacea

Habitat wise the cetaceans of Indian waters are of three types *viz.*, riverine form, coastal estuarine and brackish water form and marine. Among 31 cetaceans of Indian waters as many as 14 species of cetacean have been reported from the rivers, estuaries, brackish and coastal waters of Indian limits but only five species *viz.*, *Orcaella brevirostris, Sousa chinensis, Stenella attenuata, Platanista gangetica, Neophocaena phocaenoides* could only be treated as true wetland species. Though *Globicephala macrorhynchus* was reported from the Hugli near Srerampur and Salt Lake near Kolkata (Agrawal and Alfred, 1999; Chakraborty, 2004) but in the recent past it is not available in its earlier distributional range and the Salt Lake is now reclaimed partly and there is no connection with river Hugli or Bay of Bengal. Another 8 species *viz.*, *Orcinus orca, Pseudorca crassidens, Stenella longirostris, Balaena mysticetus, Balaenoptera edeni, B. musculus, Megaptera novaeangliae* are occasional or accidental visitors of the coastal wetlands and often seen to stranded on the coasts.

Family: Delphinidae

Species 2: *Delphinus delphis* Linnaeus; Common Name: Common Dolphin.

Found in the coastal waters of both Arabian Sea (Goa and Malabar coast) and Bay of Bengal (Chennai coast).

Conservation Status: IWPA- Schedule II; CITES- Appendix II; IUCN- LC

Species 3: *Globicephala macrorhynchus* Gray; Common Name: Indian Pilot Whale.

Found in Bay of Bengal in Indian limits and earlier reported from the river Hugli and Salt Lake, Kolkata. Blyth in the year 1850 collected one specimen from Salt Lake, North 24 Parganas (Chakraborty, 2004).

Conservation Status: IWPA- Schedule II; CITES- Appendix II; IUCN- DD

Species 4: *Orcinus orca* (Linnaeus); Common Name: Killer Whale.

The species was recorded only from Gujarat coast (Baroda) of Indian mainland and north of Andaman Island.

Conservation Status: IWPA- Schedule II; CITES- Appendix II; IUCN- DD

Table 10.1: List of Essentially Wetland Mammals of India

Sl.No.	Name	EC	WC	ES	BR	FW	END	Status IUCN (2012)
	Order: Sirenia							
	Family: Dugongidae							
1.	*Dugong dugon* (Muller)	+	+	–	–	–	–	VU
	Order: Cetacea							
	Family: Delphinidae							
2.	*Delphinus delphis* Linnaeus	+	+	–	–	–	–	LC
3.	*Globicephala macrorhynchus* Gray	+	–	–	–	–	–	DD
4.	*Orcinus orca* (Linnaeus)	–	+	–	–	–	–	DD
5.	*Pseudorca crassidens* (Owen)	–	+	–	–	–	–	DD
6.	*Orcaella brevirostris* (Gray)	+	–	+	+	–	–	VU
7.	*Sousa chinensis* (Osbeck)	+	+	+	–	+(?)	–	NT
8.	*Stenella longirostris* (Gray)	–	+	–	–	–	–	DD
9.	*Stenella attenuata* (Gray)	+	+	+	–	–	–	LC
	Family: Platanistidae							
10.	*Platanista gangetica* (Roxburgh)	–	–	+	–	+	–	EN
	Family: Balaenidae							
11.	*Balaena mysticetus* Linnaeus	–	+(?)	–	–	–	–	LC
	Family: Balaenopteridae							
12.	*Balaenoptera edeni* Anderson	+	+	–	–	–	–	DD
13.	*Balaenoptera musculus* (Linnaeus)	–	+	–	–	–	–	EN
14.	*Megaptera novaeangliae* (Borowski)	–	+(?)					LC
	Family: Phocoenidae							
15.	*Neophocaena phocaenoides* (G. Cuvier)	+	+	+	–	–	–	VU

EC: East Coast; WC: West Coast; ES: Estuary; BR: Brackish; FW: Freshwater; END: Endemic; EN: Endangered; VU: Vulnerable; NT: Near Threatened; DD: Data Deficient; LC: Least Concern.

Species 5: *Pseudorca crassidens* (Owen); Common Name: False Killer Whale.

A chiefly deep water oceanic species but in India reported from the coasts of Kozhikode, Thiruvananthapuram, Gulf of Cambay, Cape Comorin, Port Blair and Campbell Bay. Probably it was reported in a much deeper water depth than coastal wetland depth specified by IUCN.

Conservation Status: IWPA- Schedule II; CITES- Appendix II; IUCN- DD

Species 6: *Orcaella brevirostris* (Gray); Common Name: Irrawady Dolphin.

It prefers inshore water and estuaries and recorded from Visakhapatnam (Andhra Pradesh), Chilka Lake (Odisha), Hugli river and Sunderbans (West Bengal), Krishna river and estuary (Andhra Pradesh).

Conservation Status: IWPA- Schedule II; CITES- Appendix II; IUCN- VU

Species 7: *Sousa chinensis* (Osbeck); Common Name: Indo-Pacific Hump-backed Whale.

It is an inhabitant of coastal and inshore waters, recorded from NE Andaman, Visakhapatnam, Thiruvananthapuram, Kozhikode, Mumbai, and mouth of Hugli river and also from a tributary of Ganga river in Jalpaiguri district of West Bengal (Alfred *et al.*, 2002).

Conservation Status: IWPA- Schedule II; CITES- Appendix II; IUCN- NT

Species 8: *Stenella longirostris* (Gray), Common Name: Spinner Dolphin.

In Indian limits, it is reported from Malabar Coast and frequently caught by gill netting in southern India.

Conservation Status: IWPA- Schedule II; CITES- Appendix II; IUCN- DD

Species 9: *Stenella attenuata* (Gray); Common Name: Pantropical Spotted Dolphin or Malay Dolphin.

It moves freely in both coastal and offshore waters and in Indian limits, reported from Sunderban estuary of West Bengal.

Conservation Status: IWPA- Schedule II; CITES- Appendix II; IUCN- LC

Family: Platanistidae

Species 10: *Platanista gangetica* (Roxburgh); Common Name: Ganges River Dolphin.

It is a true wetland dolphin, reported from the rivers Ganga, Bhaghirathi, Gagra, Gandak, Jamuna, Kosi, Dudwa, Banas, Sone, Chambal, Hugli, Tista, Brahmaputra and its tributaries, and even in Dihang, Buri Dihang, Luhit and Kulsi of NE India.

Conservation Status: IWPA- Schedule I; CITES- Appendix I; IUCN- EN

Family: Balaenidae

Species 11: *Balaena mysticetus* Linnaeus; Common Name: Greenland Right Whale or Bowhead. This is not a coastal wetland species, recorded once from Gujarat coast (Alfred *et al.*, 2002).

Conservation Status: CITES- Appendix I; IUCN- LC

Family: Balaenopteridae

Species 12: *Balaenoptera edeni* Anderson; Common Name: Bryde's Whale.

It is found in both off and close to the shore in Arabian Sea and Bay of Bengal. An individual was stranded at Digha coast, East Medinipur district of West Bengal in the month of December 12, 2013 (Venkataraman *et al.*, 2015).

Conservation Status: IWPA- Schedule II; CITES- Appendix I; IUCN- DD

Species 13: *Balaenoptera musculus* (Linnaeus); Common Name: Blue Whale.

It is not a true coastal species but in Indian limits reported from the coasts of Mangalore, Cochin, Tuticorin, Kozhikode, South Kanara, Surat, Okha and Lakhswadweep. Most probably it was observed in much deeper water than the depth of specified coastal wetland. Recently, a 42 feet long individual was stranded alive at Revdanda Coast, 17 km south of Alibaug near Mumbai (Maharashtra) on June 24, 2015.

Conservation Status: IWPA- Schedule II; CITES- Appendix I; IUCN- EN

Species 14: *Megaptera novaeangliae* (Borowski); Common Name: Hump-back Whale.

It is also not a true coastal wetland species but there is a calving area at the western coast and reported from Thiruvananthapuram and Gulf of Mannar (Agrawal and Alfred, 1999).

Conservation Status: IWPA- Schedule II; CITES- Appendix I; IUCN- LC

Family: Phocoenidae

Species 15: *Neophocaena phocaenoides* (G. Cuvier); Common Name: Black Finless Porpoise.

It is found in the coastal waters and estuaries and reported from the coasts of Mumbai, Goa, North and South Kanara, Kozhikode and Thiruvananthapuram in western India. In eastern coast it is reported from Chennai and Hugli estuary up to Kolkata.

Conservation Status: IWPA- Schedule I; CITES- Appendix I; IUCN- VU

Wetland Dependent and Wetland Associated Mammals

The term wetland dependent and wetland associate are very much overlapping. In true sense the wetland dependent animals lives on wetland flora and fauna and seldom leave the habitat. But the foraging area of the wetland associated species is much higher than the dependent species as well as not fully dependent on wetland for their food and shelter. A wide variety of mammals occurs in Indian freshwater wetlands but none of them occurs in large numbers. It has already been mentioned that many mammalian species adjusted themselves in a particular wetland habitat that they are not regarded as wetland mammal in other habitat. Certain mammals *viz.*, tiger, cheetal, wild boar, rhesus macaque, leopard cat, Bengal fox, Indian grey mongoose, small Indian mongoose are almost wetland dependent or associate in Sunderban Mangals, while majority of them could not be regarded as wetland mammal in true sense. The mammals like elephant, gaur, sambar, and wild dog live as wetland associate along the vicinity of the Periyar Lake in the Periyar Wildlife Sanctuary but they could not be regarded as wetland mammals.

The wetlands of Keoladeo Ghana National Park represent 21 species of mammals (Alfred and Nandi, 2001) of which few ungulates *viz.*, blackbuck, cheetal, sambar, nilgai and certain carnivores like Bengal fox, Indian grey mongoose, small Indian mongoose, jackal are wetland associated species in this particular habitat but not so in other bio-geographic zones.

The Little Rann of Kutchch in its saline flats supports a sizeable population of the Indian wild ass, chinkara and blackbuck. As the distributional range of Indian wild ass is confined to the salt marsh of Little Rann thus it may be considered as wetland associate but in fact in summer months it moves usually around the thorny scrub jungle and in monsoon, when the Rann is flooded, it moves both through the marshes and highlands (Prater, 1971). Similarly the Kiang (*Equus kiang*) moves freely in the marshes and meadows around high altitude brackish lakes of Ladakh and Sikkim (Nandi, 2001) as well as in the drier regions also. Thus it may be considered as wetland associated species. Taking in to account all these anomalies, in this article, the mammals have been selected as wetland dependent/wetland associate as they spent either some part of their life in marshes or depend on wetland for their food, shelter and foraging area. Among the Indian land mammals only 23 species under 6 orders and 11 families could be considered as wetland dependent/associate. Brief accounts of their distributional range in India and conservation status are enumerated below.

Order: Carnivora

Family: Mustelidae

Species 1: *Aonyx cinereus* (Illiger); Common Name: Oriental Small-clawed Otter.

It commonly inhabits streams, rivers, lakes, flooded rice fields, creeks and estuaries. It is distributed in the states of Bihar, Himachal Pradesh, Karnataka, Kerala, Sikkim, Uttar Pradesh, West Bengal and NE India.

Conservation Status: IWPA- Schedule I; CITES- Appendix II; IUCN- VU

Species 2: *Lutra lutra* (Linnaeus); Common Name: Common Otter.

Mountain streams and lakes are its favourite haunts in hilly terrain. It is distributed from Jammu and Kashmir, Punjab and Himachal Pradesh east through Himalaya to NE States including West Bengal and Sikkim. It is also reported from Maharashtra in west and Pondicherry, Tamil Nadu and Karnataka in south (Hussain, 2013).

Conservation Status: IWPA- Schedule I; CITES- Appendix I; IUCN- NT

Species 3: *Lutrogale perspicillata* (I. Geoffroy Saint-Hilaire); Common Name: Smooth Coated Otter.

An otter of plains even adapted to live in dry Deccan Plateau and NE India. It is distributed all over the country except high Himalayas (Hussain, 2013).

Conservation Status: IWPA- Schedule II; CITES- Appendix II; IUCN- VU

Family: Felidae

Species 4: *Prionailurus viverrinus* (Bennett); Common Name: Fishing Cat.

It inhabits forests in the Himalaya, swamps, brackish lakes, estuaries and mangroves in plains. Found in the states of Andhra Pradesh, Assam, Karnataka, Kerala, Maharashtra, Odisha, Tamil Nadu, Uttar Pradesh and West Bengal.

Conservation Status: IWPA- Schedule I; IUCN- EN

Species 5: *Felis chaus* Schreber; Common Name: Jungle Cat.

It inhabits dry, open parts of the country, keeping more to grassland, scrub jungle and reedy banks of rivers and marshes. Mukherjee (2013) identified this cat as water dependent species inspite of its distribution in drier parts. It is found in Gujarat, Jammu and Kashmir, Madhya Pradesh, Rajasthan, Sikkim, Uttar Pradesh, West Bengal and in peninsular India south to Krishna river.

Conservation Status: IWPA- Schedule II; CITES- Appendix II; IUCN- LC

Species 6: *Prionailurus bengalensis* (Kerr); Common Name: Leopard Cat.

Mostly a forest cat but also found in marshes, mangroves and near villages also. It is distributed in north-western, central, NE states and West Bengal.

Conservation Status: IWPA- Schedule I; CITES- Appendix I; IUCN- LC

Family: Herpestidae

Species 7: *Herpestes urva* (Hodgson); Common Name: Crab-eating Mongoose.

A very good swimmer and diver, more aquatic than any other mongoose. It is found in NE states and West Bengal.

Conservation Status: IWPA- Schedule IV; CITES- Appendix III; IUCN- LC

Species 8: *Herpestes vitticollis* (Bennett); Common Name: Striped-necked Mongoose.

A typical forest animal haunts by the banks of rivers and frequenting swamps and flooded rice fields. It is distributed only in Western Ghats in parts of Kerala, Karnataka and Tamil Nadu.

Conservation Status: IWPA- Schedule IV; CITES- Appendix III; IUCN- LC

Species 9: *Herpestes javanicus palustris* Ghose; Common Name: Indian Marsh Mongoose.

This species is reported to be endemic and restricted to the wetlands of West Bengal in eastern India (Mudappa, 2013). More precisely it is restricted in the marshes of Howrah, North and South 24 Parganas (Agrawal *et al.*, 1992).

Conservation Status: IWPA- Schedule IV; CITES- Appendix III (*javanicus*); IUCN- EN

Order: Insectivora

Family: Soricidae

Species 10: *Chimarrogale himalayica* (Gray); Common Name: Himalayan Water Shrew.

It inhabits montane rivers and streams of Himalaya in the states of Himachal Pradesh, Jammu and Kashmir, Sikkim, Uttar Pradesh and West Bengal.

Conservation Status: IUCN- LC

Species 11: *Nectogale elegans* Milne-Edwards; Common Name: Sikkim Water Shrew.

This shrew is fully adapted to aquatic life, lives in montane rivers and streams at altitude of 900 -3000m (Corbet and Hill, 1992); distributed only in Sikkim (Alfred and Chakraborty, 2002; Jiang and Hoffman, 2013) in Indian limit. Wroughton (1916) reported the species from the Darjeeling district of West Bengal.

Conservation Status: IUCN- LC

Species 12: *Feroculus feroculus* (Kelaart); Common Name: Kelaart's Long-clawed Shrew.

It inhabits hilly swamps and marshes upto 2400 m in southern India on the border of Kerala and Tamil Nadu and possibly distributed upto Palni Hills (Jiang and Hoffman, 2013).

Conservation Status: IUCN- EN

Order: Primates

Family: Cercopithecidae

Species 13: *Macaca fascicularis* (Raffles); Common Name: Crab-eating Macaque or Nicobar Long-tailed Macaque.

It lives in Katchal and Nicobar group of islands in the swamps and often visits creeks and seashore.

Conservation Status: IWPA- Schedule I; CITES- Appendix II; IUCN- LC

Order: Perissodactyla

Family: Equidae

Species 14: *Equus hemionus* Pallas; Common Name: Indian Wild Ass.

It lives in the grass covered (*bets*) salt marshes of Little Rann of Kutchch and when the *bets* are inundated during monsoon, it moves to higher elevation and also wading freely from one island to another.

Conservation Status: IWPA- Schedule I; CITES- Appendix I; IUCN- EN

Species 15: *Equus kiang* Moorcroft; Common Name: Kiang.

It lives in the vicinity of brackish water lakes of Ladakh (Jammu and Kashmir) and Sikkim.

Conservation Status: IWPA- Schedule I; CITES- Appendix II; IUCN- LC

Family: Rhinocerotidae

Species 16: *Rhinoceros unicornis* Linnaeus; Common Name: Great One-horned Rhinoceros.

It prefers swamp and grass jungles but also moves to wood jungles up ravines and low hills. Found in Assam and West Bengal and introduced in Dudhwa National Park, Uttar Pradesh.

Conservation Status: IWPA- Schedule I; CITES- Appendix I; IUCN- VU

Order: Artiodactyla

Family: Suidae

Species 17: *Porcula salvania* Hodgson; Common Name: Pygmy Hog.

It is distributed in Assam *Terai* in swampy or riverside grass jungles; extremely rare.

Conservation Status: IWPA- Schedule I; CITES- Appendix I; IUCN- CR

Species 18: *Sus scrofa* Linnaeus; Common Name: Wild Boar.

It is distributed throughout India. In true sense not a wetland species but moves freely in swamps and riverside grass jungles but the subspecies of Andaman and Nicobar islands prefers to live near swamps.

Conservation Status: IWPA- Schedule I for *S.s. andamanensis,* others Schedule III; IUCN- LC

Family: Cervidae

Species 19: *Axis porcinus* (Zimmermann); Common Name: Hog Deer.

It prefers to live in grass jungles by the river banks and open grass plains adjacent to swamp and distributed from Himachal Pradesh, Punjab through Uttar Pradesh east to NE States and West Bengal.

Conservation Status: IWPA- Schedule III; IUCN- EN

Species 20: *Rucervus duvaucelii* (G. Cuvier); Common Name: Swamp Deer or Barasingha.

The north Indian race *duvauceli* is completely swamp dwelling and seldom seen out of water which is distributed north of river Ganges in the *Terai* forest of Uttar Pradesh. The NE race *ranjithsinhi* is distributed in the Brahmaputra plains of Assam in the proximity of water. Another race *branderi* is distributed south of river Ganges in Madhya Pradesh and prefers to live in open grassland and not a wetland species.

Conservation Status: IWPA- Schedule I; CITES- Appendix I; IUCN- VU

Table 10.2: List of Wetland Dependent/Associate Mammals of India

Sl.No.	*Name*	*WD*	*WA*	*END*	*Status IUCN (2012)*
	Order: Carnivora				
	Family: Mustelidae				
1.	*Aonyx cinereus* (Illiger)	+	–	–	VU
2.	*Lutra lutra* (Linnaeus)	+	–	–	NT
3.	*Lutrogale perspicillata* (I. Geoffroy Saint-Hilaire)	+	–	–	VU
	Family: Felidae				
4.	*Prionailurus viverrinus* (Bennett)	+	–	–	EN
5.	*Felis chaus* Schreber	–	+	–	LC
6.	*Prionailurus bengalensis* (Kerr)	–	+		LC
	Family: Herpestidae				
7.	*Herpestes urva* (Hodgson)	+	–	–	LC
8.	*Herpestes vitticollis* (Bennett)	+	–	–	LC
9.	*Herpestes javanicus palustris* Ghose	+	–	+	EN
	Order: Insectivora				
	Family: Soricidae				
10.	*Chimarrogale himalayica* (Gray)	+	–	–	LC
11.	*Nectogale elegans* Milne-Edwards	+	–	–	LC
12.	*Feroculus feroculus* (Kelaart)	+	–	–	EN
	Order: Primates				
	Family: Cercopithecidae				
13.	*Macaca fascicularis* (Raffles)	+	–	–	LC
	Order: Perissodactyla				
	Family: Equidae				
14.	*Equus hemionus* Pallas	–	+	–	EN
15.	*Equus kiang* Moorcroft	–	+	–	LC
	Family: Rhinocerotidae				
16.	*Rhinoceros unicornis* Linnaeus	–	+	–	VU
	Order: Artiodactyla				
	Family: Suidae				
17.	*Porcula salvania* Hodgson	–	+	–	CR
18.	*Sus scrofa* Linnaeus	–	+	–	LC
	Family: Cervidae				
19.	*Aix porcinus* (Zimmermann)	–	+	–	EN
20.	*Rucervus duvaucelii* (G. Cuvier)	+	+	–	VU
21.	*Rucervus eldii* (McClelland)	+	–	–	EN
	Family: Bovidae				
22.	*Bubalus arnee* (Kerr)	–	+		EN
	Order: Rodentia				
	Family: Muridae				
23.	*Bandicota indica* (Bechstein)	+	+	–	LC

WD: Wetland Dependent; WA: Wetland Associate; END: Endemic; CR: Critical; EN: Endangered; VU: Vulnerable; NT: Near Threatened; LC: Least Concern.

Species 21: *Rucervus eldii eldii* (McClelland); Common Name: Manipur Brow-antlered Deer.

Only a small herd finds shelter in the floating swamps (*Phumids*) of Loktak Lake in Manipur.

Conservation Status: IWPA- Schedule I; CITES- Appendix I; IUCN- EN

Family: Bovidae

Species 22: *Bubalus arnee* (Kerr); Common Name: Water Buffalo or Wild Asian Buffalo.

It prefers to live in tall grass jungles with reed beds adjacent to swamp or river banks or river flats and distributed in Brahmaputra valley of Assam, also in Arunachal Pradesh and Meghalaya and probably in West Bengal; south of Ganges in the Bastar district of Chhattishgarh, adjoining Odisha and Maharashtra.

Conservation Status: IWPA- Schedule I; CITES- Appendix III; IUCN- EN

Order: Rodentia

Family: Muridae

Species 23: *Bandicota indica* (Bechstein); Common Name: Large Bandicoot Rat. It is distributed almost throughout India except high Himalaya. The population of southern West Bengal is much water dependent than the population of any other region and makes burrows and lives at the banks of tanks (Chakraborty, 1991, 1992). This population takes aquatic molluscs as their primary food and crabs and fish also (Chakraborty and Chakraborty, 1999). The population other than southern West Bengal takes usual food like other rats but fond of animal food and prefers to live near water and even in the sewage system in the cities.

Conservation Status: IWPA- Schedule V; IUCN- LC

Situation in West Bengal

In West Bengal, about 54 natural and nine manmade wetlands are there (Biswas and Trishal, 1993) and in addition to these there are numerous small water bodies including ponds, puddles etc. The two categories combine to cover an area of about 344527 ha which is almost 8.5 percent of the total wetland area of India (Sanyal *et al.*, 2012). So far, 21 mammalian species under 6 orders and 13 families (Table 10.3) are identified as wetland mammals in West Bengal which is 5 percent of the total mammalian species of India and 55.26 percent of the total wetland species of the country. The rich wetland mammalian diversity in the state is due to its varied physiographic condition which includes Eastern Himalayan region at the north, vast Gangetic Plain from the south of the Himalayan region, western highland, the eastern extension of the Chhotanagpur Plateau and distinct deltaic region with Sunderban mangroves at the southern most area. Nandi *et al.* (2013) enlisted 17 wetland species in West Bengal which included *Nectogale elegans* also. But Alfred *et al.* (2002) and Jiang and Hoffman (2013) did not confirm their presence in West Bengal. Long back, Wroughton (1916) reported the species from Darjeeling Himalaya

Table 10.3: List of Wetland Mammals of West Bengal

Sl.No.	*Name*	*Status (IUCN) 2012*	*EWS*	*WD*	*WA*	*District-wise Distribution*
	Order: Cetacea					
	Family: Delphinidae					
1.	*Globicephala macrorhyncha* Gray	DD	+	–	–	14, 16
2.	*Orcaella brevirostris* (Gray)	VU	+	–	–	16
3.	*Sousa chinensis* (Osbeck)	NT	+	–	–	8, 16
4.	*Stenella attenuata* (Gray)	LC	+	–	–	16
	Family: Platanistidae					
5.	*Platanista gangetica* (Roxburgh)	EN	+	–	–	4,6,7,12,13,14,16
	Family: Balaenopteridae					
6.	*Balaenoptera edeni* Anderson	DD	+	–	–	11
	Family: Phocoenidae					
7.	*Neophocaena phocaenoides* (Cuvier)	VU	+	–	–	16
	Order: Carnivora					
	Family: Mustelidae					
8.	*Aonyx cinereus* (Illiger)	VU	–	+	–	4,5,7,8,14,16
9.	*Lutra lutra* (Linnaeus)	NT	–	+	–	1,4,5,7,8,12,14,16
10.	*Lutrogale perspicillata* (I. Geoffroy Saint-Hilaire)	VU	–	+	–	3,4,8,14,16
	Family: Felidae					
11.	*Prionailurus viverrinus* (Bennett)	EN	–	+	–	4,6,8,14,16
12.	*Felis chaus* Schreber	LC	–	–	+	1-17
13.	*Prionailurus bengalensis* (Kerr)	LC	–	–	+	1-3,5-10,14-17

Contd...

Table 10.3–*Contd...*

Sl.No.	*Name*	*Status (IUCN) 2012*	*EWS*	*WD*	*WA*	*District-wise Distribution*
	Family: Herpestidae					
14.	*Herpestes urva* (Hodgson)	LC	–	+	–	5,8
15.	*Herpestes palustris* Ghose	EN	–	+	–	6,14,16
	Order: Insectivora					
	Family: Soricidae					
16.	*Chimarrogale himalayica* (Gray)	LC	–	+	–	5
	Order: Perissodactyla					
	Family: Rhinocerotidae					
17.	*Rhinoceros unicornis* Linnaeus	VU			+	8
	Order: Artiodactyla					
	Family: Suidae					
18.	*Sus scrofa* Linnaeus	LC	–	–	+	4,8,13-16
	Family: Cervidae					
19.	*Aix porcinus* (Zimmermann)	EN	–	–	+	5,8
	Family: Bovidae					
20.	*Bubalus arnee* Kerr	EN	–	–	+	8
	Order: Rodentia					
	Family: Muridae					
21.	*Bandicota indica* (Bechstein)	LC	–	+	+	1-4,6-17

EWS: Essentially Wetland Species; WD: Wetland Dependent; WA: Wetland Associate.

District Code: 1: Bankura; 2: Bardhaman; 3: Birbhum; 4: Kolkata; 5: Darjiling; 6: Haora; 7: Hugli; 8: Jalpaiguri; 9: Koch Bihar; 10: Maldah; 11: Medinipur (East and West); 12: Mursidabad; 13: Nadia; 14: North 24-Parganas; 15: Purulia; 16: South 24-Parganas; 17: Dinajpur (North and South).

and Agarwal *et al.* (1992) included the species in West Bengal fauna. During the last 100 years there is no record of this species from Darjeeling Himalayan region, while the species are found in Sikkim Himalaya adjacent to Darjeeling Himalaya. Thus, the same has not been included in the wetland fauna of West Bengal. Though Indian elephant spent some time of a day along the marshes or rivers in North Bengal forests but not dependent for food etc. on those wetlands. Thus it could not be considered as wetland mammal. Maximum numbers of wetland mammals are found in South 24 Parganas followed by North 24 Parganas and Jalpaiguri districts (Table 10.3). Among the wetland mammals of India about 50 percent are threatened (Tables 10.1 and 10.2) and the same is little more in West Bengal *i.e.* 57 percent (Alfred *et al.*, 2001, 2006). As regards to the endemicity, there is only one wetland mammal (*Herpestes javanicus palustris*) distributed in certain niches of West Bengal. Thus conservation of those habitats is a must for conserving the species. It is obvious that more wetland species were there in the last century even. Four charismatic mammals *viz.*, *Bubalus arnee*, *Cervus duvaucelii*, *Rhinoceros javanicus* and *Muntiacus muntjac* were there in Suderbans even in the 19th century.

Threats

The main threats to the wetland mammals are encroachment of wetlands for agriculture and human settlement, damming in the upper stretches of the rivers, land erosion in the hilly areas, siltation and pollution load, which leads to depletion of faunal diversity including mammalian fauna for the lack of their shelter, food, breeding and foraging ground.

Conservation

For conservation of wetland resources including its mammalian fauna certain steps have already been taken up.

- ☆ Declaration of Protected areas like National Park, Sanctuaries, Biosphere Reserves.
- ☆ Identification of Ramsar Site.
- ☆ Establishment, notification and proper implementation of Indian Wildlife (Protection) Act, 1972.
- ☆ Signing the CITES treaty.
- ☆ Establishment of Wetland Wing with Department of Environment and Forests and Climate Change under Government of India.
- ☆ Establishment of Wetland Wing with State Biodiversity Board.
- ☆ Species oriented conservation programme.

Inspite of the above steps the most essential is the public awareness and effective legislation practice. Until and unless the people's participation and political willingness for conservation takes place nothing could be fruitful and more and more species will be enlisted under 'Extinct' category.

Summary

Wetland mammals reside permanently or temporarily in aquatic habitat for food, shelter and foraging ground. In India, there are altogether 38 wetland mammals so far been identified of which 15 are essentially wetland mammal, another 23 are either wetland dependent or associate. Among the 15 essentially wetland or aquatic mammals, 1 belongs to the order Sirenia under the family Dugongidae and another 14 are grouped under 5 families of the order Cetacea. Among 38 Indian wetland mammals, 21 species is distributed in West Bengal which is 55.26 percent of the total wetland mammal and 5 percent of Indian mammalian fauna. So far, only 1 endemic wetland mammal has been recognized in India which is distributed in West Bengal only. About 50 percent of Indian wetland mammals are threatened as per IUCN (2012) which is little more *i.e.* 57 percent in West Bengal.

References

Agrawal VC, Das PK, Chakraborty S. *et al.*, 1992. Mammalia. In *Fauna of West Bengal, Part 1. State Fauna Series 3*. Zoological Survey of India, Kolkata. pp. 27-169.

Agrawal VC and Alfred JRB. 1999. *Handbook of Whales, Dolphins and Dugong from Indian Seas*, i-iv, Zoological Survey of India, Kolkata. pp. 1-150.

Alfred JRB and Nandi NC. 2000. Faunal Diversity in Indian Wetlands, *ENVIS Newsl*, Zoological Survey of India, 6(2).

Alfred JRB, Das AK and Sanyal AK. 2001. *Ecosystem of India*, ENVIS Centre, Zoological Survey of India Kolkata. pp. 1- 410.

Alfred JRB and Nandi NC. 2001. Wetlands: Freshwater. In. *Ecosystem of India*, (Alfred JRB, Das AK and Sanyal AK. eds). ENVIS Centre, Zoological Survey of India, Kolkata. pp. 165-193.

Alfred JRB, Sinha NK and Chakraborty S. 2002. Checklist of Mammals of India. *Rec. zool. Surv. India*, Occ. Paper No. 199 : 1-289.

Alfred JRB and Chakraborty S. 2002. Endemic mammals of India. *Rec. zool. Surv. India, Occ. Paper* No. 201: 1-37.

Alfred JRB, Das AK and Sanyal AK. 2006. *Animals of India: Mammals*. ENVIS Centre, Zoological Survey of India, Kolkata. pp. 236.

Biswas DK and Trishal CL. 1993. Initiatives for conservation of wetlands in India, In. *Biodiversity Conservation: Forests, Wetlands and Deserts* (Frame B, Victor J and Joshi Y. eds.), New Delhi. pp. 47-53.

Chakraborty R. 1991. Burrow system of the Large Bandicoot Rat, *Bandicota indica* (Bechstein) (Rodentia : Muridae) in West Bengal. *Rec. Zool. Surv. India*, 89(1-4): 105-122.

Chakraborty R. 1992. Large Bandicoot Rat, *Bandicota indica* (Bechstein). In. *Rodents in Indian Agriculture*, Vol. 1. (Prakash I and Ghosh PK. eds.) Scientific Publishers, Jodhpur.

Chakraborty R and Chakraborty S. 1999. Feeding behaviour of the Large Bandicoot Rat, *Bandicota indica* (Bechstein) [Rodentia : Muridae]. *Rec. Zool. Surv. India*, 97(2): 45-72.

Chakraborty R. 2004. Catalogue of Mammalian Exhibits of the Indian Museum. *Rec. Zool. Surv. India*, Occ. Paper, 219 : 1-99.

Corbett GB and Hill JE. 1992. *The mammals of Indo-Malayan Region. A Systematic Review.* Oxford University Press, Oxford.

Hussain SA. 2013. Otters. In. *Mammals of South Asia* (Johnsingh AJT and Manjrekar N. eds.). pp. 499-521.

IUCN 2012. IUCN Red List of Threatened Species, IUCN, Gland, Switzerland.

Jiang X and Hoffman R. 2013. Insectivores. In. *Mammals of South Asia* (Johnsingh AJT and Manjrekar N. eds.). pp. 1-51.

Mudappa D. 2013. Herpestids, Viverrids and Mustelids. In. *Mammals of South Asia* (Johnsingh AJT and Manjrekar N. eds.). pp. 471-498.

Mukherjee S. 2013. Small Cats. In : *Mammals of South Asia* (Johnsingh AJT and Manjrekar N. eds.). pp. 531-540.

Nandi NC. 2001. Wetlands: Brackishwater. In. *Ecosystem of India*, (Alfred JRB, Das AK and Sanyal AK. eds). ENVIS Centre, Zoological Survey of India, Kolkata. pp. 195 -217.

Nandi NC, Das AK and Dey A. 2013. *Wetland Faunal Diversity in West Bengal*. West Bengal Biodiversity Board and Nature Book India, New Delhi. pp. 1-256.

Prater SH. 1971. *The Book of Indian Animals.* 3rd (Revised) Edition, Bombay Natural History Society. pp. 1-324.

Sanyal AK, and Alfred JRB, Venkataraman K. *et al.*, 2012. *Status of Biodiversity of West Bengal*. Zoological Survey of India, Kolkata. pp. 1-969.

Venkataraman K, Chattopadhya A and De JK. 2015. Report on the occurrence of a Brayde's Whale on Digha coast, West Bengal. *ENVIS Newsl*, Zoological Survey of India, Vol. 21: 12-14.

Wroughton RC. 1916. Scientific Results from the mammal Survey. *J. Bombay Nat. Hist. Soc.*, 24: 644-649.

Chapter 11

Indian Sundarban: Faunal Diversity and their Ecological Adjustment

Pranabes Sanyal

School of Oceanographic Studies, Jadavpur University, Kolkata

Introduction

Sundarban - the land of beauty, is an abode for diverse life forms and at the same time one of the worst victims of global climate change. It is the largest contiguous tidal mangrove forest in the world covering two neighboring countries – India and Bangladesh (Blasco and Aizpuru, 1997). The major part of the forest (6,000 km^2) lies in Bangladesh and the rest (4,263 km^2) in the state of West Bengal in India. The Indian part of Sundarban spreads across 104 islands with an area of 9,630 km^2, bounded by river Hooghly on the west to Ichamati-Raimangal in the east, Bay of Bengal in the south and Dampier-Hodges line in the north and is spread over the two districts namely, South 24-Parganas and North 24 Parganas. It also includes about 5000 km^2 of reclaimed land over 52 islands. The name Sundarban may have been derived from the "Sundari" (*Heritiera fomes*) trees that are found in this forest as climax species. The geographic location of Sundarban is between 21°31′N and 22° 31′N latitudes and 88°10′ E and 89°51′E longitudes. It lies within the vast delta on the Bay of Bengal formed by the confluence of Ganga, Brahmaputra and Meghna rivers across southern Bangladesh and West Bengal, India and is the largest 'prograding' delta of the world. It is intersected by a complex network of tidal waterways, mudflats and small islands of salt-tolerant mangrove forests. As part of the world's largest delta, formed from sediments deposited by three great rivers and covering the Bengal Basin, the land has been moulded by tidal action, resulting in a distinctive physiography. A major portion of the Indian Sundarban (2,585 km^2) has been designated as Sundarban Tiger Reserve in the year 1973 under "Project Tiger" by the Ministry of Environment and Forest, Government of India and was declared as a World Heritage Site by UNESCO in 1987. It was declared as a Biosphere Reserve on the recommendation of the Man and Biosphere Committee by UNESCO in 1989.

Unique Features

Sundarban possess several unique features which are all of standalone importance.

- It is the only mangrove tiger-land of the planet.
- Largest mangal diversity of the globe having 95 intertidal species.
- Largest prograding delta of the earth.
- Home of some unusual species which are normally found at the foothills like small clawed otter (*Aonyx cinerea*) and leopard cat (*Prionailurus bengalensis*).
- Contributes to the fisheries of entire eastern Indian coast.
- Purifies the untreated mixed sewage load of 1100 MLD from Kolkata and surrounding areas (Akhand, 2012).
- Sundarban protects the metropolis of Kolkata and suburbs from the fierce annual high gale from Bay of Bengal.
- The forest also stabilizes the coastal belt and protects the low-lying country from tidal surges, cyclones and other natural calamities. Research has indicated that mangroves can absorb 30-40 per cent of total force of a cyclone and reduce the force of powerful waves before crashing inland.

Sundarban provides a fascinating maze of on-going ecological processes as it represents the process of delta formation and the subsequent colonization of the newly formed deltaic islands and associated mangrove communities. These processes include monsoon rains, flooding, delta formation, tidal influence and plant colonization. Ecologically this unique Biosphere Reserve includes Sundarban National Park (1700 km^2), 3 Sanctuaries *viz.* Sajnekhali Wildlife Sanctuary (362 km^2), Halide Island Wildlife Sanctuary (6 km^2) and Lothian Island Wildlife Sanctuary (38 km^2) and are notable for its rich biodiversity and to provide vital ecological services and goods. It is estimated that the Sundarban is home to 95 intertidal mangal plants constituting 63 per cent of the total mangrove forests of India containing more than 10 percent of mammals and 25 percent of birds found in India, together with the emblematic species - the Bengal Tiger.

Floral Diversity

Mangroves can be considered aquatic when they are partially submerged in water during high tide.They are able to tolerate salinity and tidal waves because of the unique adaptive features like presence of stilt roots, root buttress, pneumatophores, thick leathery wax coated leaves to reduce transpiration. Most of them are well adapted both for land as well as aquatic habitats. The Indian Sundarban is the most viable gene pool of mangal species of the world having 95 Intertidal species (Sanyal, 2008) of which two species the 'Sundari' (*Heritiera fomes*) and 'Golpata' (*Nypa fruticans*) are included in IUCN Red Data Book as 'Endangered', while another 10 species are included under 'Vulnerable' category. The Plant 'Chakkaora' (*Sonneratia caseolaris*) is one of the seven edible mangrove fruit trees, having special capacity of purifying the sewage borne metallic pollutants (Akhand, 2012), its nutritive value being better than that of common apple and called 'Mangrove Apple' (Haldar, 2015). There are edible high protein algae growing on mudflats (*Ulva lactuca*) and ponds (*Enteromorpha intestinalis*). Its exceptional biodiversity includes a wide range of flora; 334 plant species belonging to 245 genera and 75 families, 165 algae and 13 orchid species as well as about a dozen of non-orchid non-fern epiphytes (Barik *et al.*, 2011). Naturally occurring plant named 'Hargoza' (*Acanthus ilicifolius*) are used to cures tiger bite injury.

Faunal Diversity

The tidal mangrove forests is in reality a mosaic of islands of different shapes and sizes, perennially washed by brackish water, shrilling in and around the endless and mind-boggling labyrinths of water channels and supports exceptional biodiversity in its terrestrial, aquatic and marine habitats, ranging from micro to macro flora and fauna. Sundarban is of universal importance for globally endangered species including the royal Bengal tiger, Ganges and Irrawaddy dolphins, estuarine crocodiles and the critically endangered and endemic river terrapin (*Batagur baska*). Sundarban supports an exceptional level of biodiversity in faunal assemblage too in both the terrestrial and marine environment, including significant population of globally endangered cat species, the royal Bengal tiger. Notable faunal diversity of Sundarban was last assessed by the Zoological Survey of India to be 1586 species including 26 endangered ones. Horseshoe crab (*Carcinoscorpius rotundicauda* and *Tachypleus gigas*) - the primitive endangered invertebrate with great medicinal value are readily available in this unique habitat. They produce lectin which is considered to be the base material for the preparation of life saving medicine to treat leukemia. The Horseshoe lysate prepared from their primitive blue blood are used to determine the expiry dates of medicines. The first report of the Hemichordate - *Saccoglossus* from mangroves was also made from Sundarban (Sanyal, 2012).

Mammals

Sundarban is the only Tiger Reserve of India which does not harbour leopards. However, it is home to 3 more cats, the endangered fishing cat (*Prionailurus viverrina*), leopard cat (*P. bengalensis*) and jungle cat (*Felis chaus*). Presence of small clawed otter (*Aonyx cinerea*) and the critically endangered hill species smooth coated otter (*Lutrogale perspicillata*) are noticed from this area. Most common land mammals mangrove habitat are chital, wild boar, rhesus monkey, jackal and palm civet. Among aquatic mammals of Irrawaddy dolphin *Orcella brevirostris* is common which replaced commonly seen Gangetic dolphin *Platanista gangetica* due to increasing river water salinity. Tiger and dolphin are considered as the indicator species of two contrasting environments, the land and water respectively.

Reptiles

A total of 59 species of reptiles have been recorded from the area. The largest estuarine crocodile *Crocodylus porosus* which is often reach more than 20 ft length is the remarkable one. Salvator lizard (*Varanus salvator*) the largest monitor lizard of India is also available. The endangered Indian python (*Python molurus*) has become a bit uncommon now a days in the Indian side, but recently in Bangladesh it was photographed gripping a spotted deer. Large sized king cobra is not uncommon and is even devoured by tiger. The endemic reptile, hard shelled river terrapin (*Batagur baska*) is found at the south eastern corner of Indian Sundarban at Mechua estuary and its breeding programme has been very successful. Besides, four species of marine turtle including the most beautiful hawksbill turtle (*Eretmochelys imbricata*) along with olive ridley turtle, green turtle, and logger head turtle are also documented.

Amphibians

Annandale (1907) reported only 3 species of amphibians from Sundarban, but now 8 species have been found. These are skipper frog (*Euphlyctis cyanophlyctis*), Six toad green pond frog (*E. hexadactylus*), bull frog (*Hoplobatrachus tigrina*), cricket frog (*Fejervarya limnocharis limnocharis*), tree frog (*Polypedates maculatus*), narrow mouthed frog (*Microhyla ornata*), grey balloon frog (*Uperodon globulosus*) and common Indian toad (*Duttaphrynus melanostictus*)

Birds

A total of 217 species of birds have been reported from Indian Sundarban. The list includes giant sized birds like, goliath heron (*Ardea goliath*), white bellied sea eagle, lesser adjutant stork etc. Besides, 52 species of migratory birds and 8 species of kingfishers are abound in Sundarban. The elusive spoon bill sandpiper was last reported near Kakdwip 10 years back.

Vulnerability

As a result of increased vulnerability of this unique ecosystem, the level of salinity in both surface water and ground water system is increasing. This increased salinity could have multiple impacts on the socio-economic development potential of the islands, including that on agriculture and fisheries. A changing salinity in the ecosystem could also impact both structure and distributional pattern of flora and fauna with cascading implications for conservation of biodiversity in the mangrove ecosystem. During the last century, important mammalian species like java rhinoceros, great Indian one horned rhinoceros, soft ground barasinga deer, hog deer, wild buffalo and gharial became extinct from Sundarban. The barking deer population became restricted to Hallide Island till eighties and became extinct thereafter. However, salinity of the water bodies has continued to decrease towards the east. This shift is due to tectonic subsidence of the Bengal basin between the 10th and 12th centuries, as well as another neo-tectonic movement during 16th century and the continued gradual eastward tilting of the underlying crust. The average salinity of water and soil therefore, decreases markedly from west to east. The area has three main hydrological zones, 1) Brackish (Oligohaline) with salinity level less than 6250 micromhos having sundari (*Heritiera fomes*) as the dominant species, 2) Moderately Saline (Mesohaline) with salinity level between 6250 and 12500 micromhos having gewa (*Excoecaria agallocha*) as the principal species and 3) Saline (Polyhaline) having goran (*Ceriops decandra*) as the dominant species.

Agricultural practices in Sundarban is a problematic issue. Productivity of crop never exceeds 2.5 ton/ha while the Indian average is 3.5 ton/ha. In winter, irrigation was possible only in 30 per cent of lands in the past and that has been dwindling day by day due to rising salinity of ground water. The age old system of irrigation are of digging shallow (3m deep) linear canals storing the rain water. Pond irrigation is also difficult since they dry up after monsoon and are often stalinized by inundation caused due to storm surges. Post *Aila* situation of 2009 turned to be a menace on this count. As such, chili is a major winter crop over most of the Sundarban, while water melon and betel leaf are successful at sandy locations (Sanyal, 1997).

Sundarban is one of the major global estuaries, which are likely to be badly affected by the accelerated rate of sea-level rise, as a direct consequence of climate change. John Milliman (1989) first pointed this aspect through his landmark publication in the journal "AMBIO". Now in the report of IPCC the concern on the effect of accelerated sea-level rise in Sundarban has been elaborately expressed. Apart from this, recently the tiger straying in villages has increased many folds due to the increased salinity of creek waters in the coastal islands. Tigers drink creek water, so they are migrating towards north near inhabited islands to become allured by the village cattle comparatively easier and concentrated more to prey upon.

The huge sediment load (1570 million tons/year) carried by Ganga-Bramhaputra river system to Sundarban (McDowell, 1995) causes constant delta subsidence and many ascribe this as the primary reason for considering rate of sea level rise in Sundarban as higher than the global average rate of rise (2mm per year). In the Indian Sundarban the rate of sea-level rise is 3.14 mm/y at Sagar and 5 mm/y at Pakhiralay. As a result, many families of already submerged Islands like Lohachara and Supuribhanga have become climate refugees. The Ghoramara Island has also been submerged more than 80 per cent with the remnant population precariously surviving under the real threat of losing their homeland. Some 7500 climate refugee families have settled on Sagar Island so far.

Conclusion

Sunderban is susceptible to the risks of rise in sea level, enhanced salinity level, and vagaries of extreme weather uncertainties as a result of climate change. Besides, the demographic load (824/sq km) has been exerting too great a pressure on Sundarban ecosystem to overcome the ill effects of climate change looming large before us. The unique biodiversity of this region will be seriously affected unless the various projects including Project Tiger and Sundarban Development Projects are being implemented properly so as to overcome the ecological imbalance being caused by climate change.

Summary

Sundarban is the largest contiguous mangrove chunk on globe and is the only "Mangrove Tiger land" of the world. Mangal diversity of flora is also towering with 95 intertidal species and 1586 faunal species. It is presently the only abode of critically endangered hard shelled river terrapin *Batagur baska* and the breeding ground of the endangered and high value bio-medically important living fossil horseshoe crab

(*Tachypleus gigas* and *Carcinoscorpius rotundicauda*). At the same time, Sundarban delta is one of the worst victims of climate change. The unique biodiversity will be seriously affected unless proper attention will be paid to mitigate the problems with much sincerity and devotion.

Acknowledgments

I sincerely acknowledge the contributions of my scholars Dr. A. Akhand, and Smt. J. Barik for their technical help and cooperation.

References

Akhand A, Chanda A, Sanyal P and Hazra S. 2012. Pollution loads of four heavy metals in water, sediment and benthic organisms in the Kulti river of Sundarban fed by the Metropolitan sewage. *Jr. Nature Environment and Pollution Technology.*, 11(2) : 153-156.

Annandale N. 1907. Notes on the freshwater Fauna of India. IX. Description of new Freshwater sponges from Calcutta, with a record of two known species from the Himalayas and a list of the Indian forms. *J. Proc. Asiat. Soc. Beng.*, 3: 15-26.

Barik J, Sanyal P and Hazra S. 2011. Sustainable development of algal culture in relation to socio-economy in Sundarbans. *Jr. Netaji Institute for Asian Studies.* XXIX (2): 1-8.

Blasco F and Aizpuru M. 1997. Classification and evolution of mangroves of India. *Tropical Ecology*, 38(2) : 357-374.

Milliman J, Broadus DJM and Frank G. 1989. Environmental and economic implications of rising sea level and subsiding deltas. The Nile and Bengal examples. *Ambio*,18(6) : 340-345.

McDowell MC. 1995. *The Development of the river Hooghly for navigation.* Commemoration volume of Port of Calcutta, 125 years. pp. 69.

Halder R, Ghosh P, Chatterji S and Sanyal P. 2015. Assessment of nutritional potential and food safety evaluation in mice and rat by the fruit of mangrove plant, *Sonneratia apetala* of Sundarban, India. *International Journal of Advanced Research*, 3: 363-383.

Sanyal P.1997. Land-use dynamics of Bengal coast. *Jr. Banabithi*, June: 15-18.

Sanyal P, Mukhopadhyay A and Das I. 2008. Sundarban-the greatest mangal diversity of the planet. *J. Ind. Soc. Coastal Agric. Res.*, 26(1): 58-62.

Sanyal P. 2012. Climate change affecting wildlife with special reference to West Bengal. In. *Coping With Disasters* (Desai M and Haldar S eds). Abhijit Publications, New Delhi. pp. 269-276.

Chapter 12

Bioresources and Socio-economic Developmental Plans (SDPs) in Wetland Ecosystem: A Case Study

Debnath Palit[1], Ambarish Mukherjee[2] and Santanu Gupta[3]

[1]Department of Botany, Durgapur Government College, Durgapur
[2]Department of Botany, Burdwan University, Burdwan
[3] SRF, PG Department of Conservation Biology, Durgapur Government College, Durgapur

---

Introduction

Wetlands are diverse ecosystems that link people, wildlife and environment in special and interdependent ways through the essential life-support functions of water (Maltby and Barker, 2009).Wetlands include a variety of habitats such as marshes, swamps, river banks, river flood plains, lakes, ponds and so on (Keddy, 2010; Hassal and Anderson, 2015) and forms an important environment for aquatic, semi-aquatic and moisture loving floral and faunal associations (LaSalle *et al.*, 1991; Diaz-Paniagua *et al.*, 2010; Bagella *et al.*, 2011). According to IUCN (2004) the total number of known species worldwide is about 1.5 million, while many more are yet to be discovered. Although most of the taxonomic groups of temperate zones have been fully recorded, the biodiversity data of rich tropical zones is far from complete.

Macrophytes are referred to as water plants, as well as *amphiphytes* and/or amphibian plants, consisting mainly of aquatic and wetland vascular plant species (Hannus and Von Numers, 2008; QingLing *et al.*, 2012; Prajeesh *et al.*, 2014) belonging to Pteridophytes and Angiosperms and exclude filamentous algae. Macrophytes play an important role in aquatic ecosystems (Choi *et al.*, 2014; O'Brien *et al.*,2014; Human *et al.*, 2015) and contribute to the general fitness and diversity of a healthy aquatic ecosystem by acting as indicators for water quality and aiding in nutrient cycling (Cvetkovic and Chow-Fraser, 2011; Beck, 2014; Achieng *et al.*, 2014; Phillips, 2015). Thus studies on wetland macrophytes have started gaining importance (Takamura *et al.*, 2003; Chambers *et al.*, 2008; Nishihiro *et al.*, 2014) as systematic stock taking of biodiversity is presently given top most priority at the same time these plants have implications with functional values of wetlands (Rahman and Hasegawa, 2011; Bornette and Puijalon, 2011; Jiang *et al.*, 2014). A fairly good number of birds inhabit wetlands for nesting, feeding and roosting and are dependent on wetland for their very survival. However, wetlands in India, are facing tremendous anthropogenic pressures (Prasad *et al.*, 2002) few studies have been carried out on the status and diversity of wetland birds in this eco region (Mukherjee and Gupta, 2012), study of wetland avifauna is also gaining importance.

Fishes are one of the important elements in the economy of many nations as they have been a stable item in the diet of Indians. Study on fish fauna in India has been well investigated through some good and novel studies pioneered by Talwar and Jhingran (1991), Kar and Dey (2002).

Wetland management generally involves activities that can be conducted with, in, and around wetlands, both natural and man-made, to protect, restore, manipulate, or provide importance for their functions and values. Effective wetland management requires knowledge on a range of wetland subjects and proper baseline information can only help a decision maker to evaluate wetland resources, to determine their functions, values, and services, to assess risks, and prioritize protection. There are several studies concerning the management of wetlands with special reference to socioeconomic development of the concerned area in various part of the globe (Turner, 1991; Maltby, 1991; Roggeri, 1995; Johnston *et al.*, 2013;Gray *et al.*, 2013;Namaalwa*et al.*, 2013; Eppink *et al.*, 2014 and Smardon, 2014).

The present investigation was, therefore, undertaken to study the species composition of aquatic macrophytes, wetland birds and fishes along with the scope of wetland management in twenty different wetlands of Birbhum district, West Bengal as a model. Detail survey data and biological attributes were used to address two main objectives: i) to analyze the composition of aquatic life (macrophytes, wetland birds and fishes) in these freshwater wetlands and ii) to prepare a list of SDPs required in wetlands of Birbhum District for proper management and sustainable use.

Materials and Methods

Study Area

Birbhum is one of the smallest districts in West Bengal representing the unique drought prone, dry regions of eastern India and is located between 23°32′30″ and 24°35′ 0″ North latitude and 88°01′ 40″ and 87°05′ 25″ East. East longitude extending over an area of 4545.00 sq km. The district is well drained by a number of rain fed rivers and rivulets running in nearly every case from west to east with a slight southerly inclination. There are only two principal rivers, *viz.* the Mor and the Ajay, the latter forming the southern boundary. The geological formations represented in Birbhum are Archaean genesis, the Gondwana system, laterite and Gangetic alluvium. The predominant soil types are old-alluvium and red laterite. Soil is acidic with the pH varying from 5.0 to 6.5.

The climate of the district is generally drought prone, dry, mild and healthy. The hot weather usually last from the middle of March to the middle of the June, the rainy season from the middle of June to the middle of October, and the winter from middle of October to the middle of March. As a rule, the wind is blown from south east in summer and from the north west in winter. This region is blessed with a good number of fresh water wetlands harboring a great variety of aquatic macrophytes. For the present study a total of twenty freshwater wetlands of Birbhum district (Figure 12.1) having unique geographical representativeness, pattern of aquatic life and multipurpose usages by local stakeholders were selected. Almost all parts (at block level) of Birbhum district were surveyed from 2010 to 2014.

Documentation of Wetland Bioresources

Observations on the aquatic macrophytes, wetland birds and piscifaunal occurrence at 20 wetlands distributed over 19 Blocks in Birbhum District were carried out through frequent field surveys at seasonal intervals during pre-monsoon (April-May), monsoon (August-September) and post-monsoon (December-January). The annual assemblage (presence or absence) of macrophyte species were noted and plants were identified using standard literature by Cook (1996). The birds were photographed by CANON Power Shot (SX510HS) and identified using standard field book (Grimmett *et al.*, 2011). Fishes were identified upto the species level, with the help of standard keys, books (Talwar and Jhingran, 1991) and online databases (Fish Base, 2014). Site specific relative abundance (here after will be noted as three categories: a) Most abundant: +++ (fish observed in >50 per cent of study sites); b) Less abundant: ++ (25-50 per cent) and c) Rare: + (<25 per cent) were derived for wetland birds and fishes. Threat category and population trend of these communities were adopted from IUCN Red List of Threatened Species (IUCN, 2014).

Survey on Wetland Management and SDPs

Structured questionnaire survey method among different stakeholders of each wetland was used to document different SDPs. In this context, a detailed stated preference study was carried out using RIS (Ramsar, 2010) and NLCP guidelines (MoEF, 2008) to devise the Socioeconomic Developmental Plans (SDPs). All the statistics have been carried out using XLSTAT (Addinsoft, 2010) statistical package.

Results and Discussion

Macrophytes Assemblage

During the entire study period, a total of 26 families and 57 species (Table 12.1a) of macrophytes were recorded; Cyperaceae (17 per cent), Lemnaceae (15 per cent) and Poaceae (15 per cent) showed higher abundance followed by Nymphaeceae (8 per cent), Pontederiaceae (7 per cent), Hydrocharitaceae (6 per cent), Araceae (5.5 per cent), Asteraceae (5 per cent) and Amaranthaceae (4 per cent) with moderate abundance. Lower abundant families include Trapaceae (3 per cent), Nelumbonaceae

(3 per cent), Marsileaceae (3 per cent), Apiaceae (2 per cent), Salviniaceae (2 per cent) Ranunculaceae (1 per cent), Acanthaceae (1 per cent) and Polygonaceae (1 per cent) etc. Uncommon families (less than 1 per cent abundance) were represented by Euphorbiaceae, Convolvulaceae, Papilionaceae, Potamogetonaceae. The plant families are categorized under three major wetland plant types – emergent (65.5 per cent), floating (25 per cent) and submerged (9.5 per cent). The trend further confirms the earlier observation made by Palit and Mukherjee (2007) and Palit and Gupta (2012).

Table 12.1(a): List of Macrophytes Documented from Wetlands of Birbhum District

Family	*Scientific Names of Macrophyte Species*
Acanthaceae	*Hygrophila difformis*[b]
Alismataceae	*Sagittaria guyanensis*[a]
Amaranthaceae	*Alternanthera paronychoides*[b], *Alternanthera sessilis*[b], *Amaranthus spinosus*[c]
Apiaceae	*Centella asiatica*[b]
Aponogetonaceae	*Aponogeton natans*[a]
Araceae	*Pistia stratiotes*[d]
Asteraceae	*Enydra fluctuans*[a]
Ceratophyllaceae	*Ceratophyllum demersum*[a]
Convolvulaceae	*Ipomoea aquatic*[a], *Ipomoea carnea*[a], *Ipomoea obscura*[a]
Cyperaceae	*Bulbostylis barbata*[b], *Cyperus castaneus*[b], *Cyperus compressus*[b], *Cyperus difformis*[c], *Cyperus pilosus*[d], *Eleocharis dulcis*[a], *Fimbristylis dichotoma*[a], *Fimbristytis aestivates*[a], *Schoenoplectus articulatus*[d]
Euphorbiaceae	*Croton bonplandianum*[a]
Hydrocharitaceae	*Hydrilla verticillata*[d], *Ottelia alismoides*[a], *Vallisneria spiralis*[a]
Lemnaceae	*Lemna acquinoctialis*[d], *Spirodela polyrrhiza*[d]
Lentibutariaceae	*Utricularia gibbosa*[a], *Utricularia stellaris*[a]
Menyanthaceae	*Nymphoides indica*[c]
Nelumbonaceae	*Nelumbo nucifera*[c]
Nymphaeceae	*Nymphaea nouchali*[d], *Nymphaea pubescens*[c]
Onagraceae	*Lindwigia perennis*[b]
Papilionaceae	*Aeschynomene aspera*[a], *Alysicarpns monilifer*[a]
Poaceae	*Coix lachryma-jobi*[a], *Digittaria ciliaris*[c], *Digittaria longifolia*[b], *Echinochloa colona*[c], *Echinochloa crussgalli*[d], *Eragrostis gangetica*[a], *Eragrostis tenella*[a], *Leptochloa chinensis*[d], *Oryza rufipogon*[b], *Phragmites karka*[a]
Polygonaceae	*Polygonum barbatum*[a], *Polygonum orientale*[a], *Polygonum plebeium*[a]
Pontederiaceae	*Eichhornia crassipes*[d], *Monochoria vaginalis*[a], *Monochoria hastata*[a]
Potamogetonaceae	*Potamogeton nodosus*[a]
Ranunculaceae	*Ranunculus seleratus*[b]
Salviniaceae	*Salvinia cuculata*[b]
Trapaceae	*Trapa bispinosa*[c]

Code used for relative abundance of macrophyte species: a: 0-1 per cent, b: 1-2 per cent, c: 2-3 per cent, d: >3 per cent.

Table 12.1(b): List of Wetland Birds Species Documented from Wetlands of Birbhum district.

Family	*Scientific Names of Wetland Birds Species*
Alcedinidae	*Alcedo atthis*[a], *Halcyon smyrnensis*[a]
Anatidae	*Anas crecca*[b], *Anas strepera*[b], *Aythya fernia*[c], *Dendrocygna bicolor*[a], *Dendrocygna javanica*[a], *Nettapus coromandelianus*[b]
Ardeidae	*Ardea purpurea*[a], *Ardeola grayii*[a], *Bubulcus ibis*[a], *Egretta garzetta*[a]
Charadriidae	*Charadrius dubius*[b], *Charadrius alexandrinus*[b]
Ciconiidae	*Anastomus oscitans*[a]
Jacanidae	*Hydrophasianus chirurgus*[c], *Metopidius indicus*[a]
Motacillidae	*Motacilla alba*[a], *Motacilla flava*[a], *Motacilla citreola*[a]
Phalacrocoracidae	*Phalacrocorax niger*[a]
Rallidae	*Amaurornis phoenicurus*[b], *Gallinula chloropus*[a], *Porphyrio porphyrio*[b], *Fulica atra*[b]

Code used for site specific relative abundance (SSRA) of wetland bird species: a: Most abundant: +++ (>50- per cent); b: Less abundant: ++ (25-50 per cent), c: Rare: + (<25 per cent).

Table 12.1(c): List of Fish Species Documented from Wetlands of Birbhum District

Family	*Scientific Names of Fish Species*
Cypriniformes	*Hypopthalmicthyes molitrix*[c], *Aristichthysnobills nobilis*[c], *Labeo bata*[a], *Catla catla*[a], *Cyprinus carpio*[a], *Ctenopharyngodon idella*[b], *Labeo calbasu*[b], *Cirrhinus mrigala*[a], *Amblypharyngodon mola*[b], *Puntius puntio*[b], *Puntius chola*[b], *Puntius guganio*[b], *Labio rohita*[a], *Puntius sarana*[b], *Puntius ticto*[b]
Cyprinodontiformes	*Aplocheilus panchax*[a]
Mugiliformes	*Liza parsia*[a]
Osteoglossiformes	*Notopterus notopterus*[b]
Perciformes	*Anabas testudineus*[c], *Trichogaster chuna*[b], *Chanda nama*[b], *Channa gachua*[b], *Channa punctatus*[b], *Channa striata*[b], *Oreochromis niloticus niloticus*[b], *Oreochromis mossambicus*[a], *Badis badis*[b]
Siluriformes	*Sperata aor*[b], *Mystus vittatus*[b], *Clarious batrachus*[b], *Heteroneustes fossilis*[b], *Lates calcarifer*[b], *Wallago attu*[b]
Synbranchiformes	*Mastacembelus favus*[c], *Macrognathus pancalus*[c]

Code used for site specific relative abundance (SSRA) of wetland bird species: a: Most abundant: +++ (>50- per cent); b: Less abundant: ++ (25-50 per cent), c: Rare: + (<25 per cent).

Wetland Birds Composition

Twenty five species of wetland birds belonging to 20 genera and 9 families were recorded as listed in Table 12.1b. The family Anatidae dominated the wetland bird community of the study area, represented by 6 species and accounted for 24 per cent of the total number of wetland bird species of the studied wetlands (Figure 12.2). Among the recorded species, 15 (60 per cent) were Common (+++), 8 (32 per cent) Uncommon (++) and 2 (8 per cent) were Less common (+) species according to SSRA (Figure 12.3). Population trend of these water bird species revealed 10

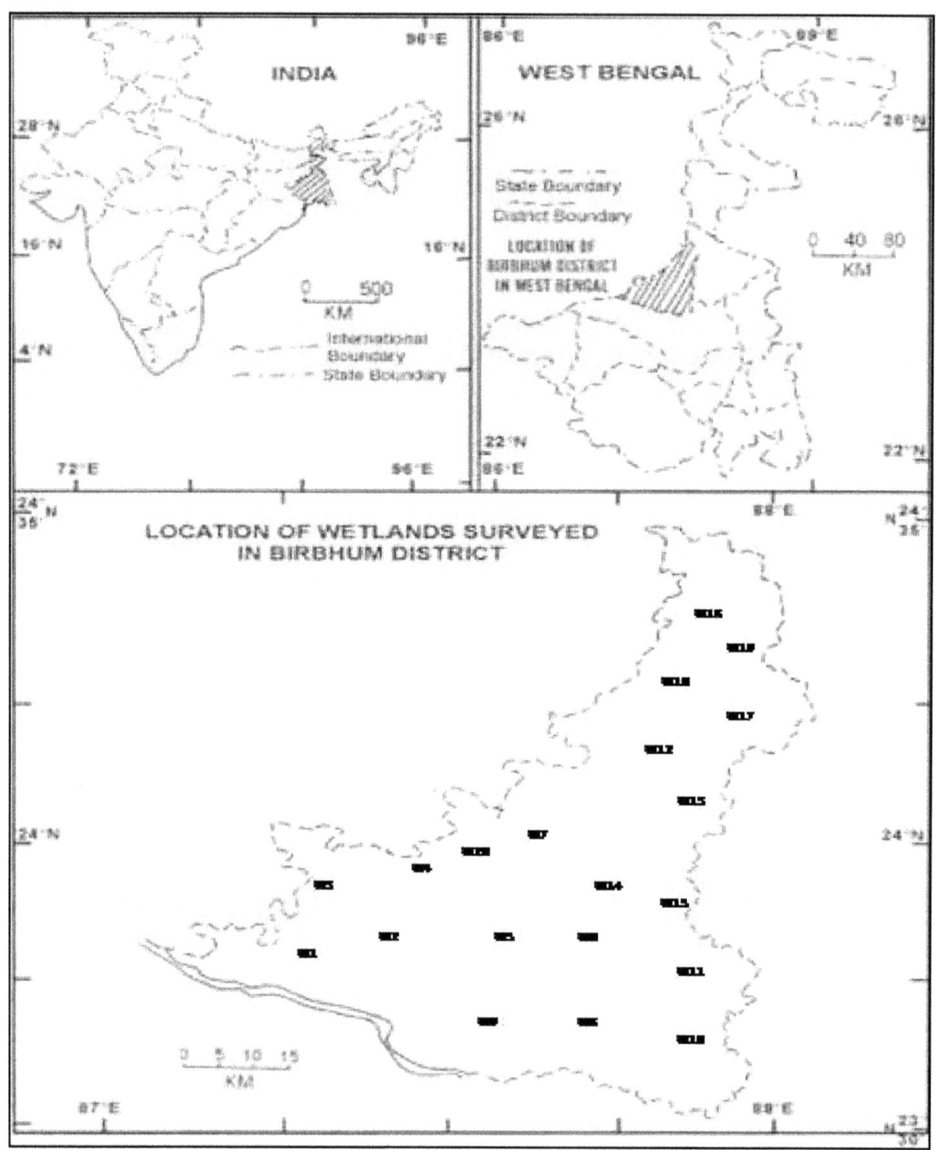

Figure 12.1: Location of Study Sites in Birbhum District, West Bengal, India.

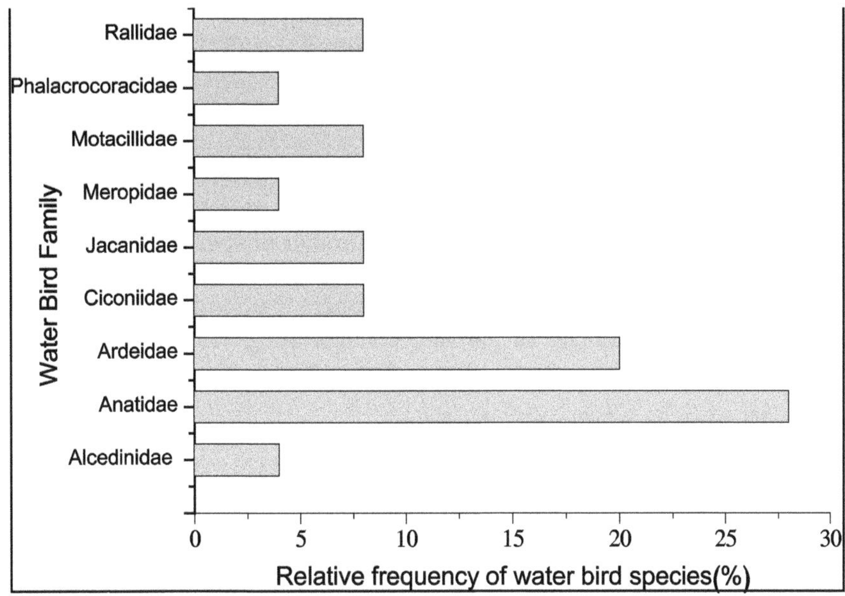

Figure 12.2: Bar Graph of Avian Families as Represented by Relative Frequency of Wetland Bird Species in Wetlands of Birbhum District.

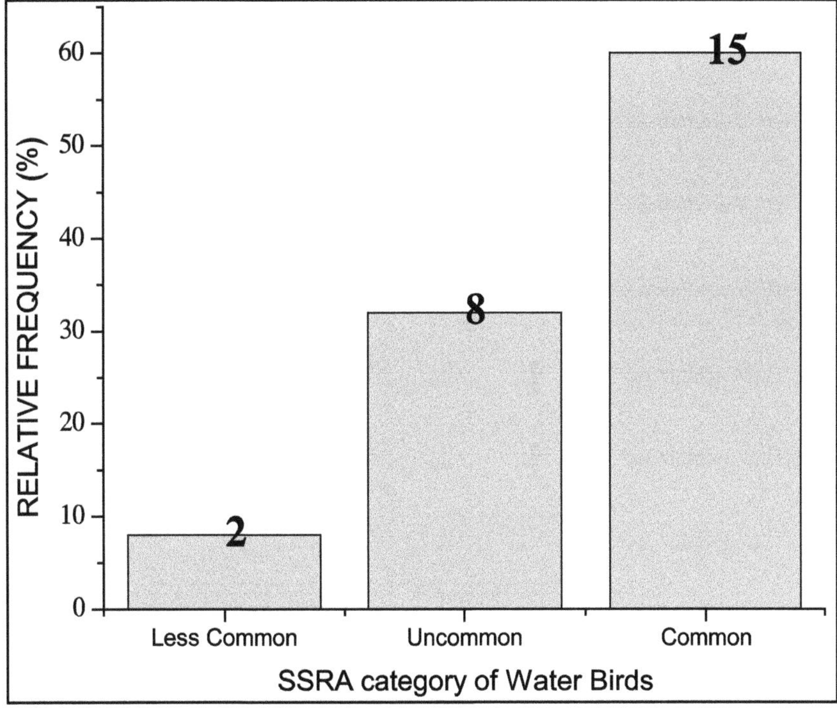

Figure 12.3: Bar Graph Representing the Relative Frequency of SSRA Category of Avian Species in Wetlands of Birbhum District (Species number is labeled).

species (40 per cent) with unknown (UN) status followed by 9 species (36 per cent) with decreasing (DE), 3 species (12 per cent) with stable (ST) and 3 species (12 per cent) with increasing (IN) status.

Piscifaunal Status

During the study period, thirty five fish species, under 25 genera, 16 families and 7 orders were collected and identified from the wetlands of Birbhum District, West Bengal, India (Table 12.1c). Maximum number of species was found under the order Cypriniformes (15) where as Osteoglossiformes, Mugiliformes and Cyprinodontiformes comprised minimum number (1) among seven orders (Figure 12.4).The relative frequencies of fish species belonging to the SSRA category is depicted in Figure 12.5.

SDPs in Wetlands of the Study Area

Figure 12.6 enumerates different SDPs for wetlands in the study area and some preferable management practices according to the mode of implementation along with public interest *i.e.* public demand have been enlisted. Diversion and treatment of sewage, weed removal, reduced usage of water for agricultural practices are important considerations in general. A majority of wetlands in urban area must be prohibited for further construction in wetland premises and a ban on wetland filling must be imposed on priority basis as also recommended by earlier worker (Islam

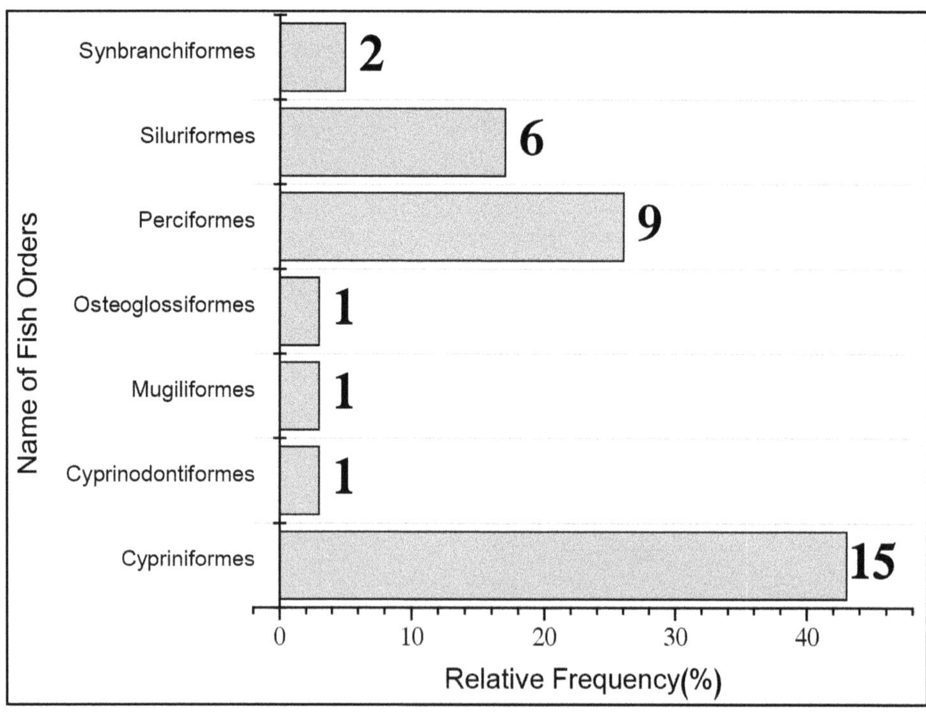

Figure 12.4: Bar Graph of Observed Fish Orders in Birbhum Wetlands of Birbhum District.

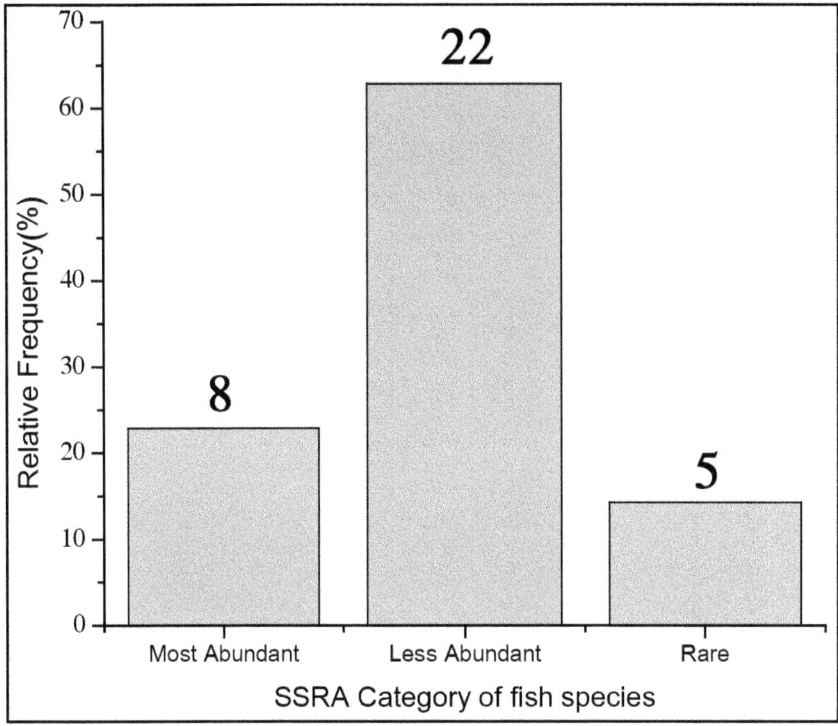

Figure 12.5: Column Graph of SSRA Categories of Observed Fish Species in Wetlands of Birbhum District.

and Kitazawa, 2013) from other parts of the globe. While, majority of wetlands in the vicinity of rural villages need bank protection, desiltation/dredging and improvement of water quality. In some specific wetlands there is a malpractice of adding unlimited poultry manure to wetland water which causes further problems to aquatic life. During winter months, there are some instances where wetland premises become horrible due to frequent picnic activities, where destruction of natural habitat was common leading to loss of biodiversity. Removal of plastic waste and other solid waste along with introduction of picnic revenue system will be the key developmental measures for those wetlands. Similar observations were also documented by MoEF (2009, 2010), Aber *et al.*(2012) and Cools *et al.* (2013) in their earlier studies. In some instances, ownership dispute may exist, which may be minimized through negotiations in consultation with local bodies and administration.

Since major portion of the study area are drought prone, village people depend mostly on these wetlands for their day to day activity, construction of proper bathing ghats in the wetland premises is a popular demand. Beside, rural livelihood depends greatly on culture of fish fauna, and as a food source it has immense importance. However, the fish culture practices suffering from lack of proper funding, which needs immediate attention. Thus formation of "effective and planned" societies are of urgent need for overall sustenance of wetlands. Among other SDPs, promotion of

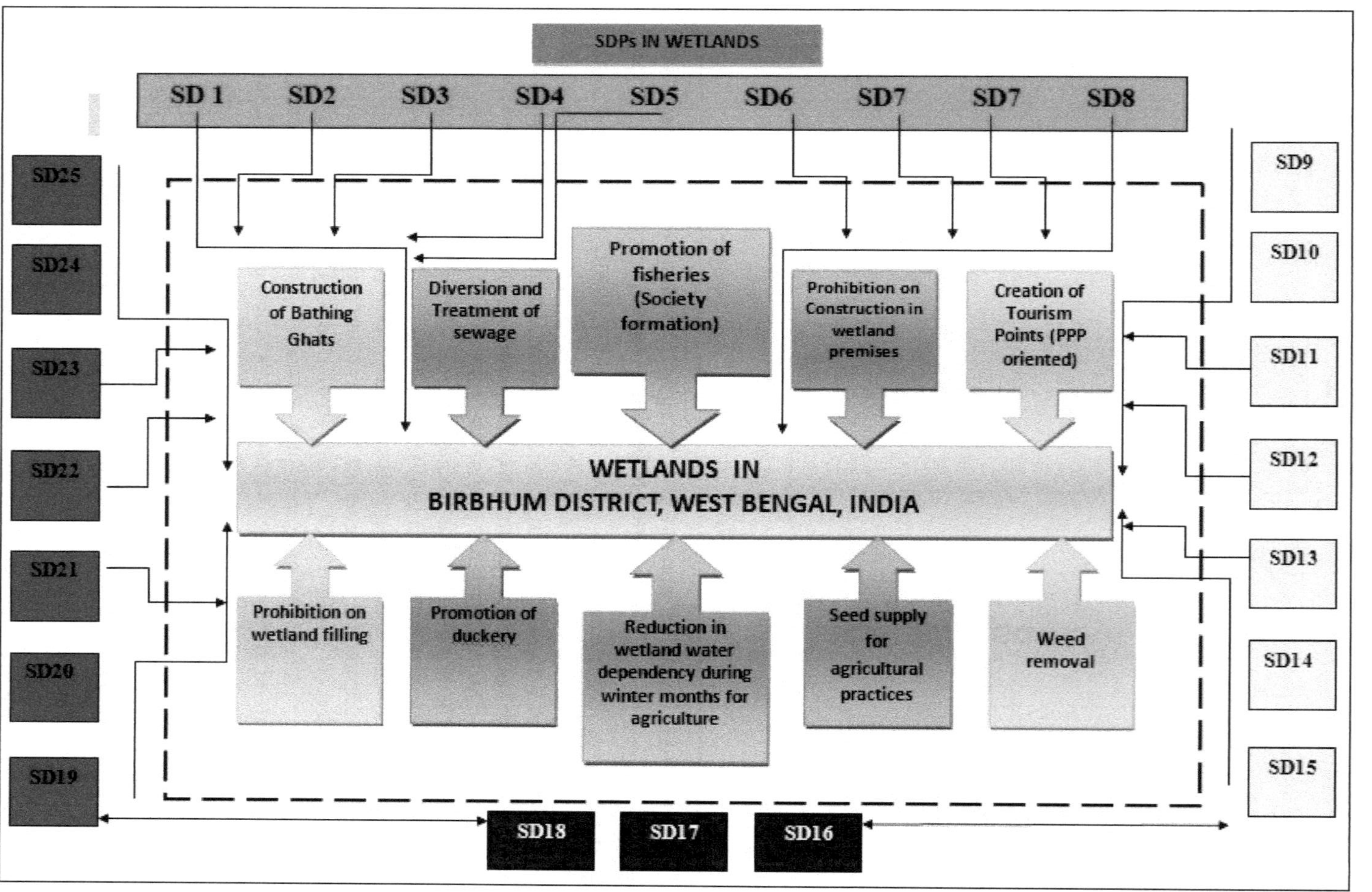
SDPs IN WETLANDS
SD 1
SD2
SD3
SD4
SD5
SD6
SD7
SD7
SD8
SD25
SD24
SD23
SD22
SD21
SD20
SD19
SD9
SD10
SD11
SD12
SD13
SD14
SD15
Construction of Bathing Ghats
Diversion and Treatment of sewage
Promotion of fisheries (Society formation)
Prohibition on Construction in wetland premises
Creation of Tourism Points (PPP oriented)
WETLANDS IN
BIRBHUM DISTRICT, WEST BENGAL, INDIA
Prohibition on wetland filling
Promotion of duckery
Reduction in wetland water dependency during winter months for agriculture
Seed supply for agricultural practices
Weed removal
SD18
SD17
SD16

Figure 12.6: A Schematic Diagram on the Major SDPs for Wise Use of Wetlands in Birbhum District.
[*Source*: Ecological studies and structured questioner Survey on 20 wetlands (block wise) during 2010 to 2014.]

Code used: SD1: Desiltation/Dredging; SD2: Bank Protection; SD3: Minimize Ownership dispute; SD4: Improvement of Water quality; SD5: Banning on unhygienic wetland use; SD6: Invasive species control; SD7: Prohibition of washing of automobiles; SD8: Controlled Picnic activity; SD9: Check on uncontrolled dumping of poultry manure in wetland; SD10: Mass awareness of stakeholders; SD11: Creation of Wetland Protection Group (at village level); SD12: Deep tube Well construction; SD13: Regular Wetland water quality monitoring (exclusively for bathing/cleaning of utensils); SD14: Promotion of Tourism (GoWB supported); SD15: Floriculture (GoWB supported); SD16: Establishment of marketing channels for useful plant and animal resources; SD17: Enhanced poultry farming in adjacent fallow land; SD18: Biodiversity Documentation and R and D Projects; SD19: Enhancement of Livlihood dependency over wetland resources; SD20: Harvest of Macrophytes; SD21: Planned irrigation; SD22: Vermiculture of organic matter removed from wetlands during management; SD23: Modernization of Aqua: Parks; SD24: Raising of gardens around wetlands to promote aesthetic value and general; SD25: Construction of STP.

tourism activities, poultry farming, floriculture practices are the leading ones which need proper attention. Besides, formation of wetland protection group and mass awareness programs for local stakeholders over wetland bioresources and their possible role in the improvement of socio-economic status for the locals should be given priority at least in this part of country.

Summary

Biodiversity encompasses the grand total of all the different life forms of our planet. This study represents an indepth assessment of key biological components and management scenario in 20 wetlands of Birbhum district, West Bengal, India during 2010 to 2014. Total 26 families and 57 species of macrophytes, 25 bird species and 35 species (16 families and 7 orders) of fishes were observed. Cyperaceae (17 per cent), Lemnaceae (15 per cent) and Poaceae (15 per cent) encompass higher abundance among macrophytes. Anatidae represented by 6 species, dominated the wetland bird community and maximum number of fish species were catalogued under the order Cypriniformes (15). Promotion of fisheries, traditional agricultural practices, duckery, tourism activities, poultry farming, floriculture are some key Socioeconomic Developmental Plans (SDPs); ensuring sustainable use of these wetlands through financial support. Our significant findings include formation of wetland protection group and formulation of mass awareness programmes for local stakeholders concerning wetland bioresources which may lead to wise use of wetlands in this drought prone region.

References

Aber JS, Pavri F and Aber SW. 2012. Conservation and Management: Wetland Planning and Practices. *Wetland environments: A Global Perspective*,218-230.

Achieng AO, Raburu PO, Okinyi L and Wanjala S. 2014. Use of macrophytes in the bioassessment of the health of King'wal Wetland, Lake Victoria Basin, Kenya. *Aquatic Ecosystem Health and Management*, 17(2): 129-136.

Addinsoft SARL. 2010. XLSTAT software, version 10, Addinsoft, Paris, France.

Bagella S, Gascón S, Caria MC. *et al.*, 2011. Cross-taxon congruence in Mediterranean temporary wetlands: vascular plants, crustaceans, and coleopterans. *Community Ecology*, 12(1): 40-50.

Beck MW, Tomcko CM, Valley RD and Staples DF. 2014. Analysis of macrophyte indicator variation as a function of sampling, temporal, and stressor effects. *Ecological Indicators*, 46: 323-335.

Bornette G and Puijalon S. 2011. Response of aquatic plants to abiotic factors: a review. *Aquatic Sciences*, 73(1): 1-14.

Chambers PA, Lacoul P, Murphy KJ and Thomaz SM. 2008. Global diversity of aquatic macrophytes in freshwater. *Hydrobiologia*, 595(1): 9-26.

Choi JY, Jeong KS, Kim SK *et al.*, 2014. Role of macrophytes as microhabitats for zooplankton community in lentic freshwater ecosystems of South Korea. *Ecological Informatics*, 24: 177-185.

Convention on Wetlands of International Importance especially as Waterfowl Habitat. Ramsar (Iran), 2 February 1971. UN Treaty Series No. 14583.

Cook CDK. 1996. Aquatic and Wetland Plants of India. Oxford University Press, Oxford, New York, Delhi.

Cools J, Johnston R, Hattermann FF. *et al.*, 2013. Tools for wetland management: Lessons learnt from a comparative assessment. *Environmental Science and Policy*, 34: 138-145

Cvetkovic M and Chow-Fraser P. 2011. Use of ecological indicators to assess the quality of Great Lakes coastal wetlands. *Ecological Indicators*, 11(6): 1609-1622.

Diaz-Paniagua C, Fernandez-Zamudio R, Florencio M *et al.*, 2010. Temporary ponds from Doñana National Park: a system of natural habitats for the preservation of aquatic flora and fauna. *Limnetica*, 1(29): 41-58.

Eppink FV, Brander LM and Wagtendonk AJ. 2014. An initial assessment of the economic value of coastal and freshwater wetlands in West Asia. *Land*, 3(3): 557-573.

Fish Base. 2014. Paris, France.www.fishbase.org

Gray MJ, Hagy HM, Nyman JA and Stafford JD. 2013. Management of Wetlands for Wildlife. In. *Wetland Techniques*. Springer, Netherlands. pp. 121-180

Grimmett R, Inskipp C and Inskipp T. 2011. *Birds of the Indian Subcontinent, Helm Field Guides*. Oxford University Press. UK.

Hannus JJ, and Von Numers M. 2008. Vascular plant species richness in relation to habitat diversity and island area in the Finnish Archipelago. *Journal of Biogeography*, 35(6): 1077-1086.

Hassall C and Anderson S. 2015. Storm water ponds can contain comparable biodiversity to unmanaged wetlands in urban areas. *Hydrobiologia*, 745(1): 137-149.

Human LR, Snow GC, Adams JB. *et al.*, 2015. The role of submerged macrophytes and macroalgae in nutrient cycling: a budget approach. *Estuarine, Coastal and Shelf Science*, 154: 169-178.

Islam MN and Kitazawa D. 2013. Modeling of freshwater wetland management strategies for building the public awareness at local level in Bangladesh. *Mitigation and Adaptation Strategies for Global Change*. 18(6): 869-888.

IUCN 2004. IUCN Red List of Threatened Species, IUCN, Gland, Switzerland.

IUCN. 2014. The IUCN Red List of Threatened Species. Version 2014. 3 <http://www.iucnredlist.org>.

Jiang X, Song X, Chen Y and Zhang W. 2014. Research on biogas production potential of aquatic plants. *Renewable Energy*, 69: 97-102.

Johnston R, Cools J, Liersch S. *et al.*, 2013. WETwin: a structured approach to evaluating wetland management options in data-poor contexts. *Environmental Science and Policy*, 34: 3-17.

Kar D and Dey SC. 2002. On the occurrence of advanced fry of *Hilsa (Tenuolosa) ilisha* (Hamilton- Buchanan) in Chatla Haor seasonal wetland of Assam. *Proc. Zool. Soc.* Kolkata, 55(2): 15-19.

Keddy PA. 2010. *Wetland ecology: principles and conservation*. Cambridge University Press,. Cambridge.

Kler TK 2002. Bird species in Kanjali wetland. *Tiger Paper*, 39(1): 29-32.

LaSalle MW, Landin MC and Sims JG. 1991. Evaluation of the flora and fauna of a *Spartina alterniflora* marsh established on dredged material in Winyah Bay, South Carolina. *Wetlands,* 11(2): 191-208.

Maltby E and Barker T. 2009. The wetlands Handbook, 1st edition. Blackwell Publishing, Oxford.

Maltby E. 1991. Wetland management goals: wise use and conservation. *Landscape and Urban Planning*, 20(1): 9-18.

Ministry of Environment and Forests. 2008. Guidelines for National Lake Conservation Plan. National river Conservation directorate, Ministry of Environment and Forests. Government of India, New Delhi. pp. 7-8.

Ministry of Environment and Forest. 2009. National Wetland Conservation Programme: Guidelines for Conservation and Management of Wetlands in India (Revised on 12.06.2009). Conservation and Survey Division, Ministry of Environmnet and Forests, Government of India, New Delhi. pp. 3-4.

Ministry of Environment and Forests. 2010. Conservation and management of Lakes: An Indian perspective. National river Conservation directorate, Ministry of Environment and Forests. Government of India, New Delhi. pp. 3.

Mukherjee A and Gupta S. 2012. Assessment of avifaunal diversity of Santragachi Wetland, West Bengal, India. *Ecology, Environment and Conservation*, 18(2): 357-362.

Namaalwa S, Funk A, Ajie GS and Kaggwa RC. 2013. A characterization of the drivers, pressures, ecosystem functions and services of Namatala wetland, Uganda. *Environmental Science and Policy*, 34: 44-57.

Nishihiro J, Akasaka M, Ogawa M and Takamura N. 2014. Aquatic vascular plants in Japanese lakes. *Ecological Research*, 29(3): 369-369.

O'Brien JM, Lessard JL, Plew D. *et al.*, 2014. Aquatic macrophytes alter metabolism and nutrient cycling in lowland streams. *Ecosystems,* 17(3): 405-417.

Palit D and Gupta S. 2012. Seasonal changes in limnological parameters and macrophyte diversity associated with wetlands in Birbhum district, West Bengal, India. *Ind. J. Plant Sci.,* 1 (2-3): 97-115.

Palit D and Mukherjee A. 2007. An inventory of wetlands in Birbhum District, West Bengal and their successional characteristics. *Environment and Ecology,* 25 (I): 173-176.

Phillips DP, Human LRD and Adams JB. 2015. Wetland plants as indicators of heavy metal contamination. *Marine Pollution Bulletin,* 92(1-2): 127-132.

Prajeesh P, Narayanan MR and Kumar NA. 2014. Diversity of vascular plants associated with wetland paddy fields (Vayals) of Wayanad district in Western Ghats, India. *Annals of Plant Sciences*, 3(5): 704-714.

Prasad SN, Ramachandra TV and Ahalya N. *et al.*, 2002. Conservation of wetlands of India- A review. *Tropical Ecology*, 43(1): 173-186.

QingLing W, XiaoLan X, Jing Y. *et al.*, 2012. Plant diversity and ecosystem health assessment of Sanyang wetland in Wenzhou, Zhejiang. *Journal of Zhejiang University (Agriculture and Life Sciences)*, 38(4): 421-428.

Rahman MA and Hasegawa H. 2011. Aquatic arsenic: phytoremediation using floating macrophytes. *Chemosphere*, 83(5): 633-646.

Ramsar. 2010. Information Sheet on Ramsar Wetlands (RIS), 2009-2014 version, Available for download from http://www.ramsar.org/doc/ris/key_ris_e.doc

Roggeri H. 1995. *Tropical freshwater wetlands: a guide to current knowledge and sustainable management*. Kluwer Academic Publishers.

Smardon R. 2014. *Wetland Ecology: Principles and Conservation*. 813-817

Takamura N, Kadono Y, Fukushima M. *et al.*, 2003. Effects of aquatic macrophytes on water quality and phytoplankton communities in shallow lakes. *Ecological Research*, 18(4): 381-395.

Talwar PK and Jhingran A. 1991. *Inland fishes of India and adjacent countries*. Oxford and IBH publishing company, New Delhi.

Turner K. 1991. Economics and wetland management. *Ambio*, 59-63.

Part II

Wetlands: Ecosystem Service Provider

Chapter 13

Biomonitoring Potential of Aquatic Insects in Freshwater Ecosystem: An Overview

Srimoyee Basu[1], K.A. Subramanian[2] and Goutam Kumar Saha[1]

[1]Entomology and Wildlife Biology Research Laboratory, Department of Zoology, University of Calcutta, Kolkata
[2]Zoological Survey of India, Kolkata

Introduction

Life is dependent on water, a valuable component for the survival of all living organisms. The total water source of the earth is approximately 1340 million cubic km, however, less than 3 per cent of that constitute the freshwater ecosystem (Katona, 2008). The inland freshwater encompasses a diverse array of ecosystems as varied as lakes and rivers, ponds and streams, temporary puddles, thermal springs and even pools of water that collect in the leaf axils of certain plants (Subramanian and Sivaramakrishnan, 2007). Freshwater habitats can be classified into three major types such as, lentic or stagnant water systems, lotic or running or flowing water systems and wetlands. Each of these ecosystems has its own set of distinctive ecology and biological community. Freshwater is an invaluable as well as a finite natural resource to man's varied activities (Pradhan, 2005), such as for drinking and personal hygiene, religious activities, fisheries, agriculture (irrigation and livestock supply), industrial production, hydropower generation and recreational activities such as bathing and fishing (Pradhan, 2005). As a result of increased human demand for water (Arthington *et al.*, 2006; Dudgeon *et al.*, 2006), the freshwater habitats are being subjected to increased levels of human disturbance around the world (Saunders *et al.*, 2002) and as a matter of fact, conservation aspects of freshwater ecosystems and its biodiversity deserves special attention.

Freshwater Biodiversity

Although, freshwater encompasses a small fraction of world's water resource, they contain a disproportionately high fraction of the world's biodiversity (Subramanian and Sivaramakrishnan, 2007). According to Dudgeon *et al.* (2006), it supports at least 100,000 species out of approximately 1.8 million, almost 6 per cent of all described species. As for example, 10,000 of the 25,000 (40 per cent) known fish species are freshwater forms (Lundberg *et al.*, 2000). If amphibians, aquatic reptiles, mammals are added to the fish diversity, it becomes one third of all vertebrate species that are confined to freshwater. Incidentally, the comprehensive knowledge on the total biodiversity of freshwater ecosystems is incomplete, particularly in case of invertebrates, microbes etc. The recent assessment of global freshwater animal species diversity by Balian *et al.* (2008) showed that a total of approximately 126,000 species have been recognized from freshwater habitats globally, of which 60.4 per cent are insects, 14.5 per cent are vertebrates, 10 per cent crustaceans, 1.6 per cent rotifers, 1.4 per cent annelids, 1 per cent platyhelminthes, 1.4 per cent nematodes, 5 per cent and 4 per cent are archanids and molluscs respectively. Adequate data on the diversity of most of the microfauna of freshwater ecosystems do not exist and due to this reason it can be said, freshwater hotspots receive less attention than their terrestrial counterparts (Myres *et al.*, 2000). As a matter of fact freshwater biodiversity is the over-riding conservation priority for biodiversity protection and freshwater ecosystem management.

The value of inland waters gets increased as ecosystems become more stressed and their goods and services are declined. Freshwater biodiversity has a particular importance for indigenous people in many parts of the world, who depend upon aquatic products for their livelihood as well as sustenance. In recent days, threats to freshwater biodiversity are the major problem. Review of literature (Allan and Flecker 1993; Naiman and Turner, 2000; Jackson *et al.*, 2001; Malmquist and Rundle, 2002; Dudgeon *et al.*, 2006) suggest that the major threats identified to freshwater faunal diversity at a global scale are over extraction and overexploitation of bioresources, water pollution from domestic and industrial point sources and excessive nutrient enrichments, flow modification due to water impoundment as a result of construction of dams, habitat degradation, introduction and invasion of exotic species. Along with this, environmental changes and climate changes occurring at a global scale are superimposed upon all of these threat categories. Hence, protection of freshwater biodiversity is a conservation challenge and is needed for the interests of all nations and governments. Conservation of intact freshwater bodies and their biodiversity remains a priority. One of the major option to conserve freshwater biodiversity is the biomonitoring of freshwater habitats to measure their eco-health.

Concept of Biomonitoring

The term biomonitoring is used for the purpose of gathering of biological data in both the laboratory and in the field for making some sort of assessment or in determining whether regulatory standards and criteria are being met in that particular ecosystem. According to Roux *et al.* (1993) different biomonitoring techniques are used and can be subdivided into a number of categories as follows-

1. Bioassessment that is based on ecological surveys of the functional or structural aspects of biological communities.
2. Toxicity bioassays which are laboratory based methodology for investigating and predicting the effect of chemicals on test organisms.
3. Behavioural bioassays which explore sub-lethal effects on fish or other species when exposed to contaminated water, usually as on site, early warning systems.
4. Bioaccumulation studies that monitor the uptake and retention of chemicals in the body of an organism and the consequent effects higher up the food chain.

Along with this, standard physico-chemical water quality methods are also required for assessing water quality health as it provide information on water quality at a particular spatial unit during the time of sampling. It cannot provide historical information on water quality. This is important for assessment of heavy metals or pesticidal contamination. Though biomonitoring can not entirely replace standard physico-chemical water quality methods (Subramanian and Sivaramakrishnan, 2007), this biomonitoring tools provide historic information into the water quality. Hence, standard physico-chemical water quality methods need to be carried out in conjunction with biomonitoring tool for evaluating health of an ecosystem. This has resulted in the development of a number of relatively simple and rapid assessment techniques by which biological and other data can be presented numerically. These techniques are referred to as 'indices'. One important method to monitor changes in the ecosystem is the use of indicator taxa, that are species or higher taxonomic groups whose parameters such as density, presence or absence of infant survivorship are used as a tool for measuring of ecosystem condition (Hilty and Merelender, 2000).

Bioindicators: Tools for Biomonitoring

The term 'bioindicator species' was first coined by Kolkwitz and Marsson in the year 1908 during their research on impact of organic pollution on aquatic organisms. Thornton *et al.* (1994) defined it as "Characteristics of the environment that provide quantitative information on the condition of ecological resources, the magnitude of stress or the exposure of a biological component to stress". According to Markert *et al.* (1997) "A bioindicator is an organism (or part of an organism, or a community of organisms) that contains information on the quality of the environment (or part of environment)". Indicator taxa have been used to evaluate toxicity levels, abundance of specific resources, levels of biodiversity, target taxa status, endemism levels and ecosystem health (Croonquist and Brooks, 1991; Kremen *et al.*, 1993; Faith and Walker, 1996). Thus the use of indicator taxa is important to biological conservation. On the other hand, a biological indicator is a taxa or taxon selected based on its sensitivity to a particular attribute, and then assessed to make inferences about the attribute. Bioindicators are evaluated through presence or absence, condition, relative abundance, reproductive success, community structure (*i.e.* composition and diversity), community function *i.e.*, trophic structure or any combination thereof (Landres *et al.*, 1988).

Criteria for an Ideal Bioindicator

Though various methods exist for water quality assessment through the use of bioindicator taxa in the aquatic ecosystems, the baseline information about the details on the attributes of the identified indicator taxa are listed below -

1. Taxonomic status: Taxonomic soundness is necessary for indicator taxa. The species should be identified up to the family or genus using taxonomic keys provided in the taxonomic literature, at least, for that particular group.
2. Tolerance: A more tolerant taxon would not show any measurable change as a result of small or medium impacts. So the indicator taxon should be selected based on their ability to withstand a broad range of human impacts.
3. Correlation to other biota: This category indicates whether changes in each indicator taxon have been correlated with ecosystem changes.
4. Wide or cosmopolitan distribution: A taxon's distribution can be categorized in different ways like Local, Regional and Global.
5. Migratory habit: It is necessary to have the information on whether the taxon has a defined seasonal shift in non-contiguous habitats in any part of the taxon's range or not.
6. Trophic level: High trophic level was defined as carnivorous taxa of which adults are mostly predaceous in nature. Medium trophic level indicates omnivorous and carnivorous taxa of which adults were also potential prey to other species and low trophic level taxa included only herbivorous species of which adults are potentially prey to other species.
7. Specialization: The information is needed for the taxon if they are fond of a particular habitat or food (monophagous or oligophagous).
8. Reproductive rates: The selected taxa should have high reproductive rates.
9. Easy for sampling
10. Well-known ecological characteristics
11. Suitability for laboratory experiments and
12. High sensitivity to environmental stress

The biological indicators are currently used globally and promoted by numerous conservation agencies to assess human impacts, including the World Conservation Union, World Conservation Monitoring Centre (UNEP), U.S. Environmental Protection Agency (USEPA), as well as the Nature Conservancy, World Wide Fund for Nature (WWF), Friends of the Earth (FOE) and Greenpeace (IUCN, 1989; USEPA, 2002).

Assessment of Bioindicator Potentiality

The indicator value of a species is expressed as the degree (per cent) to which it fulfills the criteria of specificity and fidelity within any particular group of site. The association between each species and site group may be determined independently by both of the clustering procedure and of the other species and the significance of each indicator species would be determined using a site randomization procedure. Earlier, the biotic index approach by many European programmes integrates the indicator species concept with elements of diversity. Biotic index is simply a "scoring system" as it gives scores to taxonomic groups on assumed tolerance of the taxa to pollution and habitat disturbance (Cairns and Pratt, 1933). Currently for biomonitoring, the Biological Monitoring Working Party (BMWP) scores (Armitage *et al.*, 1983, Walley and Hawkes, 1997) and the Average Score Per Taxon (ASPT) is frequently used (Sivaramakrishnan *et al.*, 1996; Subramanian and Sivaramakrishnan, 2007). Basic biodiversity indices and biomonitoring scores can be calculated by putting the data in matrix. Pivot table function of MS Office Excel is useful in creating data matrix (Subramanian and Sivaramakrishnan, 2007). Basic data analysis can be done by using free software in Windows platform to estimate basic biodiversity parameter programs such as PAST, ESTIMATE-S, PC-ORD and BIODIVERSITY PRO etc.

Assessment using Aquatic Organisms

Assessment of biological condition is essential for effective ecosystem conservation, which is also true for freshwater ecosystems (Rosenberg and Resh, 1993; Simon, 2003; Palmer *et al.*, 1998; Cao *et al.*, 2002). The complexity of ecosystems has forced conservation biologists to develop alternative methods to monitor changes that would be too costly or difficult to measure directly (Landres *et al.*, 1988; Meffe and Carroll, 1994; Hilty and Merelender, 2000). One such cheapest method used for biomonitoring of aquatic ecosystem is the use of indicator taxa. The goal of monitoring ecosystem health is to identify the chemical, physical and biological changes due to human impacts. These assessments are mainly based on the measurements, interpretation and identification of different biotic indicators that are partially based on different ecological concepts and statistical techniques. Hence, incorporation of biological assessment into Government policy and regulations is needed and it would be a sensible step for different agencies and departments responsible for the management of aquatic ecosystems health (Yoder and Rankin, 1996; Metzeling *et al.*, 2006). The inland waters are inhabited by different vertebrate and invertebrate species. Among the invertebrates, the sponges, flatworms, molluscs, polychaete and oligochaete worms, crustaceans, insects are the most diverse group of organisms found in freshwater. Hence, aquatic insects are the most promising agents to understand how accurately they reflect the actual biological attributes and prove themselves as an ideal bioindicator. The different taxa show differential sensitivities that taxa exhibit in response to single or mixed disturbances. Although many taxa may decrease in abundance with increasing stress and their relative sensitivities usually differ (Lenat, 1988). Moreover, other taxa can increase in abundance with stress. The organisms that could be used as indicators of ecological condition in aquatic ecosystems are the aquatic macroinvertebrates for obvious reasons (Rosenberg and Resh, 1993; Metzeling *et al.*, 2006) as they are sensitive to stress and can reveal the effects of perturbation and habitat alteration (Arimoro *et al.*, 2007). These current monitoring techniques used to detect one or more of these changes to identify water quality problems at a site (Sivaramakrishnan *et al.*, 1996).

Aquatic Insects as Biomonitoring Agents

Among the freshwater animal taxa, the community structure of aquatic insects may be useful in assessing the water quality because of their high abundance, high birth rates with short generation time, large biomass in freshwater habitats. Insects are the most diverse group of organisms in freshwater. Estimation on the global number of aquatic insect species derived from the fauna of North America, Australia and Europe is about 45000 of this about 5000 species are known to inhabit inland wetlands of India (Subramanian and Sivaramakrishnan, 2007). Infact, aquatic insects constitute a dominating group of the benthic, limnotic and littoral fauna of lentic ecosystems and can be used assertively as efficient bio-indicators for bio monitoring of eco-health of freshwater ecosystem. They have evolved various morphological and physiological adaptations for living in aquatic environment. Modifications such as flattening of body, streamlining, reduction of projecting structures, suckers, friction pads, hooks, silk and sticky secretions are known to be existed in different groups who are usually found in lotic or running waters (Subramanian and Sivaramakrishnan, 2007). Other than this, they possess air tubes to obtain atmospheric oxygen (water bugs and dipteran larvae), haemoglobin pigments (chironomid larvae), cutaneous and gill respiration (immature stages of most aquatic insects), air bubbles (beetles and some bugs), plastron respiration that helps them to stay longer under water, legs with swimming hairs (water bugs). These morphological adaptations are closely related to behavioral adaptation such as, to avoid water current. Very few aquatic insects have adapted to a completely submerged life cycle. A major problem of being submerged is the respiration. As a result they have adapted some morphological or physiological adaptations. Some insects such as mayfly, caddies fly, stone fly, depend upon the dissolved oxygen (DO). The variation in environmental factors also affects the emergence pattern of different aquatic insects. One of the most important factors affecting aquatic life is temperature. In tropical region, many species show continuous emergence throughout the year. Some species undergo diapause in the form of egg, larvae, or pupae that help them to overcome unfavorable conditions.

Survey on Aquatic Insects Used as Bioindicator

Insects that are aquatic and semi-aquatic either throughout their livelihood and part of their life and can be used as bioindicator or biomonitoring agents are summarized in Table 13.1.

Table 13.1: Aquatic Insects Reported as Bioindicator of Aquatic Ecosystem

Order	*Examples*	*References*
COLLEMBOLA	*Podura aquatica*	Venkatesan, 1991
ORTHOPTERA	Tridactylidae (pigmy mole crickets)	Venkatesan, 1991
HEMIPTERA	important families are Belostomatidae, Notonectidae, Corixidae, Nepidae, Gerridae, Hydrometridae etc.	Thirumalai,1999; Subramanian and Sivaramakrishnan, 2007
EPHEMEROPTERA	*Ephemerella granle Baetis* sp., *Caenis* sp. etc.	Subramanian and Sivaramakrishnan, 2007
ODONATA	Family- Gomphidae, Aeshnidae etc. of Sub order Anisoptera (Dragonfly) and Stream Jewels, Torrent Darts of Suborder Zygoptera (Damselflies) etc.	Venkatesan,1991; Subramanian and Sivaramakrishnan, 2007
PLECOPTERA	Stoneflies eg., *Pteronarcys* sp., *Neoperla* sp. etc.	Venkatesan,1991; Subramanian and Sivaramakrishnan, 2007
MEGALOPTERA	*Sialis rotunda*	Venkatesan,1991
NEUROPTERA	Spongilla flies, *e.g. Climacia* sp.	Venkatesan,1991
TRICHOPTERA	Caddisflies. *e.g. Lepidostoma* sp., *Hydropsyche* sp., *Macronema* sp. etc.	Subramanian and Sivaramakrishnan, 2007
LEPIDOPTERA	*Aulocodes* sp., *Acentropus niveus* etc.	
HYMENOPTERA	*Caraphractus cinctus*	Venkatesan,1991
COLEOPTERA	*Cybister* sp.(Dytiscidae),*Dineutus*sp.(Gyrinidae), *Eubrianax* sp. (Psephenidae) etc.	Subramanian and Sivaramakrishnan, 2007
DIPTERA	*Culex* mosquito larvae, *Chironomous* sp. (Chironomidae), *Simulium* sp. (Simulidae), *Philorus* sp. (Blephoriceridae) etc.	Venkatesan, 1991, Subramanian and Sivaramakrishnan, 2007

Aquatic Hemiptera as Potential Bioindicator of Wetland Ecosystems of Himalaya and Sub-Himalaya Regions of West Bengal: A Case Study

The aquatic and semi-aquatic Hemiptera occupy a broad spectrum of aquatic habitats and are adapted to a broad variety of niches (Spencer and Andersen, 1994). Because of their abundance and richness in the mountain streams as well as in the stagnant freshwater bodies, the water bugs justifiably represent a major component of aquatic life which are not only the important member of food web of freshwater ecosystem, but also biomonitoring agents to evaluate health of aquatic ecosystem and intend to protect. As they require specific habitats, they are vulnerable to the loss of physical integrity of the aquatic ecosystem. Hence, their diversity can be correlated with the changes in physical integrity. Consequently, these features allow them to be used as bioindicator. They are commonly known as water bugs, can be distinguished from the other groups of aquatic insects by their unique feature of having piercing-sucking mouthparts, which has suctorial function. In India, 16 families of water bugs are found and each family differs considerably according to their habitat preferences with morphological and physiological adaptations. They have significant economic importance as bio indicators and good number of studies has been conducted to determine bugs association with environmental parameters. For example, the suitability of water striders of the family Gerridae as heavy metal indicators has been documented, whereas corixids are considered as one of the useful indicators of water quality (Nummelin *et al.*, 1998).

Study Design

Study Area

The present study was conducted in the two districts of West Bengal, mainly in the Darjeeling District, which comprises of the 'Darjeeling Himalaya', falling under Eastern Himalaya and the Himalayan foot hills and also in the Jalpaiguri district, which mainly consists of the Terai region or Sub-Himalaya region of West Bengal.

Methodology

Systematic stratified random sampling was carried out for all the families of aquatic and semi-aquatic Hemuptera and other aquatic insects in different wetlands ranging from ponds, lakes, waterfalls, hill streams, rivers, riffles, runs, roadside seeps, stagnant small pools within forests, irrigation canals such as canal within agricultural or paddy field, canal in the tea garden, canal besides dam etc. of Darjeeling Himalaya and Jalpaiguri sub-Himalaya region of West Bengal. A total of 50 sampling sites were surveyed to fulfill the objective. The collected samples were preserved in 70 per cent ethyl alcohol in small to large Borosil glass vials depending on their body size. The vials are labeled with the proper data such as locality, date and name of collector in the field. The collected samples of water bugs were brought to laboratory and were examined under stereoscopic binocular microscope (Leica M205A) and identified using standard taxonomic literature.

Data Organization

The data were organized using MS Excel 2007 with corrected spelling of samples to avoid problem of 'pseudo taxa'. The data matrix is used to assign the BMWP (Biological Monitoring Working Party) scores across 50 sampling sites.

Assigning BMWP (Biological Monitoring Working Party) scores: The determination of BMWP scores was based on the standard table of Armitage *et al.* (1983) and revised scoring system (Table 13.2) by Walley and Hawkes (1997). For calculation of BMWP score, identification of family is sufficient. The BMWP scores usually equal to the sum of the tolerance scores of all macro invertebrate families in the sample. The scores for various spatial scales can be obtained by summing the individual scores of all families present. Score values for individual families reflect their pollution tolerance based on the current knowledge of distribution and abundance. A higher BMWP score is considered to reflect a better water quality. Pollution intolerant families have high BMWP scores, while pollution tolerant families have low scores (Subramanian and Sivaramakrishnan, 2007). The abundance of species of aquatic and semi-aquatic Heteroptera found in different water quality sites are calculated by using MS Excel 2007. The BMWP score was calculated for the 50 sampling sites across the study area. The water quality parameters such as dissolved oxygen (DO), pH, salinity, turbidity, conductivity were also investigated for all the sampling sites.

Table 13.2: Biological Monitoring Working Party Score Revised by Walley and Hawkes (1997)

BMWP Score	*Category*	*Interpretation*
0-10	Very poor	Heavily polluted
11 to 40	Poor	Polluted or impacted
41-70	Moderate	Moderately impacted
71-100	Good	Clean but slightly impacted
>100	Very good	Unpolluted, unimpacted

Observations

The Biomonitoring Working Party score (BMWP) calculated for all of the 50 sampling sites shows that the site 1 of Rishi River and Rabijhora of Darjeeling district, Kalipur wetland of Gorumara National park of Jalpaiguri have highest BMWP scores with highest family richness. As per the standard table of BMWP scoring system (Walley and Hawkes, 1997), the Rishi River, located between West Bengal and Sikkim border of Darjeeling district at an altitude of 554 m can be referred to as a very good and unpolluted sites with the score of 102 and 13 families are found here. Whereas, the Kalipur wetland within Gorumara National Park of Jalpaiguri and Rabijhora of Darjeeling are also good sites, more or less clean, but slightly impacted. In the Kalipur wetland within Gorumara National Park both Nepomorpha and Gerromorpha are found. The sites such as Buri Torsha river of Bish Khutia, situated between the North Khairabari Reserve Forest and South Khairabari Reserve Forest, Murti river near Medla camp within Gorumara National Park, Sikhiajhora of Buxa Tiger Reserve, wetland within Chapramari National Park, Poro river, Raidhak river of Jalpaiguri and Ghoshpukur dighi, Rellykhola near Teesta Bazar, stagnant pool beside Rishi River of Darjeeling are with the BMWP scores ranging between 41-70 and hence these sites are moderately impacted. A total of 29 sites are determined as polluted sites with the BMWP scores ranges between 11- 40. Though few of these are located within the forest zone and the study indicates that these areas are still affected with several human activities like habitat degradation, intentional forest firing, pisciculture for increasing human demand of nearby villages, uncontrolled grazing in the forest zone. The sites like Bagdogra Sanyasithan tea garden near Siliguri, canal between Gava Ganga and Kamala tea garden, Bhimbhar dighi near Sayedabad tea garden, Manjukhola within Phuguri tea estate of Darjeeling are mainly polluted with the pesticidal runoff used in those tea gardens. The three sites which have streams or runs flowing across village are Gourjanjhora near Mal, Teesta canal near Gajaldoba Teesta barrage, Chel river near Gorubathan and are also polluted and impacted with human anthropogenic activities. In the study area of Darjeeling and Jalpaiguri region of West Bengal, among the 50 sites, 7 sites are found to be heavily polluted and hence very poor sites with <10 BMWP scores. Those are Chaitanyajhora within Buxa Tiger Reserve, Jhora within Chilapata forest near Mendabari beat, Mal river, Stagnant pool in the North Khairabari Reserve forest near Madarihat of Jalpaiguri district and Teesta river near Chitre Bridge, stream near Pulbazar, Jhora near Manebhanjang of Darjeeling.

Identification of Hemipteran Species as Bioindicator

To determine about the indicator species of aquatic and semi-aquatic Hemiptera, the study on the abundance of few species was carried out. A few species like *Mesovelia vittigera*, *Anisops breddini*, *Enithares* spp., *Gerris (G.) nepalensis* are only restricted to lentic ecosystems and may also be used to indicate the water quality. The species like *Ptilomera (P.) laticaudata* (Figure 13.1), *Chimarrhometra orientalis* (Figure 13.2), and *Heterobates rihandi* (Figure 13.3) were totally absent in very good and good water quality, but showed highest abundance in the poor water quality. However, rate of abundance decreases in the very poor water quality. Another sensitive species which was relatively common in the lotic habitat of Darjeeling and Jalpaiguri region is the *Ptilomera (Proptilomera) himalayensis*. The result (Figure 13.4) shows that they were extremely abundant in the clean sites with good water quality. The other lotic species which was very sensitive to water quality changes is the *Amemboa mahananda*. This species (Figure 13.5) was found in the very good to moderate water quality and rate of abundance was highest in the moderately impacted areas. But, this is totally absent in the poor and very poor quality of waters which are under serious threats. *Anisops breddini* is not found in the very good to good sites (Figure 13.6) and it is prevalent in the moderately impacted and polluted sites where some degree of fishing has been noticed. The most widespread species found in both lentic and lotic ecosystems of the study area are *Hydrometra greeni*, *Neogerris parvulus*, and *Limnogonus (L.) fossarum fossarum* of Gerromorpha and *Micronecta* spp. of Nepomorpha. *Hydrometra greeni* (Figure 13.7) is absent in the unpolluted and clean sites with good water quality. While, it is highly prevalent in the poor, polluted and moderately impacted sites. The other species like *Limnogonus (L.) fossarum fossarum* (Figure 13.8) and *Neogerris parvulus* (Figure 13.9) are most

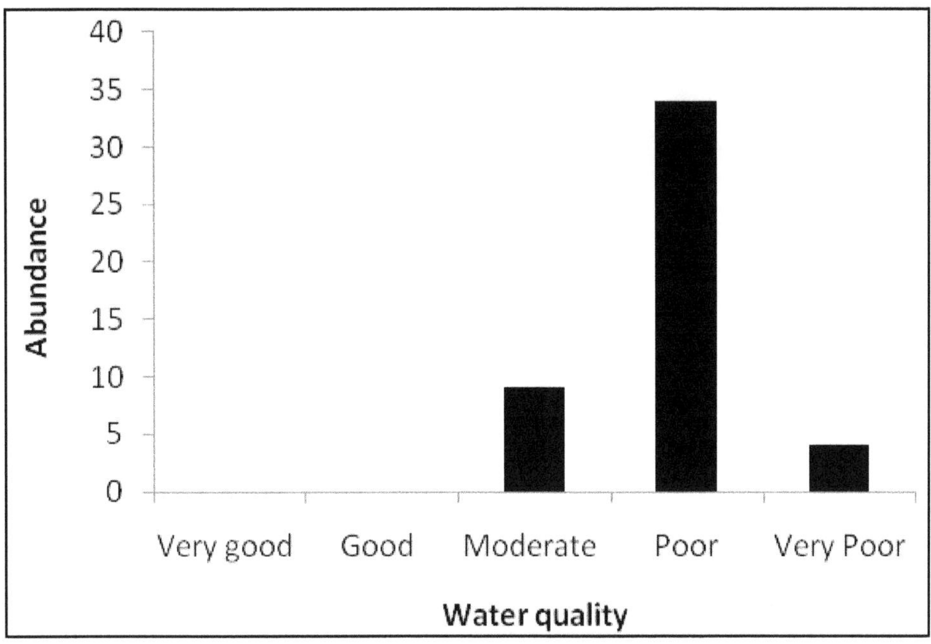

Figure 13.1: Abundance of *Ptilomera (P.) laticaudata.*

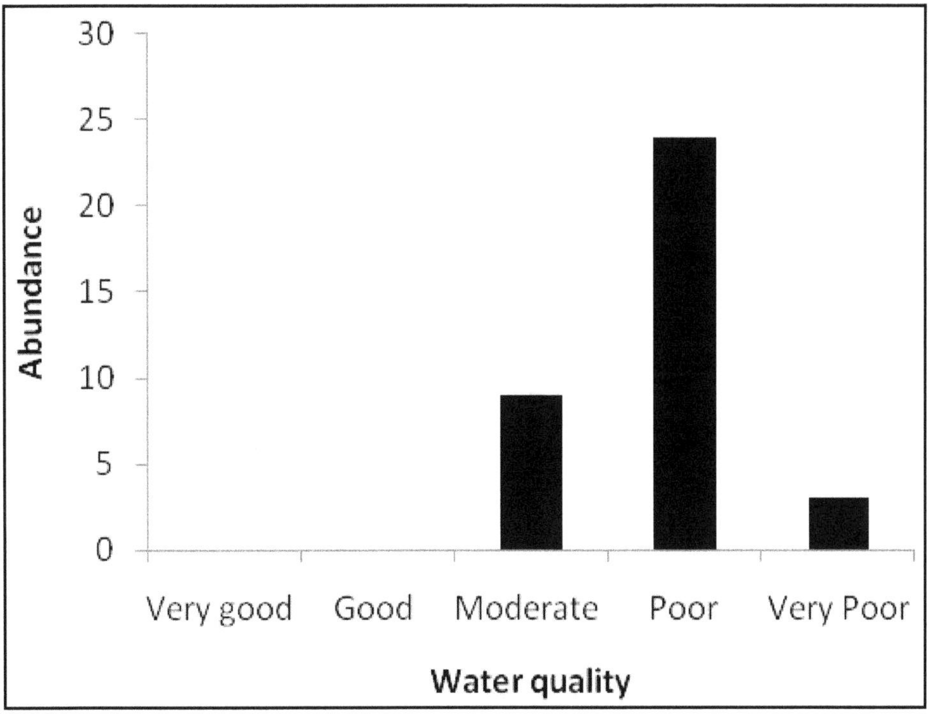

Figure 13.2: Abundance of *Chimarrhometra orientalis.*

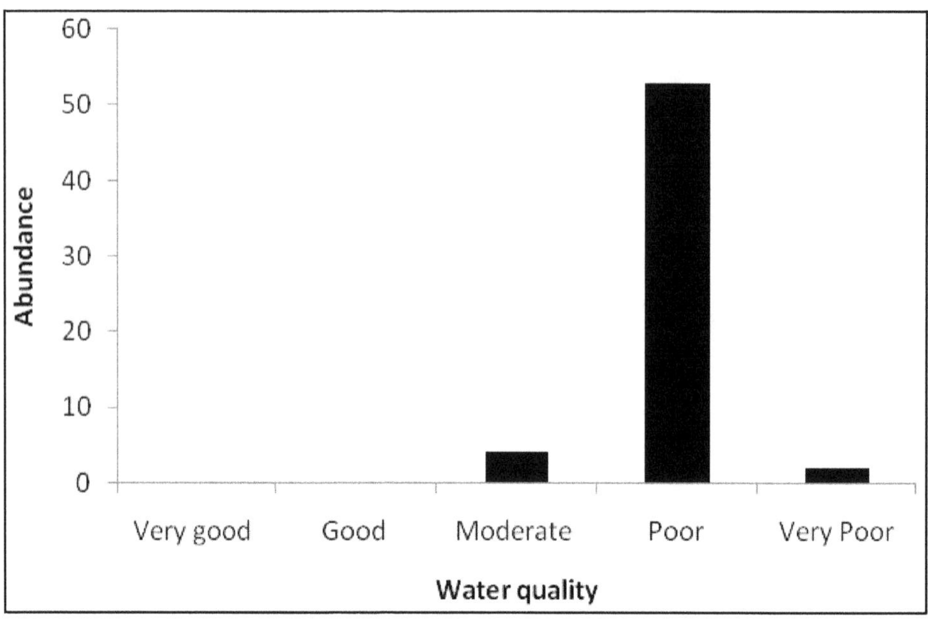

Figure 13.3: Abundance of *Heterobates rihandi.*

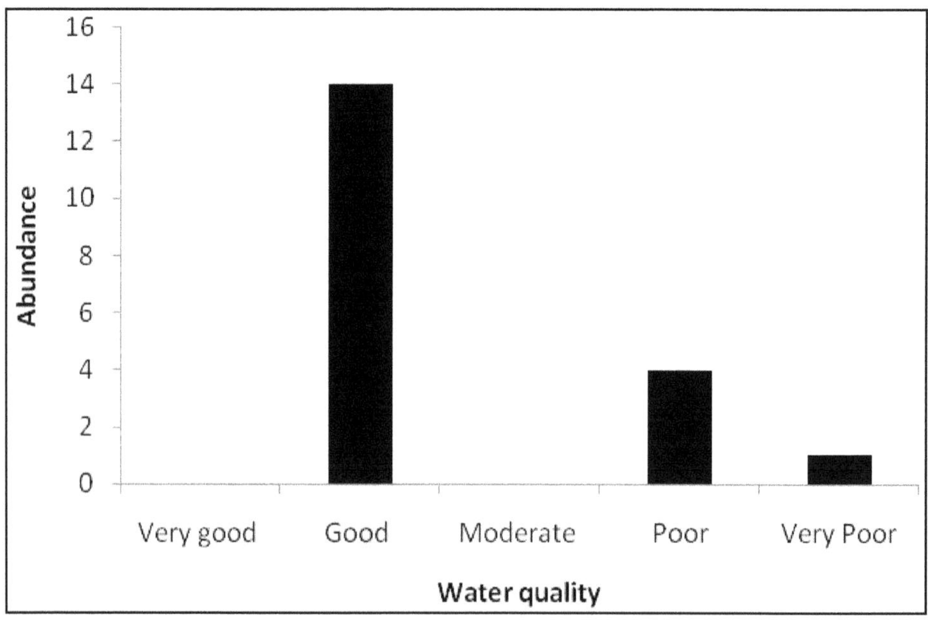

Figure 13.4: Abundance of *Ptilomera (P.) himalayensis.*

abundant in the moderately impacted sites and totally absent in the very good and very poor sites. The species which are mostly prevalent in the stagnant ecosystems of the study region are *Mesovelia vittigera* (Figure 13.10), *Anisops breddini*, *Gerris (G.) nepalensis* and the species of the genus *Enithares* (Figure 13.11). Among the semi-

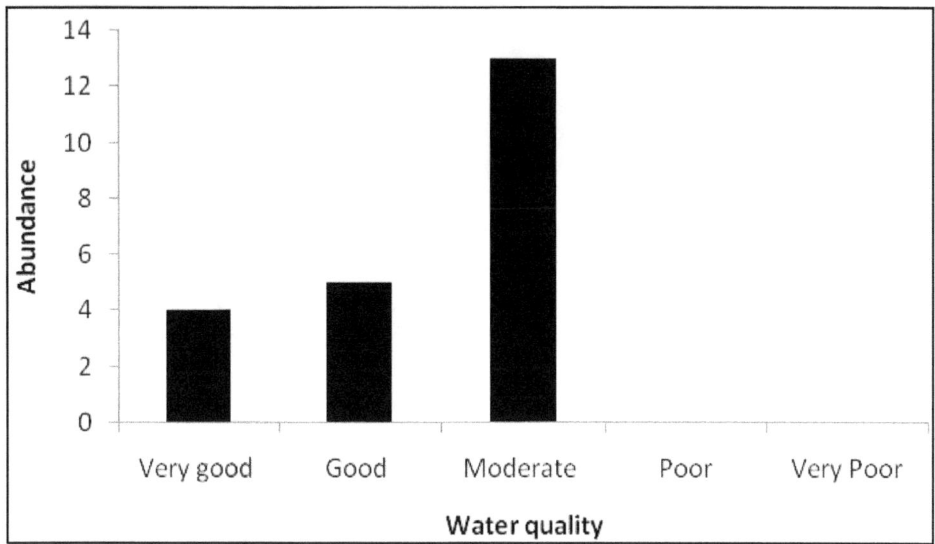

Figure 13.5: Abundance of *Amemboa mahananda*.

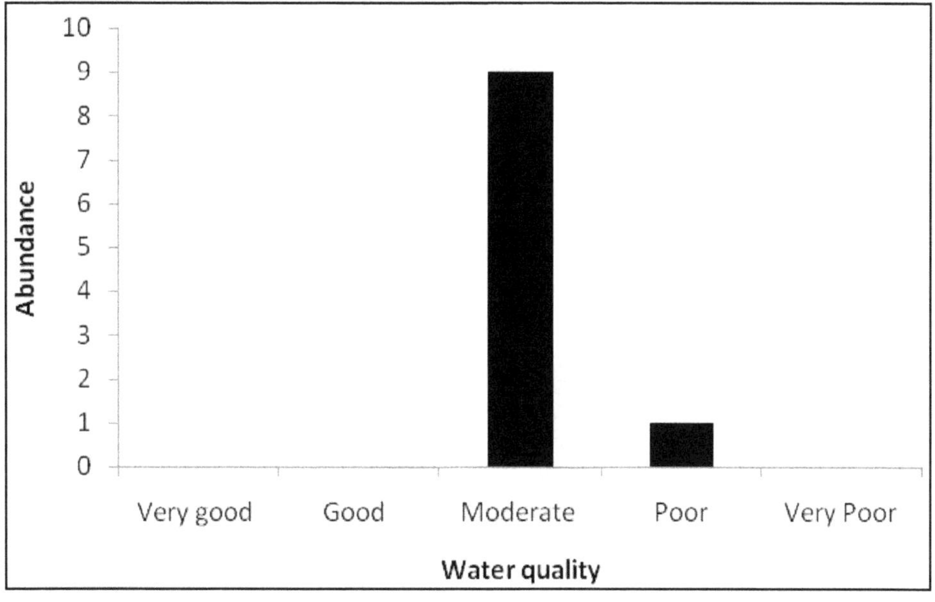

Figure 13.6: Abundance of *Anisops breddini*.

aquatic bugs, *Gerris (G.) nepalensis* (Figure 13.12) is present in higher numbers in the clean lentic sites with good quality of water and its abundance decreases as the impact or pollution load become higher and completely absent in the heavily polluted sites. The abundance of *Ranatra varipes* group (*R. v. varipes* and *R. v. atropha*) (Figure 13.13) and *Diplonychus annulatus* (Figure 13.14) are highest in the polluted

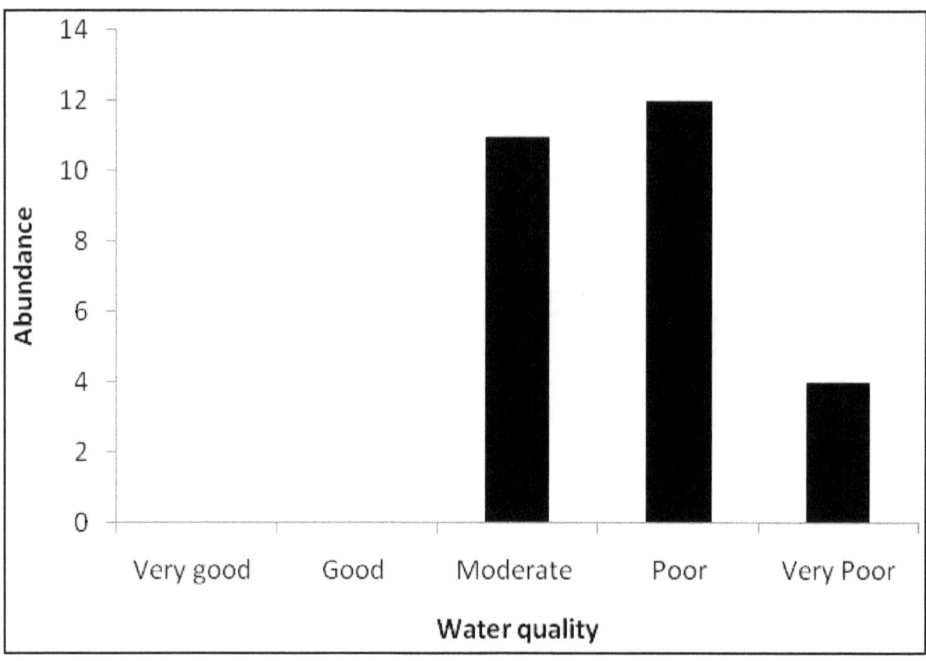

Figure 13.7: Abundance of *Hydrometra greeni.*

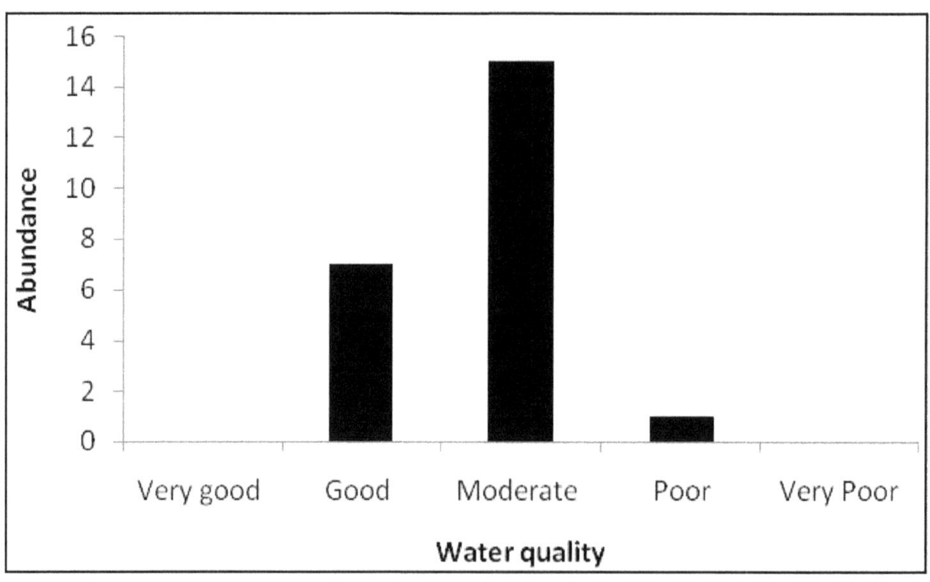

Figure 13.8: Abundance of *Limnogonus (L.) fossarum fossarum.*

sites with poor water quality and present in moderate quantity in the moderately polluted sites. They are totally absent in the very good and good sites. The list of species as indicators of water quality is provided in the Table 13.3.

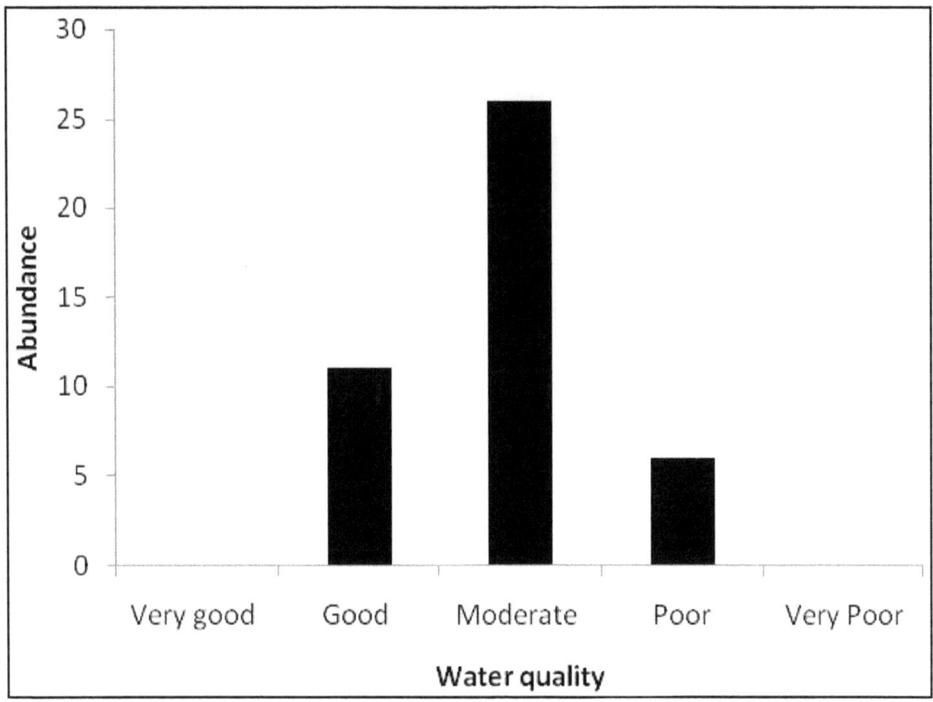

Figure 13.9: Abundance of *Neogerris parvulus.*

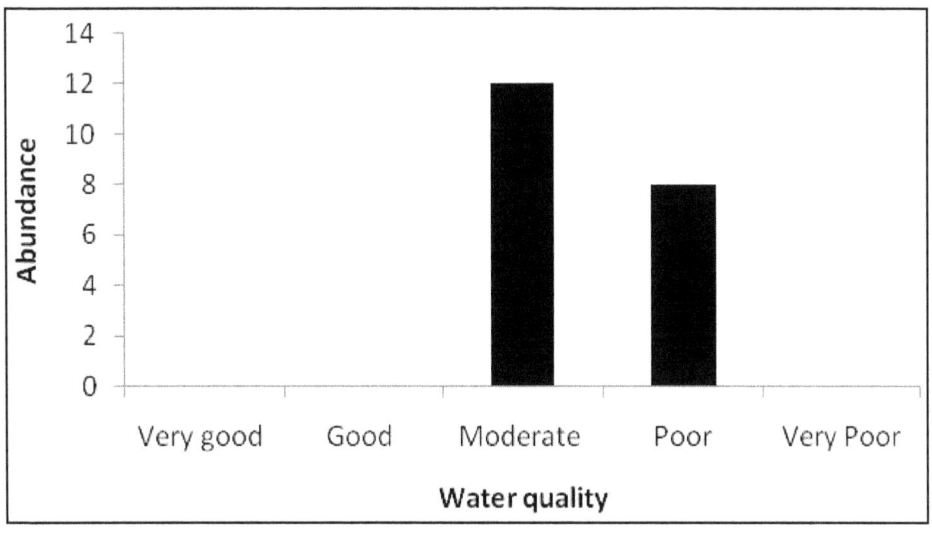

Figure 13.10. Abundance of *Mesovelia vittigera.*

Inference

A total of 1030 examples belonging to 47 species of aquatic and semi-aquatic Hemiptera were collected from the 50 sampling sites. The polluted sites which are

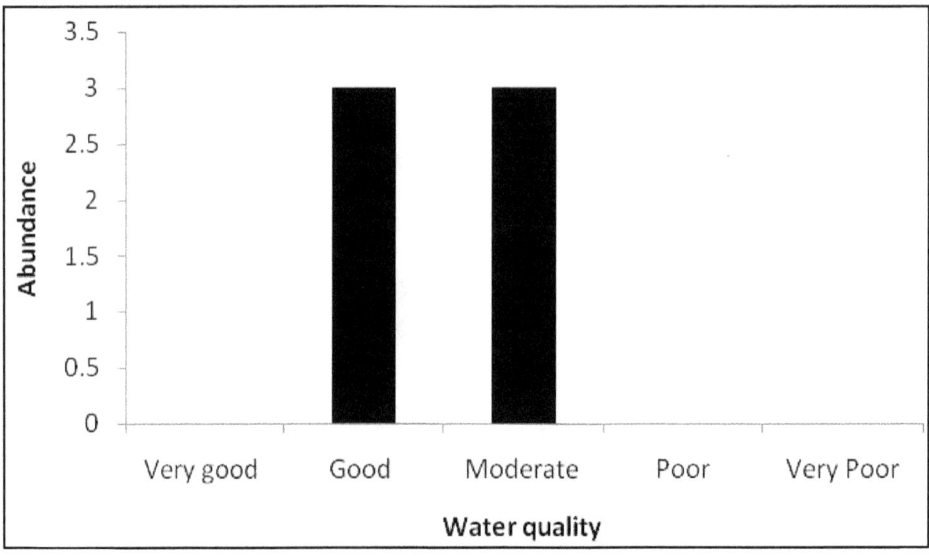

Figure 13.11: Abundance of *Enithares* spp.

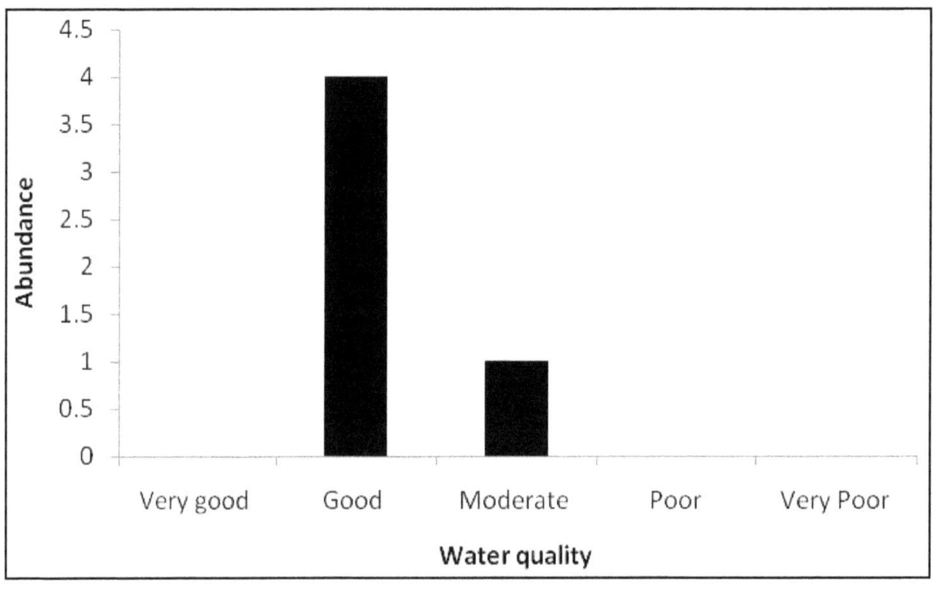

Figure 13.12: Abundance of *Gerris (G.) nepalensis.*

designated as poor sites were mainly dominated by *Rhagovelia* spp., *Diplonychus annulatus, Ranatra varipes varipes, R. varipes atropha, Mesovelia vittigera, Hydrometra greeni, Heterobates rihandi, Chimarrhometra orientalis, Ptilomera (P.) laticaudata, Ptilomera (P.) assamensis*. Based on abundance and frequency of occurrence, the species of aquatic and semi-aquatic Hemiptera best represented in the very good sites were *Amemboa mahananda, Gerris (G.) nepalensis, Ptilomera (P.) himalayensis, Enithares* sp.

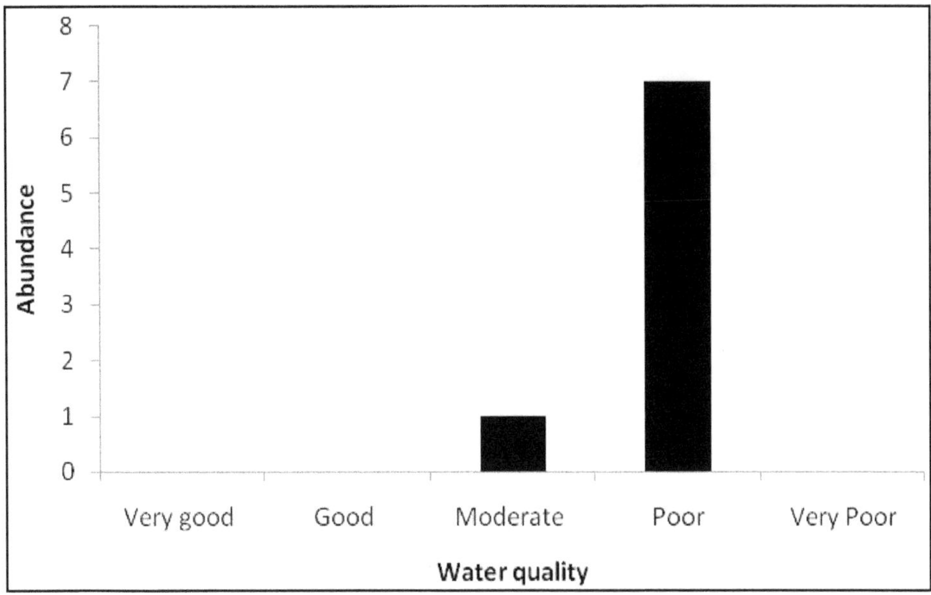

Figure 13.13: Abundance of *Ranatra varipes* Group.

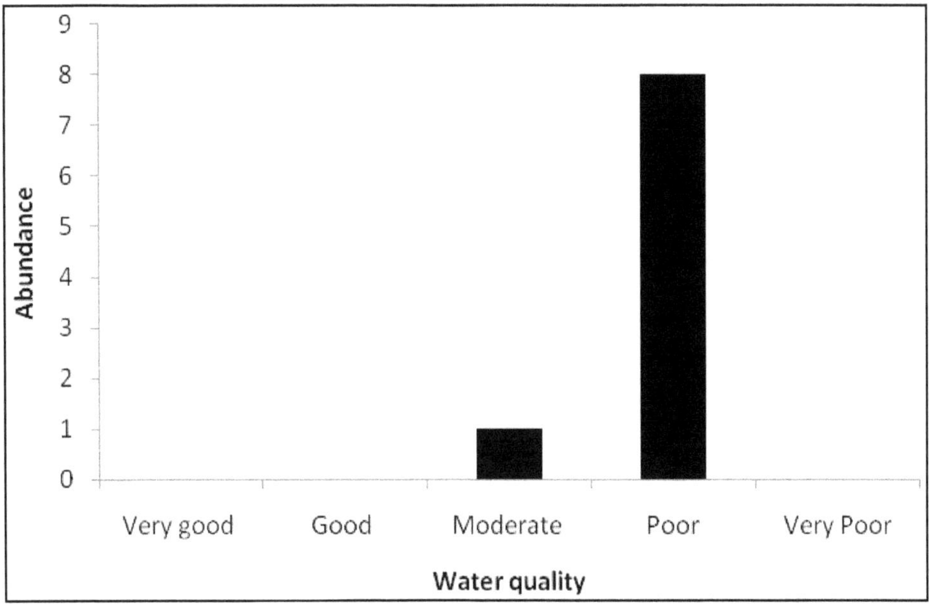

Figure 13.14: Abundance of *Diplonychus annulatus*.

and with their high abundance in clean sites they form the basic core taxa expected in healthy aquatic ecosystems. They can be referred to as pollution sensitive species. The pollution tolerant species identified for the running water bodies are *Ptilomera (P.) assamensis, P.(P.)laticaudata, Heterobates rihandi, Metrocoris* spp., *Chimarrhometra*

Pollution Tolerant Species
Ptilomera (P.) assamensis
Chimarrhometra orientalis
Hydrometra greeni
Ranatra varipes varipes
Aquarius adelaides
Diplonychus annulatus
Pollution Sensitive Species
Amemboa mahananda
Enithares sp.
Gerris (G.) nepalensis

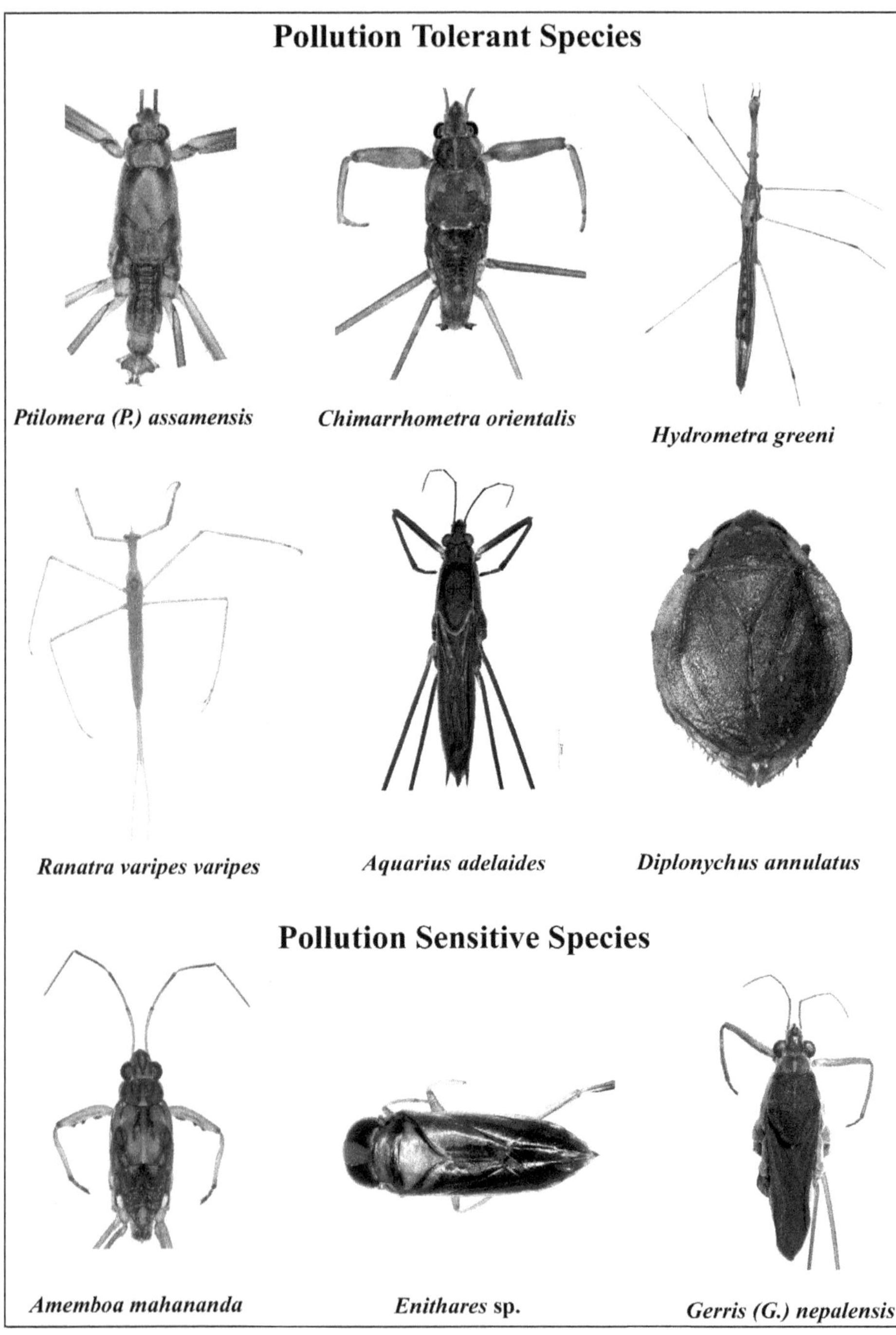

Figure 13.15

orientalis, Limnogonus (L.) fossarum fossarum, Hydrometra greeni. However, for the stagnant water bodies, the species like *Ranatra varipes, Diplonychus annulatus, Micronecta* spp., *Anisops breddini* are determined as pollution tolerant species.

Table 13.3: Aquatic Hemipteran Species Identified as Potential Indicators of Water Quality

Sl.No.	Water Quality Category	Species
1.	**Very good**	*Amemboa mahananda*
2.	**Good**	*Ptilomera (P.) himalayensis, Enithares* spp., *Gerris (G.) nepalensis, Amemboa mahananda*
3.	**Moderate**	*Neogerris parvulus, Hydrometra greeni, Enithares* spp., *Amemboa mahananda, Micronecta* spp., *Anisops breddini, Mesovelia vittigera, Limnogonus (L.) fossarum fossarum*
4.	**Poor**	*Ptilomera (P.) assamensis, Metrocoris* spp., *Ptilomera (P.) laticaudata, Chimarrhometra orientalis, Ranatra varipes, Diplonychus annulatus, Rhagovelia* spp, *Hydrometra greeni, Heterobates rihandi, Microneca* spp., *Anisops breddini, Mesovelia vittigera,*
5.	**Very Poor**	*Ptilomera (P.) assamensis, Metrocoris* spp., *Hydrometra greeni, Ptilomera (P.) laticaudata, Heterobates rihandi, Ptilomera (P.) himalayensis, Chimarrhometra orientalis*

Conclusion

Aquatic macroinvertebrates are usually regarded as good bioindicator than vertebrates as they respond very quickly to environmental changes (Kremen *et al.*, 1993). They are the most frequently used taxa for environmental monitoring (Hellawell, 1986). They are long-lived, easy to collect and their collection is inexpensive. But, the disadvantages of studying this group are their identification is laborious, time consuming and needs expertise. Biological monitoring requires continuous assessments of water quality. These monitoring techniques have attracted attention now a days because the organisms give response to the toxicity of the environment. They respond to the stress by their presence or absence, variation in abundance, population or community structure. Although traditionally, water quality monitoring techniques have focused on physical and chemical measurements of water quality, biomonitoring is an important tool to assess the ecosystems' health and provide valuable baseline information to formulate proper and effective management strategies.

Summary

The freshwater biodiversity and inland waters constitutes a valuable natural resource. Recently, the freshwater ecosystems are experiencing declines in biodiversity and hence conservation priority is the foremost challenges to the scientists. However, the conservation and management of freshwater ecosystems are critical to the human interest, nation and Government. The major threats identified to the freshwater biodiversity are over-exploitation, flow modification, water pollution, introduction of exotic species and habitat degradation. Hence, to conserve the freshwater ecosystems and to detect which ecosystems are under threat, biomonitoring using aquatic macroinvertebrates are promising. Among the macroinvertebrates, many species of aquatic bugs are proved to be a good bioindicator such as, *Enithares* sp. (Purkayastha and Gupta, 2013), *Metrocoris* sp., *Ranatra* sp., *Anisops* sp, *Sigara* sp., *Microvelia* sp., *Mesovelia* sp. (Chakrabarty *et al.*, 2014), *Gerris* sp., *Anisops* sp., *Limnometra* sp., *Ranatra* sp. (Gupta *et al.*, 2013), *Aquarius adelaides* (Pal *et al.*, 2012). They respond to the stress of ecosystems either by changing their abundance, community structure, presence or absence or by consuming the heavy metal toxins like cadmium, mercury etc. Hence, they can be used as an important bioindicator of aquatic ecosystems.

Acknowledgements

The authors are grateful to the Director, Zoological Survey of India and Head, Department of Zoology, University of Calcutta for their help and constant encouragements. We also thank Miss Swarnali Mukherjee, Research Scholar, Entomology and Wildlife Biology laboratory, University of Calcutta, for helping us in various ways. Special thanks are also due to Miss E. Jehamalar, Research Associate, Zoological Survey of India, Kolkata, for helping us in identification of species.

References

Allan JD and Flecker AS. 1993. Biodiversity conservation in running waters. *Bioscience*, 43 (1): 32- 43.

Arimoro FO, Ikomi RB and Iwegbue CMA. 2007. Water quality changes in relation to Diptera community patterns and diversity measured at an organic effluent impacted stream in southern Nigeria. *Ecological Indicators*, 7: 541-552.

Arthington D, Gessner AH, Kawabata MO. *et al.*, 2006. Freshwater biodiversity: importance, threats, status and conservation challenges. *Biological Review*, 81: 163-182.

Armitage, PD, Moss D, Wright JF and Furse MT. 1983. The performance of a new biological water quality score system based on macroinvertebrates over a wide range of unpolluted running-water sites. *Water Research*, 17 (3): 333-347.

Balian EV, Segers H, Léveèque C and Martens K. 2008. The Freshwater Animal Diversity Assessment: an overview of the results. *Hydrobiologia*, 595: 627-637.

Cairns J and Pratt JR. 1993. A history of biological monitoring using benthic macroinvertebrates. In: Freshwater *Biomonitoring and Benthic Macroinvertebrates* (Rosenberg DM and Resh, VH eds.). Chapman and Hall, New York. pp. 10-27.

Cao Y, Larsen DP, Hughes RM. *et al.*, 2002. Sampling efforts affect multivariate comparisons of stream assemblages. *Journal of the North American Benthological Society*, 21: 707-714.

Chakravorty PP, Sinha M and Chakrabarty SK. 2014. Impact of industrial effluent on water quality and benthic macroinvertebrate diversity in freshwater ponds in Midnapore district of West Bengal, India. *Journal of Entomology and Zoology Studies*, 2 (2): 93-101.

Croonquist MJ and Brooks RP. 1991. Use of avian and mammalian guilds as indicators of cumulative impacts in riparian-wetland areas. *Environmental Management*, 15(5): 701-714.

Dudgeon D, Arthington AH, Gessner MO. *et al.*, 2006. Freshwater Biodiversity: Importance, Threats, Status and Conservation Challenges. *Biological Reviews*, 81: 163-182.

Faith DP and Walker PA. 1996. How do indicator groups provide information about the relative biodiversity of different sets of areas?: on hotspots, complementarity and pattern-based approaches. *Biodiversity Letters*, 3: 18-25.

Gupta S, Dey S and Purkayastha P. 2013. Use of aquatic insects in water quality assessment of ponds around two cements factories of Assam, India. *International Research Journal of Environment Sciences*, 2 (7): 15-19.

Hellawell JM. 1986. *Biological Indicators of Freshwater Pollution and Environmental Management.* Elsevier Applied Science Publishers, London.

Hilty J and Merenlender A. 2000. Faunal indicator taxa selection for monitoring ecosystem health. *Biological Conservation*, 92: 185-197.

IUCN. 1989. From Strategy to Action: The IUCN Response to the Report of the World Commission on Environment and Development. IUCN, Gland, and Cambridge, UK.

Jackson RB, Carpenter SR, Dahn CN. *et al.*, 2001. Water in a changing world. *Ecological Applications*, 11: 1027-1045.

Jansson A. 1977. *Micronecta* (Heteroptera, Corixidae) as indicators of water quality in two lakes in southern Finland. *Annales Zoologici Fennici*, 14: 118-124.

Katona JN. 2008. The environmental significance of bioindicators in sewage treatment. *Acta Polytechnica Hungarica*, 5 (3), 117-124.

Kolkwitz R and Marsson M. 1908. Ökologie der flanzliche saprobien. *Berichte der Deutschen Botanischen Gesellschaft*, 26: 505-519.

Kremen C, Colwell RK, Erwin TL. *et al.*, 1993. Terrestrial arthropod assemblages: their use in conservation planning. *Conservation Biology*, 7: 796-808.

Landres PB, Verner J and Thomas JW. 1988. Ecological uses of vertebrate indicator species: a critique. *Conservation Biology*, 2: 316-328.

Lenat DR. 1988. Water quality assessment of streams using a qualitative collection method for benthic macroinvertebrates. *Journal of North American Benthological Society*, 7: 222-233.

Lundberg G, Kottelat M, Smith GR. *et al.* (2000). So many fishes, so little time: an overview of recent ichthyological discovery in continental waters. *Annals of the Missouri Botanical Gardens*, 87: 26-62.

Malmquist B and Rundle S. (2002). Threats to the running water ecosystems of the world. *Environment Conservation*, 29: 134-153.

Markert B, Oehlmann J and Roth M. 1997. General aspects of heavy metal monitoring by plants and animals. In. Environmental biomonitoring - exposure assessment and specimen banking. (Subramanian G, Iyengar V, eds.) *ACS Symposium Series*, Vol. 654. Washington DC: American Chemical Society.

Meffe GK and Carroll CR. 1994. *Principles of Conservation Biology*. Sinauer Associates, Inc., Sunderlans, MA. pp. 600.

Metzeling L, Tiller D, Newall P. *et al.*, 2006. Biological objectives for the protection of rivers and streams in Victoria, Australia. *Hydrobiologia*, 572: 287-299.

Myres N, Mittermeier R, Mittermeier, GC. *et al.*, 2000. Biodiversity hotspots for conservation priorities. *Nature*, 403: 853-858.

Naiman RJ and Turner MG. 2000. A future perspective on North America's freshwater ecosystems. *Ecological Applications*, 10: 958-970.

Nummelin M, Lodenius M and Tulisalo E. 1998. Water striders (Gerridae: Heteroptera) as indicators of heavy metal pollution. *Entomologica Fennica*, 8: 185-191.

Pal A., Sinha DC and Rastogi N. 2012. *Gerris spinolae* Lethierry and Severin (Hemiptera: Gerridae) and *Brachydeutera longipes* Hendel (Diptera: Ephydridae): Two Effective Insect Bioindicators to Monitor Pollution in Some Tropical Freshwater Ponds under Anthropogenic Stress. *Psyche*, 2012: 1-10.

Palmer MA, Bernhardt ES, Allan, JD. *et al.*, 1998. Auditing the Earth: the value of the world's ecosystem services and natural capital. *Environment*, 40: 23-27.

Pradhan B. 2005. *Water quality classification model in the Hindu-Kush Himalayan region: the Bagmati River in Kathmandu valley, Nepal.* Submitted to ICIMOD: 1-76.

Purkayastha P and Gupta S. 2013. Estimating ecosystem health of shallow water pond in Lower Irongmara, Barak Valley, Assam, India using ASPT, SPI and BMWP score. *International Research Journal of Biological Sciences*, 2 (8): 1-4.

Rosenberg DM and Resh VH. 1993. *Freshwater Biomonitoring and Benthic Macroinvertebrates*. Chapman and Hall, London.

Roux A, Hilgers A, de Feraudy H. *et al.*, 1993. Auroral kilometric radiation sources: *In situ* and remote observations from Viking. *Journal of Geophysical Research*, 98(A7): 11657-11670.

Saunders DL, Meeuwig JJ and Vincent ACJ. 2002. Freshwater protected areas: Strategies for Conservation. *Conservation Biology*, 16 (1): 30-41.

Simon TP. 2003. Biological response signatures: Toward the detection of cause-and-effect diagnosis in environmental disturbance. In. *Indicator Patterns Using Aquatic Communities* (Simon TP. ed.), CRC Press LLC, Boca Raton, Florida, USA. pp. 3-12.

Sivaramakrishnan KG, Hannaford JM and Resh VH. 1996. Biological Assessment of the Kaveri River Catchment, South India, and Using Benthic Macroinvertebrates: Applicability of Water Quality Monitoring Approaches Developed in Other Countries. *Int. J. Eco. and Env.Sci*, 32: 113-132.

Spencer JR and Andersen NM. 1994. Biology of water striders: interactions between systematic and ecology. *Annual Review of Entomology*, 39: 101-128.

Subramanian KA and Sivaramakrishnan KG. 2007. Aquatic Insects for Biomonitoring Freshwater Ecosystems-A Methodology Manual. Ashoka Trust for Ecology and Environment (ATREE), Bangalore, India. pp. 31.

Thornton KW, Saul GE and Hyatt DE. 1994. Environmental Monitoring and Assessment Program: Assessment Framework. EPA/620/R-94/016. U.S. Environmental Protection Agency, Office of Research and Development: Research Triangle Park, NC.

Thirumalai G. 1999. Aquatic and semi-aquatic Heteroptera of India. *Indian Association of Aquatic Biologists*, 7: 74.

USEPA. 2002. Biological indicators of watershed health. Last updated June 13, 2002, last accessed July 27, 2002. http://www.epa.gov/bioindicators/

Venkatesan P.1991. Aquatic insects- at a glance. In. *Perspectives of aquatic entomology-manual*, Chapter 2 (Venkatesan P. ed). Loyola College, Madras.

Walley WJ and Hawkes HA. 1997. A computer-based reappraisal of the Biological Monitoring Working Party score system incorporating abundance rating, site type and indicator value. *Water Research*, 31 (2): 201-210.

Yoder CO and Rankin ET. 1996. The role of biological indicators in a state water quality management process. *J. Env. Mon. Assess.*, 51 (1-2): 61-88.

Chapter 14

A Narrative on the Theories and Empirical Evidences of Intraguild Predation

Shreya Brahma, Gautam Aditya and Goutam Kumar Saha

Entomology and Wildlife Biology Research Laboratory, Department of Zoology, University of Calcutta, Kolkata

Introduction

The theoretical perspectives and applied examples of intra-guild predation (IGP) system has been reviewed in recent years by several workers (Wissinger, 1992; Rosenheim *et al.*, 1999; Blumenshine and Hambright, 2003; Vance-Chalcraft and Soluk, 2005) including trophic supplements (Daugherty *et al.*, 2007), to sustain such systems. The heteropteran bugs are voracious feeder of different aquatic organisms including the mosquitoes, tadpoles (Babbitt and Jordan, 1996), and smaller detritivore heteropterans (Hampton and Gilbert, 2001) of freshwater ecosystems. These organisms have been used as models to study different ecological theories related to predation mechanisms in arthropods and insects in particular. Many of the heteropteran bugs bear the potential to regulate mosquito abundance (Saha *et al.*, 2010, 2012, 2014). It has been noted that the notonectids *Buenoa* (Hampton *et al.*, 2000) and *Notonecta* (Blaustein, 1998) can act as intra-guild predators. Evaluation of intra-guild predation in variety of insects like aphids (Snyder and Ives, 2003), dragonfly nymphs (Wissinger and McGrady, 1993; Suutari *et al.*, 2004; Crumrine and Crawley, 2003; Crumrine, 2005) and water scorpion (Ohba and Swart, 2009) are known, which determine the population level of different prey species in ecological communities (Langellotto and Denno, 2004). The success of intra-guild predators can be affected by morphs of the intermediate predators (Lucus *et al.*, 1997) and functional and numerical responses may not be appropriately reflecting the ability of predators (Lester and Harmsen, 2002). Habitat structure of ricefields augments aggregation of mosquitoes and other insects (Lawler and Dritz, 2005), which might show intra-guild predation system. In the context of mosquito biocontrol using natural predators it is expected that the intra-guild prey may influence the

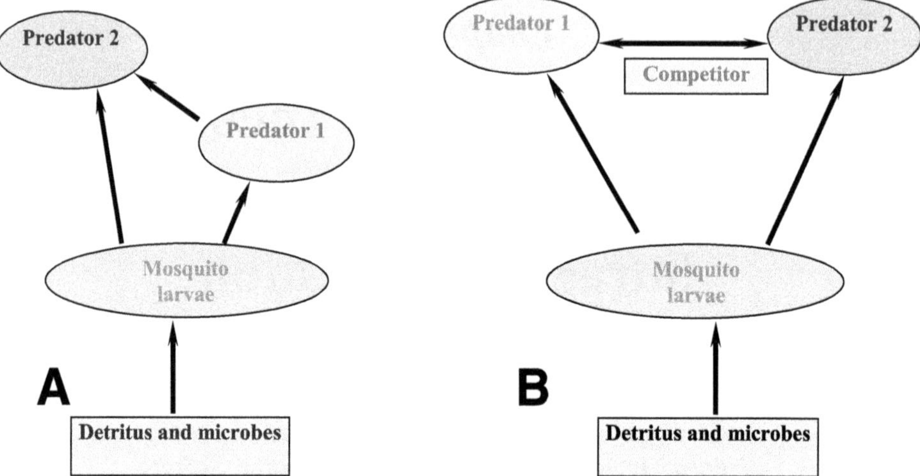

Figure 14.1: Two Basic Models of Intra-guild Predation (Rosenheim *et al.*, 1995). In A, the two predators are occupying different trophic levels, but share the same prey. While in B, the two predators are competitors to one another. The present HYPOTHESIS is aimed at estimation of the effect of the predators on the target prey population, with the extensions (Blaustein and Chase, 2007; Juliano, 2009) represented below C.

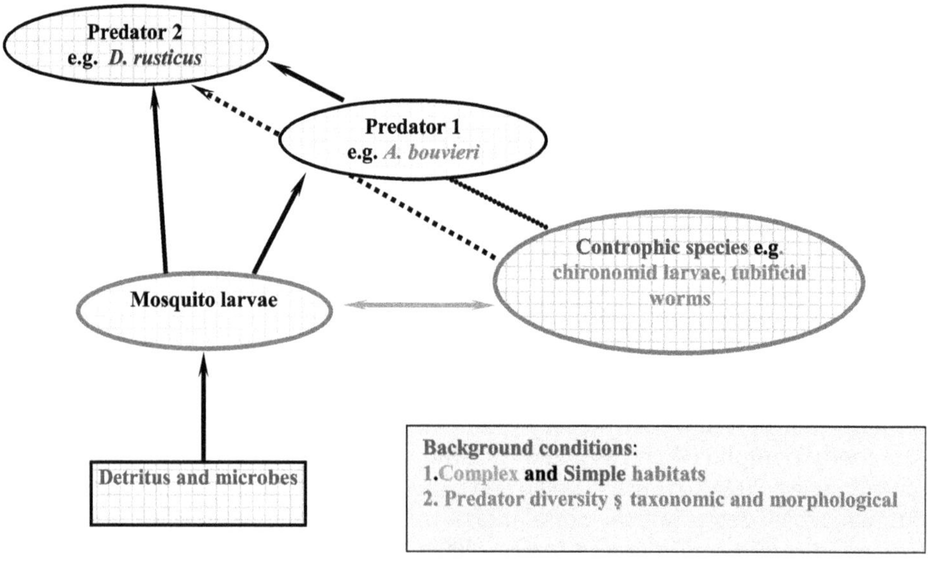

efficiency of regulation by the predators (Blaustein and Chase, 2007; Juliano, 2009). The aquatic hemipterans considered in this compilation are established predators regulating the abundance of different aquatic organisms (Gorai and Roychoudhuri, 1962; Jana *et al.*, 2009; Hazarika and Goswami, 2010). The belostomatid, notonectid and the nepid bugs prey upon mosquito immature (Nishi and Venkatesan, 1989; Aditya *et al.*, 2004; Sivagnaname, 2009). The preference for mosquitoes in presence of alternative prey is known to be density and frequency dependent (Aditya *et al.*, 2005). These water bugs are opportunistic predators (Saha *et al.*, 2010) and can be utilized in regulation of mosquitoes in wetlands and different mosquito larval habitats. However, intra-guild predation employing these predators or other predators of economic importance has not been noted in Indian context.

The Intraguild Predation: Theory and Empirical Evidences

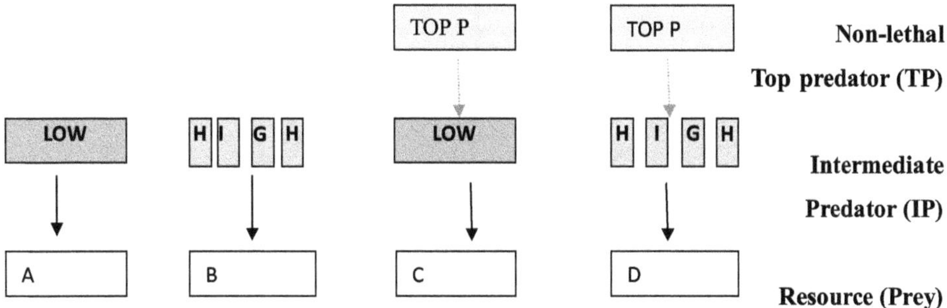

Figure 14.2: Illustrations of the Four different Systems in GP.

(A) Low intermediate predator diversity, (B) high intermediate predator diversity (C) low intermediate predator diversity with non-lethal predator present, and (D) high intermediate predator diversity with non-lethal predator present. The arrows represent predicted impact of predators on consumers -the arrow thickness showing the strength of the impact. Dashed arrows represent non-lethal impacts (trait-mediated indirect interaction).

In Figure 14.2, the impact of the intraguild predator and top predator on the prey resource is depicted with varying density levels and the trophic links. The impacts are generally trait-mediated changes in the behaviour and morphology of the prey resource, since the density variations in the top predator is not assumed. However, such density mediated impact can be seen in case of prey resource, provided ample variations in the prey density occur such that the predator is able to induce changes in the population level of the prey. Contrast to these, the presence of two predators sharing the common prey imparts effect at the level of the life history traits such that the modifications become adaptive. This is illustrated among several predator and prey taxa particularly in agroecosytems like cotton cultivation and the aquatic ecosystems consisting of water bugs and odonate larvae. The implications of such predator prey interactions are recognized as indirect interactions, when the species interactions in the mosquito larval habitats are considered.

In case of intraguild predation the impact of the predators – IG predator and top predator can induce defenses on the shared prey as has been seen in *Daphnia* where the presence of multiple predator leads to the development of spines. The prey taxa revives its defense mechanism with higher vigour and intensity once the

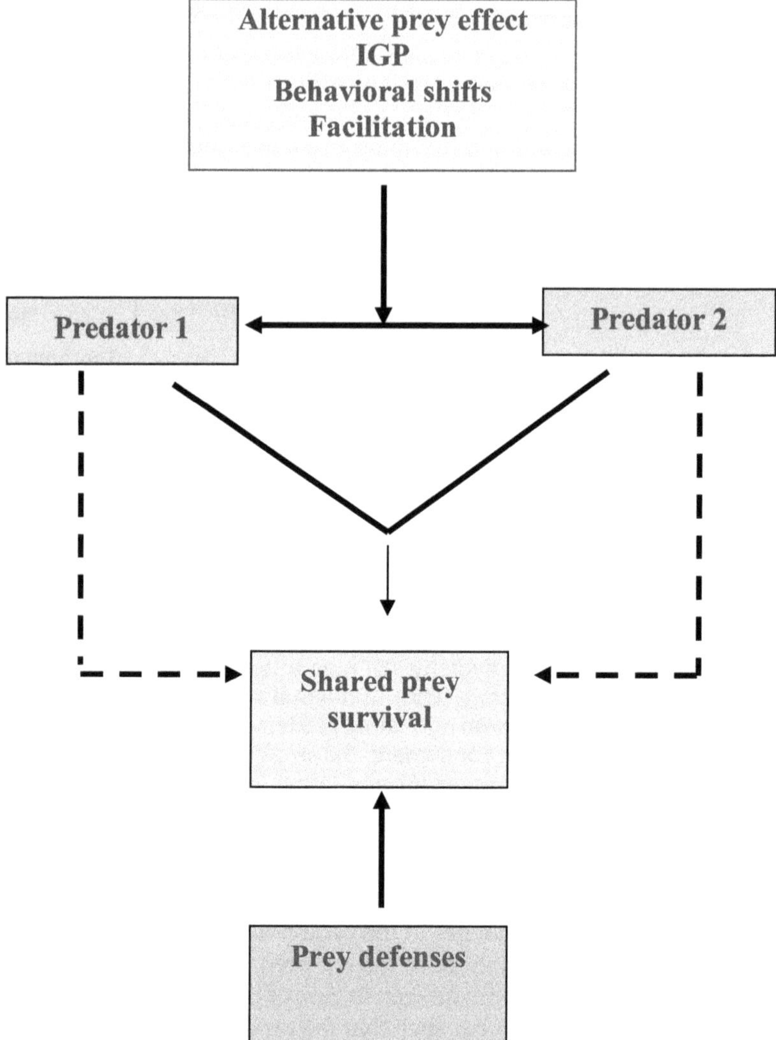

Figure 14.3: Conceptual Overview of the Effects of Two Predators on Shared Prey Survival.

Predators 1 and 2 both have an independent effect on prey survival when alone (dashed lines). When both predators are present (solid lines), the effect may depend on the interaction between predators. Whether risk reduction or enhancement occurs may depend on the occurrence and magnitude of factors such as IGP and behavioral shifts made under the threat of IGP by IG prey. Prey defenses can promote survival but may conflict with multiple predators and decrease survival (adapted from Crumrine and Crowley, 2003).

intermediate predator and top predator population increases in number (Figure 14.3). As a whole the indirect interaction between these taxa at different trophic levels may exhibit apparent competition, apart from direct competitions. An alternate prey can diffuse the competition as well as augment the shared prey population if present. This has been elaborated in the review illustrating the impact of several prey taxa in the mosquito larval habitats where the presence of the alternative prey favours the increase of the mosquito prey population (Blaustein and Chase, 2007).

As shown in Table 14.1, the controphic species can share both predators and resources. Here, the impact of the controphic species on the target species depends on a balance between competition for resources and apparent competition through shared predators.The presence of a controphic species and predator can benefit the target species if the former is more susceptible to predation, but harm the target species if the former is less susceptible to predation. The developmental response in terms of growth of the top predator and intermediate predators can result from consumption of the target and alternative prey. Intraguild predation may also induce increase in abundance of the inferior competitors due to reduced abundance of a superior competitor caused by a top predator.

Table 14.1: Direct and indirect Interactions at Various Trophic Levels. Schematic diagrams of the modules by which species can both directly and indirectly affect a target species. Here, we focus specifically on controphic species and the way that they interact with each other, resources, and predators. One-way arrows point to the consumer. Two-way arrows designate a mutually non consumptive negative interaction (as stated in Blaustein and Chase, 2007)

TYPES OF INTERACTIONS	Interference competition	Exploitative competition	Apparent competition	Indirect mutualism	Intraguild predation	Keystone predation
Predator (PR)			PR		PR	PR
Controphic Species (A and B)	A ↔ B	A B	A B	A B	A	A B
Resources (R, R1 and R2)		R		R1 ↔ R2	R	R

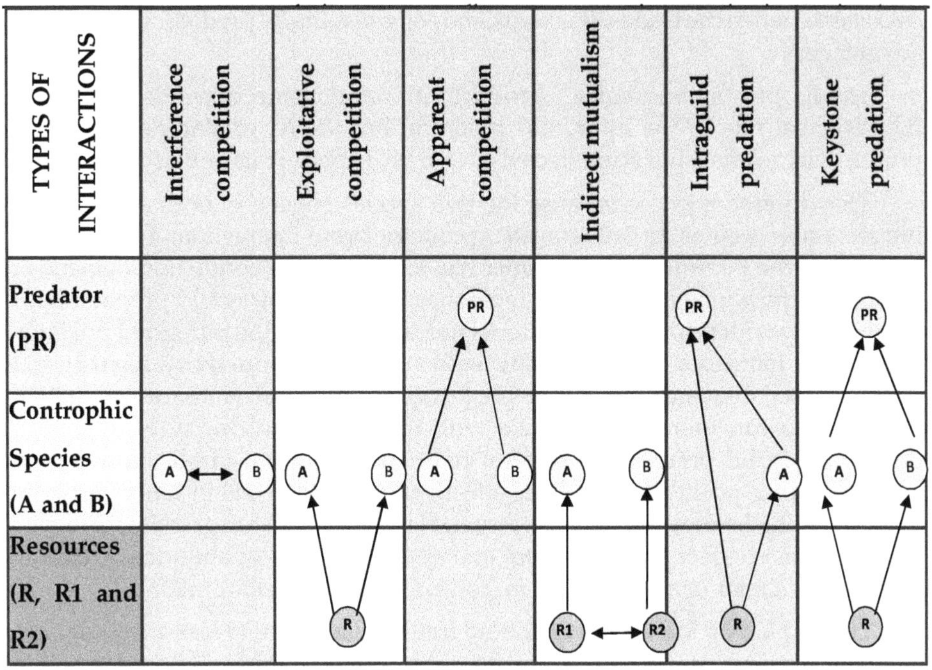

- **Controphic species**: Two or more species sharing the same trophic level but not necessarily the same functional food groups.
- **Apparent competition**: Two or more prey species that share the same predator have a mutually negative interaction because they contribute to increased predator populations and thus to ultimately higher predation intensities on each prey species.
- **Apparent mutualism**: Two or more prey species sharing the same predator have reduced predation intensity because the predator can feed on these alternative prey.

IGP in Different Taxa: Theory and Empirical Evidences

At the community level the impacts of the intraguild predators can be viewed through the analyses of the food web as shown in the Figure 14.4, where an insect species assemblage in terrestrial ecosystem has been assessed in terms of food web properties to evaluate the potential of indirect interactions among the controphic species in terms of intraguild predation.

The various other forms of intraguild predation can be viewed as a conceptual model depicting the indirect interactions involving three trophic levels. Two key observations on the influence of dispersal on the coexistence and abundance-productivity relationships of two species engaged in intraguild predation are:

Dispersal enhances coexistence when a trade-off between resource competition and IGP is strong and/or when the IG prey has an overall advantage and impedes coexistence when the trade-off is weak and/or when the IG predator has an overall advantage.

The IG prey's abundance - productivity relationship depends crucially on the dispersal rate of the intraguild predator, but the IG predator's abundance-productivity relationship is unaffected by its own dispersal rate or that of the IG prey.

This difference arises because the two species engage in both a competitive interaction as well as an antagonistic (predator-prey) interaction. The intraguild prey being the intermediate consumer has to balance the conflicting demands of resource acquisition and predator avoidance, while the intraguild predator has to contend only with resource acquisition (Figure 14.5). Thus, the intraguild predator's abundance increases monotonically with resource productivity regardless of either species' dispersal rate, while the intraguild prey's abundance-productivity relationship can increase/decrease with increasing productivity depending on the intraguild predator's dispersal rate. The important implication is that a species' trophic position determines effectiveness of dispersal in sampling spatial environmental heterogenecity. The dispersal behavior of a top predator is likely to have a stronger effect on coexistence and spatial patterns of abundance, than the dispersal behavior of an intermediate consumer (Amerasekare, 2006).

The impact of the larvae of the migratory dragonfly (*Tramea lacerata*) on a common resident dragonfly species (*Erythemis simplicicollis*), and on damselflies, through manipulative field experiments revealed the following (Wissinger and

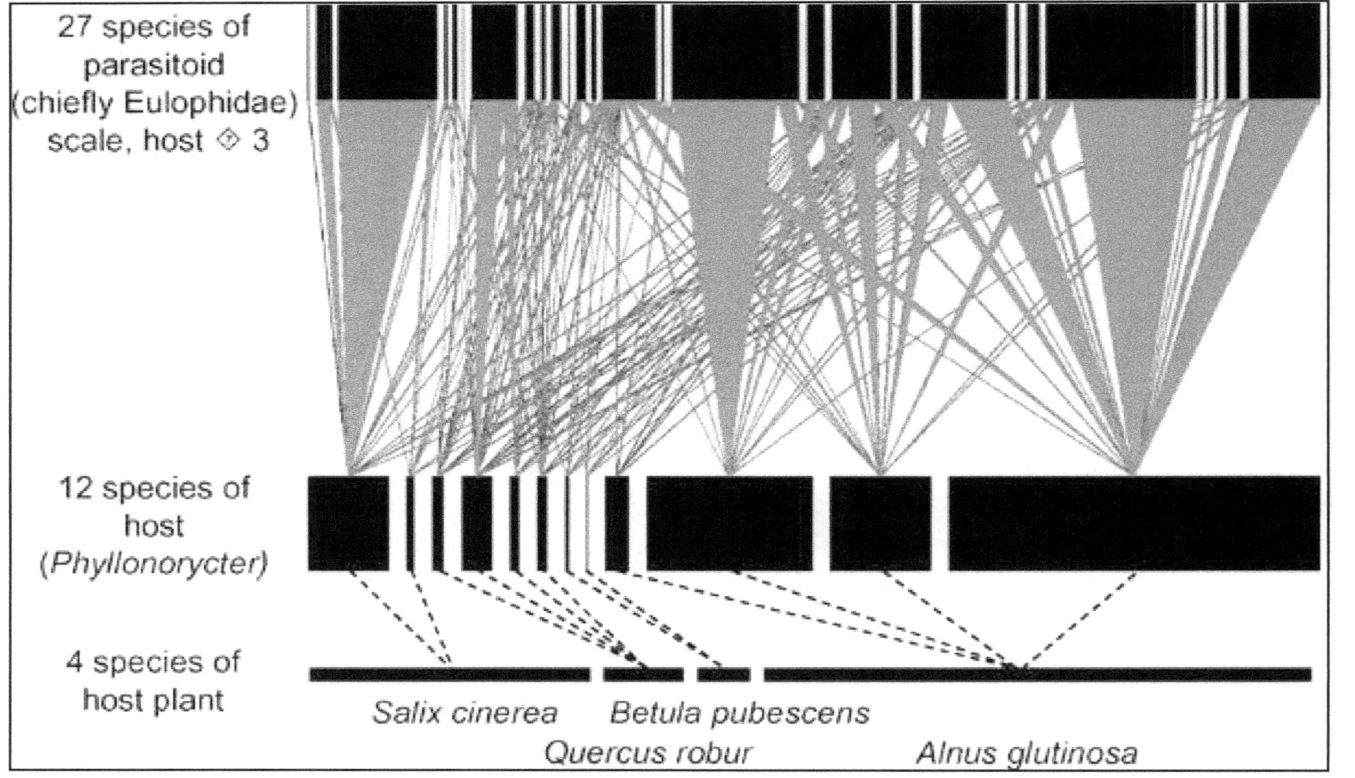

Figure 14.4: An Example of a Quantitative Food Web including the Possibilities of Intraguild Competition.

The web describes a community of closely related leafminers (Lepidoptera: Gracillariidae) that are attacked by a range of hymenopterous parasitoids. The thickness of the bars in each register depicts relative density. All the leafminers are monophagous (at this site), whereas most parasitoids attack multiple hosts. The relative importance of each host is represented by the "wedges" linking the two trophic levels (After van Veen *et al.*, 2006).

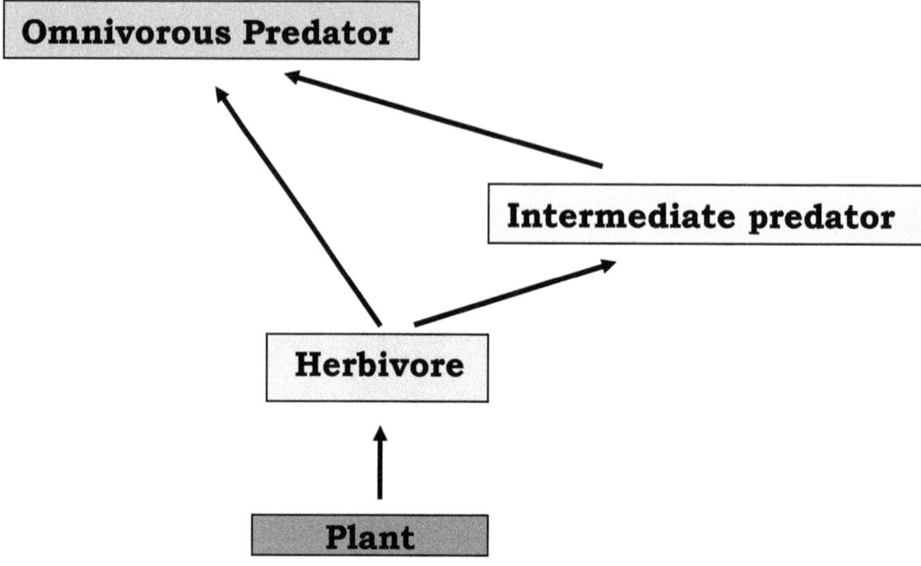

Figure 14.5: Trophic Web of the Arthropod Community Represented in the Simulation Model.

McGrady, 1993). The combined effects of the two predators were not additive. In order to determine the causes of the non-additive predation rates of damselflies, in single predator treatments, the dragonfly consumption rates of the damselflies were compared to those on presence of heterospecifics and conspecifics, with their menta surgically modified to prevent prey capture (Wissinger and McGrady, 1993). Demented *Tramea* reduced the consumption rates of *Erythemis* to less than half of that observed when *Erythemis* foraged alone. *Erythemis* numbers were also reduced by *Tramea* predation. *Erythemis* had neither effect on *Tramea*. Both of the negative

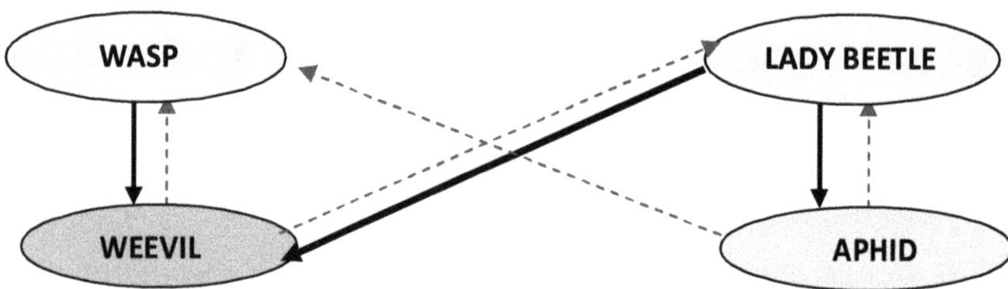

Figure 14.6: Indirect Interactions in Utah Alfalfa Fields.

(1) parasitism of the alfalfa weevil by the host-specific wasp *Bathyplectes curculionis* may be enhanced by the adult wasp's consumption of pea aphid honeydew; (2) weevil survivorship may be depressed when aphids are present and promote aggregation of lady beetles, as these primarily aphidophagous predators will feed on both prey species (short-term apparent competition); and (3) weevil parasitism by the wasp may be undercut by lady beetle consumption of aphids (through a three-link chain of direct interactions). The relative thickness of arrow lines for lady beetles reflects relative preferences.

effects of *Tramea* on *Erythemis* will have indirect positive effects on Damselflies. The behavioural component (reduced *Erythemis* foraging rate) should be more important than the "trophic link" (reduced *Erythemis* numbers) indirect effect. Together with these indirect positive effects will allay, but not completely compensate for, the direct negative effects of *Tramea* predation on damselflies (Wissinger and McGrady, 1993). These results illustrates that an asymmetric potential for intraguild predation can lead to asymmetries in interference competition and to non additive effects on prey mortality. The addition or removal of predators that interact in this manner to or from communities should have only a small net effect on prey, because of compensating direct and indirect effects. In some cases, the combined effects of them or more predators on shared prey are:

1. Additive-without any indirect effect.
2. More than additive-indication of facilitation among predators and/or resource enhancement.
3. Less than additive-indicating negative interactions between predators (interference, intraguild predation) and/or resource depression.

It has been predicted that reduced foraging rates of larval dragonflies delay emergence as adult which increases the likelihood of asymmetric competition and intraguild predation in the subsequent generation.

In case of *Tramea* and *Erythemis* combination, the interactions will be:

1. As intraguild predators, with asymmetric benefit to *Tramea* both in terms of energy gain and the elimination of a competitor, and
2. As interference competitors, with the negative effect manifested as reduced rate of prey consumption for *Erythemis*. Larger species have been observed to function as superior interference competitors.

In another experiment using habitat complexity as factor to regulate intraguild predations, in terrestrial community the following were observed.

The structural complexity of habitats is an important factor influencing natural enemy abundance and food web dynamics in invertebrate based communities (Figure 14.6). A meta analysis of published research work reveals that increased habitat structure results in a large and significant increase in natural enemy abundance (Langellotto and Denno, 2004). Similarly, decreasing habitat structure significantly diminished natural enemy abundance. Manipulation of detritus amount at the habitat spatial scale had the strongest effect on natural enemy abundance. Most guilds of natural enemies were significantly affected when the structural complexity of the habitat was altered. The possible reasons for such effect of vegetation structure/habitat complexity on natural enemy aggregation may be:

1. Refuge from intraguild predation.
2. More effective prey capture.
3. Access to alternative resources (like alternative prey, pollen or nectar).

The basal resources mediate top-down impacts on herbivores, and provide encouragement that manipulations of habitat complexity can be made in agro-ecosystems that will enhance the effectiveness of the natural enemy complex for more effective pest suppression. Natural enemies may aggregate in complex structured habitats because they:

1. Encounter more abundant prey.
2. Gain refuge from predation.
3. Are able to locate and capture prey more effectively.
4. Encounter a more favourable microclimate.
5. Gain access to alternative resource.

Experimental studies have shown that the intraguild predation of hemipteran predators by wolf spiders is vastly reduced in litter-rich habitats compared to litter-free ones. Likewise hunting spiders gain refuge from conspecific cannibals in structurally complex habitats (Langellotto and Denno, 2004).

The presence of intermediate predators suppresses the consumer population, and this effect (suppression) tended to increase, with increased intermediate predator density when the top predator is absent. In the presence of top predator, increased intermediate predator diversity showed the opposite effect on the consumers compared to the absence of top predator, *i.e.* decreased suppression of consumers with increased diversity. Thus, the loss of intermediate predator species weakened or strengthened predator-prey interactions depending on if the top predator is present or not, while loss of top predator only strengthened prey-predator interactions. The loss of a predator species, therefore, may render different, but perhaps predictable effect on the functioning of a system depending on from which trophic level it is lost and on the initial number of species in that trophic level (Jonsson *et al.*, 2007).

Increased predator diversity weakens predator-prey interactions, in a system of multiple lethal predators. In case of fishes present as top predators, the diversity of intermediate predators (larval dragonflies) weakened (Jonsson *et al.*, 2007).

Introguild predation occurs when a member of the guild preys upon another member of the same guild. IGP is a taxonomically widespread interaction within communities and exists broadly in nature. In Japanese wetlands, adults of *Laccotrephes japonensis* younger nymph of *Kirkaldyia deyrolli* and tadpoles are regarded as intraguild predator, intraguild prey and common prey respectively (Ohba and Swart, 2009). In terrestrial ecosystems many predators exhibit trophic level omnivory that involves consumption of both herbivores and other predators. This renders predator function to be indeterminate, which means a predator can operate from either the third or fourth trophic level and have opposite effects on herbivore and plant populations. A simulation model produces 4 predictions (Rosenheim and Corbett, 2003):

1. Actively foraging predators may be effective regulators of sedentary herbivore populations

2. Sit-and-wait predators are unlikely to suppress populations of sedentary herbivores, but may act as omnivorous top predators, suppressing populations of widely foraging intermediate predators and thereby increasing herbivore densities.
3. Among widely foraging predators attacking a common herbivore prey, predators that are large relative to their body size of their prey will be more mobile and thereby more vulnerable to predation by sit– and-wait omnivores, compared to predators that are similar in size to their prey
4. Widely foraging omnivores are unlikely to disrupt herbivore population suppression generated by intermediate predators and may instead enhance herbivore suppression. This is opposite to the impact of sit-and-wait predators.

Susceptibility of *Aeshna viridis* larvae to the IGP by similar sized larva of *A. grandis* and *A. juncea* was evaluated under laboratory conditions (Suutari *et al.*, 2004). It was observed that, *A. viridis* is susceptible to IGP and interference competitipn. The larvae of *A. grandis* dominated the middle and outer portion of *Stratiotes aloides*, a macrophyte exploited by *A. viridis* for oviposition substrate and evading fish predation. In absence of *A. grandis* larvae, *A. viridis* colonized the middle and outer parts of the rosettes. Possibly the asymmetric predation between odonate larvae of equal size can be intense, and that both IGP and interference competition impact *A. viridis*. Natural habitat complexity diminishes their impact, the IGP would nevertheless influence the distribution of *A. viridis* in *S. aloides* waters and restrict microhabitat use. IGP and interference are frequent among size-structured assemblages of odonate larvae and is characterized by size and density dependence occurring routinely in the shallow, littoral zone of eutrophic lakes and ponds (Suutari *et al.*, 2004).

Intraguild predation can influence top-down control of prey populations-either herbivores or detritivores and can impact the community structure that may eventually influence the food web dynamics. Keeping this in view, understanding of the context in which the impact of intraguild predation will be more prominent is essential to approaching population and community dynamics. It has been shown through this study on mirids wolf-spiders and plant hopper system of intraguild predation that multiple predators are more effective in suppressing herbivores in structured vegetation. This implies that habitat management may be adopted to augment the overall effectiveness of the predator complex in minimizing antagonistic interactions among natural enemies (Finke and Denno, 2002).

The intraguild predator *Tytthus vagus* (mired bug; Hemipters: Myridae) and the generalist predator the wolf spider *Pardosa littoralis* (Araneae: Lycosidae) interacted antagonistically reducing the predation pressure on the phloem feeding plant hopper *Prokelisia dolus* and *Prokelisia marginata* (Hemiptera: Delphacidae) in structurally simple habitats. In structurally complex habitats this antagonistic interaction was dampened by providing a refuge for mirids from spider predation, thereby increasing the combined effectiveness of these predators in suppressing plant hopper population. This was observed to be same in natural conditions, where enhanced co-occurrence of these predators under complex habitat conditions in the field (Finke and Denno, 2002). This provides evidence that in salt marsh vegetation diminished the occurrence of intra guild predation and increased overall enemy impact on the shared herbivore prey, when the mirids and wolf spiders are considered as prey (Finke and Denno, 2002).

Niche overlap indices do not address the potential for 'intraguild predation' in assemblages in which potential competitors also prey each other. Many species concurrently interact as competitors and intraguild predators and that this complex pattern of interspecific interaction plays an important role in the organization of many communities. Information about the potential for intraguild competition and predation is important for formulating strategies for regulation of target organisms. Thus an index of the opportunity for competition (IOC), revealing the frequency with which similar size classes of two species encounter each other, and index of the opportunity for intraguild predation (IOP) calculates the frequency of encounters among disparate size classes of the same two species (Wissinger, 1992).

Using this approach to measure overlap, competition and IGP in size structured population of dragonfly nymphs revealed that:

1. Some species with high overlap values should interact mainly as competitors, others mainly as intraguild predators, and many as both competitors and predators.
2. Subtle differences in phenology and/or size specific shifts in habitat distribution can lead to the potential for asymmetric interspecific competitions.
3. Some species with low pair wise IOP and IOC values are nonetheless vulnerable to the effects of diffuse competition or intraguild predation.
4. Seasonal segregation reduces competitive overlap but at the same time increases the opportunity for intraguild predation.

Conclusion

IGP can be viewed as a specific form of omnivory, where multiple consumers feed on each other and on a shared prey. This mechanism of trophic interaction is of particular relevance to the study and application of conservation of threatened species and post species regulation. Although several studies have emphasized the dynamics of three-species IGP, but the potential of other factors like exterior interactions have been overlooked in those models (Daugherty *et al.*, 2007). Three forms of feeding outside the IGP module by intraguild predators (*i.e.* trophic supplementation) affect the dynamics of the predators both IG predators and IG prey and their shared resource. Availability of alternative prey, supply of donor controlled resource and predator plant feeding affected the dynamics of IGP models.

It was seen that coexistence is possible without the IG prey being a superior competitor for the original shared resource if the IG prey effectively exploit one of the types of trophic supplements. Supplements to IG predator restricted the potential for co-existence while supplements to IG prey ameliorate the disruptive effects of the

IG predator on the suppression of the shared resource thereby promoting effective control of the resource in the presence of both predators. These propositions have been evaluated through mathematical models.

References

Aditya G, Bhattacharya S, Kundu N. *et al.*, 2004. Predatory efficiency of the water bug *Sphaerodema annulatum* on mosquito larvae (*Culex quinquefasciatus*) and its effect on adult emergence. *Bioresource Technology*, 95: 169-172.

Aditya G, Bhattacharya S, Kundu N and Saha GK. 2005. Frequency dependent prey selection of predacious water-bugs on *Armigeris subalbatus* immatures. *Journal of Vector Borne Diseases*, 42: 9-14.

Amarasekare P. 2006. Productivity, dispersal and the coexistence of intraguild predators and prey. *Journal of Theoretical Biology*, 243: 121-133.

Babbitt KJ and Jordan F. 1996. Predation on *Bufo terrestris* tadpoles: effects of cover and predator identity. *Copeia*, 1996(2): 485- 488.

Blaustein L. 1998. Influence of predatory backswimmer, *Notonecta maculata* on invertebrate community structure. *Ecological Entomology*, 23 (3): 246-252.

Blaustein L and Chase JM. 2007. Interactions Between Mosquito Larvae and Species that Share the Same Trophic Level. *Annual Review of Entomology*, 52: 489-507.

Blumenshine SC and Hambright KD. 2003. Top-down control in pelagic systems: a role for invertebrate predation. *Hydrobiologia*, 491: 347-356.

Crumrine PW and Crowley PH. 2003. Partitioning Components of Risk Reduction in a Dragonfly-Fish Intraguild Predation System. *Ecology*, 84: 1588-1597.

Crumrine PW. 2005. Size structure and substitutability in an odonate intraguild predation system. *Oecologia*, 145: 132-139.

Daugherty MP, Harmon JP and Briggs CT. 2007. Trophic supplements to intraguild predation. *Oikos*, 116: 662-677.

Finke DL and Denno RF. 2002. Intraguild Predation Diminished in Complex-Structured Vegetation: Implications for Prey Suppression. *Ecology*, 83: 643-652.

Gorai AK and Raychoudhury DN. 1962. Food and feeding habits of *Anisops bouvieri* Kirkaldy. *Proceedings of Indian Science Congress*, Part 3: 522-523.

Hampton SE, Gilbert JJ and Burns CW. 2000. Direct and indirect effects of juvenile *Bueona macrotibialis* (Hemiptera: Notonectidae) of zooplankton of a shallow pond. *Limnology and Oceanography*, 45: 1006-1012.

Hampton SE and Gilbert JJ. 2001. Observations of insect predation on rotifers. *Hydrobiologia*, 446/447: 437- 444.

Hazarika R and Goswami MM. 2010. Aquatic Hemiptera of Gauhati University, Guwahati, Assam, India. *J. Threatened Taxa*, 2: 778-782.

Jana S, Pahari PR, Dutta TK and Bhattacharya T. 2009. Diversity and community structure of aquatic insects in a pond in Midnapore town, West Bengal, India. *Journal of Environmental Biology*, 30(2): 283-287.

Jonsson M, Johansson F, Karlsson C and Brodin T. 2007. Intermediate predator impact on consumers weakens with increasing predator diversity in the presence of a top-predator. *Acta Oecologica*, 31: 79-85.

Juliano SA. 2009. Species interactions among larval mosquitoes: context dependence across habitat gradients. *Annual Review of Entomology*, 54: 37-56.

Langellotto GA and Denno RF. 2004. Responses of invertebrate natural enemies to complex-structured habitats: A meta-analytical synthesys. *Oecologia*, 139: 1-10.

Lawler SP and Dritz DA. 2005. Straw and winter flooding benefit mosquitoes and other insects in a rice agroecosystem. *Ecological Applications*, 15: 2052-2059.

Lester PJ and Harmsen R. 2002. Functional and numerical responses do not always indicate the most effective predator for biological control: an analysis of two predators in a two-prey system. *Journal of Applied Ecology*, 39: 455-468.

Lucus E, Coderre D and Brodeur J. 1997. Instar-Specific Defense of *Coleomegilla maculata* Lengi(Col: Coccinelidae): Influence on attack success of the intra-guild predator *Chrysoperla rufilabris* (Neur: Chrysopidae). *Entomophaga*, 42(1/2): 3-12.

Nishi R and Venkatesan P. 1989. Predation ingestion rate and its bearing on prey death rate in *Anisops bouvieri* Kirkaldy. *Journal of Entomological Research*, 13 (1-2): 140-145.

Ohba S and Swart CC. 2009. Intraguild predation of water scorpion *Laccotrephes japonensis* (Nepidae: Heteroptera). *Ecological Research*, 24: 1207-1211.

Rosenheim JA and Corbett A. 2003. Omnivory and the indeterminacy of predator function: can knowledge of foraging behavior help? *Ecology*, 84: 2538-2548.

Rosenheim JA, Limburg DD and Colfer RG.1999. Impact of Generalist Predators on a Biological Control Agent, *Chrysoperla carnea:* Direct Observations. *Ecological Applications*, 9: 409-417.

Saha N, Aditya G, Saha GK and Hampton SE. 2010. Opportunistic foraging by heteropteran mosquito predators. *Aquatic Ecology*, 44: 167-176.

Saha N, Aditya G, Banerjee S and Saha GK. 2012. Predation potential of odonates on mosquito larvae: implications for biological control. *Biological Control*, 63 (1): 1-8.

Saha N, Aditya G and Saha GK. 2014. Prey preferences of aquatic insects: potential implications for the regulation of wetland mosquitoes. *Medical and veterinary entomology*, 28 (1): 1-9.

Sivagnaname N. 2009. Selective and frequency dependent predation of aquatic mosquito predator *Diplonychus indicus* Venkatesan and Rao (Hemiptera: Belostomatidae) on immature stages of three mosquito species. *Entomological Research*, 39: 356-363.

Snyder WE and Ives AR. 2003. Interactions between Specialist and Generalist Natural Enemies: Parasitoids, Predators, and PEA Aphid Biocontrol. *Ecology*, 84: 91-107.

Suutari E, Rantala MJ, Salmela J and Suhonen J. 2004. Intraguild predation and interference competition on the endangered dragon fly *Aeshma viridis*. *Oecologia*, 140: 135-139.

Vance-Chalcraft HD and Soluk DA. 2005. Multiple predator effects result in risk reduction for prey across multiple prey densities. *Oecologia*, 144: 472- 480.

van Veen FJF, Morris RJ and Godfray HCJ. 2006. Apparent competition, quantitative food webs and the structure of phytophagous insect communities. *Annual Review of Ecology and Systematics*, 51: 187-208.

Wissinger SA. 1992. Niche overlap and the potential for competition and intraguild predation Between Size-Structured Populations. *Ecology*, 73: 1431-1444.

Wissinger SA and McGrady J. 1993. Intraguild Predation and Competition Between Larval Dragonflies: Direct and Indirect Effects on Shared Prey. *Ecology*, 74: 207-218.

Chapter 15

Plankton Mediated Carbon Cycling Process in Sewage-Fed Fisheries

Parthiba Basu[1], Sarmistha Saha[1,2] and Tapan Saha[2]

[1]*Ecology Research Unit, Department of Zoology, University of Calcutta, Kolkata*

[2]*Institute for Environmental Studies and Wetland Management, Kolkata*

Introduction

Ecologically prudent city sewage removal and management has become a global necessity. The surface water pollution has assumed a critical magnitude in the metropolitan areas of South Asia, especially in Nepal, Bangladesh and India, due to the high loads of waste disposal into the short stretches of rivers (Karn and Harada, 2001; Bassi *et al.*, 2014). In India, most of the cities have expanded ahead of the municipal area and have no sewage and sewerage management service and smaller towns are worse in sewage management (Kamyotra and Bhardwaj, 2011).

In this context, wastewater aquaculture can be envisaged as an efficient way to address city sewage management. In wastewater aquaculture, waste is transformed to usable resource. In China, utilization of house and farm wastewater to produce phytoplankton in some fish ponds integrating aquatic macrophytes conservation and irrigation in cultivation fields, took place in 1960s and came up with ecological and economic benefits (Yan and Zhang, 1994). In Sweden, Stensund wastewater aquaculture was built in 1992 as a greenhouse mesocosm in a northern climate (Guterstam, 1996) to manage wastewater as a resource by using ecological engineering at the experimental level. In Hungary, sewage water has been used for healthy growth of fish in oxidation ponds (Olah *et al.*, 2003). In Munich, Germany fishermen apply secondary effluent that is chemically treated sewage to their fishpond covering approximately 230 ha area producing 100 to 150 tons of fish pre year (Olah *et al.*, 2003). Compared to these, East Kolkata Wetland (EKW), Kolkata, India has been the largest and oldest in transforming sewage water for fish culture where fishermen has been practicing wastewater fish culture since 1930 (Ghosh, 1993,1999) and commercially produce 30,000 tons of fish per year (Saha *et al.*, 2014).

Wetlands in general, serve as reservoirs of carbon and play a critical role in the global carbon cycle and consequently affect climatic factors at both global and regional scale (Kulinski and Pempkowiak, 2008). As a large reservoir of carbon, the hydrosphere plays a very important role in the global carbon cycle (Joos *et al.*, 1996). Air-water CO_2 exchange which is governed by a number of physical parameters and processes, plays a vital role in this cycle. In general wetlands are considered as net heterotrophic systems which frequently act as the source of atmospheric CO_2 (Bauer *et al.*, 2013). It is estimated globally that surface water of wetlands has a CO_2 flux of 1.4 Pg C per year (Raich and Potter, 1995). The two major processes - photosynthetic uptake and respiratory release, balance the degree of CO_2 flux. Photosynthesis and respiration are dependent upon different aspects like allochthonous input of labile organic matter, temperature, availability of sunlight, water residence time, metabolism rates and nutrient load (Smith and Hollibaugh, 1993; Howland *et al.*, 2000; Wang and Veizer, 2000; Abril *et al.*, 2002; Ahad *et al.*, 2008).

In open waters, phytoplanktons are the main producer of organic carbon and it accounts for approximately 30 per cent of the world's primary production (Falkowski, 1980). Phytoplanktons regulate the recycling of dissolved and volatile carbon at the air-water interface (Amon and Benner, 1996) and act as CO_2 sink. CO_2 uptake at water surface is positively correlated with primary productivity, phytoplankton biomass and surface irradiance (Platt *et al.*,1990). Absorbed CO_2 is transformed into organic carbon through phytoplankton and through grazing it is transferred to zooplankton which acts as main food source for fish, thus planktons in wetlands act as biological pumps for carbon which enters into the aquatic food chain. Dead planktons in the food chain are one of the major sources of carbon in soil sediment that add to the particulate organic carbon (POC) pool. Dissolved organic carbon (DOC) is also one of the major components of food webs (Packard *et al.*, 2000) as it is the main source of energy for microbial metabolism (Tranvik, 1992) and is a major component of photosynthetic release (Ducklow and Carlson, 1992). Changes in DOC concentrations along with the temperature, depth and seasonal variation, the amount and proportion of autochthonous and allochthonous sources have been studied in various types of aquatic ecosystems (Wetzel, 2001; Dafner and Wangersky, 2002; Sugiyama *et al.*, 2004). DOC forms complexes with metal ions, and can affect particle accumulation and sedimentation rates by controlling the surface charge of colloids and suspended matter in water through adsorption (Kretzschmar *et al.*, 1997). Mineralization of DOC through microbial degradation ultimately enriches the soil organic carbon pool (Jumars *et al.*, 1989, Amon and Benner,1996). Export of particulate organic carbon (POC) leads to the net sequestration of CO_2 through sedimentation. Regardless of the significant role of the carbon cycle in the Earth's biogeochemistry, until recently studies from southern Asia have been few (Borrett *et al.*, 2012) and all the available studies were done in either estuarine or in marine ecosystem. Moreover, there is dearth of basic information on carbon cycling in freshwater systems.

All the measures which have been taken to manage sewage throughout the world were mostly of experimental basis for a short stretch of time or within a small area whereas the sewage-fed fish culture in EKW is probably the oldest and worlds' largest recycling system that has remained active commercially for over past eighty years. This provides an opportunity to assess how such a perturbed system has sustained over the years and such assessment would provide important clue to replicate this system elsewhere as a multipurpose sewerage recovery and carbon sequestration strategy. Tracking carbon flow pathway through the plankton community would be an important primary step towards understanding the ecosystem process of this unique system.

In view of this, the present study aims to assess the ultimate fates of different forms of carbon within the bio-geochemical carbon cycle in the sewage-fed fisheries system. We attempted to identify the main environmental factors that act as forcing functions to regulate the plankton communities and physical processes of the carbon cycle in this freshwater system. A conceptual carbon flow model for this system is also attempted. Since the sewage-fed wetland system receives high quantity of organic matter, we hypothesize that inland sewage-fed fisheries would entrap more carbon in the form of dissolve inorganic carbon (DIC) and sequester more amount of soil organic carbon (SOC) thereby acting as an efficient carbon sink.

Materials and Methods

Study Site

We took water and soil samples from fisheries of East Kolkata Wetland (EKW), located on the eastern fringes of Kolkata (Pradhan *et al.*, 2008). The span of EKW is about 12,500 hectare area. Within EKW 5,852.14 hectare is covered by water bodies of which approximately 3,798 hectare are used for fisheries (Ghosh, 1993; Sarkar *et al.*, 2009). EKW is situated in between latitudes 22°25' to 22°40' North and longitudes 88°20' to 88°35' East, between the levee of river Hooghly on west to Kultigang in the east and the land-ocean boundary of Bay of Bengal at south (Figure 15.1). These wetlands provide habitat for variety of arthropods, molluscs, fishes and birds. Local economy is mainly controlled by the commercial fish production from EKW.

We have categorized study ponds into two groups, East Kolkata Wetlands ponds and Control ponds. EKW ponds are situated within EKW Ramsar site. We have collected samples from nine different ponds of this area. Six ponds which we have categorized as control ponds are situated near EKW Ramsar site area. These control ponds do not receive any sewage water and are free from fisheries activities. All study ponds were sampled from 2011 to 2013 at four seasons – winter, pre-monsoon, monsoon and post-monsoon.

Climate Zone

EKW is in the wet tropical climatic zone, with distinct seasonal climatic changes. The seasons are characterized as pre-monsoon (April-June) with average high temperature ranging from 30–41°C and minimum precipitation; monsoon (July-September), when 70-80 per cent of annual rainfall occurs, post-monsoon

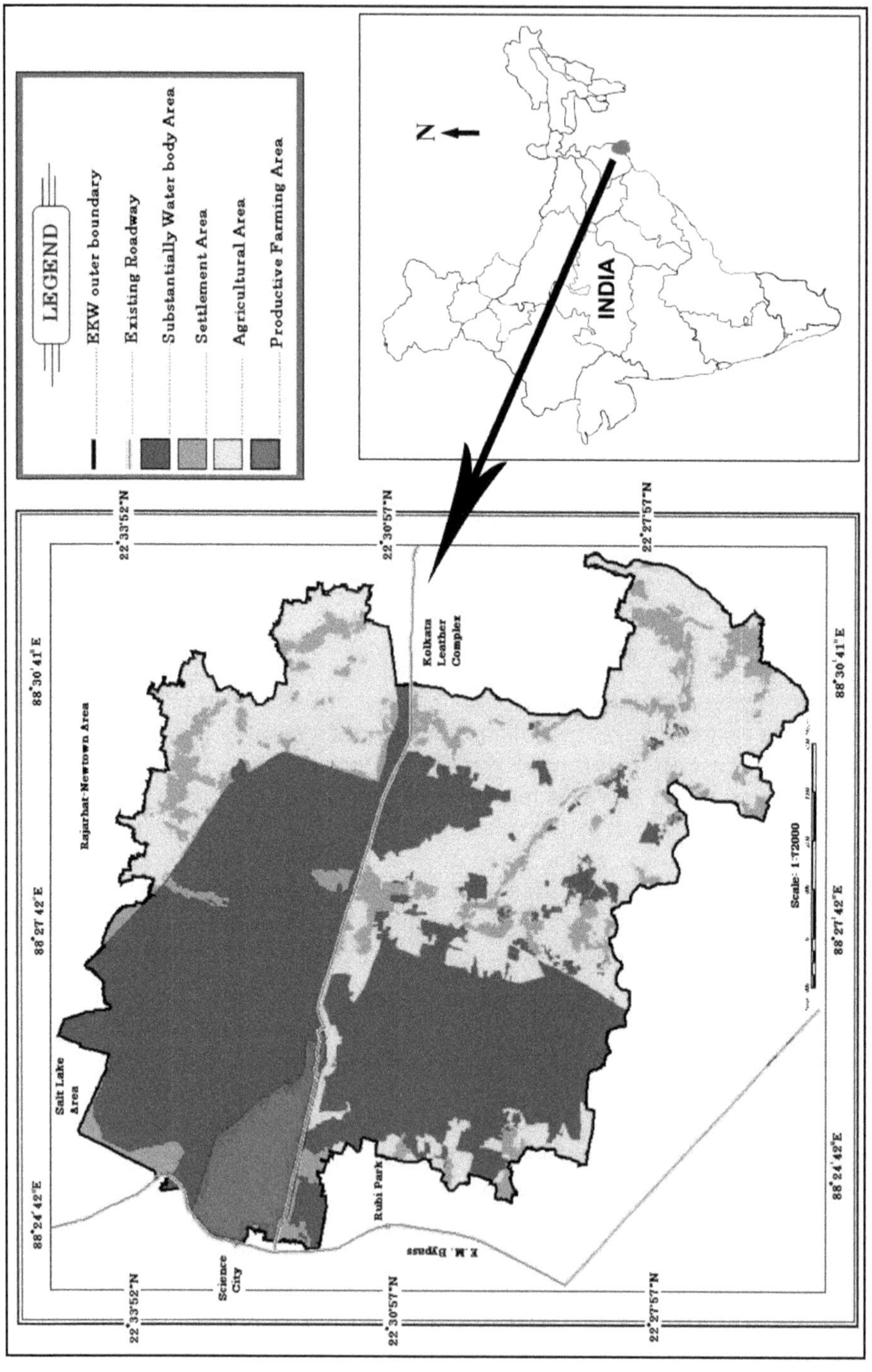

Figure 15.1: Study Site - East Kolkata Wetlands (Maiti *et al.*, 2012).

(October-December), with average 25° C temperature, negligible rainfall and winter (January-March) with temperature below 23° C and almost no rainfall. Here in India monsoon is usually dominated by southwest winds while the average humidity is about 80 per cent and more or less uniform throughout the year.

Phytoplankton and Zooplankton Sampling

Using plankton nets, we have sampled planktons from surface water. Phytoplankton samples were collected by sieving 25 lts. of water using bolting silk plankton net (No.25) and were preserved with 4 per cent formaldehyde. Phytoplanktons were identified from keys provided by Fritsch (1948), Bold and Wynne (1978), Palmer (1980) and Gupta (2005). Forty litres of water were sieved through zooplankton net (No.20) to collect zooplankton samples and were preserved with 70 per cent alcohol. Sedgewick-Rafter Counting Cell Slide (100 x 1mm squares) was used to obtain the number of organisms per litre of water. Wet and dry weights were measured for zooplankton and the corresponding carbon content was estimated following Friedler *et al.* (2003). Gross primary production, Net primary production and community respiration were measured in situ by the light and dark bottle method following Trivedy and Goel (1984).

Analysis of Physico-chemical Parameters

We have collected water and soil samples both from the fishery ponds of EKW and the control ponds. EKW fish ponds are shallow with 3 to 4 ft. depth where as control ponds are natural ponds with 20 to 25 ft. depth. Water samples from both type of ponds were collected from subsurface region at ~ 1 ft. depth and were analyzed for parameters like water temperature, water pH, alkalinity, dissolved organic carbon (DOC), particulate organic carbon (POC), dissolved oxygen (DO), dissolve bicarbonate (DBC), dissolved carbon dioxide (DCO_2 or $CO_2(aq)$), biological oxygen demand (BOD), dissolved inorganic carbon (DIC) and carbon from dead soil organisms (DSO). We have collected soil samples from the ponds' bottom and analyzed them to measure pH, organic carbon (SOC) and inorganic carbon (SIC). Soil organic and inorganic carbon contents were measured following standard methods (Bold and Wynne, 1978). Soil pH was measured seasonally from a saturated soil–water paste (Trivedy and Goel, 1984; Gupta, 2002). In the field, we have used glass bottles to collect water samples for estimation of DOC and POC and preserved them on ice in the dark and moved to the laboratory. Water samples were filtered through Whatman 42 filter paper and the filtrate was used to determine POC. Total alkalinity for water samples were estimated by titration method as suggested by Trivedy and Goel (1984). Dissolved carbon dioxide (DCO_2) and dissolved bicarbonate (DBC) was assessed following standard methods (APHA, 1985, 1992). Dissolved inorganic carbon (DIC) was estimated following a standard method (Greenberg *et al.*, 1992). All the chemical parameters were calculated in ppm unit. We have used portable sample analyzer (Multi 340 i/SET) to digitally assess water temperature and pH in the field and measured DO by Winkler's iodometric method (APHA, 1985, 1992).

Statistical Analysis

Kolmogorov-Smirnov's Test was done to check normality of the data. We have verified homoscedasticity of the residuals using Levene's test. All the data sets were found to follow normal distribution and hence we used parametric statistics. To assess the relationships between environmental factors and different forms of carbon in EKW water, we have used Pearson Correlation, Multiple Regression, Redundancy Analysis (RDA), Multiple Comparison using Tukey HSD test, one-way and multi-way ANOVA and compared them with the results of the same in control ponds. For RDA, Monte Carlo permutation test was done after 1000 permutations after randomizing the data. In case of multiple regressions, we have performed stepwise backward analysis using predictor variables where non-significant factors ($p > 0.05$) were eliminated and best models were selected depending on the beta values. Data analysis was done using SPSS 14.0 software (windows evaluation version) and CANOCO 5. To prepare the conceptual map of carbon cycling in the sewage-fed fisheries of EKW, we have used mathematical software STELLA 9.0.2.

Results

Statistical analysis showed significant difference in BOD, WpH, Alk, SpH, SOC, SIC, DCO_2, DIC, DSO, and DOC between the sewage-fed fisheries and control ponds (Table 15.1). There is also significant seasonal differences in regulatory factors *e.g.*, WT, DO, WpH, SpH and different forms of carbon *e.g.*, SOC, DCO_2, DIC and DSO.

Table 15.1: Multi-way ANOVA of different Environmental Factors and Carbon Forms. Environmental factors are water temperature, water pH, DO, BOD, alkalinity, soil pH and carbon forms are SOC, SIC, DBC, DCO_2, DIC, DOC, and DSO. (All abbreviations are referred in the text).

Variables	*R Square (adj.)*	*F*	*p*
WT	0.990	319.03	p<0.001
DO	0.881	25.39	p<0.001
BOD	0.720	9.43	p<0.001
WpH	0.850	19.59	p<0.001
ALK	0.915	36.39	p<0.001
SOC	0.856	19.90	p<0.001
SIC	0.129	1.48	0.241
SpH	0.991	368.91	p<0.001
DBC	0.913	35.55	p<0.001
DCO_2	0.971	110.63	p<0.001
DIC	0.957	74.94	p<0.001
DOC	0.704	8.83	p<0.001
DSO	0.849	19.89	p<0.001

For EKW ponds, best combined multiple regression model of *regulatory* physico-chemical parameters which are the main forcing functions that explained the fluctuation in SOC was WT + WpH + SpH + DO ($R^2 = 0.901$; $F = 26.146$, $p<0.01$). DCO_2 + DIC was also the best combined multiple regression model of carbon compartments that explained SOC ($R^2 = 0.661$; $F = 23.456$; $p<0.01$) in EKW ponds (Table 15.2). There were significant seasonal variation in both SOC ($R^2 = 0.856$; F

= 19.90, p<0.001) and DSO values (R^2 = 0.849; F = 19.89, p<0.001) in EKW ponds (Table 1). Seasonal fluctuation in DSO follows values of SOC in both the systems of EKW fisheries and control ponds.

Fluctuation in SIC pool in EKW ponds is significantly explained by the phytoplankton load (R^2 = 0.749; F = 69.492; p<0.01). Fluctuations in zooplankton load was best explained by WT in control ponds (R^2 of 0.808; F = 42.163; p<0.01) while for EKW ponds it was best explained by combined multiple regression model WT + Alk + SIC + DCO_2 (R^2 = 0.927; F = 22.233; p<0.01) (Table 15.2).

Table 15.2: Summaries of Linear Regression Models (Combined model) between Zooplankton Loads, Gross Primary Production (GPP), DOC and SOC with different Regulatory Factors and Carbon Forms. Only the best models are represented in the current table. Water temperature (WT), alkalinity (Alk), soil inorganic carbon (SIC), dissolved carbon dioxide (DCO_2), Water pH (WpH), dissolved oxygen (DO), soil pH (SpH), dissolved inorganic carbon (DIC) on observed zooplankton load, gross primary production and organic carbon forms of water and soil.

Response Variables	*Best Model*	*F*	*R^2 (Adjusted)*	*P Value*
Zooplankton load	WT+Alk+SIC+DCO_2	22.233	0.927	p< 0.01
GPP	WpH+DCO_2	24.266	0.669	p< 0.01
DOC	Alk+DO+WpH	31.148	0.797	p< 0.01
SOC	WT+WpH+SpH+DO	26.146	0.901	p< 0.01
SOC	DCO_2+DIC	23.456	0.661	p< 0.01

Within fishery ponds of EKW, fluctuations present in gross primary production (GPP) were best explained by the combined multiple regression model WpH+DCO_2 (R^2 = 0.669; F = 24.266; p<0.01) (Table 15.2). GPP was found to be highest in monsoon in EKW ponds whereas it reached the peak in pre-monsoon in control ponds lowering in winter. We have found higher rates of Community respiration (Cresp.) during post-monsoon season in control ponds where as it remained almost constant throughout the year in EKW. Fluctuations in DOC was best explained by the combined multiple regression model Alk + DO + WpH (R^2= 0.797; F = 31.148; p < 0.01) (Table 15.2).

Dissolved carbon dioxide and other dissolve forms of inorganic carbon show higher concentration in EKW water than control ponds. There was significant seasonal variation in DCO_2 values (R^2 = 0.971; F = 110.63, p<0.001) (Table 15.1) Seasonal variation in DIC was lower in both EKW and control pond water compared to DCO_2. Values of DCO_2 reach the peak at winter season in both EKW and control ponds.

Phytoplankton carbon count was highest (0.506 mg L^{-1}) in monsoon at sewage-fed fisheries while for the control ponds it is highest in pre-monsoon (0.091 mg L^{-1}). Zooplankton carbon content varied from 0.032 to 0.09 mg L^{-1} in EKW and 0.006 to 0.015 mg L^{-1} in control ponds, showing higher values during post-monsoon to winter and lower in monsoon. There was significant negative correlation between phytoplankton and zooplankton carbon content in both EKW (Pearson correlation, r = –0.837, p<0.01) and control ponds (Pearson correlation, r = -0.878, p<0.01).

There was significant seasonal variation in DO values (R^2 = 0.881; F = 25.39, p<0.001) (Table 15.1) and particularly between monsoon and post-monsoon season in both EKW and control ponds (Tukey Post hoc test, p<0.001). There was significant seasonal variation observed in SpH values (R^2 = 0.991; F = 368.91, p<0.001) (Table 15.1). Higher values of soil pH are recorded during winter (7.03 for EKW and 5.07 for control ponds) and lower values during pre-monsoon (6.8 in EKW and 4.8 in control ponds). Water pH values remained almost same across all seasons in control ponds and little variation was present in the EKW ponds. DO value increases from winter to post-monsoon in EKW whereas in control ponds, it is highest in monsoon and remain same in all other seasons.

We have done triplot diagram from RDA with sampling sites, plankton communities and significant explanatory variables (alkalinity, water temperature at four seasons). The RDA analysis clearly distinguishes the EKW ponds from the control ponds (Figure 15.2). First two major axis of RDA significantly account for 95.3 percent variation. Alkalinity significantly explains 84.4 per cent variation of total explained variation (F = 27.1, p = 0.02) and water temperature significantly contributes 13.8 per cent variation (F = 14.3, p = 0.02). Correlation matrix of response and explanatory axis scores and correlation of two significant explanatory variables on first two main axes are described in Table 15.3. Explanatory variable alkalinity (alk) is in axes 1 and water temperature (WT) in axes 2.

Table 15.3: RDA Correlation Matrix: Response and Explanatory Axis Scores and Correlation of Two Significant Explanatory Variables on First Two Main Axes. Explanatory variables are water temperature (WT) and alkalinity (alk).

	Resp Ax1	Resp Ax2	Resp Ax3	Resp Ax4	Expl Ax1	Expl Ax2	Expl Ax3
Resp Ax1	1.000						
Resp Ax2	-0.023	1.000					
Resp Ax3	-0.118	0.482	1.000				
Resp Ax4	-0.104	-0.253	-0.000	1.000			
Expl Ax1	0.988	-0.000	0.000	-0.000	1.000		
Expl Ax2	-0.000	0.872	-0.000	-0.000	0.000	1.000	
Expl Ax3	0.000	0.000	0.000	0.000	0.000	0.000	0.000
Expl Ax4	0.000	0.000	0.000	0.000	0.000	0.000	0.000
WT	-0.129	-0.865	0.000	0.000	-0.131	-0.991	0.000
alk	0.953	-0.230	0.000	-0.000	0.965	-0.263	0.000

Position of phytoplankton and zooplankton load on first two major axes of RDA, percent variation of phyto and zooplankton loads explained by significant explanatory variables and the variance are described in Table 15.4. Variation in phytoplankton load (93.86 per cent) and zooplankton load (96.73 per cent) are best explicated by explanatory variables (Table 15.4).

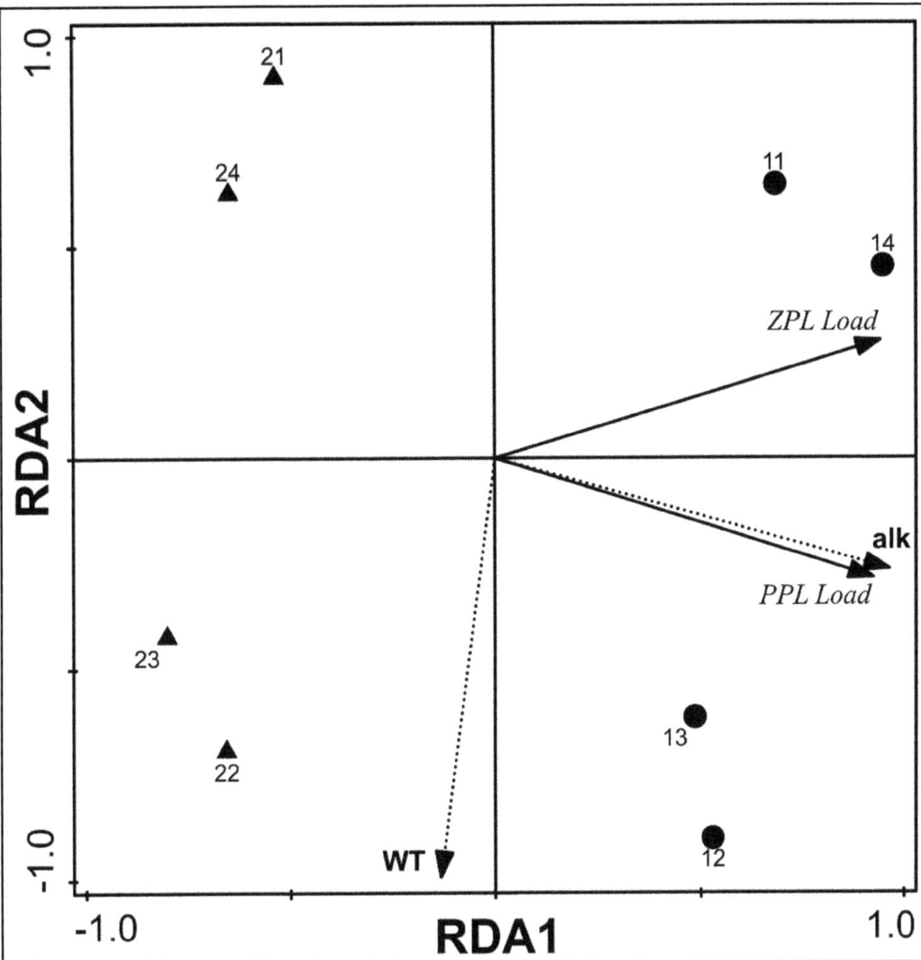

Figure 15.2: RDA Plot Representing the Study of Physical and Biotic Parameters in EKW and Control Pond (Sampling site EKW ponds are represented by solid black circles and numbered as 1 whereas sampling site Control ponds are represented by solid black triangles and numbered as 2; seasons are denoted as: winter = 1, pre-monsoon = 2, monsoon = 3 and post-monsoon = 4).

Table 15.4: Position of Response Variables on Two Major Axes, Variation Explained by Significant Explanatory Variables and Variance Plot of RDA

	AXIS 1	*AXIS 2*	*Variation Explained by Explanatory Variables*	*Variance*
PPL Load	0.9263	-0.2840	93.8646	0.9987
ZPL Load	0.9434	0.2781	96.7336	1.0013

Discussion

Assessment of SOC and the factors that regulate it is crucial to understand the nutrient flow dynamics in the fishery ponds. Major component of carbon cycling pathways in aquatic ecosystem are plankton communities through which abiotic forms of carbons are transformed and transferred to higher trophic levels (Ray, 2008; Mukherjee *et al.*, 2012). Phytoplankton load (indiv./L) varies seasonally along with zooplankton load in both EKW and control ponds. Carbon content of phytoplankton is the main source of particulate organic carbon (POC) present in water and inorganic carbon of soil (SIC). Dead floating and submerged flora including phytoplankton and fauna (zooplankton, macroinvertebrates and fishes) is the prime source of soil organic carbon (SOC) within these aquatic systems (Ghosh, 1993; Mukherjee *et al.*, 2012). During monsoon, runoff from adjacent localities also adds up to the SOC pool in both EKW and control ponds.

Various factors *e.g.*, soil texture; temperature and rainfall etc. are the prime regulators of the observed spatial and temporal variation in SOC (Sebastian and Chacko, 2006). Of the different factors, the dynamics of SOC mineralization is strongly influenced by temperature that regulates the microbial mineralization process (Insam, 1990; Kirschbaum, 1995; Winkler *et al.*, 1996). Temperature plays an important role in controlling both the amount and turnover time of SOC mineralization (Singh and Gupta, 1977; Raich and Schlesinger, 1992; Lloyd and Taylor, 1994; Kirschbaum, 1995; Kätterer *et al.*, 1998). Higher soil temperature was observed in pre-monsoon compared to post-monsoon which causes higher microbial activity (Gupta, 2002).

Soil pH (SpH) is also one of the important physical factors that govern the soil system and thereby control conversion of one form of soil carbon to another (APHA, 1985, 1992) In the present study, SpH shows significant correlation with SIC values in both EKW and control ponds, which is expected in an anaerobic condition of the EKW and the control ponds (Sarkar *et al.*, 2009). Due to anaerobic degradation, CO_2 is trapped by alkaline earth metals forming calcium carbonate and magnesium carbonate which increase the SpH (Wetzel, 2001). However, the EKW ponds have higher SpH compared to the control ponds since the former receives larger doses of organic nutrients in the form of sewage. SpH, along with POC, DSO and DCO2 also significantly explains fluctuations of SIC throughout the year. SIC shows significant negative correlation with DCO_2 and high positive correlation with DSO and POC in both EKW and control ponds. Partially degraded POC is converted to DCO_2 and is entrapped by the alkaline earth metals of soil (APHA, 1985, 1992). This makes the soil environment alkaline which regulates mineralization process within the soil (Wetzel, 2001).

DBC, DCO_2 and DIC values are higher in EKW ponds compared to control ponds. DBC and DCO_2 both contribute to DIC. The DCO_2 pool is mostly formed by diffusion from the atmosphere, respiration by the biotic community of the aquatic ecosystem and microbial degradation of soil organic carbon which releases free CO_2 (Wetzel, 2001). Soil pH influences conversion processes in anaerobic system as in the present system (Bolin *et al.*, 1979; Sarkar *et al.*, 2014). Under acidic condition, a fraction of SIC contributes to DIC which is then transformed to DCO_2. In ponds, an anoxic state prevails at soil-water interface prompting conversion of SIC into free CO_2. Water pH (WpH) and DCO_2 are inversely related and higher values of CO_2 (aq.) are observed in winter with lower values during monsoon.

The present study also shows significant negative correlations between SOC and DCO_2, and between DIC and DOC in EKW ponds. Photosynthesis causes uptake of CO_2 (aq.) and dissolved inorganic carbon causes increase in algal biomass. Degradation and subsequent deposition of this algal biomass increases soil organic matter and DOC.

In the present study, phytoplankton carbon is highest during pre-monsoon and monsoon. Phytoplankton blooms essentially in the monsoon (Clark and Flynn, 2000; Tortell and Morel, 2002; Cassar *et al.*, 2003) which consequently assumes a significant role in organic carbon cycle. The dynamics of the DOC pool follow inputs and outputs from various sources through different processes like leaching, mineralization, microbial degradation, decomposition and polymerization in the sewage-fed aquatic system. Secretion from submerged plants or animal body, cells of phytoplankton as exudates directly supply to the DOC pool (Ahn *et al.*, 2009). Easily leachable organic matter from dead organisms and microbial activity upon these dead and decaying matter supplement DOC pool. The availability of DOC in water depends on the amount of organic carbon leached from SOC. Due to microbial degradation DOC is released from POC and at higher pH; POC is dissolved to form DOC. Since, EKW is a sewage-fed system, the sewage canals contain a significant amount of DOC of allochthonous origin (Ghosh *et al.*, 1990). DOC in turn is mineralized to DIC and polymerizes to form POC depending upon the physicochemical and environmental regulating factors like DO and temperature. Other than DOC, other main sources of POC generation are sinking phytoplankton, excretion, dead and decaying parts of zooplankton and their fecal matter. Sewage also infuses a little fraction of POC into the system.

WpH values are dependent on temperature of surrounding environment. In the tropics, temperatures are higher during pre-monsoon (March-June) and lower during winter (December to February). Microbial activity amplifies with the rising temperature in pre-monsoon and decreases with the lower temperature in winter (Ahn *et al.*, 2009). DOC shows significant negative correlation with WpH and DO. Photosynthesis results in oxygen generation which elevates DO level in water and uptake of dissolved carbon dioxide increases WpH level. In presence of ample sunlight, high temperature, DO and higher pH, the rate of aerobic degradation of DOC increases during photosynthesis.

DBC largely regulates the alkalinity of water. DCO_2 shows high negative correlation with all the environmental regulating factors such as WpH, water temperature (WT), DO and alkalinity at both EKW and control ponds. This is natural since these same factors also enhance photosynthesis process for which dissolved carbon dioxide is one of the raw materials. DIC values are three to four folds higher in EKW ponds compared to control ponds. DIC also shows significant negative correlation with the above factors in EKW ponds but not in the control ponds where it shows positive correlation with alkalinity and no relation with the other factors. This is indicative of a higher rate of photosynthesis in EKW ponds compared to control ponds. Due to high photosynthesis rate, DIC is used as an extra source of CO_2 other than DCO_2, therefore generating higher volume of algal biomass in EKW ponds. However, in control ponds, DCO_2 is the main carbon form to be used for photosynthesis while, WpH plays the main environmental regulator.

Phytoplankton bloom in pre-monsoon (control ponds) and monsoon (EKW ponds) is not synchronized with increased zooplankton population size which increases during winter and post-monsoon season. Increased zooplankton community in winter and post monsoon graze upon phytoplankton causing a decline in phytoplankton load.

Alkalinity significantly explains highest variation in zooplankton and phytoplankton load in our studied system where significant and positive regulation was observed in the phytoplankton community. Alkalinity buffers pH changes in pond water. It improves phytoplankton productivity by stabilizing pH and increase pond fertility by escalating nutrient availability (Wurts and Durborow, 1992). Plankton communities, mostly zooplankton community, are sensitive towards temperature, and are negatively regulated. Temperature acts as a detrimental factor for zooplankton population as the enzyme activity is optimum at around 30°C but inactivate at higher temperature (Dutta *et al.*, 2006). At higher temperature, enzymatic activity breaks down and causes zooplankton cell death. In EKW, depth of fish pond is very low causing fast increase in water temperature which may hampers the enzymatic activity in zooplankton cells. This may be the reason behind negative regulation of temperature on zooplankton community. Both zooplankton and phytoplankton load increased towards managed East Calcutta fish ponds where alkalinity was also increased due to input of high amount of nutrient (soluble forms of nitrogen and phosphorus) from sewage in EKW ponds. Phosphates are one of the common bases (Wurts and Durborow, 1992) present in high quantity in EKW water which increases the alkalinity here.

Considering all the biological and physico-chemical factors and regulatory physical processes, and analyzing their relationship explained above, a conceptual model with pathways of carbon with respect to the phyto and zooplankton communities have been developed that explains the nutrient flow dynamics in the EKW pond system (Figure 15.3).

In the above conceptual model of carbon cycling pathways of sewage-fed ponds of EKW and natural control ponds, phytoplankton and zooplankton loads have been considered as biotic factors. Factors like temperature, pH, DO, alkalinity etc; that control nutrient cycling in the system are denoted as forcing function as they regulate the transformation of carbon particles at different trophic levels from one form to other. In the conceptual map forcing functions are represented by the circles. Red arrows denote the effect of forcing functions over the physical processes. Physical processes occur at different trophic levels and are included in this conceptual model as photosynthesis, respiration, parasitism, predation, grazing, mineralization, polymerization, microbial degradation or mineralization, diffusion, dissociation, leaching and sedimentation. The main compartments are PPL, ZPL, POC, DOC, SOC, DBC, and DCO_2. These carbon compartments are represented by

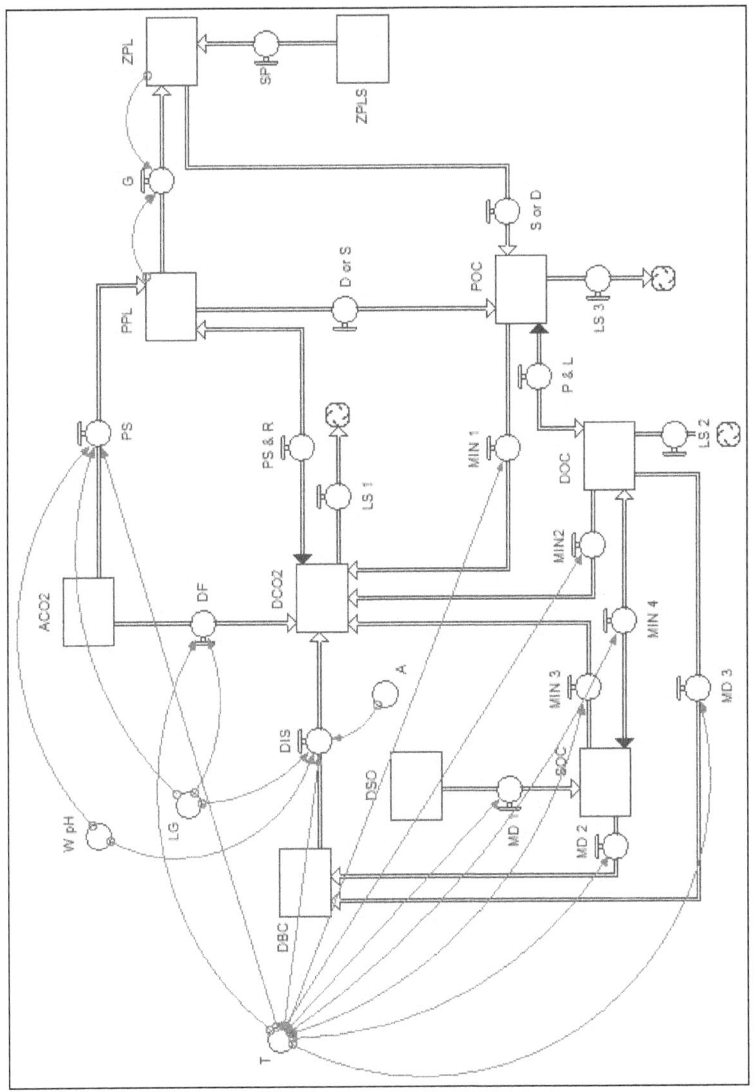

Figure 15.3: Conceptual Diagram of Carbon Cycling Pathways in the Fisheries of EKW and Control Pond Systems.

Abbreviations–A: Alkalinity, A CO_2: Aerial carbon dioxide, D: Death/decay, DBC: Dissolved bicarbonate, DCO_2: Dissolve CO_2, DF: Diffusion, DOC: Dissolve organic carbon, DIS: Dissociation, DSO: Input from dead soil organisms, G: Grazing, L: Leaching, LG: Light, MIN 1, 2, 3, 4: Mineralization, MD 1, 2, 3: Microbial degradation, P: Polymerization, PA: Parasitism, POC: Particulate organic carbon, PPL: Phytoplankton, PR: Predation, PS: Photosynthesis, R: Respiration, S: Sedimentation, SP: Self predation, SOC: Soil organic carbon, T: Temperature, W pH: Water pH, ZPL: Zooplankton, ZPLS: Zooplankton Species, LS 1,2,3: Loss to the system.

rectanglar boxes in the conceptual map. Connecting pipes between different carbon compartments in our model stand for different physical process mediating flow of carbon from one compartment to another. Direction of the connecting pipes (uni-flow and bi-flow) here signify the course of carbon flow between compartments. The seven compartment model considers that aerial carbon dioxide diffuse in water in presence of ample sunlight and high temperature to form dissolved carbon dioxide. Phytoplankton produces glucose through photosynthesis process. The main source of carbon is DCO_2 but at suitable environmental condition, when rate of photosynthesis is higher, other forms of DIC as DBC are also used as an extra source of carbon for photosynthesis. Unlike the studied sewage-fed system, previous studies and structural model of carbon cycle from adjacent Hooghly estuary and mangrove show litter as the main source of carbon (Mukherjee *et al.*, 2012). The present study shows, water pH, temperature and light are the physical environmental factors regulating the process of photosynthesis. Zooplanktons are directly dependent upon phytoplankton for nutrition. Through grazing, carbon is transferred to zooplankton from phytoplankton whereas self-predation causes transfer of carbon within different zooplankton species. Both phyto and zooplankton form POC by death and sedimentation. Polymerization and leaching help in interconversion between POC and DOC. Through mineralization DOC is converted to SOC and vice versa. Carbons from dead soil organism add up to increase SOC. Mineralization or microbial degradation causes transformation of SOC and DOC to different forms of DIC. In estuary of Sunderban delta, surface water remains undersaturated with respect to DO and large quantities of DCO_2 are formed (Ghosh *et al*, 1991). In both estuaries and inland wetlands, temperature regulates microbial degradation process (Smith and Hollibaugh, 1993; Ray, 2008). In our study, partial loss of carbon occurs into the system from DOC, POC and DCO_2. DCO_2 dissociates into DBC. Factors like WpH, light, temperature and alkalinity regulate this dissociation process. Phyto and zooplankton community respiration add to increase DCO_2 which is again used up in photosynthesis. Plankton communities are of immense importance as they convert abiotic carbon particles into biotic form and make it available to other heterotrophs and thus keep the carbon cycle continuing.

Summary

The present study clearly indicates that cycling of organic carbon is an integrated process for both the soil and water of East Kolkata Wetlands system. Alkalinity, water temperature act as major forcing functions controlling phytoplankton communities and different physical processes in water; whereas soil pH is the main factor controlling fluctuations of soil inorganic carbon. Plankton communities play important role in transfer and transformation of organic and inorganic carbon forms like soil organic carbon, particulate organic carbon, carbon from dead soil organism, soil inorganic carbon and dissolve CO_2. Phytoplankton and zooplankton carbon count in sewage-fed system is near about five times more than normal ponds. Gross primary production values are higher in sewage-fed system than normal ponds at all four seasons throughout the year whereas values of community respiration remain overall same in both systems causing higher net primary productivity in sewage-fed fisheries. Mineralization of one carbon form like soil organic carbon to soil inorganic carbon, dissolve inorganic carbon and free CO_2 is governed by microbial activity which is regulated by soil pH, water pH, water temperature and alkalinity. Dissolve CO_2, dissolve organic and inorganic carbon, soil organic and inorganic carbon values are higher in the sewage-fed fisheries than natural ponds. Dissolve and particulate organic carbon mineralize to soil organic and inorganic carbon - through this process large amount of carbon are sequestered within this sewage-fed wetland systems. Carbon from dead soil organisms enhances both the soil organic carbon and particulate organic carbon pool in soil and water consecutively. The conceptual model of carbon cycling presented here, based both on our result as well as secondary information obtained from existing literature, summarizes the major regulating factors in carbon cycle and we hope this in turn would help in the formation of appropriate policies for management, conservation and better sustenance of the single Ramsar site of the state of West Bengal- the East Kolkata Wetlands.

Acknowledgments

We are grateful to the Director of Institute of Environmental Studies and Wetland Management (IESWM) for facilities provided. We are thankful to Dr. Phanibhusan Ghosh for his expert guidance in chemical analysis. We are also thankful to Mr. Anirudha Sen, Mr. Debaditya Kumar and Mr. Soumik Chatterjee for helping us in different aspects. We thank Dept. of Zoology, University of Kolkata where all data analysis was done. This work is supported in-aid by funding agency Department of Environment, Government of West Bengal, India ['EN/P/1858/T-VIII-2/004/2009'].

References

Abril G, Noguueira M, Etcheber H. *et al.* 2002. Behaviour of organic carbon in nine contrasting European estuaries. *Estuarine, Coastal and Shelf Science*, 54: 241-262.

Ahad JME, Barth JAC, Ganeshram RS. *et al.* 2008. Controls on carbon cycling in two contrasting temperate zone estuaries: The Tyne and Tweed, UK. *Estuarine, Coastal and Shelf Science*, 78: 685-693.

Ahn MY, Zimmerman AR, Comerford NB. *et al.* 2009. Carbon Mineralization and Labile Organic Carbon Pools in the Sandy Soils of a North Florida Watershed. *Ecosystems*, 12: 672-685.

Amon RMW and Benner R. 1996. Bacterial Utilization of Different Size Classes of Dissolved Organic Matter. *Limnol. Oceanogr.*, 41: 41-51.

APHA. 1985. *Standard method for the examination of water and waste water*, (18th edition), American Public Health Association Washington DC.

APHA. 1992. *Standard method for the examination of water and waste water*, (18th edition), American Public Health Association Washington DC.

Bassi N, Kumar MD, Sharma A and Pardha-saradhi P. 2014. Status of wetlands in India: A review of extent, ecosystem benefits, threats and management strategies. *Journal of Hydrology*, 2: 1-19.

Bauer JE, Cai WJ, Raymond PA. *et al.* 2013. The changing carbon cycle of the coastal ocean. *Nature*, 504: 61-70.

Bold HC and Wynne MJ. 1978. *Introduction to the Algae – Structure and Reproduction*, Prentice Hall of India Private Limited, New Delhi.

Bolin B, Degens ET, Duvigneaud P and Kempe S. 1979. The global biogeochemical carbon cycle. In: *The Global Carbon Cycle* (1st edition) Bolin B, Degens ET, Kempe S, Kenter P. (Eds.). John Wiley and Sons: Chichester , UK. pp. 1-56.

Borrett SR, Christian RR and Ulanowicz RE. 2012. Network Ecology (Revised) In: *Encyclopedia of Environmetrics* (2nd edition). El-Shaarawi AH, Piegorsch WH (Eds.). John Wiley and Sons: Chinchester, UK. pp.1767-1772.

Cassar N, Laws EA and Bidigare RR. 2003. Bicarbonate uptake by Southern Ocean phytoplankton. *Global Biogeoche. Cycles* 18 Online serial: GB2003, doi: 10.1029 / 2003GB002116.

Clark DR and Flynn KJ. 2000. The relationship between the dissolved inorganic carbon concentration and growth rate in marine phytoplankton. *Proc. Royal Soc. London Series*. B, 267: 953-959.

Dafner EV and Wangersky PJ. 2002. A brief overview of modern directions in marine DOC studies Part II—Recent progress in marine DOC studies. *J. Environ. Monit.*, 4: 55-69.

Ducklow HW and Carlson CA. 1992. Oceanic bacterial production. *Advances in Microbial Ecol.*, 12: 113-181.

Dutta TK, Jana M, Pahari P and Bhattacharya T. 2006. Effect of temperature, pH and salt on amylase in *Heliodiaptomus viduus* (Gurney) (Crustacea: Copepoda: Calanoida). *Turk J Zool.*, 30: 187-195.

Falkowski PG. 1980. *Primary Productivity in the Sea*. Plenum Press, New York

Friedler E, Juanico M and Shelef G. 2003. Simulation model of wastewater stabilization reservoirs. *Ecol. Eng.*, 20: 121-145.

Fritsch FE. 1948. *The Structure and Reproduction of the Algae*. Cambridge University Press, Cambridge.

Greenberg AE, Clesceri LS and Eaton AD. 1992. *Standard Methods for Examination of Water and Wastewater. American Public Health Association*. Washington, DC, USA.

Ghosh D. 1993. Towards sustainable development of the East Calcutta Wetlands, India. Towards the wise use of wetlands. Wise use project. Switzerland, Gland: Ramsar Convention Bureau. 220-236.

Ghosh D. 1999. Waste water utilization in East Kolkata Wetlands from local practice to sustainable option. Urban Waste Expertise Programme (UWEP) Occasional Paper, Gouda, The Netherlands: WASTE.

Ghosh PB, Singh BN, Chakroborty C. *et al.* 1990. Mangrove litter production in a tidal creek of Lothian island of Sundarbans, India. *Ind. J. Mar. Sci.*, 19: 292-293.

Ghosh SK, De TK, Choudhoury A and Jana TK. 1991. Oxygen deficiency in Hooghly estuary, east coast of India. *Ind. J. Mar. Sci.*, 20:216-217.

Gupta PK. 2002. *Methods in Environmental Analysis: Water, Soil and Air*. Agrobios (India), Jodhpur, India, 203-217.

Gupta RK. 2005. *Algal Flora of Dehradun District Uttaranchal*. Botanical Survey of India. India.

Guterstam B. 1996. Demonstrating ecological engineering for wastewater treatment in a nordic climate using aquaculture principles in a greenhouse mesocosm. *Ecological Engineering*, 6(1-3):73-97.

Howland RJM, Tappin AD, Uncles RJ. *et al*. 2000. Distributions and seasonal variability of pH and alkalinity in the Tweed Estuary, UK. *Sci. Total Environ.*, 251-252:125-138.

Insam H. 1990. Are the soil microbial biomass and basal respiration governed by the climatic regime? *Soil Biol. Biochem.*, 22: 525-532.

Joos F, Raynaud D and Wigley T. 1996. Radiative forcing of climate change. In: *Climate Change 1995*, Cambridge University Press, Cambridge, UK. pp. 76-86.

Jumars PA, Penry DL, Baross JA. *et al*. 1989. Closing the microbial loop: dissolved carbon pathway to heterotrophic bacteria from incomplete ingestion, digestion and absorption in animals. *Deep-Sea Res.*, 36: 483-495.

Kamyotra JS and Bhardwaj RM. 2011. Municipal wastewater management in India. India infrastructure report 2011. IDFC. Oxford, 299-311.

Karn SK and Harada H. 2001. Surface Water Pollution in Three Urban Territories of Nepal, India, and Bangladesh. *Environmental Management*, 28(4): 483-496.

Kätterer T, Reichstein M, Andrén O and Lomander A. 1998. Temperature dependence of organic matter decomposition: a critical review using literature data analyzed with different models. *Biol. Fert. Soils.*, 27: 258-262.

Kirschbaum MF. 1995. The temperature dependence of soil organic matter decomposition and the effect of global warming on soil organic C storage. *Soil Biol. Biochem.*, 27: 753-760.

Kretzschmar R, Hesterberg D and Sticher H. 1997. Effects of adsorbed humic acid on surface charge and flocculation of kaolinite. *Soil Sci. Soc. Am. J.*, 61: 101-108.

Kulinski K and Pempkowiak J. 2008. Dissolved organic carbon in the southern Baltic Sea: Quantification of factors affecting its distribution. *Estuarine, Coastal and Shelf Science*, 78: 38-44.

Lloyd J and Taylor JA. 1994. On the temperature dependence of soil respiration. *Functional Ecol.*, 8:315-323.

Maiti SK, Saha S and Saha T. 2012. Planktonic diversity of river Bidhyadhari during high and low tides near Malancha Ghat in West Bengal, India. *Phytotaxonomy*, 12: 160-164.

Mukherjee J, Roy M, Ray S. *et al*. 2012. Mechanism of transformation of various forms of carbon and cycling pathways in the Hoogly estuarine system. *Estuaries: Classification, Ecology and Human Impacts*. Nova science publishers Inc., NY. 6: 1-27.

Olah J, Sharangi N and Datta NC. 2003. City sewage fish ponds in Hungary and India. *Aquaculture*, 54: 129-135.

Packard T, Chen W, Blasco D. *et al*. 2000. Dissolved organic carbon in the Gulf of St. Lawrence. *Deep-Sea Res.*, 47: 435-459.

Palmer CM. 1980. *Algae and Water Pollution*. Castle House Publication Ltd. England.

Platt T, Sathyendranath S and Ravindran P. 1990. Primary production by phytoplankton: analytic solutions for daily rates per unit area of water surface. *Proc. Royal Soc. London*, 241B: 101-111.

Pradhan A, Bhaumik P, Das S. *et al*. 2008. Phytoplankton diversity as indicator of water quality for fish cultivation. *American Journal of Environmental Sciences*, 4 (4): 406-411.

Raich JW and Schlesinger WH. 1992. The global carbon dioxide flux in soil respiration and its relationship to vegetation and climate. *Tellus*, 44B: 81-99.

Raich JW and Potter CS. 1995. Global patterns of carbon dioxide emissions from soil. *Global Biogeochemical Cycles*, 9 (1): 23-26.

Ray S. 2008. Comparative study of virgin and reclaimed islands of Sundarban mangrove ecosystem through network analysis. *Ecological Modeling*, 215: 207-216.

Saha S, Basu P and Saha T. 2014. Size Does Matter: Role of Large Size Wetlands in the Diversity of Wetland-Dependent Bird Species of East Kolkata Wetlands. In. *Current Perspectives in Natural Resource Management*, (Mahata C. ed) pp. 49-67.

Sarkar S, Ghosh PB, Mukherjee K. *et al*. 2009. Sewage treatment in a single pond system at East Kolkata Wetland, India. *Water Sci. Technol.*, 60 (9): 2309-2317.

Sarkar S, Ghosh PB, Sil AK and Saha T. 2014. Suspended particulate matter dynamics act as a driving force for single pond sewage stabilization system. *Ecological Engineering*, 69: 206-212.

Sebastian R and Chacko J. 2006. Distribution of organic carbon in tropical mangrove sediments (Cochin, India). *Int. J. Environ. Studies.*, 63: 303-311.

Singh JS and Gupta SR. 1977. Plant decomposition and soil respiration in terrestrial ecosystems. *Botanical Rev.*, 43: 449-526.

Smith SV and Hollibaugh JT. 1993. Coastal metabolism and the oceanic organic carbon cycle. *Rev. Geophysics.*, 31: 75-89.

Sugiyama Y, Anegawa A, Kumagai T. *et al*. 2004. Distribution of dissolved organic carbon in lakes of different trophic types. *Limnol.*, 5: 165-176.

Tortell PD and Morel FMM. 2002. Sources of inorganic carbon for phytoplankton in the eastern subtropical and equatorial Pacific Ocean. *Limnol. Oceanogr.*, 47: 1012-1022.

Tranvik LJ. 1992. Allochthonous dissolved organic matter as an energy source for pelagic bacteria and the concept of the microbial loop. *Hydrobiol.*, 229: 107-114.

Trivedy RK and Goel PK. 1984. *Chemical and biological methods for water pollution studies*. Environmental publications.

Wang X and Veizer J. 2000. Respiration–photosynthesis balance of terrestrial aquatic ecosystems, Ottawa area, Canada. *Geochim. Cosmochim. Acta*, 64: 3775–3786.

Wetzel RG. 2001. *Limnology: Lake and River Ecosystems* (3rd ed.). Academic Press, San Diego, California, USA.

Winkler JP, Cherry RS and Schlesinger WH. 1996. The Q10 relationship of microbial respiration in a temperate forest soil. *Soil Biol. Biochem.*, 28: 1067–1072.

Wurts WA and Durborow RM. 1992. Interactions of pH, Carbon dioxide, Alkalinity and Hardness in Fish Ponds. SRAC publication no. 464.

Yan J and Zhang Y. 1994. How wetlands are used to improve water quality in China. In. *Global Wetlands: Old World and New* (Mitsch, WJ. ed.). Elsevier, Amsterdam. pp. 369-376.

Chapter 16

Heavy Metal Accumulation in Zooplankton and Fish in a Sewage-Fed Wetland

Subinooy Mondal and Apurba Ratan Ghosh

Department of Environmental Science, The University of Burdwan, Burdwan

Introduction

Aquatic environment is constantly gaining burden of heavy metal accumulation with the advancement of agricultural technique, drainage of sewage from municipality and discharge of industrial wastes (Khare and Singh, 2002; Jayakumar and Paul, 2006). Besides, there are a number of activities such as mining, smelting of metal ores, industrial and automobile emissions, battery recycling units, painting industries, and application of insecticides, sewage and fertilizers contribute to the high levels of toxic metals in the system (Woodling *et al.*, 2001; Patra *et al.*, 2005). Water bodies of many pisciculture units from different parts of India are loaded with heavy metal contamination from low to alarming concentrations (Das and Kaviraj, 2000). Faunal community, particularly zooplankton forms an important trophic level for grazing and nutrient regeneration for juvenile and adult fishes. Thus, regulate the composition and productivity of entire fish communities (Werner and Hall, 1979; Arrhenius, 1996; Hakala *et al.*, 2003). The trend of accumulation of heavy metals in different occasions was studied by different authors, as Bahnasawy *et al.* (2011) showed the concentration of different metals in water, plankton, and fish tissues in the order of Zn > Cu > Pb> Cd in the coastal water as well as offshore waters of the Bay of Bengal, where the average concentration of trace metals in zooplankton followed the same sequence like, Fe > Zn > Cu > Ni > Co > Cd > Pb (Rejomon *et al.*, 2008). The physicochemical characteristics of a wetland exert a great effect upon biota and affects the uptake and accumulation of metals in fish (Playle *et al.*, 1992; Kock *et al.*, 1996; Yang and Chen, 1996). However, there are very few reports on accumulation of heavy metals in zooplankton and fish in the food chain of the sewage-fed wetlands. The present study has been designed to assess the current limnological status of the East Kolkata Wetland (EKWL), and the influence of different physicochemical or environmental variables towards accumulation of heavy metals (Cu, Cd and Pb) in water, zooplankton and different parts of gastrointestinal tract of *Labeo bata*, a fast growing species of EKWL.

Materials and Methods

Analysis of Limnological Parameters

Water samples, zooplankton and fish tissues were collected season-wise from three wetland ponds of East Kolkata Wetland designated as M1, M2, M3 and one sewage carrying canal M4 in order to cover the maximum areas of the water bodies of experimental site. Water samples were collected from epilimnion zone from three spots of each M1, M2 and M3 of wetlands and marked as M1A, M1B, M1C; M2A, M2B, M2C; and M3A, M3B, M3C respectively and also from M4, a canal which directly carries the sewage of Kolkata metropolitan city (Figure 16.1); and were preserved in well stoppered polythene bottles, formerly soaked in 10 per cent nitric acid for 24 hr and rinsed with ultrapure water. Prime physico-chemical factors of water such as pH, temperature, conductivity, TDS, dissolved oxygen, COD, BOD, total hardness, calcium, total alkalinity, chloride, ammonia-nitrogen, nitrate-nitrogen, sulphate, phosphate, iron, sodium and potassium were measured as per APHA (2008).

Analysis of Heavy Metals

For heavy metal analysis filtered plankton (4 per cent formaldehyde fixed), fish tissues of *Labeo bata* (Hamilton) were collected, weighed, and put in small Erlenmeyer flasks, dried in an oven at 105°C for about 24 h, and digested by nitric acid and perchloric acid (2: 1) on a hotplate until the solution became clear. Cu, Cd and Pb concentrations in the water were determined by the extraction method (APHA, 2008). Digested tissues and preserved water samples were measured by atomic absorption spectrometer (GBC Avanta). Plankton and fish samples were prepared for heavy metal analysis according to the method described by Kalay *et al.* (1999). Blanks were maintained simultaneously with same procedure.

Results and Discussion

Limnological Study

The physico-chemical characteristics like pH, dissolved oxygen, conductivity, hardness and water temperature, and other chemical parameters were analyzed throughout the year during the period of 2010-2011 and 2011-2012 and were recorded as an average value. Limnochemical investigations were also conducted like nitrogen bound in ammonia (NH_4-N), and in nitrate (NO_3-N), as well as analysis on chloride (Cl^-) and the biological oxygen demand (BOD_3) were also conducted the results of which are shown in Table 16.1.

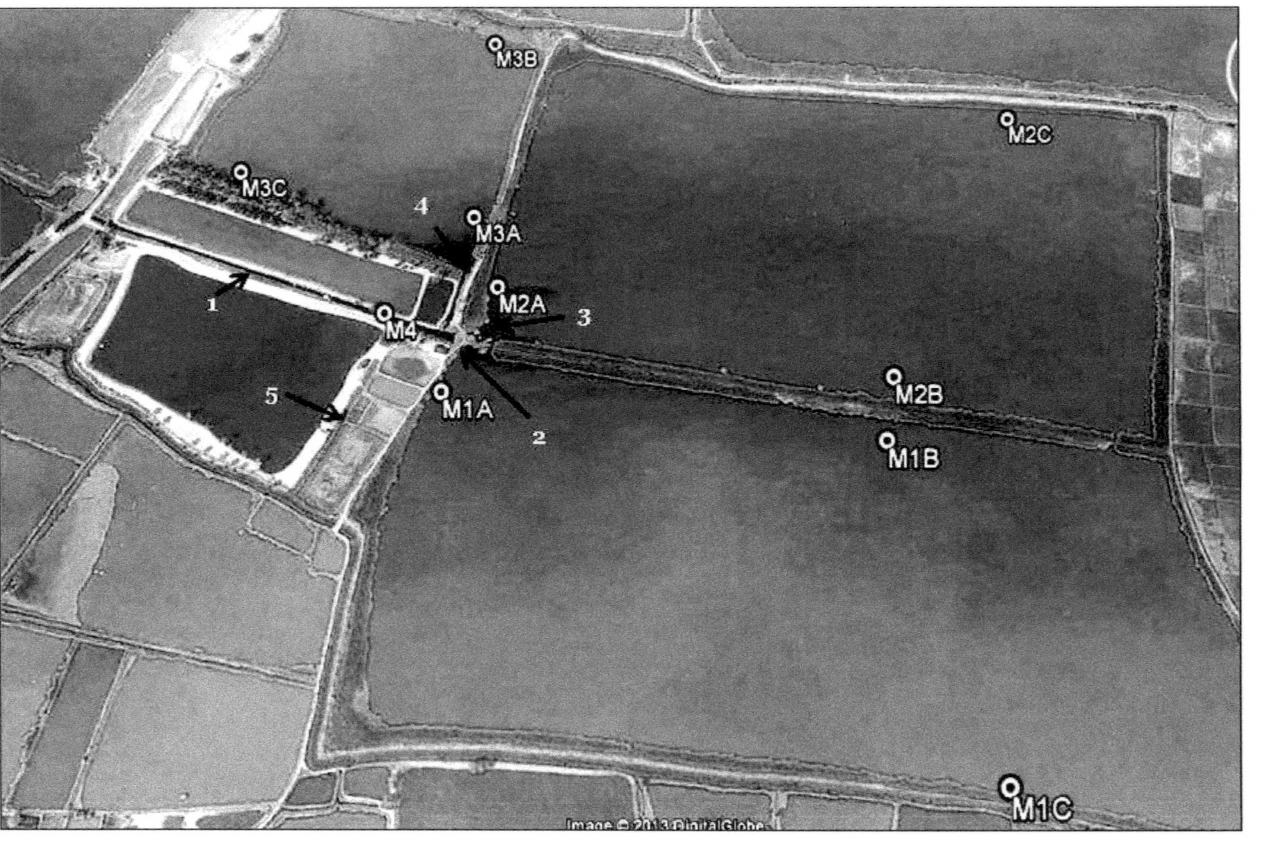

Figure 16.1: Sampling Sites: M1A, M1B and M1C of M1 pond, M2A, M2B and M2C of M1 pond, M4A, M3B and M3C of M3 pond and M4 sewage carrying canal; 1: Sewage carrying canal; 2, 3 and 4: Entrance of sewage in M1, M2 and M3 pond respectively; 5: Brooder stock pond.

Table 16.1: Analysis of Physico-chemical Parameters (Average ± SD, range) of Water and different Water Quality and Aquaculture Standards

Parameters	Sampling Sites				Drinking Water Standard		IS 13891 : 1994 Aquaculture Standard
	M1	M2	M3	M4	WHO (1997)	IS 10500 : 1995	
pH	7.75±0.10 7.64–7.82	7.69±0.14 7.54–7.80	8.22±0.25 7.98–8.48	7.47	–	6.5–8.5	6.5–8.5
Temp.	27.9±4.29 33.4–21.1	27.8±4.99 29.7–32.8	28.3±4.47 21.36–33.4	29.88	7.0–8.5(D)	–	2.0–35.0
Cond.	782.23±46.34 751.20–835.50	796.30±45.41 750.25–841.04	741.17±8.98 731.00–748.00	838.50	250–1480	2000 (MP)	–
TDS	557.97±26.89 541.65–589.00	559.87±26.47 536.00–588.33	526.00±6.87 518.50–532.00	595.00	500–1000	500–2000	–
DO	3.96±1.16 2.63–4.75	3.47±1.56 2.34–5.25	6.63±0.67 6.24–7.40	1.70	–	5 (D)	4.0 (MR)
COD	182.60±0.53 182.00–183.00	248.34±5.76 241.76–252.50	121.17±1.13 120.00–122.25	297.25	–	–	–
BOD	184.32±2.65 181.50–186.75	183.23±0.32 182.91–183.54	167.00±1.80 165.00–168.50	209.50	–	–	–
T. H.	174.27±12.37 164.00–188.00	193.05±14.12 182.00–208.96	176.17±5.06 173.00–182.00	221.50	150–500	300–600	–
Ca	121.17±8.27 115.72–130.69	136.11±5.65 129.58–139.61	122.73±9.65 116.59–133.86	195.38	75–200	75–200	–
T. A.	176.40±26.96 159.75–207.50	176.40±9.78 166.12–185.59	206.58±20.42 192.50–230.00	182.50	–	200–600	100–300
Cl	104.05±5.77 100.72–110.72	97.86±6.54 91.97–104.90	102.17±8.89 94.25–111.79	115.97	250–500	250–1000	–
NH_3–N	1.93±0.35 1.52–2.14	1.70±0.52 1.10–2.06	1.12±0.10 1.01–1.21	2.87	–	–	1.5 (MP)

Contd...

Table 16.1–*Contd...*

Parameters	*Sampling Sites*				*Drinking Water Standard*		*IS 13891 : 1994 Aquaculture Standard*
	M1	*M2*	*M3*	*M4*	*WHO (1997)*	*IS 10500 : 1995*	
NO_3–N	2.73±1.47 1.57–4.38	4.82±2.55 2.35–7.44	2.74±0.49 2.19–3.12	7.86	10–45	–	2 (MP)
SO_4	23.91±4.30 20.94–28.84	19.78±4.47 15.50–24.42	28.77±2.22 27.22–31.31	29.38	200–400	200–400	
PO_4	0.60±0.30 0.30–0.91	0.44±0.23 0.30–0.71	0.25±0.00 0.25–0.26	0.91	–	–	–
Fe	0.37±0.03 0.33–0.39	0.57±0.21 0.43–0.81	0.47±0.03 0.44–0.50	0.38	0.3–1	0.3–1	2 (MP)
Na	67.27±3.39 64.50–71.05	67.43±1.08 66.21–68.25	68.17±2.13 65.75–69.75	64.00	120–400	–	–
K	16.52±1.53 14.80–17.75	16.31±1.09 15.11–17.25	18.17±0.14 18.00–18.25	18.00	12(D)	–	–

D: Desirable; HD: Highly desirable; MR: Minimum requirement; MP: Minimum permissible; T.H.: Total hardness; T.A.: Total alkalinity.

All the units are mg/l, except Temp. (ºC), Cond (µS/cm).

The pH recorded throughout the study period ranged from 7.82 to 7.47. pH of four different sites showed an average value of 7.75 ± 0.10, 7.69 ± 0.14, 8.22 ± 0.25 and 7.47 at M1, M2, M3 and M4 sites respectively; it was lowest (7.47) from M4 site and highest (7.82) from M1 site. But the pH values always remained within the aquaculture standard (6.5-8.5) (IS 13891: 1994) range (Table 16.1).

Temperature varied with the seasons showing highest and lowest values of 33.4°C and 21.3°C in M1 and M3 respectively. The average temperature varied slightly and it was 27.9 ± 4.29°C in M1, 27.8 ± 4.99°C in M2 and 28.3 ± 4.47°C in M3 ponds and 29.88 °C in M4 (Table 16.1).

Yearly average conductivity values were found as 782.23 ± 46.34 µS/cm, 796.30 ± 45.41 µS/cm, 741.17 ± 8.98 µS/cm and 838.50 µS/cm, in M1,M2, M3, and M4 respectively with highest value in M2 (841.04 µS/cm) and lowest (731.00 µS/cm) in M3 ponds (Table 16.1).

The TDS values ranged from 541.65 mg/l to 589.00 mg/l with an average of 557.97 mg/l in M1, 536.00 to 588.33 mg/l with an average of 559.87 mg/l in M2 site and 518.50 mg/l to 532.00 mg/l, with mean value 526.00 mg/l in M3 (Table 16.1).

There was a distinct seasonal variation of dissolved oxygen concentration in the EKWL: higher value of 6.63 ± 0.67 mg/l were obtained at M3 and the lowest value at 1.70 mg/l was recorded at M4 but other ponds were according to the aquaculture standard, *i.e.*, within the acceptable limit *i.e.*, 4 mg/l due to high organic and inorganic load. Here, DO concentration of water is strongly affected by the presence of oxidisable organic matter like dissolved organic carbon (DOC), and also organic and inorganic nitrogen concentrations.

COD (297.25 mg/l) and BOD (209.50 mg/l) were high in M4 site. Here, the range of COD values at the sites of M1, M2, M3 and M4 canal were 182.00 to 183.00 mg/l, 241.76 to 252.50 mg/l, 120.00 to 122.25 mg/l and 297.25 mg/l respectively (Table 16.1). As the East Kolkata Wetland is fed mainly with metropolitan and domestic sewage so the present experimental ponds contained with high level of the organic pollution and the magnitude was in the order of: M4> M2 > M1 > M3.

The result of total hardness and calcium concentrations (Table 16.1) indicated that the maximum hardness of 208.96 mg/l was obtained in M2 (av. 193.05 ± 14.12 mg/l) while the lowest value of 164.00 mg/l was recorded in M1 (av.174.27 ± 12.37 mg/l). The result also showed that the maximum calcium concentration was obtained in M4 (195.38 mg/l) and the lowest in M1 (115.72 mg/l).

There was also a seasonal variation in the total alkalinity values, among the ponds *i.e.*, 176.40 mg/l in M1, 176.4 mg/l in M2, 206.58 mg/l in M3 and 182.50 in M4.

Total iron in every sites remained within the drinking water standard (IS 10500: 1995) (Table 16.1) as well as aquaculture standard (IS 13891: 1994) (Table 16.1).

Ammonia-nitrogen varied from 1.52 to 2.14 mg/l in M1, 1.10 to 2.06 mg/l in M2, 1.01 to 1.21 mg/l in M3 and 2.47 mg/l in M4 sites and nitrate-nitrogen was 1.57 to 4.38 mg/l in M1, 2.35 to 7.44 mg/l in M2, 2.19 to 3.12 mg/l in M3 and 7.86 mg/l in M4 sites, all these values showed higher range than the aquaculture standard. Average values of Ammonia-nitrogen and nitrate-nitrogen showed the following order of M4 > M1> M2 > M3 and M4 > M2 > M3 > M1 respectively (Table 16.1).

In the present study, phosphate concentrations were of 0.60 ± 0.30 mg/l, 0.44 ± 0.23 mg/l, and 0.25 ± 0.00 mg/l at M1, M2 and M3 ponds respectively which always remained within the optimum value as per standards.

Production of sulphate occurs during decomposition of sewage and is the common effluents of tanneries and paper mills. The concentration of sulphate in EKWL ranged between 20.94 and 28.84 mg/l in M1 (av. 23.91± 4.30 mg/l), 15.50 and 24.42 mg/l in M2 (av. 19.78 ± 4.47 mg/l) and 27.22 and 31.31 mg/l in M3 (av. 28.77 ± 2.22 mg/l) ponds but in M4 (29.38 mg/l) that was always high. Phosphate concentrations should be interpreted in conjunction with the concentrations of nitrate, total suspended solids (turbidity) and dissolved oxygen.

The average concentration of sodium was 67.27 mg/l, 67.43 mg/l, 68.17 mg/l and 64.00 mg/l in M1, M2, M3 and M4 sites respectively and potassium varied from 17.75 to 14. 80 mg/l, 15.11 to 17.25 mg/l, 18.00 to 18.25 mg/l and 18.00 mg/l in M1, M2, M3 and M4 sites respectively. The sodium concentration was recorded to be lower than the drinking water quality standard (WHO, 1997) but potassium showed higher value (Table 16.1).

Heavy Metals Study

The mean concentration of Cu, Cd, and Pb in the water samples of four sites (M1, M2, M3 and M4) of EKWL are shown in Table 16.2. The mean concentration of the metals in water was found in the following order of Cd (0.020 mg/l) < Cu (0.055 mg/l) < Pb (0.311 mg/l). This pattern of occurrence agrees with the earlier studies performed on Lake Manzala (Abdel-Baky *et al.*, 1998). Present experimental ponds receive huge quantities of sewage of Kolkata metropolitan city endorsing the concentration of cadmium (Cd) in the water of EKWL between 0.0012 and 0.0050 μg/mg (av. 0.0024 ± 0.0012 μg/mg) in summer, 0.0023 to 0.0048 μg/mg (av. 0.0034 ± 0.0008 μg/mg) in monsoon, from 0.0011 to 0.0029 μg/mg (av. 0.0017 ± 0.0005 μg/mg) in post-monsoon and 0.0010 and 0.0032 μg/mg (av. 0.0017 ± 0.0006 μg/mg) in winter. In summer, the range of concentration of lead (Pb) in water was found between 0.0165 and 0.0460 μg/mg (av. 0.0279 ± 0.0099 μg/mg), in monsoon 0.0166 and 0.0346 μg/mg (av. 0.0235 ± 0.0058 μg/mg), in post-monsoon 0.0166 to 0.0274 μg/mg (av. 0.0224 ± 0.0036 μg/mg) and in winter it ranged from 0.0017 and 0.0285 μg/mg (av. 0.0196 ± 0.0085 μg/mg). Concentration of copper (Cu) in summer varied between 0.0128 and 0.0330 μg/mg (av. 0.0243 ± 0.0062 μg/mg), in monsoon 0.0171 and 0.0461 μg/mg (av. 0.0274 ± 0.0086 μg/mg), in post-monsoon 0.0197 and 0.0371 μg/mg (av. 0.0277 ± 0.0049 μg/mg) and in winter 0.0058 and 0.0183 μg/mg (av.0.0114 ± 0.0042 μg/mg).

Generally, Cd in M4 showed higher concentration of (0.0060 ± 0.001 μg/mg) and Pb (0.550 ± 0.017 μg/mg. This site carries many pollutants in undiluted condition from different sources like, agricultural, industrial, and sewage drains. The level of metals exhibited seasonal fluctuations, highest during summer, while lowest during winter. Ali and Abdel-Satar (2005) attributed to the increased metal concentrations in the water during hot seasons (summer and monsoon) which was released from the sediment to the overlying water under the effect of both high temperature and a fermentation process resulting from the decomposition of organic matter. The seasonal variations of metals in water have also been reported in Lake Manzala (El-Safy and Al-Ghannam, 1996; Abdel-Baky *et al.*, 1998) and in Nile River (Hamed, 1998).

Table 16.2: Analysis of Heavy Metals (average ± SD) of Water, Zooplankton and different Organs of Fish (*L. bata*)

Sampling Sites	*Cd (μg/mg)*	*Pb (μg/mg)*	*Cu (μg/mg)*
		Water sample	
M1A	0.0051 ± 0.0008	0.0455 ± 0.0113	0.0131 ± 0.0055
M1B	0.0038 ± 0.0007	0.0330 ± 0.0083	0.0064 ± 0.0026
M1C	0.0033 ± 0.0006	0.0471 ± 0.0096	0.0116 ± 0.0072
M2A	0.0053 ± 0.0011	0.0486 ± 0.0046	0.0129 ± 0.0036
M2B	0.0044 ± 0.0005	0.0371 ± 0.0035	0.0098 ± 0.0029
M2C	0.0042 ± 0.0007	0.0365 ± 0.0070	0.0083 ± 0.0041
M3A	0.0048 ± 0.0006	0.0474 ± 0.0163	0.0133 ± 0.0090
M3B	0.0047 ± 0.0001	0.0444 ± 0.0156	0.0165 ± 0.0067
M3C	0.0044 ± 0.0007	0.0287 ± 0.0200	0.0126 ± 0.0066
M4	0.0060 ± 0.0019	0.0550 ± 0.0176	0.0151 ± 0.0060
		Zooplankton	
M1	0.0004	0.0021	0.0055
M2	0.0009	0.0015	0.0072
M3	0.0008	0.0024	0.0067
Average ± SD	0.0007 ± 0.0003	0.0020 ± 0.0005	0.0065 ± 0.0009
	Fish (*Labeo bata*) Tissues (Average values in wetland ponds)		
Bone	0.0026	0.0171	0.0038
Gill	0.0018	0.0016	0.0071
Liver	0.0119	0.0531	0.0208
Intestine	0.0108	0.0118	0.0084
Muscle	0.0009	0.0024	0.0038
Fin	0.0004	ND	0.0027
Scale	0.0024	0.0106	0.0487

The order of availability of metals in zooplankton was, Cu (av. 0.0007 ± 0.0003 μg/mg) > Pb (av. 0.0020 ± 0.0005 μg/mg)> Cd (av. 0.0065 ± 0.0009μg/mg) (Figure 16.2). Correlation study among Cu ($r = 0.988$), Pb (0.472) and Cd ($r = 0.817$) concentration of water and zooplankton had a positive relation (Tables 16.3–16.5). This corresponds to the same order of abundance of these metals in water, which supports the hypothesis that water is an important source of plankton contamination. Elmaci *et al.* (2007) reported that the quantity of heavy metals in plankton depends

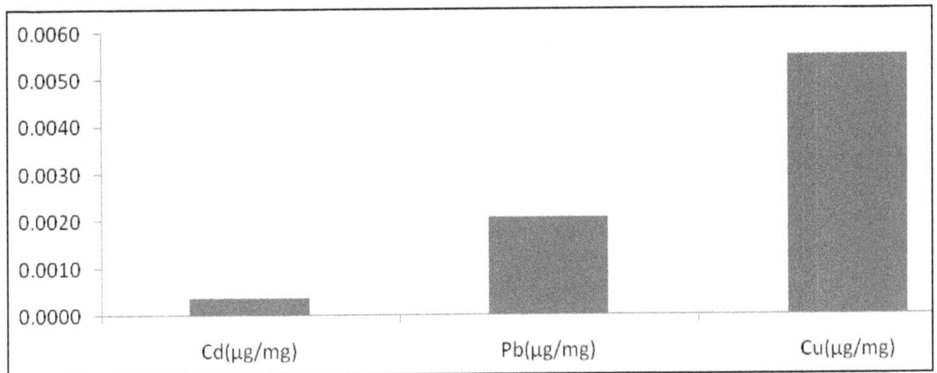

Figure 16.2: Accumulation of Metals in Zooplankton.

on their concentration in water and partially on sediment. The concentration of heavy metals in plankton has been reported to depend upon several factors, such as the productivity of the water body, the physico-chemical properties of the water, quantitative and qualitative species composition of zooplankton and phytoplankton, the capacity of heavy metal absorbance, and the season (Radwan *et al.*, 1990; Elmaci *et al.*, 2007). Compared with results of other studies, small plankton and macrozooplankton from American lakes accumulated lower levels of Cu, Zn, Cd, and Pb (Chen *et al.*, 2000). Plankton from lakes in southern Finland showed higher levels of Cu but lower levels of Cd, Zn, and Pb (Tulonen *et al.*, 2006). Compared with this study, Elmaci *et al.* (2007) recorded enormously higher concentrations of Cu, Zn, Cd and Pb.

The metal concentration in the different organs of *L. bata* showed variations (Table 16.2). Among them Pb showed highest accumulation in liver and bone, Cu mainly accumulated in gill, intestine, muscle and scale, whereas, Cd accumulated highly in fin. In the present study, average concentration of Cd ranged from 0.0004 to 0.0119 µg/mg in fish (Figure 16.3). According to Canadian Council of Ministers of the Environment (1999), Cd concentration should not exceed the standard value (5.0 µg/g) for human consumption. According to Australian standard, Cu concentration must not exceed 10 µg/g in fish (Van den Broek *et al.*, 2002) and the concentration of Cd in muscles of fish from EKWL ranged from 0.0487 to 0.0027 µg/g, which was within the permissible concentration (10 µg/g) for human consumption (USEPA,

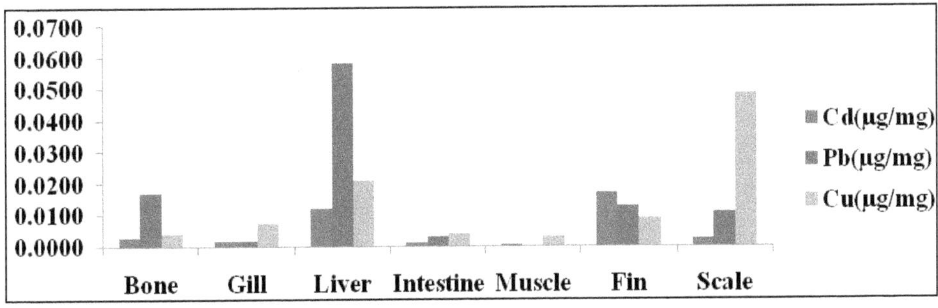

Figure 16.3: Accumulation of Metals in Fish Tissue (*L. bata*).

1999). Detail correlations among metals and fish organs are depicted in Tables 16.3–16.5.

Table 16.3: Correlation among Accumulation of Cd in different Fish (*L. bata*) Organs and Zooplankton

	Zooplankton	*Bone*	*Gill*	*Liver*	*Muscle*	*Fin*	*Intestine*	*Scale*
Bone	0.500							
Gill	0.866	0.866						
Liver	0.277	0.971	0.721					
Muscle	–0.000	0.866	0.500	0.961				
Fin	–0.756	0.189	–0.327	0.419	0.655			
Intestine	–0.778	–0.933	–0.988	–0.820	–0.629	0.176		
Scale	–0.866	0.000	–0.500	0.240	0.500	0.982	0.359	
Water	0.988	0.629	0.933	0.423	0.156	–0.645	–0.866	–0.778

Table 16.4. Correlation among Accumulation of Pb in different Fish (*L. bata*) Organs and Plankton

	Zooplankton	*Bone*	*Gill*	*Liver*	*Muscle*	*Intestine*	*Scale*
Bone	0.968						
Gill	–0.338	–0.564					
Liver	–0.206	0.047	–0.851				
Muscle	–0.554	–0.327	–0.596	0.929			
Intestine	–0.964	–0.866	0.075	0.459	0.756		
Scale	–0.149	–0.392	0.981	–0.937	–0.741	–0.120	
Water	0.472	0.235	0.670	–0.960	–0.995	–0.690	0.802

Table 16.5: Correlation among Accumulation of Cu in different Fish (*L. bata*) Organs and Plankton

	Zooplankton	*Bone*	*Gill*	*Liver*	*Muscle*	*Fin*	*Intestine*	*Scale*
Bone	1.000							
Gill	0.419	0.399						
Liver	0.866	0.855	0.817					
Muscle	0.831	0.843	–0.157	0.441				
Fin	–0.866	–0.877	0.091	–0.500	–0.998			
Intestine	–0.945	–0.938	–0.693	–0.982	–0.603	0.655		
Scale	–0.996	–0.994	–0.500	–0.908	–0.777	0.817	0.971	
Water	0.817	0.804	0.866	0.996	0.358	–0.419	–0.961	–0.866

The higher levels of Cu can be characterised on the basis of biological role in normal metabolism and the growth of plankton and fish, which cause an active uptake and storage. There were significant differences in heavy metal concentrations between seasons and fish organs. According to Shakweer (1998), the concentration of trace metals in various organs of fish showed the degree of water pollution in the aquatic environments. Ravera (2001) noticed the aquatic organisms usually received the pollutants (*e.g.*, metals) from the water or from the food and accumulated in their tissues.

The skin tissue, along with the gill tissues contains mucous layer on the outer surface. The presence of mucus is an effective barrier and important for controlling of diffusion of water pollutants across the fish epidermis (Yilmaz, 2003). There are few studies about the accumulation of metal in fish skin, although it has been considered as the first site of defence and accumulation of the foreign substances.

Muscles contain the lowest concentration of the measured metals. The present study revealed that fish muscles have a low tendency to accumulate the heavy metals to which they are exposed, which has also been supported by Yilmaz (2005). The level of heavy metals in hepatic tissue was ranked as follows: Pb> Cu>Cd. The fish liver is the site of metabolism and excretion of xenobiotic compounds with morphological alterations occurring in some toxic conditions (Rocha and Monteiro, 1999). Metals can influence the hepatic enzyme activities and can lead to pathological hepatic changes, depending on the metal type and concentration, fish species, length of exposure and other factors (Paris-Palacios *et al.*, 2000). Although the zinc and copper have a crucial role in several enzymatic processes but they are tentatively classified as highly toxic metals by Hellawell (1986) and can be bioaccumulated in aquatic organisms.

Aquatic animals can accumulate copper when exposed to the higher level of toxic concentrations (Mazon *et al.*, 2002), which may lead to redox reactions generating free radicals and thus cause biochemical and morphological alterations (Varanka *et al.*, 2001, Monteiro *et al.*, 2005).

Lead residues can cause haematological, gastrointestinal and neurological dysfunction in animals. Severe or prolonged exposure to lead can also cause chronic nephropathy, hypertension and reproductive impairment. Lead can influence the enzymatic activities, alters the cellular calcium metabolism and slows the nerve conduction (Lockitch, 1993). Cadmium is also a common environmental pollutant which is highly toxic and considered to have no biological function (Hallenbeck, 1984). It causes severe membrane integrity damage along with a consequent loss of membrane-bound enzyme activity which results in cell death (Younes and Siegers, 1984). According to other reports nonessential metals can cause anaemia both in mammals (Kostic *et al.*, 1993) and in fish (Gwozdzinski *et al.*, 1992).

Conclusion

In West Bengal, East Kolkata Wetland is an important water resource due to its dimensions and economic importance. The average values of some physico-chemical parameters like total hardness, ammonia-nitrogen, nitrate-nitrogen have higher value than the minimum requirement as well as permissible value for freshwater aquaculture at the entry points of sewage in the wetlands but subsequently shows the dilution of pollution load and finally approaching towards to satisfy the requirements of aquaculture especially pisciculture. The results of the present study clearly demonstrated that contamination with Cu, Cd, and Pb occurs due to the continuous discharge of different pollutants into it. The study revealed that the metals in fish organs were within the allowable limits for human consumption, so, it appears to be of less concerned, but it can affect the reproductive physiology of fishes as well as distribution of zooplankton which finally may cause a negative effect on fish production and abundance of natural food in near future. Great efforts along with active cooperation between different authorities are strongly needed to protect the wetland from pollution for reducing environmental risk.

Summary

Aquatic organisms especially plankton and fish are constantly exposed to different xenobiotics including metals due to human activities. This study measured the levels of Cu, Cd and Pb in water, zooplankton and in different fish tissues like, muscle, gill, intestine, liver, bone and scales of *Labeo bata* (Hamilton) from the different wetland ponds *viz.*, M1, M2, M3 and sewage carrying canal M4 of East Kolkata wetland to assess the degree of contamination with those toxic metals. The mean concentration of these metals in water was found to be in the following order: Cd (0.020 mg/l) < Cu (0.055 mg/l) < Pb (0.311 mg/l), but in fish muscles it was in the safety baseline levels. The accumulation of heavy metals in different fish tissues followed the different order like, Pb > Cu > Cd in bone and liver of the fish, Cu > Cd > Pb in gill and fin, and muscle and scale showed the order of Cu > Pb > Cd but in intestine it was Pb > Cd >Cu. The order of accumulation and concentration of metals in zooplankton was Cu > Pb> Cd. According to the Aquaculture standard (IS 13891 : 1994) the levels of limnological parameters of water of wetland pond like, pH, temperature, conductivity, TDS, salinity, COD, BOD, total hardness, calcium, total alkalinity, chloride, sulphate, phosphate, iron, sodium and potassium were within the minimum permissible ranges but in sewage carrying canal (M4) the concentration of dissolved oxygen was 1.70 mg/l; ammonia-nitrogen was 1.01- 2.87 mg/l and nitrate-nitrogen was 1.57-7.68 mg/l which exceed the minimum permissible ranges. So, it strongly advocates the need of study of risk assessment for proper usage and management practices of wealth of East Kolkata Wetland. Further study on health of freshwater fish *L. bata* may be established as a good bioassay indicator for measuring the extent of accumulation.

Acknowledgements

The authors are grateful to the Head, Department of Environmental Science, The University of Burdwan, West Bengal, India and to UGC for sanctioning a Major Research Project. The first author is grateful to the UGC, New Delhi, India for awarding the Junior Research Fellowship.

References

Abdel-Baky TE, Hagras AE, Hassan SH and Zyadah MA. 1998. Environmental impact assessment of pollution in Lake Manzalah, 1-Distribution of some heavy metals in water and sediment. *J. Egypt. Ger. Soc. Zool.*, 26: 25-38.

Ali MH and Abdel-Satar AM. 2005. Studies of some heavy metals in water, sediment, fish and fish diets in some fish farms in ElFayoum province. *Egypt. J. Aquat. Res.*, 31: 261-273.

APHA 2008. *Standard Method for the Examination of Water and Wastewater* (17th ed.), APHA. Washington, DC.

Arrhenius F. 1996. Diet composition and food selectivity of 0-group herring (*Clupea harengus* L.) and sprat (*Sprattus sprattus* (L.) in the northern Baltic Sea. *ICES J Mar Sci.*, 53: 701-712.

Bahnasawy M, Khidr Abdel-Aziz and Dheina N. 2011. Assessment of heavy metal concentrations in water, plankton, and fish of Lake Manzala, Egypt Turk. *J Zool.*, 35(2): 271-280.

CCME (Canadian Council of Ministers of the Environment). 1995. Protocol for the derivation of Canadian sediment quality guidelines for the protection of aquatic life. CCME EPC-98E. Prepared by Environment Canada, Guidelines Division, Technical Secretariat of the CCME Task Group on Water Quality Guidelines. Ottawa. [Reprinted in Canadian environmental quality guidelines, Chapter 6, Canadian Council of Ministers of the Environment, 1999, Winnipeg.]

Chen CY, Stemberger RS, Klaue B. *et al.*, 2000. Accumulation of heavy metals in food web components across a gradient of lakes. *Limnol. Oceanogr.*, 45: 1525-1536.

Das S and Kaviraj A. 2000. Cadmium accumulation in different tissues of common carp, *Cyprinus carpio* treated with activated charcoal, EDTA and single superphosphate. *Geobios*, 27: 69-72.

Elmaci A, Teksoy A, Olcay Topaç F. *et al.*, 2007. Assessment of heavy metals in Lake Uluabat, Turkey. *Afr. J. Biotech.*, 6: 2236-2244.

El-Safy MK and Al-Ghannam ML. 1996. Studies on some heavy metal pollutants in fish of El-Manzala Lake. In. *Proc. Conference on Food Borne Contamination and Egyptians Health*, Mansoura. pp.151-180.

Gwozdzinski K, Roche H and Peres G. 1992. The comparison of the effects of heavy metal ions on the antioxidant enzyme activities in human and fish *Dicentrarchus labrax* erythrocytes. *Comput. Biochem. Physiol.*,102: 57-60.

Hakala T, Viitasalo M, Rita H. *et al.*, 2003. Temporal and spatial variation in the growth rates of Baltic herring (*Clupea harengus membras*) larvae during summer. *Mar Biol.*, 142: 25- 33.

Hallenbeck WH. 1984. Human health effects of exposure to cadmium. *Cell. Mol. Life Sci.*, 40: 136-142.

Hamed MA. 1998. Distribution of trace metals in the River Nile ecosystem, Damietta branch between Mansoura city and Damietta province. *J. Egypt. Ger. Soc. Zool.* 27: 399-415.

Hellawell JM. 1986. Biological Indicators of Freshwater Pollution and Environmental Management. Elsevier Applied Science Publishers Ltd., London and New York. pp. 546.

Jayakumar P and Paul VI. 2006. Patterns of cadmium accumulation of the catfish *Clarias batrachus* (Linn.) exposed to sublethal concentration of cadmium chloride. *Veterinarski Arhiv.,* 76: 167-177.

Kalay M, Ay O and Canli M. 1999. Heavy metal concentrations in fish tissues from the Northeast Mediterranean Sea. *Bull Environ Contam Toxicol.,* 63: 668-673.

Khare S and Singh S. 2002. Histopathological lesions induced by copper sulphate and lead nitrate in the gill of fresh water fish *Nandus nandus. J Ecotoxicol Environ Monit.* 12: 105–111

Kock G, Triendl M, Hofer R. 1996. Seasonal patterns of metal accumulation in Arctic char (*Salvelinus alpinus*) from an oligotrophic Alpine lake related to temperature. *Can. J. Fish. Aquat. Sci.* 53: 780–786

Kostic MM, Ognjanovic B, Dimitrijevic S. *et al.,* 1993. Cadmium-induced changes of antioxidant and metabolic status in red cells of rats: *In vivo* effects. *Eur. J. Haematol.,* 51: 86-92.

Lockitch G. 1993. Perspectives on lead toxicity. *Clin. Biochem.,* 26: 371-381.

Mazon AF, Cerqueira CCC and Fernandes MN. 2002. Gill cellular changes induced by copper exposure in the South American tropical freshwater fish *Prochilodus scrofa. Environ. Res.* 88: 52-63.

Monteiro SM, Mancera JM, Fonta Dnhas- Fernandes A, Sousa M. 2005. Copper induced alterations of biochemical parameters in the gill and plasma of *Oreochromis niloticus. Comput. Biochem. Physiol. C.* 141: 375-383. DOI: 10.1016/j.cbpc.2005.08.002

Paris-Palacios S, Biagianti-Risbourg S and Vernet G. 2000. Biochemical and ultrastructural hepatic perturbation of *Brachydanio rerio* (Teleostei, Cyprinidae) exposed to two sublethal concentrations of copper sulphate. *Aquat. Toxicol.,* 50: 109-124.

Patra RC, Swarup D, Naresh R. *et al.,* 2005. Cadmium level in blood and milk from animals reared around different polluting sources in India. *Bull. Environ. Contam. Toxicol.,*74: 1092-1097.

Playle RC, Gensemer RW and Dixon DG. 1992. Copper accumulation on gills of fathead minnows – influence of water hardness, complexation and pH of the gill microenvironment. *Environ. Toxicol. Chem.,* 11: 381-391.

Radwan S, Kowalik W and Kowalczyk C. 1990. Occurrence of heavy metals in water, phytoplankton and zooplankton of Mesotrophic Lake in eastern Poland. *The Science of the Total Environment.,* 96: 115-120.

Ravera O. 2001. Monitoring of the aquatic environment by species accumulator of pollutants: a review. *J. Limnol.,* 60: 63-78.

Rejomon G, Balachandran KK, Nair M and Joseph T. 2008. Trace metal concentrations in marine zooplankton from the western Bay of Bengal. *Applied Ecology and Environmental Research,* 6(1): 107-116.

Rocha E and Monteiro RAF. 1999. Histology and Cytology of Fish Liver: A Review. In. *Ichthyology: Recent Research Advances* (Saksena DN. ed.). Science Publishers, Enfield, New Hampshire. pp. 321-344.

Shakweer LM. 1998. Concentration levels of some trace metals in *Oreochromis niloticus* at highly and less polluted areas of Mariut Lake. *J. Egypt. Ger. Soc. Zool.,* 25: 125-144.

Tulonen T, Pihlstrom M, Arvola L and Rask M. 2006. Concentrations of heavy metals in food web components of small, boreal lakes. *Boreal Environ. Res.,* 11: 185-194.

USEPA. 1999. Biological assessment of the Idaho water quality standards for numeric water quality criteria for toxic pollutants. U.S. EPA, Region 10, Seattle, Washington DC.

Van den Broek JL, Gledhill KS and Morgan DG. 2002. Heavy metal concentrations in the mosquito fish, *Gambusia holbrooki,* in the Manly Lagoon catchment. In. *UTS Freshwater Ecology Report 2002,* Department of Environmental Sciences, University of Technology, Sydney, Australia.

Varanka Z, Rojik I, Varanka I *et al.,* 2001. Biochemical and morphological changes in carp (*Cyprinus carpio* L.) liver following exposure to copper sulfate and tannic acid. *Comput. Biochem. Physiol. C.,* 128: 467-478.

Werner EE and Hall DJ. 1979. Foraging efficiency and habitat switching in competing sunfishes. *Ecology,* 60: 256-264.

WHO 1997. Guidelines for drinking water quality, World Health Organization, Geneva.

Woodling JD, Brinkman SF and Horn BJ. 2001. Non uniform accumulation of cadmium and copper in kidney's of wild brown trout *Salmo trutta* populations. *Arch. Environ. Contam. Toxicol.,* 40: 381-385.

Yilmaz AB. 2003. Levels of heavy metals (Fe, Cu, Ni, Cr, Pb and Zn) in tissue of *Mugil cephalus* and *Trachurus mediterraneus* from Iskenderun Bay, Turkey. *Environ. Res.,* 92: 277-281.

Yilmaz AT. 2005. Comparison on Heavy Metal Levels of Grey Mullet (*Mugil cephalus* L.) and Sea Bream (*Sparus aurata* L.) Caught in Iskenderum Bay (Turkey). *Turkey Journal of Veterinary Animal Science,* 29: 257-262.

Yang HN and Chen HC. 1996. Uptake and elimination of cadmium by Japanese eel, *Anguilla japonica,* at various temperatures. *Bull. Environ. Contam. Toxicol.,* 56: 670-676.

Younes M and Siegers CP. 1984. Interrelation between lipid peroxidation and other hepatotoxic events. *Biochem. Pharmacol.* 33: 2001-2003.

Chapter 17

Water Hyacinth: Origin, Mode of Invasion and its Role as Ecosystem Service Provider

Barnali Sarkar, Gautam Aditya and Goutam Kumar Saha

Entomology and Wildlife Biology Research Laboratory, Department of Zoology, University of Calcutta, Kolkata

---

Introduction

Freshwater ecosystems of the world face threats of invasive species that leads to alteration of the community structure and ecosystem functions. Preservation of the community structure is essential to continue with the benefits derived from the freshwater bodies, including the valuable ecosystem services. Although quarantine measures and strict legal enforcements have reduced the intentional introduction of species from the native to the exotic habitats but the invasion of certain species all through the globe leads to the degradation of the habitats to a considerable extent. Conservation and management planning are emphasized to prevent and restore the freshwater ecosystems from the ill effects of the invasive species. While ecological principle based technology has proved to be effective in reconstituting the community of the freshwater bodies, the effect of many of the invasive species still remains a concern for preservation of the ecosystems. Among the different invasive species recognized worldwide the water hyacinth (*Eichhornia crassipes* (Mart.) Solm; Pontederiaceae) is a well known example affecting millions of hectares of wetlands and similar ecosystems beyond its native range. A fair amount of studies have been conducted on the ecology, growth and biology of the species in the countries where water hyacinth has established a huge population in the wetlands superseding the native population of the macrophytes. The water hyacinth is considered as one of the worst 100 invasive species in the world with huge efforts being given for its control and concern to prevent its geographical expansion. Invasive weed species are characterized by their ability to exploit a wide range of habitats and tolerate extensive fluctuation of the habitat conditions, besides exhibiting fast growth and reproduction in comparison to indigenous weed species that are adapted to the native conditions. Thus the ability of water hyacinth to express multiple forms in response to the changes in the environmental conditions, enable evasion of the barriers towards establishment in the newly invaded habitats. Empirical and theoretical research pertaining to the invasive weed species suggests that water hyacinth exhibits plastic response to the environmental variations and establish itself as a dominant species in the invaded water bodies. While vast number of studies highlight the impact of water hyacinth as invasive species, equally good number of research, particularly during the last two decades emphasize the bioresource potential of water hyacinth. The extent of nutrient loading by water hyacinth may enable its use in managing eutrophic conditions of the water bodies apart from remediation of heavy metals and different pollutants. An alternative use of the water hyacinth as a bioremediation agent arises out of its functions in the mitigation of environmental pollution and uses in feed for livestock. Considering the possible benefits that can be derived from water hyacinth and the damage potential as invasive species, the trade off is essential before adjudicating in one of the alternative categories. The following paragraphs will provide an overview to decide the nomenclature of the water hyacinth as an invasive plant with potential benefits for bioremediation and food for livestock. Apart from the discussion on the invasive nature and the effects on the aquatic community, the role of water hyacinth in environmental bioremediation will be highlighted in the following sections.

Origin and Mode of Invasion

The weed, water hyacinth is native to South Amazonian basin, and is recognized as one of the worst, nuisance, free floating plant species that has infested almost all the continents at present. It is most prevalent in tropical and subtropical water bodies with high nutrient content due to agricultural runoff, deforestation and improper sewage management. The most alarming fact is that it may expand to higher latitude with rise in temperature (Rodriguez-Gallago *et al.*, 2004; Hellmann *et al.*, 2008; Rahel and Olden, 2008). However, it is still uncertain how, why and when the weed has been carried outside its native range. Whatever the cause of invasion may be, whether intentional and accidental, it has now been reckoned as a most successful invasive species to outcompete native vegetation, phytoplankton and even its predator weevil *Neochetina eichhorniae* and *Neochetina bruchi* (Wilson *et al.*, 2005). *Eichhornia*, by virtue of its complicated root system prevents photosynthetically active light to reach, thus blocking light penetration by submerged leaves or attached algae (Cattaneo *et al.*, 1998). Being an ornamental plant from South America it was introduced in Egypt in about 1879 and then it spread to Congo, the Nile and Lake Victoria during 1950's. It was then introduced to Asia sometimes in 1888 and then in Australia. The problem associated with this weed invasion in Lake Victoria, the second largest lake of the world was apparent since 1990s (Twongo, 1993; Harley *et al.*, 1996). By 1995, the Ugandan lake was almost infested by the weed (Matagi, 2002).

Until the infestation of this weed in the concerned countries, the effects on the native biota remained mostly illusive. The infestation of water hyacinth in the African continent was reported through the studies in Shire river, Malawi (Terry, 1991), Benin (van Thielen *et al.*, 1994), Sudan (Beshir and Bennett, 1985), that elucidated the negative impact of water hyacinth on aquaculture and riverine transportation system. In Australia, the spread of water hyacinth initiated from Darwin, Northern Territory, during 1890 and has since spread in different regions along the coasts, as well as in the adjacent Pacific islands (Burton, 2005). In Asian countries, the spread was recorded from South Asia, Middle East and Japan by the end of nineteenth century (Ueki *et al.*, 1975). In USA, the infestation of water hyacinth was recorded at the beginning of 20th century in Lousiana (Sculthorpe, 1967) followed by its rapid spread in the tropical regions along the south Atlantic and Gulf coast (Aurand, 1982; Owens and Madsen, 1995). Although restricted, the infestation of water hyacinth is also reported from the European countries. Infestation of water hyacinth led to the displacement of the local weed species in many instances, even if these were exotic in nature. Considering the spread of the water hyacinth across the continents, almost all studies recorded a general effect of the water hyacinth as invasive species, either altering the habitat or changing the species interactions in the invaded community.

Growth Features and Adaptations as Invasive Species

The genotypic constitution of this alien invader accounts for adoption of various reproductive strategies in accordance with different habitat conditions. More the number of strategies adopted to overcome adverse situations more is its rate of success of ecesis in a non-native environment (Mandryk and Wein, 2006). The rapid clonal expansion of the plant accounts for its status as successful invader. Clonal plants generally adopt two types of expansion strategies 'Phalanx' versus 'guerilla'. The guerilla strategy causes the plant to quickly infiltrate areas horizontally by means of vegetative propagation of rhizomes, stolons and adventitious roots, whereas, phalanx strategy minimizes the distance between modular units (Doust, 1981). Such expansion strategies are of adaptive significance to explore new unoccupied habitat maximizing resource utilization and minimize competition (Fahrig *et al.*, 1994). The plant occurs in two morphological forms *viz.* one with short bulbous (SB) and buoyant petioles found generally in open water or at the edges of plant mats while the other is with slender, tall, nonbulbous (TN) petioles which occurs in crowded areas. Both morphs are specialized to utilize specific niches by virtue of its phenotypic plasticity. The SB form is structurally adapted to grow horizontally to colonize a free open habitat whereas the TN form is more adapted to vertical growth thus avoiding self-shading in a crowded, competitive environment where light acts as a limiting factor. Numerous stolon, ramets and increased petiole length was observed in case of SB forms rather than TN form which is indicative of its tendency to colonize rapidly. TN form allocates more energy in nutrient sequestering and reproduction rather than colonization (Williams *et al.*, 2005). In case of TN plants, stolon elongation is a common adaptation to compensate photosynthesis at low light condition. Another important adaptation of both the forms is the presence of large, intercellular, irregular air spaces in roots, leaves and rhizome. Such air-filled canaliculies provide a constant internal, closed system where oxygen and carbon-di-oxide produced can be recycled for use in respiration and photosynthesis respectively. The epidermis of both forms unlike the terrestrial plants consists of a single layer of hexagonal cells with thin cellulose wall covered by thin cuticle, an adaptation that facilitates the survival under eutrophic condition (Qaisar *et al.*, 2005). The floating form of the weed does not carry roots with them so that it can easily spread its diverse reproductive units elsewhere blown by the water current. Accordingly, due to absence of root system, plant may need to take the help of another macrophytes or helophytes for the purpose of anchorage (Tellez *et al.*, 2008). The weed has developed an excellent root adaptation commensurate with the depth and level of water body. Whenever, water level tends to rise the macrophytes develop enormous root system whereas surprisingly, the extent of leaf surface and root mass becomes reduced when the prevailing condition just reverses (Oki and Ueki, 1984). Relatively few numbers of predators as well as competitors are there to impede infestation of the weed beyond its native range. Also, the plant has been reported to be in evolutionary most favourable symbiotic relationship with several organisms like nitrogen fixing *Azotobacter chroococcum* (Purchase, 1977), invertebrates and vertebrates (Ultsch, 1976; Auffenberg, 1980; Godley, 1982).

The presence of aerenchymatous tissue formed by a typical phellogen of epidermal or cortical origin accounts for its free floating nature and thus rapid propagation occur with water and wind current. The plant can reproduce both sexually and asexually and both modes of reproduction contribute to its pernicious capacity to invade rapidly. In mild climate, it flowers throughout the year and produces enormous seeds (Langeland and Burks, 1998) elsewhere. Vegetative reproduction takes place by means of breaking off rosettes of clonal individuals. The stolon can be easily broken by wind or wave action and such light weight floating plant parts can be transported easily from one place to another (Barrett, 1980, Langeland and Burks, 1998). The plant can produce 3.4 seeds annually (Bock, 1966) as calculated by combining the data that the number of flowering plants, time required to complete flowering *i.e.* 15 days (Penfound and Earle, 1948), total number of capsules present in each inflorescence (1-20; Gopal, 1987) and number of seeds in each capsule (3-250; Bock, 1966). Inspite of producing fewer seeds and possessing low germination rate, seedling density is supposed to be greater than that of the mature plant (Mathews, 1967). It takes nearly about 1-1.5 months to arise leaf after germination (Parija, 1934). Thus a plant weighing 10 gm with seedling density 100 m^{-2} would increase its biomass to more than 1 kgm^{-2} within two month if optimum conditions prevail. Seeds germinate within six months especially under dry condition (Ueki and Oki, 1979) and can survive 15-20 years in dormant condition. However, further research on the seedling density and seedling growth is necessary in order to predict the future growth potential of the plant.

The weed grows following logistic equation was explained by Wilson *et al.* (2005) in his model, though it cannot explain the dispersion at spatial scale satisfactorily (Higgins and Richardson, 1996). Several parameters such as light, frost, salinity, humidity, wind, water current, carbon dioxide concentration, nutrient depletion and diseases are supposed to influence the growth rate of the weed but there has been lack of research in this arena for which the extent of impact has not been assessed quantitatively as yet. Soil salinity (Olivares and Colonnelo, 2000) and low humidity (Freidel and Bashir, 1979) has been reported to limit the growth of the plant under simulated laboratory condition, though, it may not be true if considered *in situ* (Allen *et al.*, 1997). The plant can grow well with increase in temperature up to the optimum and can perpetuate well at temperature below the minimum tolerance range. Such ability to survive at wide range of temperature regime was responsible for prolific growth of the plant. Another factor which makes the plant successful invader is its heliophilicity (Francois, 1981) *i.e.* to grow under conditions of wide fluctuation in light intensity. *Eichhornia* can grow well at a pH value of 6.5-7.2 and has been found to increase pH thereby ameliorating water condition. The physiology of plant is adapted in such a way that extreme nutrient rich condition foments its growth and propagation. Nitrate contributes huge towards its growth potential. Calcium deficiency also hinders its vegetative propagation (Tellez *et al.*, 2008). Different parts of plants are modified in various ways in order to thrive under condition of highly polluted environment which ultimately has made the invasion much easier. For instance, reduction in the number of layers in hypodermis, presence of raphide crystals of various sizes in parenchymatous cells of petiole are some of the unique survival strategies of the weed to thrive from oligotrophic to eutrophic condition. The successful invasion of the plant was attributed to its ability to withstand under conditions of extreme fluctuation in temperature, nutrient supply, pH level and even toxic condition up to certain extent.

Nutrient Budget for Optimal Growth and Propagation

Growth rate of plant is positively correlated with nitrogen concentration in water (Reddy and De Busk, 1991; Carignan and Neiff, 1994) whereas, phosphorus sometimes has been reported to act as limiting factor (Srivastava *et al.*, 1994). The growth rate shows a hyperbolic relation with limiting nutrients if plotted graphically. Water hyacinth leaf has been reported to possess nitrogen seven times than that of phosphorus. Accordingly, phosphorus becomes limiting if its concentration in water becomes less than that of one-seventh of the concentration of nitrogen (Gopal, 1987; Goel *et al.*, 1989). Calcium has been reported to be principle inorganic component to regulate the growth of the plant (Tabuti *et al.*, 1998). The plant accumulates this nutrient and also oxalate differentially in different parts of the body. Seasonal variation and different life history stages influence on the distribution of calcium and oxalate in different organs of the plant. In wet season, calcium deposition was highest in leaf followed by rhizome, root, petiole and stolon in descending order. Such differential accumulation may be due to restriction of movement of the element inside phloem thus interfering with redistribution to other organs of the body (Mengel and Kirby, 1982). During dry season, Ca^{2+} localization was highest in case of rhizome followed by leaf. Soluble calcium decreases with age in all organs except the leaf (Tabuti *et al.*, 1998).

Role as Ecosystem Service Provider

The concept of 'ecosystem services' integrates ecosystem functions with the perceived value for human well-being. The presumption is that the human society is dependent on the ecosystem products and the services that help to maintain the environmental conditions as well as provides food and shelter for the human being. The components and the services of the ecosystems that enable sustenance of the environmental condition as well as the human population remains less recognized in terms of the economic value. Certain services remain intangible, though essential in sustenance of human population on the earth. Ecosystem functions refer variously to the habitat, biological or system properties or processes of ecosystems. Ecosystem goods (such as food) and services (such as waste assimilation) represent the benefits human populations derive, directly or indirectly, from ecosystem functions. For simplicity, we will refer to ecosystem goods and services together as ecosystem services. Invasive plants are often recognized as detrimental for the invaded ecosystems since they alter the community composition and the habitats to such an extent that the functions and services are affected to a large extent. In many instances, the decision on the elimination or regulation of the species in the native ecosystems depends on the cost benefit factor from economic viewpoint rather than the environmental consequences. Opposing to the prevailing concept of highlighting negative impacts of the invasive species on the generation of ecosystem services, theories have been advocated that suggest the gain in the ecosystem functions through the development of strategies that make use of the invasive plants and animals in amelioration of the ecosystem conditions. The general perceptions eluted from the scientific forum substantiated with theoretical and empirical studies over turned the consideration of the possible beneficial effects of the invasive species in amelioration of the environmental condition. Although evidences of the possible ill effects by the invasive plants and animals have been demonstrated globally, least effort has been given to recognize the beneficial effects of the prospective invaders. Possible lack of neutral judgment and lack of inclusion of the non use value of the invasive species has been the primary cause for undermining the prospective benefits derived from the invasive species. Non native species can also be a part of the conservation program or even can be used for definitive economic purposes rendering them as useful resources. Invasive plants bear intrinsic beneficial values when present in the native ecosystems, which may be continued in the invaded ecosystems. Current research suggests that water hyacinth is useful in fodder industry, biogas production, removal of pollutants from the degraded environments and in energy production. Much of the green chemistry studies use the invasive species like water hyacinth as a focal bioresource. Based on these studies and considering the biology of water hyacinth, it is imperative that this invasive species has many beneficial effects than can be included as ecosystem services for the benefit of the human well being. In the following sections the prospective use of water hyacinth as a biological resource is being highlighted to substantiate the concept of beneficial effects of invasive species in freshwater ecosystems.

A) Nutrient Removal and Water Purification

Eutrophication of freshwater is a major cause of decline in the water quality and hampers the use of the water for domestic or industrial propose. Irrespective of the source of water being natural or man-made, presence of excess amount of nitrate and phosphate along with increased total suspended solids and biochemical oxygen demand (BOD) are the characteristic of eutrophic condition of the freshwater.

Among the different strategies to purify the water and reduce the nutrient load of the concerned freshwater, the use of macrophytes has long been suggested. The choice of the characteristic plant species for the purpose is the floating weeds like water hyacinth and water lettuce. The efficacy of the macrophytes depends on the ability to exhibit a correspondence of absorption of the target nutrient inside the tissues with respect to the availability in the environment. Water hyacinth is considered as a plant of choice for the removal of the inorganic nutrients with higher efficacy than many plants of the same habitat. Water hyacinth has been used in the removal of nitrogen, phosphorus and other elements from the polluted water, thereby qualifying them as essential for the phytoremediation strategy to reduce the nutrient load in nitrogen polluted waters (Fang *et al.*, 2007). The association of the water hyacinth with eutrophic conditions has long been recognized as a basis to establish the relationship with the nutrient uptake and thus removal from the ambient water body, particularly in areas where the infestation of water hyacinth is high. Experimental evidence suggests that water hyacinth can grow well with the addition of nitrogen and phosphorus in water (Rogers and Davis, 1972; Boyd, 1976). According to one estimate, the average rate of nitrogen and phosphorus removal are over 3.4 and 0.43 kg/ha/day respectively from the ambient water (Boyd, 1976). Dried sample of water hyacinth contained 1.2 to 5.6 per cent (average of 2.5 per cent) of nitrogen, and 0.1 to 0.8 per cent (average of 0.47 per cent) of phosphorus. The protein content of water hyacinth decline with the age and the absorption ability for the nitrate and phosphate (Wolverton and McDonald, 1978). In a different study, researchers demonstrated that one hectare of water hyacinth plants under optimal conditions could absorb the average daily nitrogen and phosphorus waste produced by 800 people. On an average 1.6-3.3 mg of phosphorus and 6.6-20.8 mg of nitrogen can be absorbed by a plant of average size (Rogers and Davis, 1972). When compared with the algal system of nutrient removal, water hyacinth is far more efficient with the capacity to absorb nitrate-N and ammonia-N by 33 per cent and 11 per cent respectively (Sheffield, 1967). The growth and absorption of nutrients by the water hyacinth plants were greater under the pH of 4-7 with 20 ppm phosphorus with an increase in phosphorus absorption from 0 to 40 ppm (Haller and Sutton, 1973). The rate of photosynthesis and stomatal conductance increases with nitrogen content in leaf (Lambers *et al.*, 2000). The nitrogen content increases as leaf starts to mature reaches its peak at young stage and then declines with ageing (Center and Wright, 1991). When different parts of plant were taken into consideration, petiole was found to accumulate and release more nitrogen rather than leaf and root. The nitrogen uptake by water hyacinth is considerably affected by the density of the plant and the nitrogen content in the ambient environment with high degree of dependence on the microbes associated with the root hair system of the plants. Removal of ammonia nitrogen by water hyacinth is mainly due to bacterial nitrification and plant accumulation. Bacterial population takes refuge inside the root system of the plant, bringing about the conversion of ammonia to nitrite and from nitrite to nitrate which then may or may not be removed by denitrification. The rhizosphere as well as the phyllosphere of the plant contain bacteria like the *Nitrosomonas* and *Azotobacter* that contributes to the decomposition of the detritus and enhance the growth and the nitrogen absorption (Wei *et al.*, 2011).

Harvesting of the weed minimizes the competitive reduction of plant growth due to crowding. On the other hand, harvesting may cause bacteria at the plant root to die out thus bacterial nitrification is hindered. Accordingly, it has been suggested that 50 per cent removal of ammonia is performed by the plant while the other 50 per cent is through bacterial transformation. Plant uptake of ammonia may not be a significant way of ameliorating water quality because they intake ammonia only when it tends to grow (Hauser, 1984). The weed has been reported to transform the detritus into ammonium nitrogen especially during first two weeks which is later on nitrified into NO_2-N and NO_3-N by bacterial transformation (Tripathi and Shukla, 1991; Ismail *et al.*, 2014). Under controlled condition in laboratory, NO_3-N removal reached to a significant value when treated with water hyacinth (Tripathi and Shukla, 1991).

In comparison to the nitrate uptake, the phosphorus regulation by water hyacinth has been emphasized to a lesser extent, though the removal of excess phosphorus can be achieved by using water hyacinth in the polluted or eutrophic water (Xie *et al.*, 2004; Chen *et al.*, 2010). The growth of the plant, however, remains unaffected with high concentration of phosphorus at a pH of 4 to 8 (Haller and Sutton, 1973). Both raw and dry plant parts were useful in uploading the phosphorus from diary waste (Reddy and DeBusk, 1991) and the swine waste water (Chen *et al.*, 2010) quite efficiently than the water lettuce (Ismail *et al.*, 2014). The growth rate and nutrient storage by *Eichhornia* is positively correlated with the phosphorus concentration of the water but is limited to the biomass of the individual plant. At a concentration of 40 ppm of phosphorus in water, the leaf, stem and the root contain 8.8, 9.3 and 9.26 mg/g of phosphorous respectively, which under control condition is much low (leaf - 1.17, stem - 0.71 and root - 0.96mg/g), which represents the ability of the plant to grow uninhibited with high concentration of phosphorus in water.

Apart from nitrogen and phosphorus, water hyacinth effectively absorbs and stores potassium, calcium and magnesium from the water at slightly alkaline pH condition (Zhou *et al.*, 2007). The stem and leaf of water hyacinth absorb potassium at an appreciable rate from the water body thus reducing the turbidity of water, thereby serving as a source of potassium salts (Zhou *et al.*, 2007). Based on the ability to remove the nutrients from the ambient environment, water hyacinth is promoted as a bioresource for the purification of the waste water originating from different sources (Xie *et al.*, 2004; Zimmels *et al.*, 2006; Chen *et al.*, 2010; Jafari, 2010).

Management of domestic, agricultural and industrial sewage involves lots of investment in terms of money and manpower and becomes problematic in areas of scanty water. In essence water hyacinth can be employed in nutrient removal from the ecosystem loaded with phosphorus and nitrogen in excess. A strategy to employ the water hyacinth as a part of the bioecological engineering to remove the nutrient and purify the water quality has long been employed in the lakes of China, where, eutrophication is a major issue for pollution of freshwater lakes (Wang *et al.*, 2012, 2013). The water hyacinth in the eutrophic conditions reduced the phosphate and nitrate level considerably and the resultant growth of water hyacinth led to a good harvest for further use after drying. Growth and prospective harvest of plant biomass was highlighted in the studies of the water hyacinth in the freshwater lakes and streams in Thailand (Mahujchariyawong and Ikeda, 2001). Modeling of the growth pattern of the water hyacinth suggests that the optimum harvest should follow the maximum benefit of nutrient removal and harvest of the yield should balance the removal of nutrient from the concerned water bodies (Lorber *et al.*, 1984). Thermal effects and the hydrological retention time were important parameters that bear effects on the efficiency of the nutrient removal capacity of water hyacinth (Yi *et al.*, 2009). Subsequent studies showed potential benefit of the use of the water hyacinth as a dominant element in the wastewater management system for purification of the waste water and reduction of the nutrient load (Fang *et al.*, 2007; Yi *et al.*, 2009; Wang *et al.*, 2012, 2013). Reduction in the nutrients and the total suspended solids from the waste water was not satisfactory in few instances (Adewumi and Ogbiye, 2009) where the topographic factor and the landscape structure remained a constraint for the expected function of water hyacinth. Excepting for the reduced efficiency to purify the waste water in this specific instance, in all other instances around the world the water hyacinth has remained an effective macrophyte for water purification and reduction of the nutrient load. Accumulation of the nutrients from the ambient water reduces the load of the concerned element thereby restoring the trophic conditions, without alteration in the macroinvertebrate species assemblage significantly (Wang *et al.*, 2012). The biological process yields huge quantity of biomass that can be used for other purposes including biogas production, while the waste water is being purified through the use of water hyacinth as a prospective plant for nutrient removal.

B) Role as Ecosystem Engineer

Water hyacinth can proliferate fast, forming a dense interlocking mat over the surface of water body with enormous amount of biomass and storage of nutrients. Due to the formation of a complicated root system and dense canopy it can modify the habitat of macroinvertebrates as well as fish thus acting as ecosystem engineer (Jones *et al.*, 1997; Crooks and Khim, 1999). The hanging roots in the water column serve as ideal habitat for epiphytic invertebrates especially amphipods (Schramm *et al.*, 1987) and their density is found to be positively correlated with the spatial niche available for colonization (Schramm *et al.*, 1987). Due to dominance of epiphytic organisms, its selective predator blue gill fish (*Lepomis macrichirus*), increases in number (Schramm and Jirka, 1989). Thus alteration of fish-invertebrates food chain results. In this way, water hyacinth regulates the principle components of the habitat as well as surrounding community.

C) Role as Community Organizer to Enrich Faunal Diversity

The presence of water hyacinth in freshwater habitats influences the macroinvertebrate assemblages in a number of ways, either enhancing the species richness or reducing the number of species in the concerned habitats (Scheffer, 1998; Patel, 2012). The resultant effect leads to the changes in the species interactions that may modify the abundance of certain species. Based on the macroinvertebrate species association, the role of the water hyacinth can be best described as a community organizer. The macroinvertebrates use water hyacinth as food resource or as refuge to defend itself from the predators. The dependence on water hyacinth by different macroinvertebrates are not restricted to individuals of aquatic habitats rather different terrestrial insects also consume different vegetative parts of the plant (de Marco *et al.*, 2001), thus, justifying the evidence of species association with water hyacinth. Among the macrophytes, water hyacinth provides an abode for different macroinvertebrates population like lepidopteran, hemipteran and coleopteran insects, which vary considerably in the native and invaded zones of the geographical distribution (Wetzel, 1983). As proposed by several workers, herbivory of living macrophytes is quite common in nature (Carpenter and Lodge, 1986; Bronmark, 1989). Detritivore pulmonate *Stenophysa* sp. dominates *Eichhornia* roots followed by *Melanoides tuberculata* and *Drepanotrema anatinum*, all of which due to possession of soft shell take refuge inside the root of *Eichhornia*. This enables *Stenophysa* to avoid predatory pressure and outcompete the competetors (Lodge, 1985). According to de Marco *et al.* (2001), detritivore snail *M. tuberculata* generally dominates in sediment followed by herbivore snail *Pomacea haustorium* and *Biomphalaria tenagophila*. However, presence of *M. tuberculata* inside *Eichhornia* roots is attributed to the fact that sediments retained by water hyacinth root supported more detritivores rather than herbivorous molluscs. He also opined that the decomposition of macrophyte is enhanced by detritivorous invertebrates (Mann, 1988) which causes the decay of aquatic macrophytes fast twice as much as terrestrial plants (Webster and Benfield, 1986). Macrophytes thus exert significant influence on invertebrate population and energy budget of the ecosystem (Wetzel, 1983; Polunin, 1984; Carpenter and Lodge, 1986; Dawson, 1988). However, evidences of low preference for herbivory for water hyacinth may possibly be connected to poor nutrient content (Lamberti and Moore, 1984; Shanab *et al.*, 2010) or allelopathic interactions as has been demonstrated in chironomid larva. The larval stages of *Chironomus ramosus* exhibited changes in the polytene chromosome structure associated with extended developmental period and increased mortality (Thorat and Nath, 2010).

D) Bioresource for Manufactured Products

Water hyacinth, either as living or dead and dry condition serves as biological resource for a variety of purpose like biogas production, ethanol production, electricity generation and different manufactured products that substantiate the economic value. In many instances, the use of water hyacinth for different commercial purposes requires prior drying, which is a constraint in humid areas. The hygroscopic nature of the plant material creates difficulty in conventional drying and further processing, such that the moisture content of the plant matter needs to less than 15 per cent. The following uses of the water hyacinth substantiates them as a potential bioresource:

1. ***Paper products***: Small scale paper making industry have been successful depending on the weed in several countries like Philippines, Indonesia, India. It can be used to prepare fiber board and bituminized board for general purpose and low cost roof materials (Jafari, 2010).
2. ***Yarn, Rope and basket:*** The plant parts, like the stems are possible resource for the manufacture of ropes and yarn following drying and treatment with sodium metabisulphite. The plant parts are also utilized

for the preparation of basket and both ropes and baskets can be used as decorating articles (Jafari, 2010).

3. ***Charcoal briquetting:*** The plant can be utilized for preparing charcoal briquette that have several thermal application like boiling water, drying purpose etc. (Jafari, 2010).
4. ***An alternative substrate for production of biogas and ethanol:*** *Eichhornia* is regarded most propitious plant to produce bioethanol (Hronich *et al.*, 2008; Gao *et al.*, 2013). The plant after saccharification with sulfuric acid is fermented with yeast to produce ethanol on commercial basis (Masami, 2008; Mishima *et al.*, 2008; Kumar, 2009; Aswathy *et al.*, 2010). Malik (2007) first pointed out the capacity of the weed mass to produce biogas. The weed is subjected to fermentation in reactors by methanogens (Patel, 2012) and produces hydrogen and methane at particular temperature (Chuang *et al.*, 2011). The yield of methane may reach up to 60 per cent approximately, which can be utilized for cooking, lighting or for powering an engine to produce shaft power (Jafari, 2010). Besides, bioelectricity is generated by *Eichhornia* based fuel cell (Mohan *et al.*, 2011).
5. ***As source of anti-oxidant and anti-bacterial agent :*** Various phytochemicals present in petiole, leaves and flower extract of *Eichhornia* were found to possess antioxidant potential as evidenced by their DPPH (1,1 diphenyl 2 pierylhydrazyl) radical scavenging potential (Chantiratikul *et al.*, 2009; Aboul-Enein *et al.*, 2011) and metal ion chelating ability. HPLC performed with the methanolic extract of the plant identified naringenin as a chemical to possess antioxidant potential and anti-inflammatory potential. The extract was also reported to reduce lipid peroxidation in fish oil (Surendraraj *et al.*, 2013) and to inhibit several bacterial pathogens like *Proteus vulgaris, Salmonella typhi* and *Bordetella bronchiseptica* (Kumar *et al.*, 2014). Naringenin a major flavonoid isolated from the weed also has the potential to remove free radicals. So it can be concluded that the plant can be used as a potent source of antioxidant on commercial basis.
6. ***Feed for livestock industry:*** The enormous biomass of *Eichhornia* can be utilized as nutrient rich feed for ducks and *Clarias gariepinus* fingerlings (Aboud *et al.*, 2005). Leaves and shoot portions are digested easily than the whole plant. Better egg laying capacity was reported in case of weed fed duck due to high content of protein, vitamins and minerals (Lu *et al.*, 2008). In duck farm, the birds fed with water hyacinth enhanced the feeding behaviour, clutch size and egg laying capacity. The egg weight also increased with a simultaneous increase in shell weight though egg shape index, Haugh units, eggshell relative weight and egg-yolk color remains unaltered (Jianbo *et al.*, 2008). Nutrients in water hyacinth are accessible for both ruminants and non-ruminants (Mara, 1976). Chopped and/or ensiled *Eichhornia* can serve as a cheap nutrient enriched diet for steer calves, brood cows and nursing calves (Mara, 1976). Feeding with such weed is practiced in various parts of Southeast Asia. In countries like China, Malaysia, Indonesia, Philippines and Thailand, the processed

weed together with vegetable waste, rice bran, copra cake and salt can be a good source of nutrients for pigs, duck and pond fish. Thus it may solve the problem of malnutrition of reared animals in developing countries up to certain extent (Jafari, 2010).

7. ***Use in agriculture:*** After decomposition, the plant can be applied in agricultural field due to its high nutrient status. A mixture of *Tithonia diversifolia* and *E. crassipes* in 1: 3 ratio can improve the soil nutrient status and thus can replace use of costly chemical fertilizer (Chukwuka and Omotayo, 2008). It can also be used as a substrate of vermicompost with cow dung after treatment for a couple of days as evidenced by their appreciable decrease in C: N ratio (Gupta *et al.*, 2007). The plant can also be used in cultivation of mushroom (Oyster mushroom *Pleurotus sajor-caju)* thereby replacing the use of saw dust. Minimizing the use of saw dust can be a great concept of forest conservation plan.
8. ***Anti-microbial and anti-algal agent :*** Both the crude methanolic and TLC fractions of the weed showed protection against gram-positive pathogen *Bacillus subtilis, Staphylococcus aureus, Streptococcus faecalis* and Gram negative *Escherichia coli* as wells as against the fungus *Candida albicans* (Aboul-Enein *et al.*, 2011). Various allelopathic chemicals extracted from the weed can act as algaecide as has been proved in case of *Chlorella vulgaris* (Shanab *et al.*, 2010).
9. ***Production of secondary metabolites:*** The weed has been reported to act as source of various phytochemicals like alkaloids, phenolic and terpenoid groups (Shanab *et al.*, 2010), few flavonoids (luteolin, apigenin, tricin, chrysoeriol, kaempferol, azaeleatin, gossypetin and orientin), amino acids (methionone, valine, glutamic acid, tryptophan, tyrosine, leucine and lysine), phosphorous, organic matter and cyanide (Nyananyo *et al.*, 2007). The compound naringenin with relatively high ash content has been documented as major phenolic compound followed by myricetin (Chantiratikul *et al.*, 2009).
10. ***Isolation of cellulose nanofibers***: Sundari and Ramesh (2012) have isolated and characterized cellulose nano fibers from the weed after various chemical treatments like bleaching, alkaline and sodium chloride reactions. Nano fibers have manifold uses like preparation of composites, biodegradable thin films, adsorbents, and fibers.
11. ***Acts as Immunostimulant :*** *Eichhornia* after various treatments with and without hot water triggers immunogenicity of fresh water prawn *Macrobrachium rosenbergii*. Ingestion of processed weed for 4 months exhibited an increase in total haemocyte count, differential haemocyte count, phenoloxidase activity, respiratory bursts, superoxide dismutase activity and glutathione peroxidase activity. Such treated prawn also showed increased protection against *Lactococcus garvieae* (Chang and Cheng, 2014).
12. ***For soil amendment:*** It can serve as good fertilizer by improving the structure and composition of soil. The plant can be used to prepare potting

compost commercially in a cost effective way (Mara, 1976). Water hyacinth can be heaped 6-8 feet high and left for 3-4 month to allow decomposition to take place. The pile should be frequently turned for oxygenation. Due to good water retention capacity, the plant biomass after completion of decomposition can be used as nutrient supplement as well as soil moisturizer (Wolverton and McDonald, 1979).

13. ***Protein concentrate:*** Extraction of protein from the pulp prepared from leaf or whole shoot portion revealed the presence of an appreciable amount of crude protein, phosphorus, ash, potassium and crude fiber (Wolverton and McDonald, 1979).
14. ***Materials for handicraft and self sufficiency:*** Water hyacinth is processed in various ways to design various ornamental and beautifying articles of cottage industry. In Java, Indonesia, many items like coasters, placements, mats, shoes, sandals, bags, wallets, vases are prepared by weaving dried petioles of the plant. Thus the weed can be inexpensive raw materials with a bit of imagination and proper business planning (Patel, 2012).

Conclusion

Although an efficient ecosystem service provider, the excessive and unusual rapid growth and broad adaptive features of water hyacinth is a major concern for the continuance of the ecosystem functions of the concerned water bodies. This is particularly relevant for the situations where the plant has been recognized as an invasive species. Regulation of water hyacinth population is necessary to promote the growth and availability of the native species and to reduce the effects of invasion. The threats and degradation imposed by the plant to the mankind both ecologically, socially and commercially cannot be left unnoticed. Accordingly, several control methods have been adopted to keep the weed under ecological threshold level. In a general perspective, the regulation of the water hyacinth can be achieved through the mechanical, chemical and biological means alone or in combination as a part of the integrated control mechanism (Jones, 2001). Owing to invasion of water hyacinth in the areas concerned, the changes in the species composition of the weeds and the macroinvertebrates and dependent taxa provides a basis to understand the habitat alteration and consequential changes in the ecosystem functions. The changes in the ecosystem functions are obvious owing to the alteration in the species that had been interacting in a different way until the colonization of water hyacinth. Since water hyacinth utilizes high amount of nitrate and phosphate that facilitate its growth, the nutrient pool of the habitat may not be sufficient for the growth of other species of macrophytes. A competitive interaction may result in the reduced diversity of other species of macrophytes. In certain instances, this may lead to altered density of the periphytons as well as algal bloom. A reduced production of algae and the periphytons leads to a cascading effect on the fish and other macroinvertebrates that heavily depend on the vegetation of the water bodies. The changes can be considered as a regulation of the higher trophic levels by the water hyacinth, mimicking the more conspicuous bottom-up effect of trophic interaction. Thus the water hyacinth population can alter the nutrient condition that may subsequently alter the vegetation pattern and change the community structure indirectly. As a direct effect the community composition may be altered through the allelopathic interaction or occupancy of the physical space by water hyacinth as a result of faster growth. The growth pattern of water hyacinth is considerably faster than other macrophytes leading to the preemption of the space available for other species in the same habitat. Further reinforcement of the competitive interactions among the different macrophyte species can cause a change in the diversity pattern of the invaded areas. Theory suggests that the invasive species affect the habitats in two different ways–by altering the habitat conditions and second by altering the interactions among the constituent species. The changes in the nutrient availability and alterations in the macrophyte species composition in water hyacinth invaded habitats support this proposition as evident from the studies in different climatic conditions. Due to greater adaptability of water hyacinth to tolerate wide range of habitat conditions along with the changes in the physical and chemical conditions of the habitats, colonization and establishment in newer habitats are supposed to bring about similar changes as observed in the habitats that are already infested with water hyacinth. The regulation of the water hyacinth at the local scale may be achieved employing a varieties of management strategies, however, the spread of the plant can hardly be stopped due to the ability to disperse through the lotic ecosystems. Although the invasive effects of the water hyacinth is unwanted to many different ecosystems, recent studies demonstrate various useful applications to restore the environmental conditions. Among the many uses of water hyacinths include the use in fodder industry and nutrient supplement in aquaculture, bioremediation of metals and dyes and the removal of excess nutrients from aquatic systems. Studies have shown that water hyacinth can accumulate higher amount of the metals in the living tissues in comparison to the environment and are termed as hyper accumulators. In combination with other hyper accumulators or alone, water hyacinth can be employed for metal extraction from the aquatic ecosystems as a part of the bioremediation strategies. Degradation of water hyacinth tissues may yield different fermented products that are considered as alternative to the available fuels and thus can be considered as a source of biomass energy.

Summary

Water hyacinth is an invasive species dominating the aquatic ecosystems of different geographical areas across five continents. Although distributed mostly in the tropical regions, the growth and physiology of water hyacinth reflect the ability to withstand eutrophic conditions and incorporate substantial amount of nitrate and phosphate in the tissues. Alterations of the habitat conditions and the species interactions are long been recognized as an effect of invasion of water hyacinth. However, the ability to withstand the nutrient rich waters enabled considering this macrophyte for purification of waste water. Besides, evidence in favor of accumulation of heavy metals suggests its prospective use in phytoremediation of the polluted water. The use of water hyacinth as a supplement to cattle feed and in biogas plants indicate the multiple functional roles as a biological resource. The multiple functional roles of water hyacinth in the aquatic ecosystems and the use in commercial products qualify it as ecosystem service provider. Opinions on the use of the water hyacinth as a biological resource and the ill effects of the extensive spread in the invaded water bodies are two opposing facts that needs to be strongly judged prior to promoting them as a biological resource to enhance the ecosystem services. In the present commentary the functional roles are being elaborated with the views on the origin and spread of water hyacinth in different continents. The discussion will provide an insight to the prospective benefits that are derived from water hyacinth and facilitate the strategies towards the water hyacinth as a biological resource. In conclusion, it remains to be decided on the requirements of the stakeholders at the local scale to regulate or promote the growth and spread of water hyacinth, though the commercial value of this plant seems to be considerably high.

References

Aboud AAO, Kidunda RS and Osarya J. 2005. Potential of water hyacinth (*E. crassipes*) in ruminant nutrition in Tanzania. *Livestock Research Rural Development,* 17: 8.

Aboul-Einein AM, Al-Abd AM, Shalaby E. *et al.,* 2011. *Eichhornia crassipes* (Mart) solms. *Plant Signaling and Behavior,* 6 (6): 834-836.

Adewumi I and Ogbiye AS. 2009. Using water hyacinth (*E. crassipes*) to treat waste water of a residential institution. *Toxicological and Environmental Chemistry,* 91: 891-903.

Allen LH, Sinclair TR and Bennett JM. 1997. Evapotranspiration of vegetation of Florida: perpetuated misconceptions versus mechanistic process. *Proceedings of the Soil and Crop Science Society of Florida,* 56: 1-10.

Aswathy US, Sukumaran RK, Devi GL. *et al.,* 2010. Bio-ethanol from water hyacinth biomass: an evaluation of enzymatic saccharification strategy. *Bioresource Technology,* 101: 925-930.

Auffenberg W. 1980. Autecological notes on *Xenochrophis piscator* (Reptilia: Serpents) from Keoladeo Ghana sanctuary. *International Journal of Ecology and Environmental Sciences,* 6: 77-82.

Aurand D. 1982. *Nuisance Aquatic Plants and Aquatic Plant Management Programs in the United States,* Volume 2, Southeastern Region. The Mitre Corporation, McLean VA.

Barrett SCH. 1980. Sexual reproduction in *Eichhornia crassipes* (water hyacinth) II. Seed production in natural populations. *Journal of Applied Ecology,* 17: 113-124.

Beshir MO and Bennett FD. 1985. Biological control of water hyacinth on the White Nile, Sudan. *Proceedings of the VI International Symposium on Biological Control of Weeds,* (Delfosse ES ed.),1984, Vancouver, Canada. pp. 491-496.

Bock JH. 1966. An ecological study of *Eichhornia crassipes* with special emphasis on its reproductive biology. PhD Thesis. University of California, Berkeley.

Boyd CE. 1976. Accumulation of dry matter, nitrogen and phosphorus by cultivated water hyacinths. *Economic Botany,* 30 (1): 51-56.

Bronmark C. 1989. Interactions between epiphytes, macrophytes and freshwater snails: a review. *Journal of Molluscan Studies*, 55: 299-311.

Burton J. 2005, "*Water hyacinth Eichhornia crassipes.*" Agfact, NSW DPI, P7.6.43

Carignan R and Neiff JJ. 1994. Limitation of water hyacinth by nitrogen in subtropical lakes of the Parana floodplain (Argentina). *Limnology and Oceanography*, 39: 439-443.

Carpenter SR and Lodge DM. 1986. Effects of submerged macrophytes on ecosystem processes. *Aquatic Botany*, 24: 341-370.

Cattaneo A, Galanti G, Gentinetta S and Romo S. 1998. Epiphytic algae and macroinvertebrates on submerged and floating-leaved macrophytes in an Italian lake. *Freshwater Biology*, 39: 725-740.

Center TD and Wright AD. 1991. Age and phytochemical composition of water hyacinth (Pontederiaceae) leaves determine their acceptability to *Neochetina eichhorniae* (Coleoptera: Curculionidae). *Environmental Entomology*, 20: 323-334.

Chang CC and Cheng W. 2014. Multiple dietary administrating strategies of water hyacinth (*Eichhornia crassipes*) on enhancing the immune responses and disease resistance of giant freshwater prawn, *Macrobrachium rosenbergii. Aquaculture Research*, 1-13.

Chantiratikul P, Meechai P and Nakbanpotec W. 2009. Antioxidant activities and phenolic contents of extracts from *Salvinia molesta* and *Eichhornia crassipes. Research Journal of Biological Sciences*, 4: 1113-1117.

Chen X, Chen X, Wan X. *et al.*, 2010. Water hyacinth (*Eichhornia crassipes*) waste as an adsorbent for phosphorus removal from swine waste water. *Bioresource Technology*, 101: 9025-9030.

Chuang YS, Lay CH, Sen B. *et al.*, 2011. Biohydrogen and biomethane from water hyacinth (*Eichhornia crassipes*) fermentation: effects of substrate concentration and incubation temperature. *International Journal of Hydrogen Energy*, 36: 14195-14203.

Chukwuka KS and Omotayo OE. 2008. Effects of *Tithonia* green manure and water hyacinth compost application to nutrient depleted soil in South-Western Nigeria. *International Journal of Soil Science*, 3: 69-74.

Crooks JA and Khim HS. 1999. Architectural vs. biological effects of a habitat-altering, exotic mussel, *Musculista senhousia. Journal of Experimental Marine Biology and Ecology*, 240: 53-75.

Dawson FH. 1988. Water flow and the vegetation of running waters. In. *Vegetation of inland waters. Handbook of Vegetation Science*, (Symoens JJ ed.) Kluwer Academic Publishers, Boston. pp. 283-309.

de Marco P, Araújo MAR, Magda K. *et al.*, 2001. Aquatic invertebrates associated with the water-hyacinth (*Eichhornia crassipes*) in an eutrophic reservoir in tropical Brazil. *Studies on Neotropical Fauna and Environment*, 36 (1): 73-80.

Doust LL. 1981. Population dynamics and local specialization in a clonal perennial (*Ranunculus repens*) : I. The dynamics of ramets in contrasting habitats. *Journal of Ecology*, 69: 743-755.

Fahrig L, Coffin DP, Lauenroth WK and Shugart HH. 1994. The advantage of long-distance clonal spreading in highly disturbed habitats. *Evolutionary Ecology*, 8: 172-187.

Fang YY, Yang XE, Chang HQ. *et al.*, 2007. Phytoremediation of nitrogen polluted water using water hyacinth. *Journal of Plant Nutrition*, 30 (11): 1753-1765.

Francois J. 1981. Environmental conditions for germination and flowering of the water hyacinth *Eichhornia crassipes* (Mart.) Solms. *Proceedings of First International Conference on Wetlands*, New Delhi.

Freidel JW and Bashir MO. 1979. On the dynamics of populations and distribution of water hyacinth in the White Nile, Sudan. In. *Weed Research in Sudan*, (Koch W. ed). University of Wad Medani, Sudan. pp. 94-105.

Gao J, Chen L, Yan Z and Wang L. 2013. Effect of ionic liquid pretreatment on the composition, structure and biogas production of water hyacinth (*Eichhornia crassipes*). *Bioresource Technology*, 132: 361-364.

Godley JS. 1982. Predation and defensive behaviour of the striped swamp snake (*Regina alleni*). *Florida Field Naturalist*, 10: 31-36.

Goel PK, Khatavkar SD and Kulkarni AY. 1989. Chemical composition and concentration factors of water hyacinth (*Eichhornia crassipes*) growing in a shallow polluted pond. *International Journal of Ecology and Environmental Science*, 15: 141-144.

Gopal B. 1987. *Water hyacinth*. Elsevier Science publishers, Amsterdam. pp. 471.

Gupta R, Mutiyar PK, Rawat NK. *et al.*, 2007. Development of a water hyacinth-based vermireactor using an epigeic earthworm *Eisenia foetida*. *Bioresource Technology*, 98: 2605-2610.

Haller WT and Sutton DL. 1973. Effect of pH and high phosphorus concentrations on growth of water hyacinth. *Water Control Journal*, 11: 59-61.

Harley KLS, Julien MH and Wright AD. 1996. Water hyacinth: a tropical worldwide problem and methods for its control. *Proceedings of 2nd International Weed Control Congress*, II : 639-644.

Hauser JR. 1984. Use of water hyacinth aquatic treatment systems for ammonia control and effluent, *Journal of Water Pollution Control Federation*, 56(3): 219-225.

Hellmann JJ, Byers JE, Bierwagen BG and Dukes JS. 2008 Five potential consequences of climate change for invasive species. *Conservation Biology*, 22: 534-543.

Higgins SI and Richardson DM. 1996. A review of models of alien plant spread. *Ecological Modelling*, 87: 249-265.

Hronrich JE, Martin L, Plawsky J and Bungay HR. 2008. Potential of *Eichhornia crassipes* for biomass refining. *Journal of industrial microbiology and Biotechnology*, 35: 393-402.

Ismail Z, Othman SJ, Law KH. *et al.*, 2014. Comparative performance of water hyacinth *(Eichhornia crassipes)* and water lettuce *(Pista stratiotes)* in preventing nutrients build-up in municipal wastewater. *Clean Soil Air Water*, 42: 1-11.

Jafari N. 2010. Ecological and socio-economic utilization of water hyacinth (*Eichhornia crassipes* Mart Solms). *Journal of Applied Sciences Environmental Management*, 14: 43-49.

Jianbo LU, Zhihui FU and Zhaozheng YIN. 2008. Performance of a water hyacinth (*Eichhornia crassipes*) system in the treatment of wastewater from a duck farm and the effects of using water hyacinth as duck feed. *Journal of Environmental Sciences*, 20: 513-529.

Jones CG, Lawton JH and Chachak M. 1997. Positive and negative effects of organisms as physical ecosystem engineers. *Ecology*, 78: 1946-1957.

Jones RW. 2001. Integrated control of water hyacinth on the Nseleni/Mposa Rivers and Lake Nsezi, Kwa Zulu-Natal, South Africa. In. *Proc. 2nd meeting of the global working group for the biological and integrated control of water hyacinth.*

Kumar A, Singh LK and Ghosh S. 2009. Bioconversion of lignocellulosic fraction of water-hyacinth (*Eichhornia crassipes*) hemicellulose acid hydrolysate to ethanol by *Pichia stipitis*. *Bioresource Technology*, 100: 3293-3297.

Kumar S, Kumar R, Dwivedi A and Pandey AK. 2014. *In vitro* antioxidant, antibacterial, and cytotoxic activity and *in vivo* effect of *Syngonium podophyllum* and *Eichhornia crassipes* leaf extracts on isoniazid induced oxidative stress and hepatic markers. *Bio Med Research International*, Article ID 459452, pp.11.

Lambers H, Chaplin III FS and Pons TL. 2000. *Plant Physiological Ecology*, 2nd ed. Springer, Heidelberg, Germany.

Lamberti GA and Moore JW. 1984. Aquatic insects as primary consumers. In. *The ecology of aquatic insects*, Resh VH and Rosenberg DM eds. Praeger Publishers, New York. pp.164-195.

Langeland KA and Burks KC (eds.). 1998. *Identification and biology of non-native plants in Florida's natural areas.* UF/IFAS. pp.165.

Lodge DM. 1985. Macrophyte-gastropod associations: observations and experiments on macrophyte choice by gastropods. *Freshwater Biology*, 15: 695-708.

Lorber MN, Mishoe JW and Reddy PR. 1984. Modelling and analysis of water hyacinth biomass. *Ecological modeling*, 24: 61-77.

Lu J, Fu Z and Yin Z. 2008. Performance of a water hyacinth (*Eichhornia crassipes*) system in the treatment of wastewater from a duck farm and the effects of using water hyacinth as duck feed. *Journal of Environmental Science*, 20: 513-519.

Mahujchariyawong J and Ikeda S. 2001. Modeling of environmental phytoremediation – the case of water hyacinth harvest in Tha-Chin river, Thailand. *Ecological Modelling*, 142: 121-134.

Malik A. 2007. Environmental challenge vis a vis opportunity: the case of water hyacinth. *Environment International*, 33: 122-138.

Mandryk AM and Wein RW. 2006. Exotic vascular plant invasiveness and forest invasibility in urban boreal forest types. *Biological Invasions*, 8: 1651-1662.

Mann KH. 1988. Production and use of detritus in various freshwater, estuarine, and coastal marine ecosystems. *Limnology and Oceanography*, 33: 910-930.

Mara MJ. 1976. Estimated values for selected water hyacinth by-products. *Economic Botany*, 30 (4): 383-387.

Masami GOO, Usui IY and Urano N. 2008. Ethanol production from the water hyacinth *Eichhornia crassipes* by yeast isolated from various hydrospheres. *African Journal of Microbiology Research*, 2: 110-113.

Matagi SV. 2002. Some issues on environmental concern in Lampala, the capital city of Uganda. *Environmental Monitoring and Assessment*, 77: 121-138.

Matthews LJ. 1967. Seedling establishment of water hyacinth. *International Journal of Pest Management*, 13: 7-8.

Mengel K and Kirkby EA. 1982. *Principles of plant nutrition*, 3rd ed. International Potash Institute. Bern, Switzerland.

Mishima D, Kuniki M, Sei K. *et al.*, 2008. Ethanol production from candidate energy crops: water hyacinth (*Eichhornia crassipes*) and water lettuce (*Pistia stratiotes* L.). *Bioresource Technology*, 99: 2495-2500.

Mohan SV, Mohanakrishna G and Chiranjeevi P. 2011. Sustainable power generation from floating macrophytes based ecological microenvironment through embedded fuel cells along with simultaneous wastewater treatment. *Bioresource Technology*, 102: 7036-7042.

Nyananyo BL, Gijo A and Ogamba EN. 2007. The physicochemistry and distribution of water hyacinth (*Eichhornia crassipes*) on the river Nun in the Niger Delta. *Journal of Applied Sciences and Environmental Management*, 11: 133-137.

Oki Y and Ueki K. 1984. Adaptation of water hyacinth grown under various habitats. *Proceedings of International Conference on Water Hyacinth*, UNEP, Nairobi. pp.222-232.

Olivares E and Colonnelo G. 2000. Salinity gradient in the Manamo River, a dammed distributary of the Orinoco Delta and its influence on the presence of *Eicchornia crassipes* and *Paspalum repens*. *Interciencia*, 25: 242-248.

Owens CS and Madsen JD. 1995. Low temperature limits of water hyacinth. *Journal of Aquatic Plant Management*, 33: 63-68.

Parija P. 1934. A note of the reappearance of water hyacinth seedlings in cleared tanks. *Indian Journal of Agricultural Science*, 25: 386-391.

Patel S. 2012. Threats, management and envisaged utilizations of aquatic weed *Eichhornia crassipes*: an overview. *Reviews in Environmental Science and Biotechnology*, 11: 249-259.

Penfound WT and Earle TT. 1948. The biology of the water hyacinth. *Ecological Monographs*, 18: 447-472.

Polunin NVC. 1984. Marine fishes of the Seychelles. In. *Biogeography and ecology of the Seychelles Islands*, Stoddart DR and Junk W. eds. The Hague. pp. 171-91.

Purchase BS. 1977. Nitrogen fixation associated with *Eichhornia crassipes*. *Plant and Soil*, 46(1): 283-286.

Qaisar M, Ping Z, Rehan SM. *et al.*, 2005. Anatomical studies on water hyacinth (*Eichhornia crassipes* (Mart.) Solms) under the influence of textile wastewater. *Journal of Zhejiang University, Science*, 6B (10): 991-998.

Rahel FJ and Olden JD. 2008. Assessing the effects of climate change on aquatic invasive species. *Conservation Biology*, 22: 521-533.

Reddy KR and DeBusk WF. 1991. Decomposition of water hyacinth detritus in eutrophic lake water. *Hydrobiologia*, 211: 101-109.

Richards JH and Lee DW. 1986. Light effects on morphology in water hyacinth (*Eichhornia crassipes*). *American Journal of Botany*, 73 (12): 1741-1747.

Rodríguez-Gallego LR, Mazzeo N, Gorga J. *et al.*, 2004. The effects of an artificial wetland dominated by free-floating plants on the restoration of a subtropical, hypertrophic lake. *Lakes and Reservoirs*, 9: 203-215.

Rogers HH and Davis DE. 1972. Nutrient removal by water hyacinth. *Weed science*, 20 (5): 423-428.

Scheffer M. 1998. *Community dynamics of shallow lakes*. Chapman and Hall, London.

Schramm HL, Jirka KJ and Hoyer MV. 1987. Epiphytic macroinvertebrates on dominant macrophytes in two central Florida Lakes. *Journal of Freshwater Ecology*, 4: 151-161.

Schramm HL and Jirka KJ. 1989. Epiphytic macroinvertebrates as a food resource for bluegills in Florida Lakes. *Transactions of the American Fisheries Society*, 118: 416-426.

Sculthorpe CD. 1967. *The Biology of Aquatic Vascular Plants*. Edward Arnold, London.

Shanab SMM, Shalaby EA, Lightfoot DA and El-Shemy HA. 2010. Allelopathic effects of water hyacinth (*Eichhornia crassipes*). *PLoS One* 5(10) : e13200.**doi: 10.1371/journal. pone.0013200.**

Sheffield CW. 1967. Water hyacinth for nutrient removal. *Hyacinth Control Journal*, 6: 27-30.

Srivastava AK, Ambasht RS and Kumar R. 1994. Enhancing effect of pollution on dry matter, nitrogen and phosphorus accumulation in water hyacinth (*Eichhornia crassipes*) in river Ganga at Varanasi. *Indian Journal of Forestry*, 17: 279-283.

Sundari MT and Ramesh A. 2012. Isolation and characterization of cellulose nanofibers from the aquatic weed water hyacinth-*Eichhornia crassipes*. *Carbohydrate Polymers*, 87: 1701-1705.

Surendraraj A, Sabeena Farvin KH and Anandan R. 2013. Antioxidant potential of water hyacinth (*Eichornia crassipes*): in vitro antioxidant activity and phenolic composition. *Journal of Aquatic Food Product Technology*, 22(1): 11-26.

Tabuti JRS, Oryern-Oriqa H, Mutumba GM and Kasharnbuzi JT. 1998. Partitioning of soluble calcium in the water hyacinth *(Eichhornia crassipes* (Martius) Solms) in Lake Victoria, Uganda. *Acta Botanica*, 111: 297-302.

Téllez TR, de Rodrigo López EM, Granado GL. *et al.*, 2008. The water hyacinth, *Eichhornia crassipes*: an invasive plant in the Guadiana river basin (Spain). *Aquatic Invasions*, 3(1): 42-53.

Thorat LJ and Nath BB. 2010. Effects of water hyacinth *Eichhornia crassipes* root extracts on midge *Chironomus ramosus* larvae: a preliminary note. *Physiological Entomology*, 35: 391-393.

Tripathi BD and Shukla SC. 1991. Biological treatment of wastewater by selected aquatic plants, *Environmental Pollution*, 69: 69-78.

Twongo T. 1993. Status of water hyacinth in Uganda. In. *Control of Africa's water weeds* (Greathead A. and de Groot P. eds). Series Number CSC (93) AGR-18. pp. 55-57.

Ueki K, Ito M and Oki Y. 1975. Water hyacinth and its habitats in Japan. *Proc. 5th Asia-Pacific Weed Science Society Conference*, Tokyo.

Ueki K and Oki Y. 1979. Seed production and germination of *Eichhornia crassipes* in Japan. *Proceedings of Seventh Asian Pacific Weed Science Society Conference*, pp. 257-260.

Ultsch GR. 1976. Ecophysiological studies of some metabolic and respiratory adaptations of sirenid salamanders. In. *Respiration of amphibious vertebrates* (Hughes GM ed.). Academic Press, London.

Van Thielen R, Ajuonu O, Schade V. *et al.*, 1994. Importation, release, and establishment of *Neochetina* spp. (Curculionidae) for the biological control of water hyacinth, *Eichhornia crassipes* (Lil.: Pontederiaceae), in Benin, West Africa. *Entomophaga*, 39: 179-188.

Wang Z, Zhang Z, Zhang J. *et al.*, 2012. Large scale utilization of water hyacinth for nutrient removal in Lake Dianchi in China: the effects on the water quality, macrozoobenthos and zooplankton. *Chemosphere*, 89: 1255-1261.

Wang Z, Zhang Z, Zhang Z. *et al.*, 2013. Nitrogen removal from Lake Caohei, a typical ultra-eutrophic lake in China with large scale confined growth of *Eichhornia crassipes*. *Chemosphere*, 92 (2): 177-189.

Webster JR and Benfield EF. 1986. Vascular plant breakdown in freshwater ecosystems. *Annual Review of Ecology and Systematics*, 17: 567-594.

Wei B, Yu X, Zhang S and Gu Li. 2011. Comparison of the community structures of ammonia-oxidising bacteria and archaea in rhizosplanes of floating aquatic macrophytes *Nitrosomonas europea* and *Nitrosomonas ureae*. *Microbiological Research*, 133: 468-474.

Wetzel RG. 1983. *Limnology*. 2nd edition. Saunders College Publishing, Philadelphia, USA.

Williams AE, Duthie HC and Hecky RE. 2005. Water hyacinth in Lake Victoria: Why did it vanish so quickly and will it return? *Aquatic Botany*, 81: 300-314.

Wilson JR, Holst N and Rees M. 2005. Determinants and patterns of population growth in water hyacinth. *Aquatic Botany*, 81: 51-67.

Wolverton BC and McDonald RC. 1978. Nutritional composition of water hyacinths grown on domestic sewage. *Economic Botany*, 32(4): 363-370.

Wolverton BC and McDonald RC. 1979. The Water Hyacinth: From prolific pest to potential provider. *Ambio*, 8(1): 2-9.

Xie Y, Qin H and Yu D. 2004. Nutrient limitation to the decomposition of water hyacinth (*Eichhornia crassipes*). *Hydrobiologia*, 529: 105-112.

Yi Q, Hur C and kim Y. 2009. Modeling nitrogen removal in water hyacinth ponds receiving effluent from waste stabilization ponds. *Ecological Engineering*, 35: 75-84.

Zhou W, Zhu D, Tan L. *et al.*, 2007. Extraction and retrieval of potassium from water hyacinth (*Eichhornia crassipes*). *Bioresource Technology*, 98: 226-231.

Zimmels Y, Kirzhner F and Malkovskaja A. 2006. Application of *Eichhornia crassipes* and *Pistia stratiotes* for treatment of urban sewage in Israel. *Journal of Environmental Management*, 81(4): 420-428.

Chapter 18

Wetlands as a Potential Mosquito Larval Habitat: Boon or Curse?

Soumyajit Banerjee, Gautam Aditya and Goutam Kumar Saha

Entomology and Wildlife Biology Research Laboratory, Department of Zoology, University of Calcutta, Kolkata – 700 019

Introduction

Wetlands are diverse ecosystems that range from permanent to temporary systems, inundated or saturated, with water that is static or flowing, fresh or saline (Lambert, 2002). They are considered as a true habitat resource providing multiple useful products and services, and carrying out water purification and carbon storage in addition to their much appreciated responsibility in sequestering carbon. In the past few decades, the importance of wetlands and their benefits have been extensively highlighted throughout the globe (de Szalay and Resh, 2000; Keiper *et al.*, 2001; Pongsiri and Roman, 2007). In spite of the services provided by the wetlands, they continue to be affected and lost by several anthropogenic activities such as draining, construction, agriculture, and pollution (Mitsch and Gosselink, 2000). However, to mitigate the human induced loss to the wetlands and thus to support rehabilitation of degraded wetlands, considerable effort has been initiated, both at the global (Zedlar, 2000; Simenstad *et al.*, 2006) and national levels (Sharma *et al.*, 2011). An awareness of healthy waterways and water re-use has created the need to develop and implement economically and ecologically sustainable wastewater management strategies.

India, with varied topography and climatic regimes, and an annual precipitation of over 130 cm, supports and sustains diverse and unique wetland habitats. The natural wetlands of India comprises of the high-altitude Himalayan lakes, wetlands situated in the flood plains of the major river systems, saline and temporary wetlands of the arid and semi-arid regions and coastal wetlands. It is estimated that the wetlands alone support about 20 per cent of the Indian biodiversity profile (Deepa and Ramachandra, 1999; Johri *et al.*, 2010). The members of the different taxonomic groups depend on wetlands for their food, shelter, and nesting sites at least for a part of their lifecycle and the wetland plays a significant role in regulating the reproductive pattern of many of its inhabitants- the breeding cycles often correspond to the regular, natural flooding of the wetlands. Among the invertebrate community, the insects have evolved a great diversity of habits and colonize a broad range of ecological niches and the aquatic insects in particular play an important role in the tropho-dynamics at each level in an ecosystem. They also act as an indicator of health of the ecosystems, determining the changes in water quality parameters.

Apart from functioning as a biodiversity rich habitat, the wetlands also provide a potential biotope for several mosquito vectors or intermediate hosts of parasites causing several mosquito borne diseases (Russell, 1999; Knight *et al.*, 2003; Muturi *et al.*, 2006; Dale and Connelly, 2012).The vectors are dependent for all, or at least part, of their life-cycles on water, including freshwater ecosystems and thus wetlands play an important role in the transmission of malaria (Krishnamoorthy *et al.*, 2005; Rajavel and Natarajan, 2006; Verma and Gupta, 2013), filariasis (Gopalakrishnan *et al.*, 2014), schistosomiasis (Jauhari and Pemola Devi, 2014), fascioliasis (Ramachandran *et al.*, 2012) and several arbo-viral (arthropod-borne virus) diseases such as Rift Valley and West Nile fever. Although most of the studies and surveys conducted till date on wetlands, have focussed on the nuisance caused by the mosquitoes and have only highlighted their negative impacts on the ecosystem functioning, neglecting the positive roles they play in sustenance of ecological balance. Thus, there exists a paradox: whilst wetlands are important in providing many benefits to society, they can also be a source of nuisance. The mosquitoes too, play a double role in a wetland ecosystem- in one hand they act as vectors of several diseases and on the other hand are a potential ecosystem service provider, through the filter feeding nature of the larvae on microorganisms, phytoplankton, and organic detritus and thus playing a significant role in nutrient cycling (Covich *et al.*, 1999). The abundance of mosquitoes in the wetlands has a strong influence on the ecosystem structure and function. The wetland productivity and relative abundance of several organisms, both vertebrates and invertebrates, are dependent on the occurrence of mosquito fauna. The larval and adult mosquitoes are the source of food for a wide variety of aquatic insects, fishes, amphibians, lizards and migrating and breeding waterfowls (Mokany, 2007). The mosquitoes too play an important role as a pollinator of an orchid *Habenaria obtusata* (Kevan *et al.*, 1993). It is quite apparent from their functions that apart from the annoyance, they too form an integral and novel part of the wetland biodiversity.

Moreover, the mosquito control and management programmes have a profound long term impact on the non-target organisms, causing a sustainable loss of the wetland productivity and diversity which in turn is felt at every trophic levels of the environment. As the nature of wetlands varies within and between a country or a state to a large extent, in terms of hydroperiod and invertebrate community composition, it is quite important and rational to recognize that not all the wetlands are responsible in producing vector mosquitoes (Schäfer, 2004). Lack of information regarding the importance of mosquitoes in wetland biodiversity and their role in lying at the base of a trophic food web, have always designated them as a nuisance pest. The present review is expected to highlight the role of mosquitoes, residing in the wetlands, in maintaining the ecological balance and community sustenance in addition to and apart from the diseases they transmit.

Wetland as Mosquito Larval Habitat

The mosquito larval habitats are diverse in terms of origin and species assemblages. The natural habitats for mosquito breeding can range from the bamboo stumps (Aditya *et al.*, 2008) and phytotelmata of different forms (Yee *et al,*. 2007) at the smaller scale to the wetlands like temporary pools (Campos *et al.*, 2004; Irwin *et al.*, 2008), ponds (McDonald and Buchanan, 1981) and rice fields (Mogi *et al.*, 1999; Sunish and Reuben, 2002; Bambaradeniya *et al.*, 2004; Das *et al.*, 2006; Wilson *et al.*, 2007) at the larger scale considering the available space. In the wetlands like ponds (Overtli *et al.*, 2002) and rice fields (Balcombe *et al.*, 2005), the species assemblage is more diverse, offering more complex interactions between trophic levels (Jeffries, 2002). The vegetation in such wetlands is also a factor determining the community structure (Burdett and Watts, 2009), prey-predator interactions should also be taken into consideration for the assessment of mosquito larval habitat (Saha *et al.*, 2012). However, the relative and absolute abundance of the species assemblage in these larval habitats depend on the habitat size (Banerjee *et al.*, 2010).

The distribution and abundance of mosquito species in the wetlands of India vary geographically and temporally, between and within regions (Rajavel *et al.*, 2005a,b; Amala and Aunradha, 2011; Purkayastha and Gupta., 2012; Banerjee *et al.*, 2013a; Karthikairaj *et al.*, 2013; Bhattacharya *et al.*, 2014). Species occupy different environmental niches and are distributed variously within diverse habitats (Dash and Hazra, 2011; Balasubramanian and Nikhil, 2013; Aneesh *et al.*, 2014). The mosquito species prevalent in the wetlands, both in India and similar other geographical regions, have characteristic habitat associations and biological attributes. Thus in order to frame suitable mosquito control strategies mosquito breeding efficiency in natural and constructed wetlands, species specific bio-ecological attributes, population dynamics, direct and indirect trophic interactions that shape up the food web, guild structure should be taken into consideration to alleviate mosquito borne disease problems.

The higher level of faunal assemblages in the wetlands provides the opportunity of using natural predators in regulating mosquitoes through predation (Knight *et al.*, 2003). The predatory insects also bear an indirect effect deterring mosquito oviposition (Spencer *et al.*, 2002; Kiflawi *et al.*, 2003; Blaustein *et al.*, 2004). Coexistence of different predators of immature stages of mosquitoes in these habitats determines the mosquito productivity and the level of population regulation (Jacob *et al.*, 2006). Contrasts to these, in the smaller habitats too, although less in variety, mosquito predators like the larvae of *Toxorhynchites* spp. (Alto *et al.*, 2005), and odonate nymphs (Fincke, 1999; Yanoviak, 2001) are common and bears an impact on the mosquito productivity. In these habitats, different mosquito species coexists through niche partitioning (Gilbert *et al.*, 2008) that reduces competition. Thus it appears that species assemblage in larval habitats influences the physiological and behavioural preferences for oviposition, habitat selection of mosquitoes (Mokany and Mokany, 2006) and pupal productivity (Keating *et al.*, 2004).

If different habitats are sampled, species-specific habitat preferences of mosquitoes and their particular spatial arrangements can be assessed and the relationship of species assemblage with the habitat size can be determined (Sunahara *et al.*, 2002). Further, the associated controphic prey species can indicate the outcome of predatory effects on the target mosquito species (Bluastein and Chase, 2007) and highlight alternative control mechanisms for mosquito species related to vector borne diseases, without affecting the non target species (Willott, 2004). In Indian context, several works on the mosquito larval habitats have highlighted the species composition of mosquitoes and associated macroinvertebrates that constitute a dynamic community in the wetlands such as rice fields (Sunish and Rueben, 2002; Das *et al.*, 2006) and temporary pools (Aditya *et al.*, 2006). The population regulation of mosquitoes by different taxonomic group in these habitats indirectly confers the availability of the mosquito vectors and thus the abundance of the mosquito borne diseases.

Wetland management decisions require a detailed and fine level of information at an organism scale depending on the species present therein. However, not many works have highlighted both the values of wetlands and the costs due to the presence of mosquitoes (Mercer *et al.*, 2005). In light of the seriousness of some of the diseases involved, the increasing focus on wetland rehabilitation and creation, and the ongoing degradation of wetlands, it is important that this matter be investigated with regard to the pertinent question whether it is possible to reinstate environment without taking into consideration the values of mosquitoes with regards to wetland as their prospective breeding habitats. A proper understanding of the relationship between wetland services and values and the occurrence, abundance, and distribution of mosquito densities may provide information relevant to the development and implementation of an Integrated Vector Management (IVM) program based on adult productivity and variability.

Effect of Environmental Parameters on the Life Cycle of Wetland Mosquitoes

Mosquitoes are the single largest group of medically important pest and vector insects that are associated with the deadly diseases of human and other animals like malaria, filaria, encephalitis, dengue and several other viral diseases. The mosquitoes complete their immature stages in aquatic environment and the fitness of the adult depends on the larval development and successful eclosion of the pupae to adult. Information on the larval development and related bio-ecology of a mosquito species help to understand the population dynamics as well as to evaluate the potential of the adult mosquitoes in disease transmission. Field as well as laboratory studies on different aspects of larval development and variation in life history parameters of different mosquito species of the genus *Anopheles*, *Aedes*, *Culex*, and *Ochlerotatus* (Focks *et al.*, 1993; Agnew *et al.*, 2002; Bedhomme *et al.*, 2003; Edillo *et al.*, 2004) have been utilized to generate the models for transmission of diseases and population regulation and control of the respective mosquito species (Depinay *et al.*, 2004; Le Menach *et al.*, 2007).

The generalised lifecycle of mosquito includes aquatic sub-adult stages and the aerial adult stages. The sub-adult stage consists of eggs, larva and pupa, typical of the holometabolic insects. Starting from the egg stage to adult, each stage is differently adapted having different morphs and life history strategies, contributing to the subsequent stages in terms of the fitness values (Yee *et al.*, 2007; Lounibos, 2007; Sisodia and Singh, 2009). The fitness of the individual mosquitoes and the population as well depends on the variations that are observed in these larval stages. The development of one stage contrast to the other is dependent on diverse and different factors (Arrivillaga and Barrera, 2004; Sharma *et al.*, 2005; Mogi, 2010; Couret and Benedict, 2014). However, the life history strategy of mosquito is shaped for the r-selection with higher number of eggs than the adults or other stages (Juliano, 2007; Williams *et al.*, 2008).

During the mosquito life cycle, the immature stages (larvae and pupae) live in aquatic habitats, and the nature of the habitat is often characteristic for the species. After emergence from the pupa, the mated adult females search for a host, blood feed and oviposit; then the cycle is repeated. Behavioural patterns within these activities are, again, characteristic of species, and it is these traits that predispose those species to be pests and/or vectors of disease causing pathogens. Flight range, blood host preferences, susceptibility to pathogens, longevity, and an inherent capacity for population increase vary between species, and are important factors in determining whether a species is a candidate pest or disease vector. Accurate identification of species, and an intimate knowledge of their particular biology, is required to properly assess the hazards and risks associated with mosquito production from natural or constructed wetlands. The anthropogenic activities like land use, agricultural practices, lead to a shift in the hydrological as well as physico-chemical, and biological features of the wetlands that act as potential mosquito larval habitats, and thus affect the life cycle pattern, duration, immature survival and oviposition behaviour (Norris, 2004; Leisnham *et al.*, 2004). The suitability of a wetland to qualify as a prospective mosquito breeding site depends on complex interactions of different abiotic and biotic factors. Precipitation, temperature, humidity, soil moisture (Buckner *et al.*, 2011), in addition to competition (Reiskind and Wilson, 2008), predation (Chase and Shulman, 2009), and availability of resources (Palik *et al.*, 2006) shapes the life history strategies of the mosquitoes inhabiting the wetland thus has implications in its life cycle pattern. The addition of nutrients from various wastes, containing chemicals, animal excreta increases the density of microorganisms in the wetland water volume (Atkinson *et al.*, 2011). As the mosquitoes are filter feeders, feeding the microbes, this nutrient enrichment directly influences the mosquito oviposition preferences (Walker and Merritt, 1991). Immature mosquito abundance also increases in wetlands enriched with NH_4-N (Sanford *et al.*, 2005) and dissolved nitrates and phosphates in natural wetlands (Mercer *et al.*, 2005). Thus it is quite apparent that the hydrological, microbiological and biochemical parameters had a strong influence in determining the attributes of the life cycle pattern of the wetland mosquitoes. The mosquitoes too, respond to elevated nutrients depending on individual life history requirements (Chaves *et al.*, 2009, Nguyen *et al.*, 2011).

Effect of Wetland on Mosquito Abundance

Mosquitoes are a substantial nuisance for humans in many residential situations, and during various occupational and recreational activities in different regions of India and can cause severe discomfort to cattle and other domestic animals. This is generally caused by the presence of extensive habitat and other favourable conditions that allow the production and persistence in large numbers. The existence of a wetland within flight range may present such a mosquito hazard for a nearby community. Generally, the nature of the habitat will influence which mosquito species will colonise it and the extent of its productivity. Newly flooded, vegetated habitats without predators can provide suitable conditions for some pest species and produce large numbers of mosquitoes within a few weeks (McDonald and Buchanan, 1981). More permanently flooded habitats, with an established and diverse invertebrate and vertebrate fauna, generally produce fewer mosquitoes, although they may support a greater range of mosquito species (Russell, 1993).

The occurrences of mosquito larvae in the wetlands are often limited by several biotic factors, such as predators and competitors (Stav *et al.*, 2000; Mokany and Shine, 2003). In addition, the nature of wetland also directs the consequences of these biotic interactions. Based on the probability of retaining standing water round the year, wetlands can be divided into three classes - temporary, permanent and semi-permanent. This water retention ability of the wetlands in turn determines the species which can inhabit in those habitats and thus their interspecific interactions (Wellborn *et al.*, 1996). Permanent wetlands always retain standing water and hence predators, including fish, hemipteran bugs, can complete their life cycles and reach very high densities. Thus it is evident that the mosquito densities will be low in permanent wetlands as a result of predation (Chase and Knight 2003). Temporary wetlands are those that fill and dry every year coinciding with the monsoon and post monsoon months. Although some predators subsist in these temporary habitats (Spencer *et al.*, 1999), the efficient mosquito predators (fish, odonates) fail to thrive in these habitats due to the ephemeral nature. Alternatively, mosquito competitors, who can adapt to the predictable yearly drying of these wetlands (zooplankton), are often quite dense in the absence of predators. Thus, lower rates of emergence, in presence of high competitor density slows the rate of larval development, along with higher larval mortality and avoidance of oviposition by females results in lesser number of mosquito emergence from the temporary wetlands (Cardo *et al.*, 2011). Semi-permanent wetlands are those that retain standing water in most years, but periodically dry when precipitation and the water table are particularly low. Hence, in years prior to a drying event, predators will be common, as in permanent wetlands, because the habitat has retained standing water for several seasons, allowing sufficient time for predator colonization and so the mosquito densities will be low. However, in years following a drying event, both the predators and competitors will be rare, as neither group of species are well adapted to drying events. Furthermore, as the mosquitoes have shorter span of life cycle and thus rapid generation times relative to their predators, mosquitoes would reveal rapid population increase in semi-permanent wetlands and extensive dispersal ability among the wetlands. These events are true for the mosquitoes in India and similar other geographical regions (Bhattacharya *et al.*, 2006; Dhiman *et al.*, 2008; Gubler, 2011; Cailly *et al.*, 2011). Tropical countries exhibit specific peaks and falls in the mosquito population structure, coinciding with the rainfall pattern (Banerjee *et al.*, 2013b; Mohan *et al.*, 2014). Monsoon and post monsoon months result in water retention and clogging in different types of wetlands. The maintenance of water level supports the mosquitoes to oviposit and thus the perpetuation of the population.

With respect to human communities, there is a potential for explosive mosquito populations in highly enriched constructed wetlands (Tennessen, 1993), but an assessment of the risks presented by mosquitoes in wastewater wetlands can be difficult because of the complexity of the situation. Risk assessment is affected by the presence of different mosquito species, their access to pathogens, environmental conditions, and contact with humans. Although various factors are involved in the occurrence of mosquito-borne diseases, they are primarily density dependent; the major factor in transmission risk is mosquito abundance, and this can be evaluated by sampling the mosquito populations (Service, 1993). Comprehensive entomological monitoring and assessment should involve regular surveys of mosquito immature and concurrent routine adult collections from human dwellings and cattle sheds adjacent to the wetlands. The collection should be followed by identification of the vector mosquitoes using appropriate keys. The data recorded on the species composition in the habitats should be subjected to regression analysis to estimate the space and diversity relationship. The relative importance of the mosquito larval habitats in terms of macroinvertebrate assemblage and vector mosquito abundance should be evaluated by employing multivariate analysis. This will supplement the information required for management of container-breeding and wetland mosquitoes, affecting the species diversity at a minimum (Willott, 2004).

Further, to comment on the species specific variation and in order to frame suitable management policies, ecological characteristics of the mosquito species should be linked with characteristics of habitats and vice versa. The feeding behaviour of mosquitoes provides vital information on the presence of preferred hosts in an area, such as mammals, birds, or amphibians. Inspections for mosquito larvae in the wetlands are expected to provide an input regarding the variation in population dynamics and structure. However, habitat based entomological surveillance forms the key to frame appropriate control and management strategies to regulate the vector mosquito population, without affecting the coexisting predator taxa.

Conclusion

Wetland restoration decreases mosquito populations in two ways: by providing proper habitat for the natural enemies of mosquitoes and by preventing or reducing flooding. In India and many industrialized countries, there has been a long tradition of draining wetlands for agriculture and forestry, as well as for reducing mosquito nuisance and disease transmission (Perry and Vanderklein, 1996). During the last decades it has, however, become more common to construct new wetlands and to restore previous ones. These new and restored wetlands are often motivated by their function as nutrient and silt filters, by their production of fish and game species and by their use as stopover sites for migrating birds. Increasing biological diversity is also been used as one of the main arguments. However, biological diversity includes the diversity of all organisms, including those that may be considered negative or harmful to human societies. Thus, to reduce the risk for a backlash in the positive trend of creating and restoring wetlands, there is a need for a more balanced view with emphasis on the controversial natural inhabitants of such wetlands. The positive biological diversity of wetlands has often been assessed by studies of birds, amphibians, dragonflies and diving beetles (Chovanec and Raab, 1997; Worrall *et al.*, 1997; Melvin *et al.*, 1998; Lundkvist *et al.*, 2001). Many obvious and prevalent inhabitants such as mosquitoes and biting midges have been neglected, although they constitute a significant and important part of the biological diversity. Due to the greater prevalence of the mosquitoes, their potential for annoyance, and their function as vectors for human and animal pathogens, they are considered as the most controversial of the wetland insects. However, mosquitoes are also useful for assessing biological diversity as they include both ecological generalists and specialists, they interact with both the aquatic and the terrestrial fauna of the wetlands.

Mosquito control is an important part of the larger issue of wetlands protection and conservation. Agricultural, industrial, commercial, and residential development remain the major threats to wetlands, either as direct causes of outright habitat loss, or as primary causes of habitat degradation, much more than the abundance and diversity of the mosquitoes in the wetlands. Continued and increased legal protection of wetlands from the above forces remains the top priority, as without it, management related issues such as mosquito control will eventually become irrelevant and the much needed legislative mandates for responsible management of the nation's remaining wetlands become unnecessary.

Mosquito abundance is an important factor that determines the vectorial capacity and R_0 (the basic reproductive rate), and thus abundance is often correlated to an epidemic (Lafferty, 2009; Reiter, 2010). Source reduction denotes techniques used to reduce mosquito populations by eliminating their oviposition and immature rearing sites, and making the mosquito larval habitats less conducive to mosquito production. Long lasting approaches to source reduction are often more economical than temporary control and if effective, eliminate or drastically reduce the need to use pesticides (Newman and Reynolds, 2004).

In conclusion, small size of constructed wetlands has the advantage, from a human point of view, of fewer mosquitoes. For a wide acceptance of constructed wetlands among human neighbours, reduced production of the major nuisance species of functional group might be crucial. Monitoring of adult mosquitoes in constructed wetlands close to human settlements could help to detect potential mosquito nuisance species, as well as potential vectors of diseases. This data enables a risk assessment and will provide useful information in case mosquito control measurement would be needed as a result of changed climatic conditions.

Sorting of mosquitoes into functional groups helps to identify nuisance species and proves especially useful when presenting mosquito data to the general public. The concept is easy to understand and thus helps communication between scientists and administrative decision makers.

Summary

In spite of the wide appreciation of the services provided by the wetlands, they are accused of being playing a negative role in acting as a congenial biotope for the mosquito vectors and thus mosquito borne diseases. The wetlands serve as a habitat for a wide variety of both invertebrate and vertebrate species, thus maintaining a high biodiversity profile. The macroinvertebrate predators present in the wetlands are elements that influence the mosquito species abundance thus providing a basis for biological control. Wetland restoration and protection help to lessen the mosquito populations by providing better quality habitat for the natural predators of mosquitoes, and by helping to reduce flooding and standing water in non-wetland areas that are wet long enough to support mosquitoes - but aren't wet long enough to establish wetland vegetation and a healthy population of mosquito predators. Degradation and obliteration of the wetlands, in view of controlling mosquitoes affects the ecosystem health and ecological well being. Improper mosquito control management programmes adversely affect the wetlands' goods as well as several non-target and highly valuable organisms. Controlling mosquitoes is not simply a narrow, anthropocentric concern. It requires holistic approach of the pros and cons of application of management strategies and overview of the socio-economic aspects. Thus assessment of the species assemblages in the wetlands habitats will indicate the possible variations in the resource exploitation and trophic interactions and therefore can help to frame biological control strategies more appropriately, without altering the ecosystem balance.

Acknowledgements

The authors are thankful to the Head of the Department of Zoology, University of Calcutta, Kolkata for the facilities provided including DST-FIST. SB acknowledges the financial assistance of UGC through SAP-RFSMS fellowship, and CSIR, India, in carrying out the work.

References

Aditya G, Pramanik MK and Saha GK. 2006. Larval habitats and species composition of mosquitoes in Darjeeling Himalayas, India. *Journal of Vector Borne Diseases,* 43 (1): 7-15.

Aditya G, Tamang R, Sharma D. *et al.,* 2008. Bamboo stumps as mosquito larval habitats in Darjeeling Himalayas, India: A spatial scale analysis. *Insect Science,* 15 (3): 245-249.

Agnew P, Mallorie H, Sidobre C and Michalakis Y. 2002. A minimalist approach to the effects of density-dependent competition on insect life-history traits. *Ecological Entomology,* 27: 96-402.

Alto BW, Griswold MW and Lounibos LP. 2005. Habitat complexity and sex dependent predation of mosquito larvae in containers. *Oecologia*, 146: 300-310.

Amala S and Anuradha V. 2011. A study on diversity of mosquitoes in the forest ecosystem of Sirumalai Hills, Dindigul, Tamilnadu. *The ECOSCAN*, 1: 219-222.

Aneesh EM, Chandran T and Lakshmi KV. 2014. Diversity and vectorial capacity of mosquitoes in Kuruva Island, Wayanad District, Kerala, India. *The Journal of Zoology Studies*, 1(4): 16-22.

Arrivillaga J and Barrera R. 2004. Food as a limiting factor for *Aedes aegypti* in water storage containers. *Journal of Vector Ecology*, 29(1): 11-20.

Atkinson C, Golladay S and First M. 2011. Water quality and planktonic microbial assemblages of isolated wetlands in an agricultural landscape. *Wetlands*, 31: 885-894.

Balasubramanian R and Nikhil TL. 2013. Mosquito (Diptera: Culicidae) fauna in Alappuzha and Kottayam district of the Kerala state, South India. *Journal of Entomology and Zoology Studies*, 1 (6): 134-137.

Balcombe SR, Bunn SE, Davies PM and McKenzie Smith FJ. 2005. Variability of fish diets between dry and flood periods in an arid zone floodplain river. *Journal of Fish Biology*, 67: 1552-1567

Bambaradeniya CNB, Edirisinghe JP, De Silva DN. *et al.*, 2004. Biodiversity associated with an irrigated rice agro-ecosystem in Sri Lanka. *Biodiversity and Conservation*, 13: 1715-1753.

Banerjee S, Aditya G, Saha N and Saha GK. 2010. An assessment of macroinvertebrate assemblages in mosquito larval habitats-space and diversity relationship. *Environmental Monitoring and Assessment*, 168 (1-4): 597-611.

Banerjee S, Aditya G and Saha GK. 2013a. Household disposables as breeding habitats of dengue vectors: linking wastes and public health. *Waste Management*, 33: 233-239.

Banerjee S, Aditya G and Saha GK. 2013b. Pupal productivity of dengue vectors in Kolkata, India: implications for vector management. *Indian Journal of Medical Research*, 37: 549- 559.

Bedhomme S, Agnew P, Sidobre C and Michalakis Y. 2003. Sex-specific reaction norms to intraspecific larval competition in the mosquito *Aedes aegypti*. *Journal of Evolutionary Biology*, 26: 721-730.

Bhattacharya S, Sharma C, Dhiman RC and Mitra AP. 2006. Climate change and malaria in India. *Current Science*, 90(3): 369-375.

Bhattacharyya DR, Rajavel AR, Mohapatra PK. *et al.*, 2014. Faunal richness and the checklist of Indian mosquitoes (Diptera: Culicidae). *Check List*, 10(6): 1342-1358

Blaustein L and Chase JM. 2007. Interactions between mosquito larvae and species that share the same trophic level. *Annual Review of Entomology*, 52: 489-507.

Blaustein L, Kiflawi M, Eitam A. *et al.*, 2004. Oviposition habitat selection in response to risk of predation in temporary pools: Mode of detection and consistency along experimental venue. *Oecologia*, 138: 300-305.

Buckner EA, Blackmore MS, Golladay SW and Covich AP. 2011. Weather and landscape factors associated with adult mosquito abundance in southwestern Georgia, U.S.A. *Journal of Vector Ecology*, 36: 269-278.

Burdett AS and Watts RJ. 2009. Modifying living space: an experimental study of the influences of vegetation on aquatic invertebrate community structure. *Hydrobiologia*, 618: 161-173.

Cailly P, Balenghien T, Ezanno P. *et al.*, 2011. Role of the repartition of wetland breeding sites on the spatial distribution of *Anopheles* and *Culex*, human disease vectors in Southern France. *Parasites and Vectors*, 4: 65.

Campos RE, Fernandez IA and Sy VE. 2004. Study of the insects associated with the floodwater mosquito *Ochlerotahus albifasciatus* (Diptera Culicidae) and their possible predators in Buenos Aires province. *Hydrobiologia*, 524: 9-10.

Cardo MV, Vezzani D and Carbajo AE. 2011. Community structure of ground-water breeding mosquitoes driven by land use in a temperate wetland of Argentina. *Acta Tropica*, 119: 76-83.

Chase J and Shulman R. 2009. Wetland isolation facilitates larval mosquito density through the reduction of predators. *Ecological Entomology*, 34: 741-747.

Chase J and Knight TM. 2003. Drought-induced mosquito outbreaks in wetlands. *Ecology Letter*, 6: 1017-1024.

Chaves LF, Keogh CL, Vazquez-Prokopec GM and Kitron UD. 2009. Combined sewage overflow enhances oviposition of *Culex quinquefasciatus* (Diptera: Culicidae) in urban areas. *Journal of Medical Entomology*, 46: 220-226.

Chovanec A and Raab R. 1997. Dragonflies (Insecta, Odonata) and the ecological status of newly created wetlands – Examples for long-term bio indication programmes. *Limnologica*, 27: 381-392.

Couret J and Benedict MQ. 2014. A meta-analysis of the factors influencing development rate variation in *Aedes aegypti* (Diptera: Culicidae). *BMC Ecology*, 14: 3.

Covich AP, Margaret AP and Crowl TA. 1999. The role of benthic invertebrate species in freshwater ecosystems: zoobenthic species influence energy flows and nutrient cycling. *BioScience*, 49 (2): 119-127.

Dale PER and Connelly R. 2012. Wetlands and human health – an overview. *Wetland Ecology and Management*, 20(3): 165-171.

Das PK, Sivagnaname N and Amalraj DD. 2006. Population interactions between *Culex vishnui* mosquitoes and their natural enemies in Pondicherry, India. *Journal of Vector Ecology*, 29: 188-191.

Dash S and Hazra RK. 2011. Mosquito diversity in the Chilika lake area, Odisha, India. *Tropical Biomedicine*, 28(1): 1-6.

De Szalay FA and Resh VH. 2000. Factors influencing macroinvertebrate colonization of seasonal wetlands: responses to emergent plant cover. *Freshwater Biology*, 45: 295-308.

Deepa RS and Ramachandra TV. 1999. Impact of Urbanization in the Interconnectivity of Wetlands. *National Symposium on Remote Sensing Applications for Natural Resources: Retrospective and Perspective (XIX-XXI 1999)*, Indian Society of Remote Sensing, Banglore.

Depinay JMO, Mbogo CM, Killeen G. *et al.*, 2004. A simulation model of African *Anopheles* ecology and population dynamics for the analysis of malaria transmission. *Malaria Journal*, 3: 29.

Dhiman RC, Pahwa S and Dash AP. 2008. Climate change and malaria in India: Interplay between temperature and mosquitoes. *Regional Health Forum*, 12: 27-31.

Edillo FE, Toure YT, Lanzaro GC. *et al.*, 2004. Survivorship and distribution of immature *Anopheles gambiae* s.l. (Diptera: Culicidae) in Banambani village, Mali. *Journal of Medical Entomology*, 41(3): 333-339.

Fincke OM. 1999. Organization of predator assemblages in Neotropical tree holes: effects of abiotic factors and priority. *Ecological Entomology*, 24: 13-23.

Focks DA, Haile DG, Daniels E and Mount GA. 1993. Dynamic life-table model for *Aedes aegypti* (L) (Diptera: Culicidae): Analysis of literature and model development. *Journal of Medical Entomology*, 30(6): 1003-1017.

Gilbert B, Srivastava DS and Kirby KR. 2008. Niche partitioning at multiple scales facilitates coexistence among mosquito larvae. *Oikos*, 117(6): 944-950.

Gopalakrishnan R, Baruah I and Veer V. 2014. Monitoring of malaria, Japanese encephalitis and filariasis vectors. *Medical Journal of Armed Forces India*, 70: 129 -133.

Gubler DJ. 2011. Emerging vector-borne flavivirus diseases: are vaccines the solution? *Expert Review of Vaccines*, 10(5): 563-565.

Irwin P, Arcari C, Hausbeck J and Paskewitz S. 2008. Urban wet environment as mosquito habitat in upper Midwest. *Eco Health*, 5: 49-57.

Jacob BG, Muturi EJ, Funes JE. *et al.*, 2006. A grid-based infrastructure for ecological forecasting of Riceland *Anopheles arabiensis* aquatic larval habitats. *Malaria Journal*, 5: 91.

Jauhari RK and Pemola Devi N. 2014. Occurrence of a Snail Borne Disease, Cercarial Dermatitis (Swimmer Itch) in Doon Valley (Uttarakhand), India. *Iranian Journal of Public Health*, 43(2): 162-167.

Jeffries MJ. 2002. Evidence for individualistic species assembly creating convergent predator-prey ratios among pond invertebrate communities. *Journal of Animal Ecology*, 71: 173-184.

Johri PK, Dayal V and Johri R. 2010. Biodiversity and community structure of aquatic insects in a natural wetland (Jalesar) of district Unnao, Uttar Pradesh, India. *Journal of Experimental Zoology India*, 13(2): 363-372.

Juliano SA. 2007. Population dynamics. *Journal of the American Mosquito Control Association*, 23 (2): 265-275.

Karthikairaj K, Ravichandran N and Sevarkodiyone SP. 2013. Bio-diversity of Mosquitoes in Sub-Urban Area, Thiruthangal, Virudhunagar District, Tamil Nadu, India. *World Journal of Zoology*, 8 (3): 319-323.

Keating J, Macintyre K, Mbogo CM. *et al.*, 2004. Characterization of potential larval habitats for *Anopheles* mosquitoes in relation to urban land-use in Malindi, Kenya. *International Journal of Health Geographics*, 3: 9.

Keiper JB, Jiannino J, Sanford M and Walton WE. 2001. Biology and immature stages of *Typopsilopa nigra* (Wirth) (Diptera: Ephydridae), a secondary consumer of damaged stems of wetlands monocots. *Proceedings of Entomological Society Washington*, 103: 89-97.

Kevan PG, Tikhmenev EA and Usui M. 1993. Insects and plants in the pollination ecology of the boreal zone. *Ecological Research*, 8: 247-267.

Kiflawi M, Blaustein L and Mangel M. 2003. Predation-dependent oviposition habitat selection by the mosquito *Culiseta longiareolata*: a test of competing hypotheses. *Ecology Letters*, 6: 35–40.

Knight RL, Walton WE, O'Meara GF *et al.*, 2003. Strategies for effective mosquito control in constructed treatment wetlands. *Ecological Engineering*, 21: 211-232.

Krishnamoorthy K, Jambulingam P, Natarajan R. *et al.*, 2005. Altered environment and risk of malaria outbreak in South Andaman, Andaman and Nicobar Islands, India affected by tsunami disaster. *Malaria Journal*, 4: 32.

Lafferty KD. 2009. The ecology of climate change and infectious diseases. *Ecology*, 90(4): 888-900.

Lambert A. 2002. The Convention on Wetlands (Ramsar, 1971): An Active Player in the Fight Against Poverty. Published on the official Ramsar website (www.ramsar.org)

Le Menach A, Takala S, McKenzie FE. *et al.*, 2007. An elaborated feeding cycle model for reductions in vectorial capacity of night-biting mosquitoes by insecticide-treated nets. *Malaria Journal*, 6: 10.

Leisnham PT, Lester PJ, Slaney DP *et al.*, 2004. Anthropogenic landscape change and vectors in New Zealand: effects of shade and nutrient levels on mosquito productivity. *Eco Health*, 1: 306-316.

Lounibos LP. 2007. Competitive displacement and reduction. *Journal of the American Mosquito Control Association*, 23 (2): 276-282.

Lundkvist E, Landin J and Milberg P. 2001. Diving beetle (Dytiscidae) assemblages along environmental gradients in an agricultural landscape in southeastern Sweden. *Wetlands*, 21: 48-58.

McDonald G and Buchanan GA. 1981. The mosquito and predatory insect fauna inhabiting fresh-water ponds, with particular reference to *Culex annulirostris* Skuse (Diptera: Culicidae). *Australian Journal of Ecology*. 6: 21-27.

Melvin SL, James Wand and Webb J. 1998. Differences in the avian communities of natural and created *Spartina alterniflora* salt marshes. *Wetlands*, 18: 59-69.

Mercer DR, Sheeley SL and Brown EJ. 2005. Mosquito (Diptera : Culicidae) development within microhabitats of an Iowa wetland. *Journal of Medical Entomology*, 42: 685-693.

Mitsch WJ and Gosselink JG. 2000. *Wetlands*. 3rd ed. John Wiley and Sons, Inc., New York, NY, USA.

Mogi M. 2010. Unusual life history traits of *Aedes (Stegomyia)* mosquitoes (Diptera: Culicidae) inhabiting *Nepenthes* pitchers. *Annals of the Entomological Society of America*, 103(4): 618-624.

Mogi M, Sunahara T and Selemo M. 1999. Mosquito and aquatic predator communities in ground pools on lands deforested for rice-field development in central Sulawesi, Indonesia. *Journal of the American Mosquito Control Association*, 15: 92-97.

Mohan S, Banerjee S, Mohanty SP. *et al.*, 2014. Assessment of pupal productivity of *Aedes* and co-occurring mosquitoes in Kolkata, India. *Southeast Asian Journal of Tropical Medicine and Public Health*, 45(6): 1279-1291

Mokany A and Mokany K. 2006. Effects of habitat permanence cues on larval abundance of two mosquito species. *Hydrobiologia*, 563: 269-276.

Mokany A and Shine R. 2003. Oviposition site selection by mosquitoes is affected by cues from conspecific larvae and anuran tadpoles. *Australian Ecology*, 28: 33-37.

Mokany A. 2007. Impact of tadpoles and mosquito larvae on ephemeral pond structure and processes. *Marine and Freshwater Research*, 58: 436-444.

Muturi EJ, Shililu J, Jacob B. *et al.*, 2006. Mosquito species diversity and abundance in relation to land use in a riceland agroecosystem in Mwea, Kenya. *Journal of Vector Ecology*, 31: 129-137.

Newman LA and Reynolds CM. 2004. Phytodegradation of organic compounds. *Current Opinion in Biotechnology*, 15(3): 225-230.

Nguyen LAP, Clements ACA, Jeffery JAL, *et al.*, 2011. Abundance and prevalence of *Aedes aegypti* immatures and relationships with house hold water storage in rural areas in southern Vietnam. *International Health*, 3: 115-125.

Overtli B, Joye DA, Castella E. *et al.*, 2002. Does size matter? The relationship between pond area and biodiversity. *Biological Conservation*, 104: 59-70.

Palik B, Batzer D and Kern C. 2006. Upland forest linkages to seasonal wetlands: Litter flux, processing, and food quality. *Ecosystems*, 9: 142-151.

Perry J and Vanderklein E. 1996. *Water Quality in Natural Resource Management*. Blackwell Science, Oxford.

Pongsiri MJ and Roman J. 2007. Examining the Links between Biodiversity and Human Health: An Interdisciplinary Research Initiative at the U.S. Environmental Protection Agency. *Ecohealth*, 4: 82-85.

Purkayastha P and Gupta S. 2012. Insect Diversity and Water Quality Parameters of Two Ponds of Chatla Wetland, Barak Valley, Assam. *Current World Environment,* 7(2): 243-250.

Rajavel AR and Natarajan R. 2006. Mosquitoes of the mangrove forests of India: part 3-Andaman and Nicobar Islands, including an update on the mosquito fauna of the islands. *Journal of the American Mosquito Control Association,* 22: 366-377.

Rajavel AR, Natarajan R and Vaidyanathan K. 2005a. Mosquitoes of the mangrove forests of India: part 1-Bhitarkanika, Odisha. *Journal of the American Mosquito Control Association,* 21: 131-135.

Rajavel AR, Natarajan R and Vaidyanathan K. 2005b. Mosquitoes of the mangrove forests of India: part 2–Sundarbans, West Bengal. *Journal of the American Mosquito Control Association,* 21: 136-138.

Ramachandran J, Ajjampur S, Chandramohan A and Varghese GM. 2012. Cases of human fascioliasis in India: Tip of the iceberg. *Journal of Postgraduate Medicine,* 58: 150-162.

Reiskind MH and Wilson ML. 2008. Interspecific competition between larval *Culex restuans* Theobald and *Culex pipiens* L. (Diptera: Culicidae) in Michigan. *Journal of Medical Entomology,* 45(1): 20-27.

Reiter P. 2010. Yellow fever and dengue: a threat to Europe? *European Surveillance,* 15: 19509.

Russell RC. 1993. *Mosquitoes and mosquito borne disease in southeastern Australia.* Department of Medical Entomology, Westmead Hospital, Westmead, Australia.

Russell RC. 1999. Constructed wetlands and mosquitoes: health hazards and management options–an Australian perspective. *Ecological Engineering,* 12: 107-124.

Saha N, Aditya G, Banerjee S and Saha GK. 2012. Predation potential of odonates on mosquito larvae: implications for biological control. *Biological Control,* 63(1): 1-8.

Sanford MR, Chan K and Walton WE. 2005. Effects of inorganic nitrogen enrichment on mosquitoes (Diptera: Culicidae) and the associated aquatic community in constructed treatment wetlands. *Journal of Medical Entomology,* 42: 766 -776.

Schafer M. 2004. *Mosquitoes as part of wetland biodiversity.* Faculty of Science and Technology, University of Uppsala, Uppsala. pp. 63.

Service MW. 1993. *Mosquito ecology: field sampling methods,* 2nd edn. Elsevier Applied Science, London.

Sharma RS, Joshi PL, Tiwari KN. *et al.,* 2005. Outbreak of dengue in National Capital Territory of Delhi, India during 2003. *Journal of Vector Ecology,* 30: 337- 338.

Sharma S, Yadav RK, Saini Y and Sharma S. 2011. Water Quality Status of Pushkar Lake as a Primary data for Sustainable Development. *South Asian Journal of Tourism and Heritage,* 4(2): 184-191.

Simenstad C, Reed D and Ford M. 2006. "When is restoration not? Incorporating landscape-scale processes to restore self-sustaining ecosystems in coastal wetland restoration." *Ecological Engineering*. 26(1): 27-39.

Sisodia S and Singh BN. 2009. Variations in morphological and life–history traits under extreme temperatures in *Drosophila ananassae*. *Journal of Biosciences*, 34(2): 263-274.

Spencer M, Blaustein L and Cohen JE. 2002. Oviposition habitat selection by mosquitoes (*Culiseta longiareolata*) and consequences for population size. *Ecology*, 83: 669-679.

Spencer M, Blaustein L, Schwartz SS and Cohen, JE. 1999. Species richness and the proportion of predatory animal species in temporary freshwater pools: relationships with habitat size and permanence. *Ecology Letters*, 2: 157-166.

Stav G, Blaustein L and Margalit Y. 2000. Influence of nymphal *Anax imperator* (Odonata: Aeshnidae) on oviposition by the mosquito *Culiseta longiareolata* (Diptera: Culicidae) and community structure in temporary pools. *Journal of Vector Ecology*, 25: 190-202.

Sunahara T, Ishizaka K and Mogi M. 2002. Habitat size: a factor determining the opportunity for encounters between mosquito larvae and aquatic predators. *Journal of Vector Ecology*, 27: 8-20.

Sunish IP and Reuben R. 2002. Factors influencing the abundance of Japanese encephalitis vectors in rice field in India - II. Biotic. *Medical and Veterinary Entomology*, 16: 1-9.

Tennessen K. 1993. Production and Suppression of Mosquitoes in Constructed Wetlands *Constructed Wetlands for Water Quality Improvement*, (Moshiri G eds.), Lewis Publishers, London.

Verma S and Gupta RD. 2013. Role of Remote Sensing Using Topographic Analysis for Wetland breeding sites Encephalitis Disease Vectors in Gorakhpur district, Uttar Pradesh, India. *International Journal of Scientific and Engineering Research*, 4(12): 15-21.

Walker ED and Merritt RW.1988. The significance of leaf detritus to mosquito (Diptera: Culicidae) productivity from tree holes. *Environmental Entomology*, 17: 199-206.

Wellborn GA, Skelly DK and Werner EE. 1996. Mechanisms creating community structure across a freshwater habitat gradient. *Annual Review of Ecological Systematics*, 27: 337-363.

Williams CR, Leach KJ, Wilson NJ and Swart VR. 2008. The Allee effect in site choice behaviour of egg-laying dengue vector mosquitoes. *Tropical Biomedicine*, 25: 140-144.

Willott E. 2004. Restoring nature, without mosquitoes? *Restoration Ecology*, 12: 147-153.

Wilson AL, Watts RJ and Stevens MM. 2007. Effects of different management regimes on aquatic macroinvertebrate diversity in Australian rice fields. *Ecological Research*, 23: 565-572.

Worrall P, Peberdy KJ and Millett MC. 1997. Constructed wetlands and nature conservation. *Water Science and Technology*, 35: 205-213.

Yanoviak SP. 2001. Predation, resource availability, and community structure in Neotropical water filled tree holes. *Oecologia*, 126: 125-133.

Yee DA, Yee SH, Kneitel JM and Juliano SA. 2007. Richness–productivity relationships between trophic levels in a detritus-based system: significance of abundance and trophic linkage. *Oecologia*, 154: 377-385.

Zedlar JB. 2000. Progress in wetland restoration ecology. *Trends in Ecology and Evolution*, 15(10): 402-407.

Chapter 19

Non-Fish Food Resources of Wetlands of West Bengal, India: Prospects for Food Security

Joy Chakraborty[1], Sk. Habibur Rahaman[1,2], Dibyendu Saha[1] and Gautam Aditya[1,2]

[1]Department of Zoology, The University of Burdwan, Burdwan
[2]Department of Zoology, University of Calcutta, Kolkata.

Introduction

The wetlands are common source of different food resources exploited by human for long. Among the food resources, fishes have been studied considerably in comparison to the non-fish food resources. The non-fish food resources include varieties of macroinvertebrates like crustaceans, snails and mussels and enormous types of hydrophytes that sustain food security and livelihood of rural India and in different regions of South-East Asian countries. In contrast to the extent of studies carried out on fish resources, little effort has been made to evaluate the resource potential and extent of exploitation of the non-fish resources. The abundance of the non-fish resources in the wetlands including rice fields, and the extent of consumption, calls for a documentation of the species specific importance as food resources.

Among the macroinvertebrates, freshwater snails, mussels, prawns and crabs constitute a bulk of the diversity and contribute to the ecosystem services substantially. These macroinvertebrates contribute to the provisioning, regulating, cultural and supporting services that are constituents of human well-being. In freshwater ecosystem, snails, mussels, prawns, crabs and hydrophytes link the food web and energy transfer through various prey-predator interactions in different trophic levels and keep the pace of the network dynamics to stabilize the biotic integrity. In the context of the role of wetlands in providing resources to sustain food security and livelihood, the importance of non-fish resources are being less emphasized in comparison to the fish resources. While records suggests that snails, mussels, crustaceans and hydrophytes are common food resource in many parts of India and other Asian countries, few studies have been conducted to indicate the diversity and the extent of exploitation by the local people. In almost all instances, the prawn species other than those that are used in commercial aquaculture are harvested by the local people as supplement food resources. For the other non-fish and hydrophyte resources harvest is a rule, though few species are used in the integrated aquaculture. Besides, limited studies have been carried out to monitor and note the extent of exploitation of the non-fish resources that are equally important in the maintenance of the integrity of the wetland ecosystems. Over exploitation of these resources contributing to the ecosystem services derived from wetlands is a concern while promoting them as alternative or supplement to the conventional food resources. Nonetheless the importance of these resources in providing food security and the livelihood of the local stakeholders raises an issue to be considered for policy makers and frame strategies for sustainable harvest of these resources. An appraisal of different non-fish and hydrophyte resources of the freshwater wetlands will substantiate the proposition of the extent of diverse flora and fauna that can supplement the conventional resources exploited through aquaculture and capture fisheries where limited number of species is considered as commercially viable. Information on the supplementary species is a pre-requisite for enhancement and diversifying conventional aquaculture practice, which may be obtained from the appraisal of the non-fish food resources. In view of the importance of the aquaculture practices and the importance of the aquatic food resources, the present commentary provides a brief outline about the species encountered in the local markets with varying availability in the market of selected districts of West Bengal, India. The compiled information is expected to represent a glimpse about the extent of the resources exploited and applies to similar places in West Bengal and other areas in Indian subcontinent, though differences in the extent of species involved is expected among the different regions.

The Non-Fish Food Resources

Following visit to different local market places in rural areas of Howrah, Hooghly, North 24 Parganas, Burdwan and Birbhum districts of West Bengal, and Kolkata metropolis, the presence of different non-fish food resources were recorded during the monsoon (July-September) and post monsoon (October and November) periods of 2010 to 2014. Selected freshwater wetlands, including ponds, inundated rice fields and temporary pools and canals were surveyed for the presence of the respective hydrophytes, snails and mussels and crustaceans. The species present in both the market places and the field conditions were considered as non-fish food resources for this compilation. Owing to the differences in the sampling techniques and the extent of variations in the unit of the resources sold in the market, the quantification of the resources in the fields and market places were not considered,

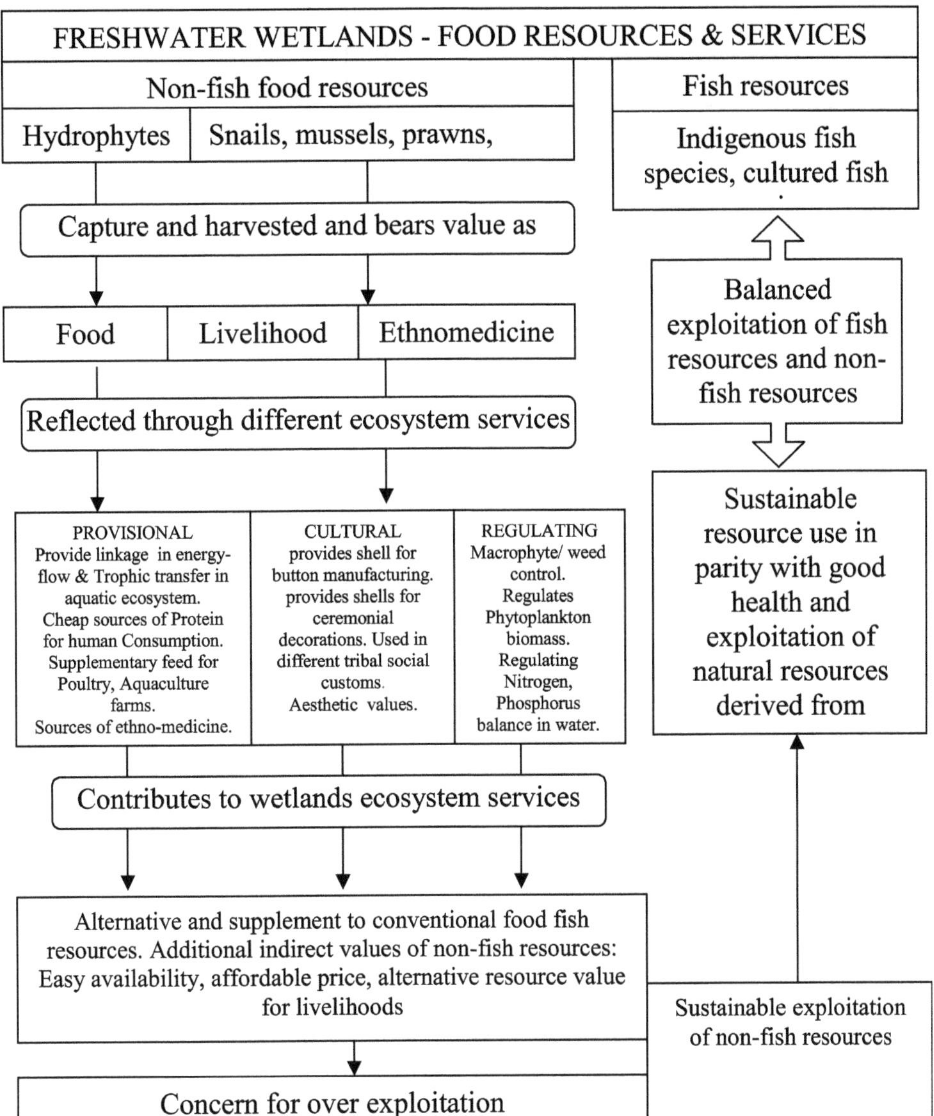

Figure 19.1: Overview of the Fish and Non-Fish Food Resources of Freshwater Wetlands and the Resource Value Contributing to the Ecosystem Services.

Although the non-fish resources are highly valued for easy availability and management against the fish resources, excess exploitation is a concern for sustainability. A balanced exploitation of both the resources is expected to sustain the wetland ecosystems of varied forms.

rather the presentation is restricted to the qualitative assessment. The species observed in the study and considered for the discussion are presented in Tables 19.1a and 19.1b, along with the abundance pattern.

Table 19.1a: The Snails, Mussels and Crustaceans Sold in the Market with Additional Medicinal Values. The abundance of the species was equal during the monsoon period though the crabs and prawns were rare during the postmonsoon period. The relative value of prawn and crabs were higher than the snails and the mussels.

Species	*Diseases*	*Method of Application for Cure of Diseases*	*Reference*
Bellamya bengalensis	Arthritis, Joint pain and rheumatism	Soup prepared from the foot of snail is used to cure these diseases.	Prabhakar and Roy (2009)
Bellamya bengalensis	Conjunctivitis, Night blindness	*Bellamya bengalensis* are collected from pond and are kept in clean fresh water in a earthen pot for night and the water is used like eye drop. This method is considered best for the cure of this disease in this region.	Prabhakar and Roy (2009)
Pila globosa	Rickets, Night blindness	Soup prepared from the eggs of *Pila* is used to cure rickets in children	Prabhakar and Roy (2009)
Lamellidens marginalis	Cardiac disease, Dehydration and giddiness	Flesh from the foot and shell powder	Prabhakar and Roy (2009)
Macrobrachium rosenberghii, M. malcomsonii, M. lamarrei, M.dayanum	Kidney dysfunction, dementia, weakness in sexual libido,	Flesh of Prawn	New, 2005; Rangappa *et al.*, 2012
Sartoriana spinigera	Diabetes, anti-carcinoma, heart attack	Crab meat	Rahaman *et al.* (2008)

Table 19.1b: The Plant Species Observed in different Market Places along with the Vegetative Parts Used as Food Resources and for Medicinal Importance. The market price of the hydrophytes were relative low than the macroinvertebrates.

Species Name	*Family*	*Purpose*	*Reference*
Sagittaria sp	Alismataceae	Underground stem used for Japanese and Chinese meat dishes.	Cook, 1996; Gichuki *et al.*, 2001; Jain *et al.*, 2007, 2011; Panda and Mishra, 2011
Sium sisarum	Umbelliferae	Cultivated for its edible roots.	– Same –
Colocasia esculenta	Araceae	The tuberous roots are low in protein and rich in starch.	– Same –
Ipomoea aquatica	Cruciferae	Leaves and stem are used as protein sources.	Mazumder *et al.*,2008; Jain *et al.*, 2011; Swapna *et al.*, 2011; *Ziv et al.*, 2012
Eleocharis dulcis	Cyperaceae	Tuber and corm used for consumption as a food in Chinese delicacy.	Jain *et al.*, 2007, 2011; Panda and Mishra, 2011
Myriophyllum aquaticum	Haloragaceae	The tips of the shoots are eaten as a vegetable.	Jain *et al.*, 2007, 2011; Panda and Mishra, 2011
Nelumbo nucifera	Nelumbonaceae	Various parts of the plant as be used in a variety of cooked and fresh dishes.	Jain *et al.*, 2007, 2011; Panda & Mishra , 2011
Nymphaea stellata	Nymphaceae	Tile seeds are eaten in India as famine food and by the poorest people regularly. are eaten cooked in India and China, and sometimes the Latin America but was introduced into tropical Asia young fruits are eaten as a salad	– Same –
Ludwigia repens.	Onagraceae	The young shoot and leaves are used as green vegetables in Thailand	– Same –
Trapa spp.	Trapacae	The nut or kernel of the spiny fruit is eaten raw or cooked or is ground into flour, which is used for various preparations	– Same –
Bacopa monnieri	Plantaginaceae	It is a medicinal Plant used for Alzheimer's disease, improving memory, anxiety, attention deficit-hyperactivity disorder (ADHD), allergic conditions, irritable bowel syndrome, and as a general tonic to fight stress.	– Same –
Centella asiatica	Apiaceae	Leaves are used for beverages and dishes	– Same –

Snails and Mussels

Freshwater snails and mussels play a vital role in the economy and tradition of West Bengal, India serving as a food of 80.81 per cent families belonging to more than 30 caste of general schedule and tribal people, of which *Bellamya bengalensis* (Chakraborty *et al.*, 2008; Mandal, 2003; Boomithan *et al.*, 2008), *Lymnaea luteola* (Prabhakar and Roy, 2009a), *Indoplanorbis exustus* (Prabhakar and Roy, 2009b) are predominant in freshwater ecosystem. Freshwater snails and mussels, relatively abundant in ricefields and ponds are the primary source of food of poor people and are considered as an alternative protein sources for human, catfish and as well as prawn farming. Freshwater snails contain high amount of fibre, fat and minerals (Eneji *et al.*, 2008) where as *B. bengalensis* provides nutrition to human and poultry as a cheap source of protein (Chakraborty *et al.*, 2008). It becomes pretty clear that molluscs are excellent sources of some required trace and minor elements required for the proper growth and development of human being (Baby *et al.*, 2010). It is also reported that snail-meat contains high amount of protein (21 per cent) higher than that of in other protein sources (Wosu, 2003) and are also used to prevent dehydration due to high content of water (73.67 per cent). *P. globosa* has been commercially farming at a large scale for supplementary feed for prawn culture, catfish farming and also for human consumption (Jahan *et al.*, 2005, 2007; Ahmed *et al.*, 2008; Barmon *et al.*, 2006; Agbogidi and Okonta, 2011). *P. globosa* and *Bellamya* spp. used as an alternative cheap source of protein for poutry farming (Mandal, 2003; Boomithan *et al.*, 2008) and for human consumption (Patil and Talmale, 2005). There are reports that the native Americans also used freshwater mollusc as a food item (Watters *et al.*, 2009). The economic and medical benefits from snail are immense as derived from their foot tissue to shells (Wosu, 2003) and are used traditionally for some ethno-medicinal purposes by some tribal and aboriginal people in North-east India. *P. globosa, Bellamya* spp. are used as a reliever of arthritis, joint pain, rheumatism, conjunctivitis, nervousness etc. The soup, prepared from eggs of *P. globosa* is used to cure rickets in children (Jamir and Lal, 2005; Prabhakar and Ray, 2009a); *P. globosa* is used for throat diseases and wound healing of poutry birds (Mavinkurve *et al.*, 2004). Composition of snails promotes fertility and assists in the cure of sterility in women (Okafor, 2001). A special form of calcium phosphate extracted from snails has been implicated in the cure of some kidney diseases, tuberculosis, diabetes, asthma, heart diseases and circulatory disorders (Abere and Lameed, 2008). Besides, snails cure hemorrhoids, prevent influenza, restore virility and enhance beauty in youth (Ayodele and Asimalowo, 1999). Snails have also been implicated in the reduction of pain and loss of blood during labour as well as in the treatment of small pox (Akinnusi, 1998). The flesh of foot of *L. marginalis* is used to cure from the cardiac diseases and the shell powder is used for the remedy of giddiness and dehydration (Prabhakar and Ray, 2009b).

Prawns and Crabs

From social and economic standpoint prawn farming will remain an important part of the rural economy. It is predominantly a cooperative and lucrative venture and does not necessitate higher education and specialized training. Farmers work together holding local management meetings, sharing information with neighbors, and in some areas working communally at harvest. *Macrobrachium rosenbergii. P. monodon* are the most popular prawn species used for commercial farming and has been exported to many parts of the world including South America and China (New, 2005). Prawns are a good source of unsaturated fat, which makes up the majority of its fat content. Unsaturated fats can help to improve blood cholesterol levels and like other fish and shellfish, are also a good source of omega-3 fatty acids, which can reduce inflammation and risk of heart disease, cancer and arthritis etc. While prawns are a low-fat food and contain many healthy fats, they are also rich in cholesterol, containing 179 milligrams per 3-ounce serving (Rangappa *et al.*, 2012; Reddy and Reddy, 2014) Prawns can help to nourish the kidneys. Chinese medicine believes the strength and weakness of a male's sexual libido, has an direct correlation with the strength and weakness of the kidneys. Therefore, those with poor kidney function should frequently consume prawns, and those who frequently eat prawns, can delay onset of aging, dementia and prolong sexual libido. Prawns can clear blood vessels and warm the meridians. Therefore, those who often suffer from cold extremities, loss of strength in the waist and hips, or heaviness of the limbs, should eat more prawns. It can help to promote blood circulation, especially on cold days.

Among the crustacean, crab is one of the main food source (Rahman *et al.*, 2008). Not only human, small vertebrate predators also feed on crabs which directs the energy flow in freshwater lentic and lotic ecosystem (Mirzadeh *et al.*, 2012). In the context of nutritional value, crabs like other shellfish has large amount of vitamins, high quality proteins and amino acids and are also rich in calcium, magnesium, copper, zinc, phosphorus and iron (Sudha Devi and Smiza, 2013; Varadharajan and Soundarapandian, 2014). It is reported that crab meat is rich in chromium which help insulin to metabolized sugar to reduce hyper-glycaemia. Crab meat also rich in selenium, is an anti-oxidant and helps to wipe out carcinogenic effects of cadmium, mercury and arsenic. Like other shellfish, crab contains omega-3 fatty acids.

Hydrophytes

Aquatic plants, hydrophytes are the major components of freshwater ecosystem related to trapping solar energy and serving as the bottom species for the movement of energy and matter in the ecosystem food webs. The hydrophytes are the power-generators of the ecosystem dynamics and play vital roles in nutrient cycling and as producers involved in carbon sequestration. In addition to the ecological roles, different species of hydrophytes are valued food resources not only for the aquatic vertebrates and macroinvertebrates but also for human and domestic animals. For instance, herbivores like otters, beaver, muskrats, turtles, moose graze on variety of aquatic plants. As a food resource, supplementing different types of minerals, vitamins and proteins, hydrophytes are highly marketable and have extraordinary demand to the local community. Hydrophytes are used as a tool for wetland management as they are indicators of water pollution. Variations in the abundance and richness of the hydrophytes can be affected by different pollutants, and thereby would reflect the nutrient status of their immediate habitat by their presence, absence and abundance and thus can be effectively used as biological indicators (Suominen, 1968; Uotila, 1971). Many macrophyte are consumed as a food and used for therapeutic purposes. *Bacopa monnieri, Centella asiatica, Enhydra fluctuans, Hygrophilla spinosa, Marsilea quadrifolia* have the most economic value.

Results and Discussion

Exploitation of freshwater wetlands for harvesting different fish and non-fish food resources is common in Indian subcontinent and other Asian countries, including the rice field ecosystems (Halwart, 2006). In many instances, at least at the local scale, the harvest of these species supplements the required food resources and secures the health and livelihood. In comparison to the conventional fish food resources, the labor and cost involved in harvesting those resources are quite low and affordable for the local stakeholders. In many instances, the consumption of these resources is made by rural poor, providing valuable calories for growth and good health. The ethno-medicinal values of these resources also increase the demand of these resources to the local stakeholders. Since the natural sources of the non-fish resources are linked to the pool of species available in the wild, the sustainable harvest is required to continue with the benefits for long term and serve as an alternative to the conventional culture and capture food fish resources. The harvest of the non-fish resources are an important and accessible alternative to the commercial aquaculture products and secure good health. On view of the multiple values of the non-fish resources, over- exploitation needs to be avoided, which is why a continuous monitoring of abundance of the resources is required. Besides, the non-fish resources needs to be harvested in a balanced way with the fish resources in order to continue with the benefits of the ecosystem services derived from freshwater wetlands.

Wetlands are important source for water purification, storage and nutrient sink, which serves in sustenance of the ecosystem functions. The constituent biotic components functions in a concerted manner to facilitate the functions of the wetlands. Hydrophytes of different forms and taxonomic identity are involved in the process of nutrient cycling and purification of the water, which may enter the water contaminated with organic wastes. The associated macroinvertebrates particularly the crustaceans and snails contribute to the regulation of the periphyton and the detritus that originate from both autochthonous as well as allochthounous sources. Interactions among the biotic components are equally crucial in maintenance of diversity and shaping the physical conditions of the wetlands like habitat complexity and heterogeneity of the conditions. It is apparent through empirical studies that the biotic components are essential for the maintenance of the wetland functions and services and thus environmental quality. Thus exploitation of the biotic resources for food and livelihood requires monitoring and sustainable harvest to continue with the benefits for future.

Summary

Freshwater wetlands serve as a source for diverse kind of plant and animal resources that are exploited as food. Among the animal resources, fish constitute a major component, though the supporting species like the freshwater prawns, crabs and snails are also harvested as a food resource and for livelihood. While majority of the studies from Indian subcontinent and from Southeast Asian countries highlight the use of the fish species as food resources, the use of the crustacean snails and hydrophytes are no less important from the point of view of food security and livelihood. An appraisal of different species of crustaceans, snails and mussels and hydrophytes of the freshwater wetlands of West Bengal, India is being made in the present commentary to highlight the resource value in sustaining food security and livelihood of local people. An approximately 20 different species of prawns and crabs, 3 different species of snails and mussels and about 20 species of hydrophytes originating from freshwater wetlands were consistently available in the markets with varying densities. A compilation of the non-fish resources and hydrophytes along with the information obtained from the available literature is being provided to highlight the resource potential for food security and livelihood. The present description is an effort to put forward the importance of the non-fish and hydrophyte resources of freshwater wetlands. The consistent presence of these resources as saleable commodity in the local markets speaks of their immense potential to substitute the conventional food resources like fish in sustaining the health and life of people at local level. Further monitoring and assessment on the exploitation pattern should be carried out to sustain the ecosystem services derived from the hydrophytes and non-fish resources.

References

Abere SA and Lameed GA. 2008. The medicinal utilization of snails in some selected states in Nigeria. In. *Proceeding of the first National conference of the Forests and Forest Products Society* (Onyekwelu JC, Adekunle VAJ and Oke DO. eds.). pp. 233-237.

Agbogidi OM and Okonta BC. 2011. Reducing poverty through snail farming in Nigeria. *Agriculture and Biology Journal of North America,* 2(1): 169-172

Ahmed N, Demaine H and Muir J. 2008. Freshwater prawn farming in Bangladesh: history, present status and future prospects. *Aquaculture Research,* 39: 806-819.

Akinnusi O. 1998. *Introduction to snail farming.* Omega Publishers, Badagry, Lagos.

Ayodele IA and Asimalowo AA. 1999. *Essentials of snails farming.* Agape Print, U.I., Ibadan.

Baby Rl, Hasan I, Kabir KA and Naser MN. 2010. Nutrient analysis of some commercially important molluscs of Bangladesh. *Journal of Scientific Research,* 2(2): 390-396.

Barmon BK, Takumi K and Osanami F. 2006. Problems and prospects of shrimp and rice prawn gher farming system in Bangladesh. *Bangladesh Development Initiative,* 8: 1-13.

Boominathan M, Subash Chandran MD and Ramachandra TV. 2008. Economic Valuation of Bivalves in the Aghanashini Estuary, West Coast, Karnataka. *ENVIS Technical Report*: 30.

Chakraborty S, Ray M and Ray S. 2008. Sodium arsenite induced alteration of haemocyte density of *Lamellidens marginalis*-an edible mollusk from India. *Clean,* 36(2): 195-200.

Cook CDK. 1996. Aquatic and wetland plants of India. Oxford University Press, New York, USA. pp. 385.

Eneji CA, Ogogo AU, Emmanuel-Ikpeme CA and Okon OE. 2008. Nutritional Assessment of Some Nigerian Land and Water Snail Species. *Ethiopian Journal of Environmental Studies and Management*, 1(2): 56-60.

Gichuki J, Guebas FD, Mugo J. *et al.*, 2001. Species inventory and local uses of the plants and fishes of the lower Sondu Miriu wetland of Lake Victoria, Kenya. *Hydrobiologia*, 458: 99-106.

Halwart M. 2006. *Biodiversity and nutrition in rice-based aquatic ecosystems. Journal of Food Composition and Analysis*, 19: 747-751.

Jahan MS, Islam MR, Rahman MR and Alam MM. 2007. Induced Breeding of *Pila globosa* (Swainson 1822) (Gastropoda: Prosobranchia) for Commercial Farming. *University Journal of Zoology, Rajshahi University*, 26: 35-39.

Jahan MS, Rahman MR, Sarker MDM and Assaduzzaman AKA. 2005. Assessment of effective physic-chemical factors for the culture of *Pila globosa. Journal of Ecobiology*, 17(3): 241-250.

Jain A, Roshnibala S, Kanjilal PB and Singh HB. 2007. Aquatic and semi-aquatic plants used in herbal remedies in the wetlands of Manipur, northeastern India. *Indian Journal of Traditional Knowledge*, 6: 346-351.

Jain A, Sundariyal M, Roshnibala S. *et al.*, 2011. Dietary use and conservation concern of edible wetland plants at Indo-Burma hotspot: A case study from North east India. *Journal of Ethnobiology and Ethnomedicine*, 7: 29.

Jamir NS and Lal P. 2005. Ethnozoological practices among Naga tribes. *Indian Journal of Traditional Knowledge*, 4(1): 100-104.

Mandal L. 2003. Snail powder - a protein source for poultry. *Feed Mix (The International Journal on Feed, Nutrition and Technology)*, 11(4): 25.

Mavinkurve RG, Shanbhag SP and Madhyastha NA. 2004. Non-marine Molluscs of Western Ghats: A status review. *Zoos' Print Journal*, 19(12): 1708-1711.

Mazumder MSA, Rahman MM, Ahmed ATA. *et al.*, 2008. Proximate composition of some small indigenous fish species (SIS) in Bangladesh. *International Journal of Sustainable Crop Production*, 3(4): 18-23.

Mirzadeh M, Kamrani E, Sojae F. *et al.*, 2012. Comparative study of the distribution the marsh crab, *Sartoriana rokitanskyi* in Hormozgan Province. *Journal of Life Science and Biomedicine*, 2(3): 92-94.

New MB. 2005. Freshwater prawn farming: global status, recent research and a glance at the future. *Aquaculture Research*, 36: 210-230.

Okafor FU. 2001. *Edible land snails: a manual of biological management and farming of snails.* Splendid Publishers.

Panda A and Misra MK. 2011. Ethnomedicinal survey of some wetland plants of South Odisha and their conservation. *Indian Journal of Traditional Knowledge*, 10: 296-303.

Patil SG and Talmale SS. 2005. A Checklist of Land and Freshwater Mollusca of Maharashtra State. *Zoo's Print Journal*, 20(6): 1912-1913.

Prabhakar AK and Roy SP. 2009a. Ethno-medicinal uses of some shell fishes by people of Kosi river basin of North-Bihar, India. *Ethnomedicine*, 3(1): 1-4.

Prabhakar AK and Roy SP. 2009b. Taxonomic diversity of shell fishes of Kosi region of North-Bihar (India). *The Ecoscan*, 2(2): 149-156.

Rahman MA, Rahman MM, Ahmed ATA. *et al.*, 2008. A survey on the diversity of freshwater crabs in some wetland ecosystems of Bangladesh. *International Journal of Sustainable Crop Production*, 3(4): 10-17.

Rangappa A, Raj Kumar T, Jaganmohan P and Reddy MS. 2012. Studies on the proximal composition of freshwater prawn *Macrobrachium rosenbergii* and *Macrobrachium malcomsonii*. *World Journal of Fish and Marine Sciences*, 4 (2): 218-222.

Reddy BS and Reddy KVS. 2014. Proximate composition of the fresh water prawn *Macrobrachium rosenbergii* in cultured and frozen stage from Nellore Coast, India. *International Food Research Journal*, 21(4): 1707-1710.

Sudha Devi AR and Smija MK. 2013. Analysis of dietary value of the soft tissue of the freshwater crab *Travancoriana schirnerae*. *Indian Journal of Applied Research*, 3(7): 45-49.

Suominen J. 1968. Changes in the aquatic macroflora of the polluted Lake Rautavesi, SW-Finland. *Annales Botanici Fennici*, 5: 65-81.

Swapna MM, Prakashkumar R, Anoop KP. *et al.*, 2011. A review on the medicinal and edible aspects of aquatic and wetland plants of India. *Journal of Medicinal Plants Research*, 5(33): 7163-7176.

Uotila P. 1971. Distribution and ecological features of hydrophytes in the polluted Lake Vanajavesi, S-Finland. *Annales Botanici Fennici*, 8: 257-295.

Varadharajan D and Soundarapandian P. 2014. Proximate composition and mineral contents of freshwater crab *Spiralothelphusa hydrodroma* (Herbst, 1794) from Parangipettai, South East Coast of India. *Journal of Aquaculture Research and Development*, 5: 2.

Watters GT, Hoggarth MA and Stansbery DH. 2009. *The Freshwater Mussels of Ohio*. The Ohio State University Press.

Wosu LO. 2003. *Commercial snail farming in West Africa - A Guide*. Ap Express Publishers, Nsukka - Nigeria.

Ziv G, Baran E, Nam S. *et al.*, 2012. Trading-off ûsh biodiversity, food security, and hydropower in the Mekong river basin. *Proceedings of National Academy of Sciences*, USA. 109: 5609-5614.

Chapter 20

Macroinvertebrate and Macrophyte Assemblages in Wetlands of North Bengal: Benefit Sharing

Dipendra Sharma, Gautam Aditya and Goutam Kumar Saha

Entomology and Wildlife Biology Research Laboratory, Department of Zoology, University of Calcutta, Kolkata

Introduction

Wetlands are considered as crucial links between the land and the water, featured by distinct ecosystem attributes and functions that render a separate identity. The hydrology, soil, vegetation, fauna and associated landscapes of wetlands carry out functions related to the flow of water, nutrient cycling, and sustenance of the biological diversity. The productivity of wetlands stands high among the different ecosystems of the world, which is one of the principal reasons for valuing wetlands as a natural capital. Irrespective of size and geographic location, wetlands serve in maintaining species diversity, supply and renew the air, water and land quality, which are valued services for sustaining the environmental quality and human well-being as well. Although the resource quality of the wetlands depends on the constituent biotic and physical components, the ecosystem services attributable to wetlands remain invariant feature irrespective of the geographical location. On a global context, wetlands are considered as highly productive ecosystems from both ecological and economical perspectives. Over the years the exploitation of wetland resources and the land use pattern of the associated landscapes have lead to the degradation and destruction of wetlands in various parts of the world. In the recent time period, the emergence of the new thoughts in terms of the natural capital and the concept of ecosystem services have set a momentum on revaluation of the resource value of the wetlands. Core to the concept of ecosystem services is the recognition of the economic and environmental values of the components of the wetlands. Incorporating the views of the Ramsar Convention and the concept of ecosystem services, surveillance of the wetlands are carried out emphasizing the biotic and abiotic resource contents and appreciation about the values of the components. Thus monitoring of the wetlands is incorporated in the conservation management programmes in almost all the countries of the world. Monitoring of wetlands thus provide an estimate of the existing components with the levels of exploitation such that the wetland health condition can be evaluated. In view of proper conservation of wetlands, monitoring and assessment of the biotic and abiotic resources are thus being highlighted and implemented.

Considering the importance and significance of monitoring wetlands for the evaluation of the quality of the services, the present compilation provides an overview of the wetlands of three districts situated on northern extremities of the state of West Bengal, popularly known as North Bengal. The topography and the land use pattern of the area remains distinct from the southern district and the tanks, ponds, temporary pools and the irrigated lands include the water bodies that qualify as wetlands. These wetlands are mostly lacustrine in origin and heavily dependent on the monsoon for the hydrological features and the hydrodynamics. The resource state of these wetlands varies with the monsoon and so are the productivity and the species diversity. Despite the significance of the wetlands, little effort has been made to monitor and evaluate the wetlands of this geographical area. Thus an initiative was taken to explore the faunal assemblage in the wetlands in North Bengal along with the physic-chemical parameters of their concerned habitat in view of their conservation, monitoring and restoration. The abundance and distribution of aquatic macroinvertebrate predators that can act as a biocontrol measures for the mosquito vector population can be revealed in spatial scale. Since, macroinvertebrates, are essential component of wetland food webs and responsible for a significant proportion of the secondary production occurring in wetlands (Davies *et al.*, 2008), an overview of the wetland macroinvertebrates will contribute to the development of ecological management through sustainable exploitation of the resources.

Materials and Methods

Study Area and Sampling Technique

The survey of wetlands had been carried out during the year 2012-2013, considering three districts of North Bengal namely Uttar Dinajpur, Malda and Coochbehar. A total of 16 wetlands (Uttar Dinajpur 4, Malda 4 and Coochbehar 8) were sampled during rainy season between June and September, initially followed by repeat sampling technique during the post-monsoon season. The comparison of area covered by water in satellite image and ground truth data can be done through pre and post-monsoon survey of wetlands. The topographic sheets are used to identify the landmarks and the route. In order to characterize the spatial location of the wetlands, the coordinates were recorded using a GPS instrument, which was further fed with the data of L-III satellite images to supplement the information recorded in the topographic sheets. The sampling sites within the wetlands were selected randomly following the principles of replication (Hurlbrt, 1984), such that the space and time intervals are maintained to qualify each of the samples as separate replicate. The random sampling was employed using an "O" frame insect net, made up of nylon net (200 μm) placed in a circular iron frame with a diameter 32.5 cm, attached with a long wooden handle of 1.2 m (Robert *et al.*, 2002; USEPA, 2002) for the collection of macroinvertebrates. The spatial scale sampling is done in three different sites of each wetland making a quadrate of an area of 1m^2 at the depth of 0.5ft and was repeatedly sampled for vegetation and aquatic organisms (Krebs, 1999). The net was browsed along the edge of the wetlands followed by dredging the sediments in both open and structured spaces consisting of macrophytes. In all instances at least three samples were considered for the collection of macroinvertebrates, particularly the aquatic insects. Simultaneous collection of the aquatic vegetations using a random quadrat of 1m^2 was made employing the same net. A multimeter environmental parameter recorder (Multi 340set®, Merck, India) was used to record the water quality parameters like pH, salinity, dissolved oxygen, temperature, conductivity of water in course of sampling. The relevant information such as history of formation of wetlands, total area and depth, socioeconomic value, uses and threats to wetlands and common/endangered species found in the wetlands were collected from the field observation and interpersonal communication with local people. Macrophytes and fringe area plants are also recorded along with their local names. The collected macroinvertebrates and macrophytes were sieved through the nylon net of 200 μm to separate the debris and mud under running tap water. Following collections, the insects and other macroinvertebrates and macrophytes were placed in plastic bags, separately, with aeration, if required. In few instances, the soft bodied insects and other macroinvertebrates were preserved appropriately in specimen containers (Tarsons®), while few specimens of each macrophytes were dried and placed in herbarium for confirmation of identification. The collected macroinvertebrates and macrophytes were brought to the laboratory for confirmation of identification and record of the data for analysis.

The macroinvertebrates were initially identified following Edmondson (1963). The identification of Hemiptera was made following Distant (1906), Coleoptera following Fowler (1912) and Odonata nymph following Kumar (2000), Subramanian (2005) and Nesemann *et al.* (2011). The identification of macrophytes upto the species level was made following Cook (1996).

Data Analysis

On the basis of the data of each collection, the species diversity (Shannon-Weiner index, Krebs,1999) was estimated for the concerned wetlands. The data obtained on the macroinvertebrates were subjected to Canonical Correspondence Analysis (CCA) (CANOCO, ter Braak and Verdonschot, 1995) to provide an overview of the ordination of macroinvertebrates in the background of the environmental factors and macrophytes. Initially a detrended correspondence analysis was applied with the matrix of macroinvertebrate count in each sample to justify the application of unimodal models. As a rule, unimodal species response model can be used provided the gradient length is greater than twice the standard deviation. Since the analysis yielded such results, further application of CCA (Canonical Correspondence

Analysis; ter Braak, 1987; ter Braak and Verdonschot, 1995; ter Braak and Šmilauer, 2002) was used to comment on the macroinvertebrate-environment relationship. The macrophytes were used as a component of the environmental variables along with the water quality parameters. The application of CCA was made following the norms (ter Braak, 1987; ter Braak and Verdonschot, 1995), where the description of the habitat preferences of the macroinvertebrates are predicted following extraction of artificial environmental gradients. The resultant ordination of the species is made as a scatter points against the environmental variables that are represented as arrows in the plot. The strength of the correlations between the environmental variables and the coordination axes is proportional to the length of the arrow (ter Braak, 1987; ter Braak and Verdonschot, 1995). The direction of an environmental arrow can be viewed as axis allowing projection of the species points in order of the species rank in terms of weighted averages in context of those environmental variables (ter Braak, 1987). The purpose of the CCA was to comply with the rapid bioassessment of the wetlands and provide a baseline information for further studies as a part of the environmental biomonitoring using the macroinvertebrates, especially the aquatic insects as key group. The data analysis was carried out following Legendre and Legendre (1998) and Zar (1999) using *XLSTAT* software (Addinsoft, 2010) with requisite samples and replicates.

Observations and Comments

The aquatic insects of the order Hemiptera, Coleoptera and Odonata dominated the macroinvertebrates sampled from the wetlands. Although numerical differences were obvious in the representative samples of wetlands of different districts, the pattern was similar in all the instances with the water bugs being the dominant predators and the snails and dipteran larvae as the prey species. A total 23 species of macroinvertebrates, and 12 species of macrophytes were recorded from the 16 wetlands as shown in Tables 20.1a and 20.1b. The characteristic features of the wetlands of North Bengal along with the macroinvertebrates observed in course of the sampling suggest little differences in the community composition (Tables 20.2–20.7). The invariant composition of the aquatic invertebrate community and the macrophytes of the concerned wetlands provide an impression of the use of the macroinvertebrates for the purpose of biomonitoring of the wetlands irrespective of the geographical location. The species diversity indices of the concerned wetlands remained within a comparable value suggesting the consistency in the richness and abundance of the constituent species (Table 20.8). The ordination of the macroinvertebrates particularly the aquatic predatory insects, against macrophytes and the water quality parameters is presented in Figure 20.1. The four canonical axes explained 80 per cent of variations in the species environment relation (Table 20.9). The total inertia was 0.556 (variance of species dispersion) and the sum of all canonical eigenvalues was 0.351, while the eigenvalue is highest in first canonical axes (0.117). The species environmental correlation is highest in 1st canonical axes (0.930). All four eigenvalues reported above are canonical and correspond to the axes that are constrained by the environmental variable. Summary of the Monte Carlo test indicate the test of significance of all canonical axes is significant ($F=1.495$; $P<0.04$). The first canonical eigenvalue ($\lambda1=0.117$) was 1.69 times the second one

($\lambda 2=0.069$) indicating a clear gradient. The first two canonical axes explains 53.5 per cent ($\lambda 1=33.5$ per cent and $\lambda 2=20$ per cent) of variance in the weighted averages of the species with respect to each environmental variables. This analysis shows a significant correlation of macroinvertebrates with the selected environmental variables in the wetlands of North Bengal, as shown in the biplot (Figure 20.1). However, correlations with axes remained poor for the macrophyte *Ipomoea carnea* while the correlation was maximum for *Salvinia* sp. The biplot shows the correlation between macroinvetebrates with macrophytes such as *Ipomoea carnea, Ipomoea aquatic, Marselia* sp., *Nymphea* sp. and *Pistia* sp. and environmental variables such as Dissolved oxygen, turbidity and salinity. However, the *Laccotrephes* sp., *Microvela* sp., *Cassida* sp., *Lymnea accuminata* and *Lammelidens* sp. show less affinity towards the macrophytes.

Table 20.1a: The List of Macroinvertebrates Sampled from the Wetlands of North Bengal

Sl.No.	*Macroinvertebrate*	*CODE*	*Sl.No.*	*Macroinvertebrate*	*CODE*
1.	*Hirudo medicinalis*	*HRM*	13.	Dragonfly nymph	DFN
2.	*Spherodema rusticus*	*RUS*	14.	Damselfly nymph	DMN
3.	*Spherodema annulatus*	*ANU*	15.	*Vivepera bengalensis*	*BEB*
4.	*Ranatra* sp.	*RAN*	16.	*Pila* sp.	*PIL*
5.	*Hydrometra* sp.	*HDM*	17.	*Lymnea luteola*	*LML*
6.	*Anisops* sp.	*ANI*	18.	*Lymnea acccuminata*	*LMA*
7.	Water strider	GER	19.	*Indopalnorbis* sp.	*IND*
8.	*Laccotrephes* sp.	*LAC*	20.	*Gabbia* sp.	*GAB*
9.	*Micronecta* sp.	*MIN*	21.	*Gyraulus* sp.	*GYR*
10	*Lethoceros* sp.	*LTH*	22.	*Lammelidens* sp.	*LLD*
11.	*Microvela* sp.	*MIV*	23.	Fish	FIS
12.	*Cassida* sp.	*CAS*	24.	Crab	CRB

Table 20.1b: The List of Macrophytes from the Wetlands of North Bengal

Sl.No.	*Macrophytes*	*CODE*	*Sl.No.*	*Hydrophytes*	*CODE*
1.	*Eichhornia crassipes*	EIC	7.	*Salvinia* sp.	SAL
2.	*Pistia stratiotes*	PIS	8.	*Azolla* sp.	AZO
3.	*Ipomoea aquatica*	IPA	9.	*Lemna minor*	LMM
4.	*Ipomoea carnea*	IPC	10.	*Lemna aequinoctialis*	LAS
5.	*Hydrilla* sp.	HRL	11.	*Nymphaea* sp.	NMP
6.	*Marsilea quadrifolia*	MAR	12.	*Vallisneria* sp.	VAL

It is evident from the observations on the wetlands of North Bengal that the aquatic insects of the orders Hemiptera, Odonata and Coleoptera dominate as predators, with abundance of different groups of macroinvertebrates as prey species. The dominance of the predatory insects in wetlands is a reflection of the prospective

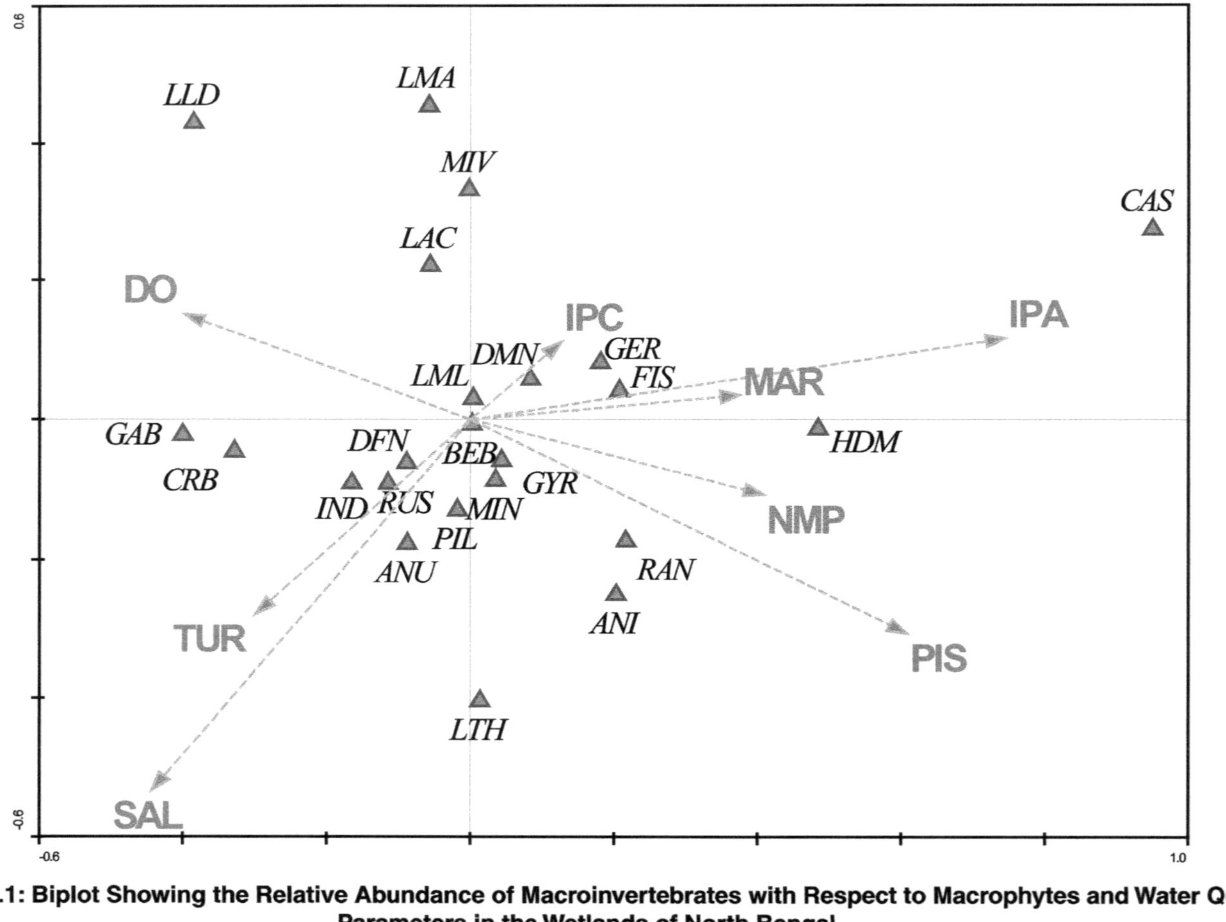

Figure 20.1: Biplot Showing the Relative Abundance of Macroinvertebrates with Respect to Macrophytes and Water Quality Parameters in the Wetlands of North Bengal.

Table 20.2: Characteristics of the Wetlands of Malda District

Characteristics	*Wetland 1*	*Wetland 2*	*Wetland 3*	*Wetland 4*
Name of the wetland	*Amritir Dhap*	*Phular bil*	*Mora kalindi bil*	*Margangi bil*
Type	Cut off mender	Cut off mender	Cut off mender	Cut off mender
Aquaveg covererage	75 per cent	40 per cent	40 per cent	65 per cent
Co-ordinate	25º01'57N88º02'47E	25°13'36N87°54'00E	25°05'41N87°58'30E	25°05'35N88°00'28E
Nature of surrounding land cover	Forest 60 per cent; Settlement 20 per cent; Agriculture 20 per cent	Forest 30 per cent; Settlement 30 per cent; Grass cover 10 per cent; Agriculture 30 per cent	Forest 50 per cent; Settlement 30 per cent; Agriculture 20 per cen	Forest 30 per cent; Settlement 30 per cent; Grass cover 10 per cent; Agriculture 30 per cent
pH	7.27	7.48	7.31	7.42
D.O and Temperature	2.59 mg/lit at 24.3º	3.11 mg/lit at 30.5ºc	1.52m g/lit at 31.9°C	3.81 mg/lit at 32.1°C
Turbidity (NTU)	4.61	1.81	1.95	10.19
Pre and post-monsoon water depth(Appx)	8ft., 20ft	30ft, 50ft.	20ft., 30ft.	12ft., 16ft.
Dominant aquatic flora	*Eichhornia* sp., *Ipomoea aquatic, Marselia quadrifolia*	*Eichhornia* sp., *Lemna major*	*Lemna major, Eichhornia crassipes, Ipomoea aquatic*	*Pistia* sp., *Marsilia* sp., *Eichhornia* sp.
Habitat for migratory birds (Yes= + and No = –)	+	+	+	+

Table 20.3: Aquatic Faunal Assemblage in Wetlands of Malda District

Sl.No.	*Specimens*	*Number of Specimen*			
		Wetland 1	*Wetland 2*	*Wetland 3*	*Wetland 4*
1.	*Hirudo medicinalis*	3	0	0	1
2.	*Spherodema rusticus*	0	1	0	0
3.	*Spherodema annulatus*	0	0	0	1
4.	*Ranatra* sp.	0	0	5	2
5.	*Hydrometra* sp.	1	0	1	0
6.	*Anisops* sp.	0	0	5	0
7.	Water strider	5	2	1	4
8.	*Laccotrephes* sp.	1	2	0	0
9.	*Micronecta* sp.	0	0	6	3
10.	*Lethoceros* sp.	0	0	0	0
11.	*Microvela* sp.	6	13	0	0
12.	*Cassida* sp.	3	0	1	0
13.	Dragonfly nymph	0	2	1	0
14.	Damselfly nymph	12	7	6	9
15.	*Vivepera bengalensis*	15	8	2	11
16.	*Pila* sp.	1	0	2	0
17.	*Lymnea luteola*	7	3	1	6
18.	*Lymnea acccuminata*	2	4	0	0
19	*Indopalnorbis* sp.	0	9	0	0
20.	*Gabbia* sp.	0	33	0	0
21.	*Gyraulus* sp.	1	4	0	10
22.	*Lammelidens* sp.	0	1	0	0
23.	Fish	2	1	1	2
24.	Crab	0	1	0	0

role in the regulation of the prey population and thus the diversity and stability of the aquatic communities. Availability of the insects and prey population is a prerequisite for maintenance of the community structure in freshwater ecosystems, which forms the basis of considering the macroinvertebrates as indicators of the ecosystem health and functioning. Certain indices like the macroinvertebrate based index of biotic integrity are employed in estimating the state of environmental conditions of the aquatic ecosystems of freshwater origin. The integrity and functions of the macroinvertebrates including the aquatic insect predators are highly dependent on the macrophyte abundance. Although the functions of the macrophytes differ with the species and ecological type, the general role as organizers and mediators of macroinvertebrate community structure remains invariant irrespective of the wetlands type and origin. Numerous studies have documented the role of the macrophytes in mediating the interactions between aquatic insect predators and

Table 20.4: Characteristics of the Wetlands of Uttar Dinajpur District

Characteristics	*Wetland 1*	*Wetland 2*	*Wetland 3*	*Wetland 4*
Name	*Chingri bil*	*Musar bil*	*Pautia bil*	Karan dighi
Type	Natural	Natural	Natural	Man Made
Aquaveg covererage	10 per cent	90 per cent	30 per cent	10 per cent
Co—ordinate	25°28'31N88°13'01E	25°28'14N88°14'10 E	25°47'10N87°56'24 E	25°47'43N87°58'30E
Surrounding land cover	Settlement 20 per cent; Agriculture 80 per cent	Settlement 20 per cent; Agriculture 80 per cent	Forest 10 per cent; Settlement 20 per cent; Grasscover 20 per cent; Agriculture 50 per cent	Forest 40 per cent; Settlement 40 per cent; Agriculture 20 per cent
pH	7.65	8.09	6.6	7.45
D.O and Temperature	4.95mg/lit at 32.1°C	3.57 mg/lit at 31.1°C	1.8 mg/lit at 30.4°C	4.21mg/lit at 31.2°C
Turbidity (NTU)	4.22	8.78	5.97	25.82
Pre and post-monsoon water depth (Appx)	4 ft, 6 ft	3 ft, 5 ft	4.5 ft, 5 ft	3 ft, 5 ft
Dominant aquatic flora	*Eichhornia* sp., *Ipomoea aquatic*, *Marselia quadrifolia*, *Nelumbo nucifera*, *Pistia stratoites*	*Eicchornia* sp., *Ipomoea carnea*, *Hydrilla verticillata*, *Vallisneria spiralis*	*Eicchornia* sp., *Ipomoea carnea*, *Nymphaea alba*, *Hydrilla verticillata*, *Chara* sp.	*Eicchornia* sp., *Marselia quadrifolia*, *Hydrilla verticillata*
Habitat for migratory birds (Yes= + and No = –)	—	—	+	+

Table 20.5: Aquatic Fauna in the Wetlands of the Uttar Dinajpur District

Sl.No.	Specimens	Number of Specimen			
		Wetland 1	Wetland 2	Wetland 3	Wetland 4
1.	*Hirudo medicinalis*	1	0	2	0
2.	*Spherodema rusticus*	0	2	3	0
3.	*Spherodema annulatus*	3	1	1	2
4.	*Ranatra* sp.	4	2	3	1
5.	*Hydrometra* sp.	1	0	2	0
6.	*Anisops* sp.	12	2	6	0
7.	*Gerris* sp.	2	1	4	5
8.	*Laccotrephes* sp.	2	1	0	1
9.	*Micronecta* sp.	1	5	2	8
10.	*Lethoceros* sp.	1	0	0	0
11.	*Microvela* sp.	6	13	0	0
12.	*Cassida* sp.	1	0	1	0
13.	Dragonfly nymph	3	1	0	2
14.	Damselfly nymph	9	11	4	7
15.	*Vivepera bengalensis*	21	34	27	14
16.	*Pila* sp.	1	2	1	4
17.	*Lymnea luteola*	7	3	1	6
18.	*Lymnea acccuminata*	0	2	0	1
19.	*Indopalnorbis* sp.	5	3	4	1
20.	*Gabbia* sp.	0	33	0	19
21.	*Gyraulus* sp.	8	3	3	2
22.	*Lammelidens* sp.	0	0	0	0
23.	Fish	2	2	3	1
24.	Crab	0	1	0	2

prey species, as well as organizers of the community. The aquatic insects and other macroinvertebrates and fish species utilize the macrophytes as refuge for safety as well sites for oviposition. The decaying macrophytes act as a source of nutrition for omnivore and detritivore macroinvertebrates. The role of macrophytes in organizing the aquatic community and the interactions of different taxa qualifies them for considering as indicator of the ecosystem health and functions. As a consequence, the macrophyte based index of biotic integrity has been formulated to monitor freshwater ecosystems. The ordination of the macroinvertebrates based on the scaling of the macrophytes portrays the respective inclination towards one or the other species of macrophytes. However, since the data is based on spatial scale studies, the fluctuations in the abundance as a function of the seasons are being concealed. It may be assumed that the intensity of fluctuations in abundance

Table 20.6: Characteristics of Wetlands of the Coochbehar District

Characteristics	*Wetland 1*	*Wetland 2*	*Wetland 3*	*Wetland 4*
Name	Buramansai bil	Nayachora bil	Panisalachari bil	Rasik bil
Type	Riverine wetland	Ox-bow lake	Cut off mender	Riverine wetland
Aquaveg covererage	80 per cent	10 per cent	35 per cent	40 per cent
Co—ordinate	26°19'18N 89°17'38E	26°16'36N 89°21'45E	26°16'48N 89°30'26E	26º24'33N 89º43'24E
Surrounding land cover	Forest 30 per cent Settlement 20 per cent Grass cover 10 per cent Agriculture 40 per cent	Forest 30 per cent Settlement 30 per cent Grass cover 10 per cent Agriculture 30 per cent	Forest 20 per cent Settlement 40 per cent Grass cover 20 per cent Agriculture 20 per cent	Forest 40 per cent Settlement 20 per cent Grass cover 20 per cent Agriculture 20 per cent
pH	6.84	7.39	7.43	7.5
D.O and Temperature	3.05mg/litat 31.4°C	3.58mg/litat 31.0°C	4.12mg/litat 30.3°C	2.37mg/litat 29.4°C
Turbidity (NTU)	4.07	12.2	3.5	10.77
Pre and post- monsoon water depth (Appx)	8ft., 12ft.	25ft., 35ft.	15ft., 20ft.	6ft., 12ft.
Dominant aquatic flora	*Azolla* sp., *Eichhornia* sp., *Marsilia quadrifolia,* *Nymphaea alba*	*Eicchornia* sp., *Salvinia* sp.	*Salvinia* sp., *Lemnaa cquinoctalis,* *Eicchornia sp.,* *Azolla pinnata*	*Eicchornia* sp., *Marselia quadrifolia,* *Salvinia cucullata*
Habitat for migratory birds (Yes= + and No = –)	+	+	+	+

Table 20.7: Characteristics of the Wetlands of Coochbehar District

Characteristics	*Wetland 5*	*Wetland 6*	*Wetland 7*	*Wetland 8*
Name	Soldanga Chara*bil*	Sonarinikutyar*bil*	Sitai*bil*	Khutemara*bil*
Type	Cut off mender	Ox-bow lake	Riverine	Ox-bow lake
Aquaveg covererage	80 per cent	88 per cent	30 per cent	30 per cent
Co—ordinate	26°15'42N89°34'34E	26°22'59N89°28'46 E	26°03'30N89°19'03E	26°10'50N89°32'31E
Surrounding land cover	Forest 30 per cent Settlement 20 per cent Grass cover 20 per cent Agriculture 30 per cent	Forest 20 per cent Settlement 40 per cent Grass cover 10 per cent Agriculture 30 per cent	Forest 40 per cent Grass cover 20 per cent Agriculture 40 per cent	Forest 20 per cent Settlement 20 per cent Grass cover 20 per cent Agriculture 40 per cent
pH	7.67	6.38	7.54	7.7
D.O and Temperature	4.56 mg/lit at 33.1°C	2.37 mg/lit at 29.4°C	5.18 mg/lit at 26.3ºC	3.23 mg/lit at 26.5ºC
Turbidity (NTU)	3.01	10.77	5.71	8.33
Pre and post-monsoon water depth (Appx)	7ft, 12ft.	6ft., 12ft.	6ft, 12ft.	7ft., 10ft.
Dominant aquatic flora	*Azolla* sp., *Eichhornia* sp., *Salvinia cucullata,* *Nymphaea alba*	*Eicchornia* sp. *Marselia quadrifolia,* *Salvinia cucullata*	*Eicchornia* sp. *Nymphaea alba,* *Lemna* sp.	*Eicchornia crassipes,* *Salvinia cucullata*
Habitat for migratory birds (Yes= + and No = –)	+	+	+	+

Table 20.8: Faunal Abundance in Wetlands (n=8) of the Coochbehar District

Sl.No.	*Specimens*	*Relative Abundance (mean ± SE)*
1.	*Hirudo medicinalis*	0.88 ± 0.39
2.	*Diplonychus rusticus*	0.5 ± 0.27
3.	*Diplonychus annulatum*	1.5 ± 0.38
4.	*Ranatra filliformis*	1.63 ± 0.42
5.	*Hydrometra* sp.	0.5 ± 0.27
6.	*Anisops* sp.	2.12 ±0.98
7.	Water strider	0.75 ± 0.49
8.	*Laccotrephes gresius*	0.75 ± 0.25
9.	*Micronecta* sp.	5.13 ± 1.28
10.	*Lethoceros* sp.	0.38 ± 0.18
11.	*Microvelia* sp.	3.25 ± 1.16
12.	*Cassida* sp.	0
13.	Dragonfly nymph	1.63 ± 0.32
14.	Damselfly nymph	6.25 ± 1.51
15.	*Bellamya bengalensis*	16.5 ± 2.1
16.	*Pila* sp.	1.12 ± 0.48
17.	*Lymnea luteola*	3.13 ± 0.72
18.	*Lymnea acccuminata*	0.75 ± 0.31
19.	*Indopalnorbis exustus*	3.5 ± 0.87
20.	*Gabbia* sp.	22.25 ± 7.86
21.	*Gyraulus* sp.	3.5 ± 0.78
22.	*Lammelidens marginalis*	0.5 ± 0.27
23.	Fish	0.88 ± 0.23
24.	*Sartoriana spinigera*	0.5 ± 0.19

may vary among the different species but the pattern may be similar. Since all the macroinvertebrates and the macrophytes considered here are dependent on the water, changes in the water quality will affect all the species, though with varying intensity. Nonetheless, the biplot portrays the relative orientation of the macroinvertebrates including the aquatic insects in accordance with the variations in the abundance of different macrophytes. The results imply that the macrophytes are crucial determinants in the abundance and richness of the macroinvertebrates that can be considered as prospective indicators of the wetland conditions. Use of macroinvertebrates of different taxonomic identity for retrieving environmental conditions has long been applied in various geographical regions, though little effort has been made in Indian context. Compared to the information on the richness and abundance of macroinvertebrates in different freshwater ecosystems in India and specifically in West Bengal, application of the indices based on the macroinvertebrates for environmental biomonitoring is still wanted. In the present

Table 20.9: The Diversity Index of Macroinvertebrates and Macrophytes District-wise and from all Wetlands of North Bengal

A: Individual district wise

District	*H' for Hydrophytes*	*H'even for Hydrophytes*
Malda	–1.7479	0.7034
Uttar Dinajpur	–1.9601	0.7888
CoochBehar	–1.6552	0.6661

District	*H' for Macroinvertebrates*	*H'even for Macroinvertebrates*
Malda	–2.6549	0.8354
Uttar Dinajpur	–2.5627	0.8064
CoochBehar	–2.3961	0.7540

B: Diversity index from all 3 districts of North Bengal

District	*H' for Hydrophytes*	*H'even for Hydrophytes*
all 3 district	–2.2141	0.8910

District	*H' for Macroinvertebrates*	*H'even for Macroinvertebrates*
all 3 district	–2.5399	0.7992

Table 20.10: The Summary of CCA of the Macroionvertebrate and Macrophytes

Summary of CCA of Macroinvertebrates of North Bengal wetlands

Axes	*1*	*2*	*3*	*4*	*Total Inertia*
Eigenvalues :	0.117	0.069	0.058	0.036	0.556
Species-environment correlations :	0.93	0.901	0.912	0.916	
Cumulative percentage variance					
of species data :	21.1	33.5	43.9	50.5	
of species-environment relation:	33.5	53.1	69.6	80.6	
Sum of all eigenvalues :					0.556
Sum of all canonical eigenvalues :					0.351
Summary of Monte Carlo test					
Test of significance of first canonical axis:			eigen value = 0.117		
			F-ratio = 1.874		
			P-value = 0.4120		
Test of significance of all canonical axes :			Trace = 0.351		
			F-ratio = 1.495		
			P-value = 0.0420		

perspective, the rapid bioassessment, though at a small scale, provided information on the use of the aquatic insects and other macroinvertebrates in environmental biomonitoring to assess the changing pattern of the biotic communities in the freshwater system. The use of macroinvertebrates including aquatic insects and the associated macrophytes in the biomonitoring of the wetlands can be a feasible way for retrieving the community conditions and ecosystem health.

Wetlands can be considered as natural capital enriched with different natural resources that provides essential ecosystem services to sustain human well being. The constituent elements of the wetlands like the macrophytes and the macroinvertebrates are essential bioresources that sustain human food security and livelihood. Species diversity of the wetlands is essential component to assess the ecosystem health condition, and thus the management of the natural resources. Recognition of ecological services provided by wetlands has stimulated renewed efforts to protect, manage, and construct them (Mitsch *et al.*, 1998; Zedler, 2006). Wetlands are created and restored for many functions, including wildlife conservation and removing pollutants (Mitsch *et al.*, 1998; Zedler, 2006). In many instances the continuing anthropogenic interferences and natural calamities causes alteration and degradation of the wetland conditions. In order to assess the changes and measure the extent of degradation and deviation of the equilibrium condition of wetland ecosystem, the use of the constituent elements can be a feasible option. The present commentary provides elementary but essential information on the common and abundant macroinvertebrate and macrophytes of the wetlands of North Bengal, which can be used in retrieving the function and health for facilitating appropriate environmental management for sustenance of the wetlands.The observed data may be useful for monitoring the wetlands of North Bengal for facilitating the ecosystem services offered by these wetlands and environmental management of the rural and urban areas surveyed. Further works in this regard would justify these conclusion and help in remediation of the degraded wetland conditions elsewhere in the state. Similar studies can be carried out on the wetlands on other districts of West Bengal, to document the extent of diverse biota that can be used as indicators of the ecosystem functions and services.

Conclusion

From the results it is evident that the surveyed wetlands of North Bengal host substantial number of macroinvertebrate that contribute to the ecosystem services provided by the wetlands. The functions of wetlands are enhanced by the presence of these macroinvertebrates that create a highly diverse condition for sustenance and maintenance of ecosystem services. The future development of the wetlands as conservation sites would require the use of these macroinvertebrates and their appropriate application to maintain integrity of the wetlands. Many of the macroinvertebrates collected from these wetlands are possible indicators of the wetland conditions and thus their use in the monitoring of the wetland conditions can be useful option to enhance ecosystem services rendered by these wetlands. The data obtained on diversity and indices of macroinvertebrates as is commonly collected in routine biomonitoring for the calculation of indices showed usefulness for the understanding of distribution trends of the species in the wetlands of North Bengal. The observed data provides a meaningful portrayal of the macroinvertebrate community in wetlands of West Bengal in general and thus this data can be useful for considering ecological restoration of degraded wetlands, elsewhere.

Summary

Wetlands are recognized as valued ecosystems providing multiple services for ecological integrity and human well-being. Consequently conservation management of wetlands is adopted to sustain the environmental quality and the ecosystem services. The valuation of wetlands as natural capital is made on the basis of the water quality and the constituent biota, which facilitates implementation of steps for conservation. Thus assessment of the biotic component of the wetlands is considered as an integral part of the monitoring wetlands. In conformation with the objectives of wetland monitoring, characteristic biota of selected wetlands of three northern districts of West Bengal were evaluated. The observations on the biota of the wetlands will provide the necessary information for continuance of the conservation efforts in the concerned geographical area. The wetlands were characterized by the consistent presence of 23 different species macroinvertebrates, fish and macrophytes with varied levels of relative abundance. In all instances the numerical abundance of the insects were observed with least variation in the macrophyte species. Using canonical correspondence analysis, the affinity of the macroinvertebrates for the macrophyte species could be deduced. The relative role of the macrophytes towards the abundance of the macroinvertebrates portrays the importance of the macrophytes as organizers of wetland invertebrate community. In view of the observations on the macroinvertebrates, fish and macrophytes, it is reasonable to conclude that the wetlands of the studied area are important source for conservation of the species concerned. Availability of the biotic resources in the wetlands confirms the importance of the wetlands in providing the essential services for the sustenance of human well-being and should be considered in the conservation of the wetlands concerned. Although systematic studies on the wetlands of the northern districts of West Bengal are yet to be accomplished, the significance for monitoring the wetlands should be initiated to document the present status and predict the future state of the wetlands so that the necessary steps for the wetland management can be justified. The present compilation may not be complete information on the wetlands of the North Bengal, but definitely provides an overview to highlight the importance of the wetlands in the districts concerned. While records reveal little or poor information about the wetlands considered in the present study, we hope that the information is truly useful for the conservation efforts and highlighting the importance of the wetlands of the concerned geographical area.

Acknowledgements

The authors acknowledge the Head, Department of Zoology, University of Calcutta for the facilities provided to accomplish the research work. DS thankfully acknowledges the help of Dr. Nitai Kundu and Mr. Tapan Saha (Senior Scientist, IESWM, Dept. of Environment, Govt. of West Bengal) for providing the opportunity for the study of wetlands of West Bengal as a Project Scientist in NWIA project.

References

Addinsoft SARL. 2010. XLSTAT software, version 9.0. Addinsoft, Paris, France.

Cook CDK. 1996. *Aquatic plant Book*. SPB Academic publishing, The Hague, Netherlands.

Davies B, Biggs J, Williams P. *et al.*, 2008. Comparative biodiversity of aquatic habitats in the European agricultural landscape. *Agriculture, Ecosystems and Environment*, 125: 1-8.

Distant WL. 1906. *The fauna of British India including Ceylon and Burma, Rhynchota*, Vol III, Taylor and Francis, London, UK. pp. 503.

Edmondson WT. 1963. *Ward and Whipple freshwater Biology*, John Wiley and Sons. New York, USA, pp. 917-940.

Fowler WW. 1912. *The fauna of British India including Ceylon and Burma. Coleoptera. (general Introduction, Cicindelidae and Paussidae)*. Taylor and Francis, London, UK. pp. 529.

Hurlbert SH. 1984. Pseudoreplication and the design of ecological field experiments. *Ecological Monographs*, 54(2): 187-211.

Krebs CJ. 1999. *Ecological Methodology*. II ed. Benjamin Cummings, CA. USA. xv : 620.

Kumar A. 2000. *Odonata (Larva), Wetland Ecosystem Series 2: Fauna of Renuka Wetland*. Zoological Survey of India, Kolkata. pp. 55-62.

Legendre P and Legendre L. 1998. *Numerical ecology*, 2nd English edn. Elsevier, Amsterdam.

Mitsch WJ, Wu X, Nairn RW. *et al.*, 1998. Creating and restoring wetlands: a whole-ecosystem experiment in self-design. *Bioscience*, 48: 1019-1030.

Nesemann H, Shah RDT and Shah DN. 2011. Key to the larval stages of common Odonata of Hindu Kush Himalaya, with short notes on habitats and ecology. *Journal of Threatened Taxa*, 3(9): 2045-2060.

Robert V, Golf GL, Ariey F and Duchemin JB. 2002. A possible alternate method for collecting mosquito larva in rice fields. *Malaria Journal*, 1: 4.

Subramanian KA. 2005. *Dragonflies and Damselflies of Peninsular India-A Field Guide*. E-Book of Project Lifescape. Centre for Ecological Sciences, Indian Institute of Science and Indian Academy of Sciences, Bangalore, India.

ter Braak CJF. 1987. The analysis of vegetation-environment relationships by canonical correspondence analysis. *Vegetatio*, 69: 69-77.

ter Braak CFJ and Šmilauer P. 2002. CANOCO reference manual and canodraw for windows user's guide: software for canonical.

ter Braak CJF and Verdonschot PFM. 1995. Canonical correspondence analysis and related multivariate methods in aquatic ecology. *Aquatic Sciences*, 57(3): 255-289.

USEPA. 2002. Methods for Evaluating Wetland Condition: Developing an Invertebrate Index of Biological Integrity for Wetlands. Office of Water, U.S. Environmental Washington, DC. EPA-822-R-02-019 Protection Agency.

Zar JH. 1999. *Biostatistical Analysis*, IV ed. New Delhi, India: Pearson Education Singapore Pvt. Ltd. (Indian Branch). pp. 663.

Zedler JB. 2006. Why are wetlands so valuable? *Arboratum leaflets*, (10): 1-2.

Chapter 21

Fish as Indicator of Metal Pollution in the Waste-Fed Ponds of East Kolkata Wetlands

Paulami Maiti* and Samir Banerjee

Aquaculture Research Unit, Department of Zoology, University of Calcutta, Kolkata

Introduction

Aquatic system is the ultimate repository for all the environmental pollutants. Of all these contaminants, heavy metals exert the most deleterious effect on the living world. These persistent, non-degradable elements accumulate in water, soil and every trophic level of the food chain leading to both ecosystem degradation and health impairment of the inhabiting organisms. Heavy metals always occur in water bodies due to natural weathering processes but with regular dumping of domestic and industrial waste, the East Kolkata Wetlands has become a huge repository of both organic and inorganic pollutants. The quantum of waste adds enormous contaminants to the environment and poison to its consumers where human being is also not spared. Concern over metal pollution, for the people of Kolkata, is due to their daily consumption of metal contaminated fish extensively grown in these wastewater wetlands. Copper and zinc are abundant in domestic waste while agricultural and industrial runoffs are loaded with excess xenobiotics of which metals form a significant part. As food is the main source of metals in terrestrial animals, a set of possible hazards thus seriously limit the use of wetland cultivated fish. Therefore, metal pollution in aquatic system warrants our attention. The impact of metal pollution depends upon the source and duration of waste

* Present Address: Lady Brabourne College, Kolkata.

input while the presence of organic and inorganic particles determines the extent of pollution and also their bio availability to the residing organisms.

Partitioning of Metals in Aquatic Ecosystem

Metal pollution of aquatic ecosystem is often more obvious in sediments, than in the overlying water as sediments are the most important sink for all the persistent pollutants that play important role in their recycling (Linnik and Zubenko, 2000). Sewage effluents contain higher concentration of both particulate and dissolved matters. This allows metals to get rapidly precipitated into the sediments by mixing with different organic and inorganic ligands associated with fine particles of silt, where these remain permanently accumulated. In sediments metals are concentrated by bacteria and other microbes that are further ingested by the benthic organisms, from where these move on to the higher trophic level especially fish. In water, metals remain either in suspension, particulate or in soluble form but the labile fraction are more harmful. Metal ions in water are readily absorbed by the planktons, which are important fish food organisms. Thus, fate of metals depends on their partitioning between dissolved and particulate phase that significantly affect the final deposition of metals in tissues (Paulson *et al.*, 1984).

Route of Entry and Fate of Metals in Fish Tissues

Aquatic organisms may acquire metals in body directly from the water via gills or through their food. Thus these non-degradable elements either accumulate or biologically magnify in the food chain, potentially affecting each trophic level. It has been observed that tissue metal level of an animal is directly related to its ambient concentration and that in its food organisms (Bervoets *et al.*, 2001). Aquatic organisms can concentrate metals in their body much higher than those found in either water or sediment. Route of metal entry in the body of fish (food, skin or gills) determine their distributional pattern (Begum *et al.*, 2009). Gut and gills are the primary routes of entry while liver and kidney are the major site for storage although metals can be significantly stored in other organs as well (Heath, 1995).

Essential metals are homeostatically regulated in the living organisms while the non-essential ones only accumulate in proportion to their concentration in the medium. Fish thus act as suitable biomarkers that indicate the extent of aquatic pollution. So food habit and the internal regulatory mechanism for metals determine the bioaccumulation pattern in fish. Some metals are preferentially accumulated over the others and their complex interaction with other metals manifest synergistic or antagonistic effects. Toxic effect of the metals depends not only on the differences in the dynamics of its distribution but also on several intracellular factors. All these interrelated factors determine the elemental concentration in the system and their relative availability, transport and toxicity to aquatic organisms.

Toxicity of metals in fish is largely influenced by season, temperature, hardness, and pH of water and the amount of organic and inorganic ligands in the aquatic environment (Kotze, *et al.*, 1999). Besides microhabitat utilization, feeding habit, age, sex of species also determine the accumulation pattern of metals in a biological system. The accumulation of metals in various intracellular ligand pools is determined by the relative rates of metal binding and release, while the rate of binding for a given ligand pool is determined by the relative binding affinity and size of the pool.

Metals as Contaminants

Metal ions and their complexes have the intrinsic property to cause injury to living organisms. These have potential mutagenic, teratogenic and carcinogenic effects and act by reacting with sulphur donor atoms of proteins, resulting in enzyme inactivation and destabilization of biomolecules (More *et al.*, 2003).

Copper and zinc are essential elements that form integral components of protein in biological system (Sivaperumal *et al.*, 2007). These have a definite nutritional and metabolic role but only at low concentration. However, at higher level toxic effects are manifested. Both these are common constituents in domestic sewage as large quantities of these metals are excreted through human faeces (Lock, 1994).

In mammals, copper deficiency leads to anemia, vascular abnormalities and depigmentation of hair. When accumulated in excess in fish, high concentration of the metal damages gills, severely impairing ion regulation and gas transfer across its membrane. It induces Wilson's disease. The metal also affects metabolism, disrupts renal function and interferes with growth rate and reproduction in aquatic animals (Shrivastava, 2009).

Zinc pollution in aquatic system occurs from galvanizing factories and those associated with synthesis of paint, plastics, rubber, cosmetics and pharmaceuticals. Besides, industries related to manufacture of electronic goods, plating, fertilizers and fungicides release huge amount of this metal in the environment. However, the stability of its compounds in aquatic system tends to be lower than that of copper, lead or cadmium. It usually remains associated with organic and inorganic compounds as hydrated ions or gets adsorbed and occluded onto them.

The metal acts as an important cofactor for more that 300 enzymes but excess of it induces gill damage, impairing respiration, leading to hepatic tissue damage, loss of equilibrium or convulsion in fish. Higher exposure to environmental zinc cause inflammatory edema in fish gills. However, zinc shows protective effect against toxicity associated with cadmium and lead (Mastan, 2014).

Lead and cadmium are non essential metals that are cumulative poisons operating at very low concentration. Lead exists in a number of valence states but majority of its salts are only sparingly soluble. The metal usually remains associated with suspended solids that form colloidal complexes with organic ligands. Road runoff, mining and effluents from smelting and battery factories generally pollute the waterways with excess of the metal. In fish, it can decrease survival and growth rates, development and metabolism, in addition to increased mucus formation (Burger *et al.*, 2002). Lead induces haematological changes leading to anaemia by inhibiting the functions of enzyme ALAD in red blood cells. In human being, it induces lesions and also affects nerves.

Cadmium compounds are mainly used in rechargeable batteries that are rarely recycled and often dumped together with household waste. Besides, electroplating, smelting factories are instrumental for increased level of the metal in the environment. Cadmium is extremely harmful and even minute concentration of the metal affects pancreas, inducing hyperplasia of secondary gill lamella, gill necrosis and skeletal deformities. It also reduces blood hemoglobin, causes kidney damage and is a general respiratory poison (Kumar and Singh, 2010).

Materials and Methods

Fish as a source of cheap protein makes it imperative to assess the level of trace elements in their tissues in view of public health importance. Moreover, various species were evaluated to determine their suitability as sentinels for monitoring of metal pollution in sewage fed system.

Study Site

The East Kolkata Wetlands (22° 27′ N,88° 27′ E) encompassing 12,500 hectares of areas, fringing the eastern part of the city are the inter-tributary, salt water marshes of the spill area of the former Bidyadhari river. In 1990, United Nations Environmental Programme, declared these wetlands as high priority conservational sites while in August 2002, the area was designated as the Ramsar Site (Kundu *et al.*, 2008).

These low lying areas daily receive city sewage, storm water runoff and effluents of about 11,000 industries where the average rate of organic loading varies between 20-70 kg per hectare. The wetlands being the largest of its kind in Asia is employed in commercial pisciculture that uses the organic load of the sewage water as fish nutrients. Waste generated in and around Kolkata are fed into the underground sewers that drain the sewage to several pumping stations. Open surface outfall channel networks act as conveyor belt distributing the untreated waste into about 308 fish ponds or *bheries.* These act as oxidation ponds where the organic load gets naturally purified. The nutrients recovered are thus recycled for pisciculture. With an annual turnover rate of 15,000 million tons of fish, the wetlands act as a resource recovery system where waste is naturally recycled into wealth providing livelihood for thousands of people (Kundu *et al.*, 2008).

Study Design

Study I : An extensive study was carried out for a period of five years for quantitative estimation of copper, zinc, lead and cadmium in the liver, kidney, muscle, skin, gills and intestine of commercial fish species widely grown in the East Kolkata Wetlands (Maiti and Banerjee, 2001, 2002, 2005, 2007, 2010, 2011, 2012). As accumulation of the metals depend largely on the trophic level and nutritional status of fish hence those selected for the study are of various food habits and occupy different niche of an pond ecosystem. Among the three Indian major carps, *Labeo rohita* is a column feeder, *Catla catla,* a surface planktivore while *Cirrhinus mrigala* is a benthivore fish. Exotic, omnivore *Oreochromis nilotica* is the most abundant species in the organically loaded water, where it is usually auto stocked. *Channa punctutus* is an endemic, predatory fish that inhabits the pond bottom. All these fishes are common choice as food for the local people.

Fish with various body weight inhabiting different niches and with varied feeding habits indicates variation in accumulation profile. The assimilation efficiency of metals in tissues depends on size, sex, feeding behavior, surface to volume ratios and concentrations of enzymes in body of fish (Authman, 2008). Thus, metal levels vary as a function of their dietary difference and body size, which is closely related to their growth and metabolism (Moriarty *et al.*, 1984). So far the present investigation is concerned, six different body weights of fish (Gr. 1 = 25 ± 5 g; Gr. 2 = 50 ± 10 g; Gr. 3 = 100 ± 20 g,; Gr. 4 = 200 ± 40 g; Gr. 5 = 400 ± 80 g; Gr. 6 = 800 ± 160 g) are chosen for the quantitative accumulation of the metals in them.

Fishes were caught, weighed, stored and transported to laboratories for quantitative analysis of the metals in their tissues. They were dissected and the different tissues were weighed accurately to 1g each for acid digestion. A modified wet digestion procedure was used to prepare biological samples for the determination of metals according to the procedure of Chernoff (1975). Samples of fish tissues were taken in Borosil hard glass test tubes and for each gram of the sample, 5 ml of conc. HNO_3 acid was added and then digested overnight at room temperature. The mixture was placed in a hot plate at 85 ± 5°C and 5 ml (Conc. Sulfuric acid: Perchloric acid at 3: 2 ratio) was then added to it. The digestion was carried out until the mixture turned into a pale transparent solution. It was cooled and filtered through an acid soaked filter paper and was adjusted to the required volume, with distilled water. Metals are detected in atomic absorption spectrophotometer (Varian AA.575).

The actual concentration of metals was calculated on the basis of total amount of the sample taken and expressed in μg/g. The accuracy of determination was checked by a standard reference material, and comparison between metal concentrations in the tissues of the fish with various body weights were performed by one-way analysis of variance ANOVA (Gomez and Gomez, 1984). Metal concentrations in water and soil were carried out following standard methods of APHA, (2005) and Nafde *et al.* (1998) respectively.

Study II: In an another experimental design, sewage input in fish ponds was stopped for six months to avoid exposure of metals in fishes. Metal level in fish tissues were quantified before and after the study period. This was designed to study or understand possible excretion of metals from the body of fish in the relatively uncontaminated environment.

Results and Discussion

In the present investigation, higher concentration of copper, zinc and cadmium are observed in the soil while lead, is more in the overlying water (Table 21.1) as compared to the maximum permissible limits for metals in the environment as per WHO standard (Table 21.2).

Copper has the strongest affinity for organic particulate matter hence gets quickly absorbed in the soil sediments. Zinc and lead have intermediate affinity for organic ligands so these are equally abundant in both water and soil fractions. Affinity of cadmium for particulate matter is however quite low. It has been observed that sediments accumulate comparatively large amount of the metal than the overlying water and so metal concentration in water increases near the pond bottom. Benthic organisms that depend on the pond substratum for food, concentrate more copper than those residing in the upper strata.

Table 21.1: Concentration of Metals in Water (Maiti and Banerjee, 2012)

	Copper	*Lead*	*Zinc*	*Cadmium*
Fresh water *(µ g/g)	Trace	Trace	Trace	Trace
Sewage Water *(µ g/g)	10.97 ± 2.87	93.33 ± 9.39	16.44 ± 5.59	2.23 ± 0.74
Pond sediment (µ g/g)	41.00 ± 5.00	32.00 ± 2.18	60.80 ± 9.11	3.48 ± 1.11

* Mean ± SD of 5 different observations

Table 21.2: Maximum Permissible Limits for Metals in the Environment (WHO Standards)

Metal	*Standard for Discharge of Metals in Inland Water (mg/l)*	*Median International Standards for Metals in Fish Flesh (µg/g wet weight)*
Copper	15.00	20.0
Lead	0.10	2.0
Zinc	15.00	45.0
Cadmium	0.01	0.3

Metals generally enter the body of fish either through food or are absorbed in ionic form through the gills. In fish, copper is taken up maximally from food that remains accumulated in fish liver. The metal induces the synthesis of metallothionein proteins in tissues especially in liver and gills. The protein maintains homeostasis of essential metals and plays an important role in detoxification of non essential ones. These proteins protect tissues from potential damage by metal toxicants. Copper is a strong inducer of metallothionein, while cadmium and zinc induces metallothionein synthesis at lower level. At higher exposure, the capacity of the metal to bind with the protein decreases.

Studies involving quantitative estimation of metals in the various fish species of East Kolkata Wetlands, indicate that all the species under study showed much higher accumulation of copper in the hepatic tissues compared to the others and it increases with increasing body weight of fish. Liver acts as a storage organ for copper that chronically removes it from other tissues (Taweel *et al.*, 2011). This reveals its distinct metabolic role in detoxification which allows other tissues to accumulate low levels of the metal. Fish depending on detritus or on live food organisms in a polluted system accumulate higher amount of copper in their intestine besides the hepatic tissue.

Copper accumulates maximally in omnivorous *Oreochromis nilotica*, predatory *Channa punctatus* and benthivore *Cirrhinus mrigala* as pond sediments in a polluted system is over laden with organically complexed copper (Figure 21.1). In contrast, *Catla catla* accumulates much less as it resides on the upper part of the water column where the metal level is low while in column feeder (*Labeo rohita*) the metal level is intermediate. Disturbances created by benthic fish at pond bottom allow metals to dissolve in water which gets available to the column feeders.

Interestingly, accumulated metals in liver are continuously excreted via bile which is however immediately reabsorbed from the intestine. Thus, only a small portion of the stored metal is actually depurated which accounts for high copper content in both liver and intestine. In *Channa punctatus* high concentration of copper has been observed in muscle, however kidney seems to be not a preferred site for the metal (Figure 21.1).

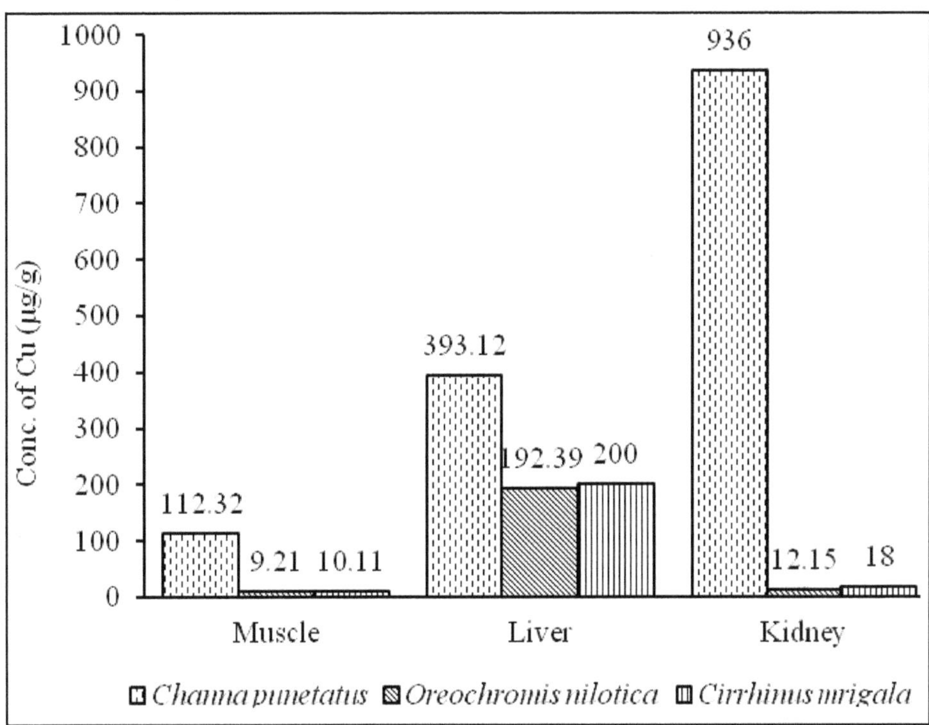

Figure 21.1: Accumulation of Copper in Fish Tissue.

Zinc is absorbed both from food and water and fish exposed to elevated levels of zinc, substantially accumulate the metal in liver and muscle. Zinc taken up by gills is actively transported across the gill membrane where it competes with calcium ions. Skin and gills being in direct contact with water also act as potential uptake and elimination sites for the metal ions. Zinc is a weak inducer of metallothionein and at chronic exposure synthesis of the protein is largely inhibited. With age, liver, gills and kidney show moderate rise in zinc level in most of the species under study. Zinc content in the tissues is positively correlated with body weight of *C. catla* and *L. rohita* which suggest that in a contaminated system uptake of metals exceeds its excretory rate.

According to our investigation, kidney shows higher affinity towards accumulation of zinc especially in *Catla catla and Channa punctatus* (Figure 21.2).

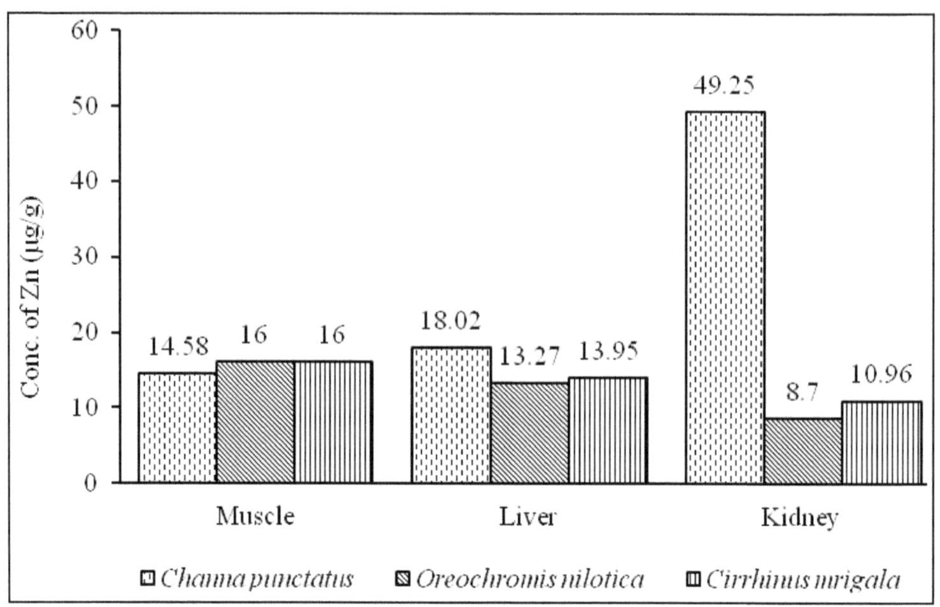

Figure 21.2: Accumulation of Zinc in Fish Tissue.

Surface planktivore, *C. catla* accumulates large amount of the metal in liver, kidney and gills that increases with body weight. This is due to its planktonic diet, as zinc is readily absorbed by the planktons. In *C. catla,* the metal accumulates maximally in gills that increases as the fish grows in size. Normoxic condition of the surface water probably pumps the available metal, which continues to accumulate in the organ (Hughes and Floss, 1978). Besides, large surface area of its gill epithelium facilitate easy uptake of the metal.

However, in benthivore *C. mrigala,* zinc was noted to be maximally accumulated in 25 g of body weight which later showed signs of depuration with age. Juvenile fishes obtain higher proportion of zinc from their planktonic diet but active homeostatic regulation in this bottom dwelling fish depurates the stored metals from the tissues.

Metals taken through food or gills are either removed by faecal excretion or by the gills. Although skin, intestine and gills act as both storage and depuration sites, the dietary zinc is generally excreted through gills. Pourang (1995) opines that kidney is the main site for accumulation and detoxification of essential metals rather than an excretory organ in teleost.

Lead and cadmium are ubiquitous environmental poisons. Water borne metals are taken up passively by fish and the final concentration is dependent on the extent and amount of exposure and accumulation strategy of the particular fish species. It is directly proportional to increased background level of contaminants.

Observation reveals that predatory *C. punctatus* show maximum accumulation of the metal in muscle and liver. In other species, its concentration is closely invariant in almost all the tissues that shows very little rise with age (Figure 21.3). Lead does

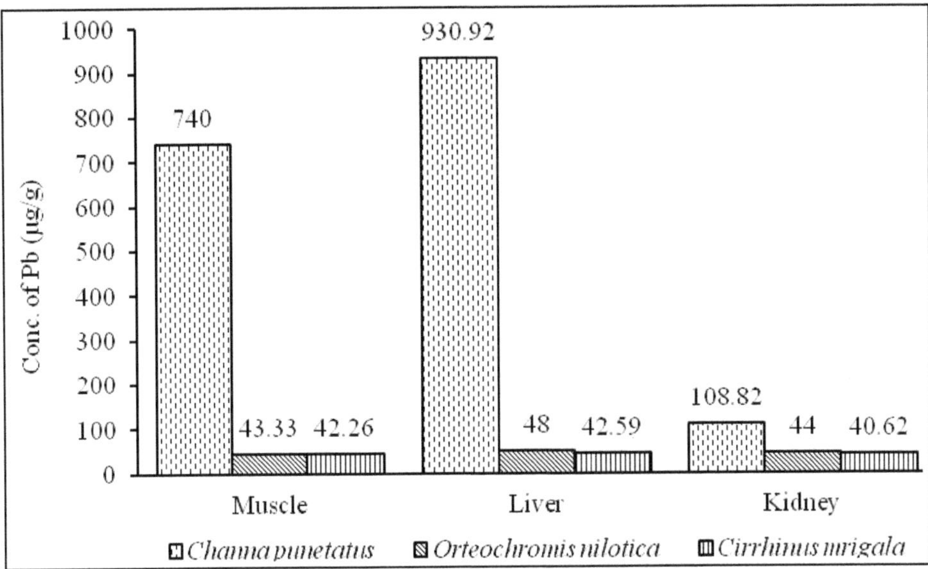

Figure 21.3: Accumulation of Lead in Fish Tissue.

not induce metallothionein synthesis and accumulates mostly in scales, kidney, bones, and liver that tends to contaminate adjacent muscular tissues (Schmitt and Finger, 1987). Skin and gills are also potential storage sites for lead. Some benthic fish accumulate more metal in their juvenile stages which is however depurated later. Younger fish with higher surface to volume ratio and higher metabolic rate are higher accumulators, compared to the larger ones.

Cadmium induces metallothionein formation at lower level. At higher ambient concentration, cadmium binding capacity of the protein decreases, considerably. Cadmium uptake is non-linear and the accumulation factor decreases with increasing background concentration. Besides the metal has lower affinity for organic ligands, and are often passively taken up by the tissues which is independent of the body size of the fish (Bohn and McElroy, 1976). However, it has been observed that concentration of cadmium in the fish increases with the metal concentration in benthic fish food organism. Its rate of accumulation and depuration in biological system is extremely low and its final concentration is dependent on the extent of environmental exposure. From the present observation, it follows that cadmium concentration in fish tissues remains low and invariant in all the species that rarely fluctuate with increasing body weight. (Figures 21.4 and 21.5) This observation is in accordance with Ekeanyanwu (2011). However, *Channa punctatus* is an exception as it significantly concentrates the metal in its muscle and kidney (Figure 21.4). Murphy *et al.* (1978) found that concentration of cadmium in fish is highly variable and are not under strict physiological control. Although cadmium is slowly accumulated in animal tissues the metal has always high concentration factor.

Heavy metal contaminated fish from East Kolkata Wetlands can predispose consumers to toxicity. Measures to eliminate metals from fish tissues require routine

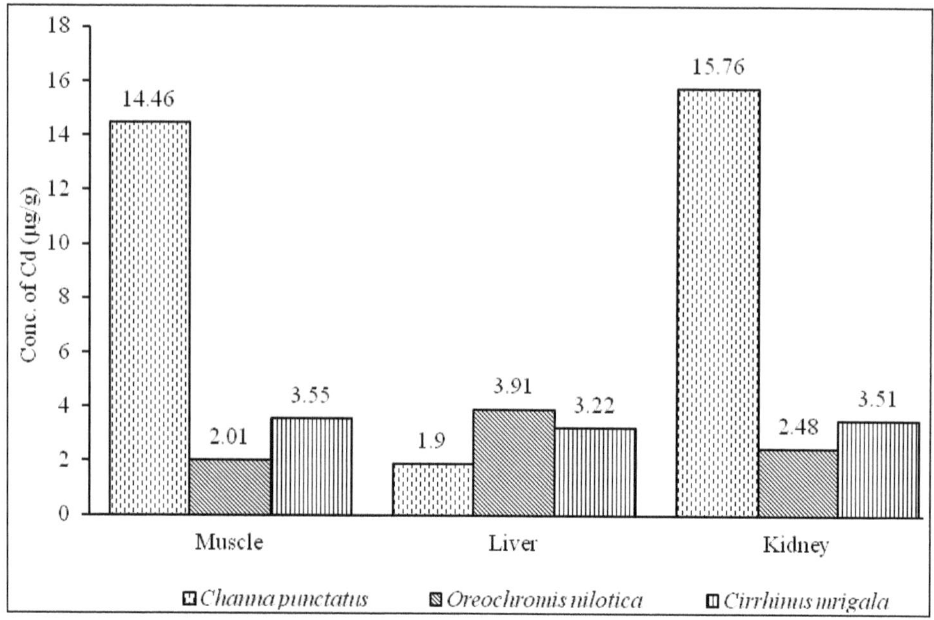

Figure 21.4: Accumulation of Cadmium in Fish Tissue.

monitoring and should be considered as a priority. Thus a study was undertaken in which sewage input in fish ponds were stopped for six months. After this period, when water and soil samples were analyzed for metal concentration in them, it was observed that copper, zinc, cadmium and lead concentration in water decreased considerably but those bound to the soil fraction did not show any appreciable change (Table 21.3). When fish tissues were analysed, copper showed complete clearance excepting in liver and intestine. Depuration rate was further slow in benthic fish due to the constant availability of food borne metal from the sediments. Elimination rate of copper from the intestinal mucosa and that of lead from gills and fins was also very slow. It has been observed that although lead accumulation in fish tissues is dependent on background level but its depuration rate is largely dependent on the overall body concentration (Schulz, 1974). Metal accumulation in any tissue is a balance between its uptake and excretion. In metal free environment rate of depuration exceeds its uptake while with high ambient concentration uptake exceeds resulting in increased body burden of metals.

Muscles of *L. rohita* and *O. nilotica* have shown rapid elimination of lead but liver and kidney of *C. catla, O. nilotica* and *C. mrigala* retained the metal. Bile eliminates sulphydryl reactive metals especially lead and cadmium through the formation of glutathione complexes (Table 21.3).

Zinc depuration from fish tissues was higher in the altered environment as mucus in gills and skin helps in considerable removal of metals. Fish showed retention of lead and cadmium in the kidney and zinc in the gills. Moreover, zinc is retained in *C. catla* even in the uncontaminated environment. Retention of metals, even at lower ambient concentration may be due to their binding in the non-

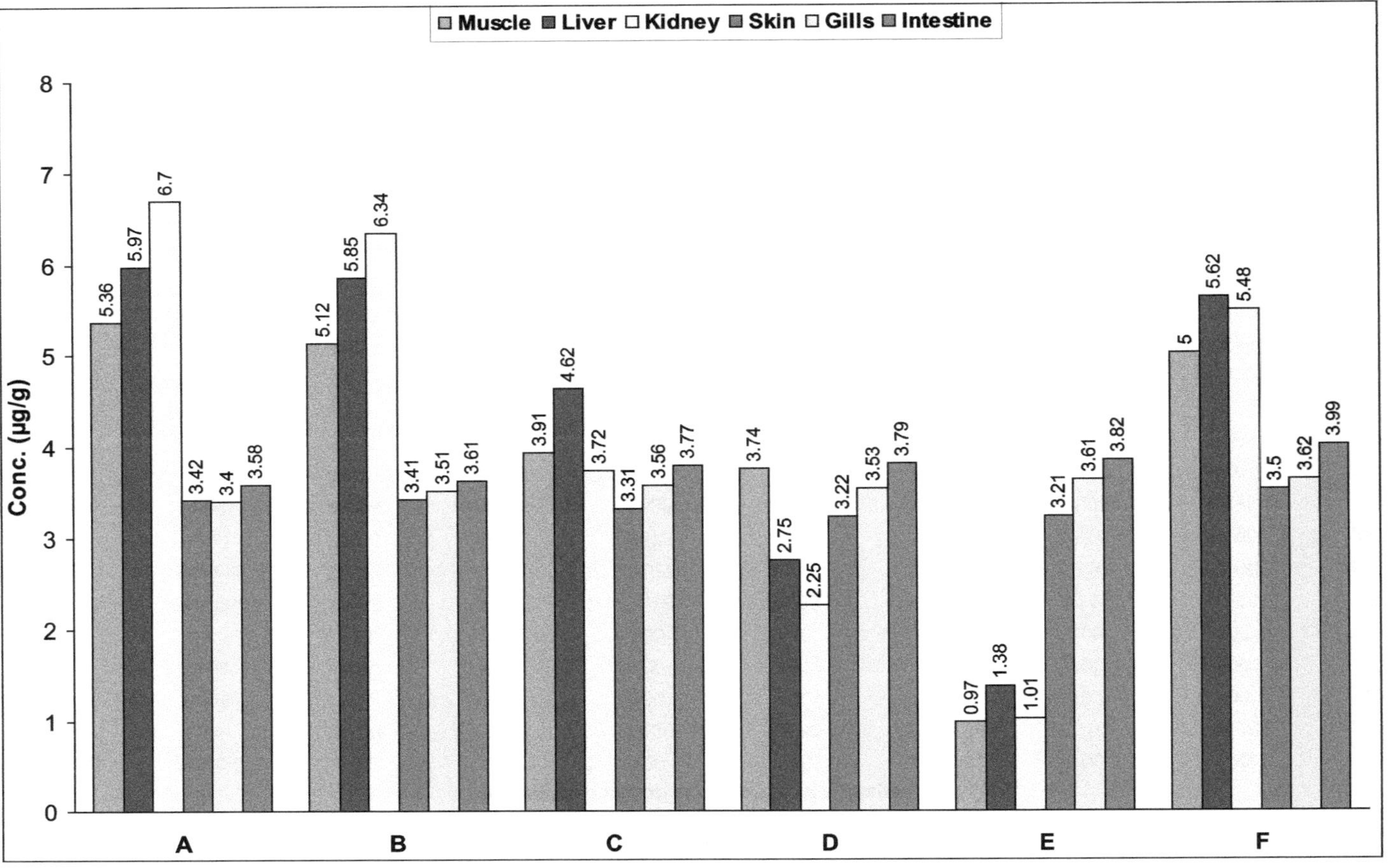

Figure 21.5: Accumulation of Cadmium in different Tissues of *Oreochromis nilotica* of Various Body Weights (A = 25 g; B = 50 g; C= 100 g; D = 200 g; E = 400 g; F = 1000 g) (Maity and Banerjee, 2010).

Table 21.3: Depuration of Metals in different Species of Fish with 50g and 800g of Body Weight

Metals	*Tissues*	*L. rohita (50g)*	*L. rohita (800g)*	*C. calta (50g)*	*C. catla (800g)*	*C. mrigala (50g)*	*C. mrigala (800g)*	*O. nilotica (50g)*	*O. nilotica (800g)*
		Percent depuration in metals in fish tissues							
Copper (μ g/g)	Muscle	100	100	100	100	100	100	100	100
	Liver	44.08	52.73	47.38	55.30	68	7.15	46.98	27.66
	Kidney	100	100	100	100	100	100	100	100
	Gills	100	100	100	100	100	100	100	100
	Skin	100	100	100	100	100	100	100	100
	Intestine	59.88	62.63	54.18	60.53	40.38	39.29	69.00	20
Lead (μ g/g)	Muscle	58.88	61.35	31.07	48.15	55.58	28.95	73.96	58.63
	Liver	56.70	49.25	29.73	42.81	37.92	21.32	35.90	24.83
	Kidney	49.06	52.14	29.38	31.75	46.09	21.71	31.85	22.51
	Gills	44.43	51.94	26.61	16.74	58.19	23.73	64.01	35.08
	Skin	52.13	56.70	22.18	30.23	61.44	31.24	28.61	46.80-
	Intestine	38.75	48.85	22.86	54.13	52.33	25.36	20.22	25.41
Zinc (μ g/g)	Muscle	44.42	58.14	30.78	18.89	51.36	54.04	63.32	44.08
	Liver	31.77	51.04	11.89	20.16	35.93	47.53	53.65	25.86
	Kidney	19.33	44.51	36.70	29.76	55.31	39.10	71.00	37.76
	Gills	12.54	27.06	14.22	8.73	27.24	39.73	33.43	44.11
	Skin	16.00	19.71	31.17	27.62	42.80	32.49	59.14	55.26
	Intestine	05.80	16.89	31.69	26.30	40.91	47.34	59.61	36.93
Cadmium (μ g/g)	Muscle	40.17	35.97	22.28	40.22	35.99	38.59	35.58	36.67
	Liver	40.72	58.26	35.89	46.94	34.47	32.92	42.86	48.33
	Kidney	41.78	30.06	43.77	43.62	37.97	35.04	21.20	42.53
	Gills	43.97	24.54	44.32	59.53	31.66	20.27	39.89	40.05
	Skin	31.67	28.48	48.72	35.19	35.61	34.16	33.14	43.14
	Intestine	10.82	28.12	43.75	50.37	35.70	34.78	44.88	52.63

(After Maiti and Banerjee, 2012).

exchangeable or slowly exchangeable pools of the body. Muscular depuration of metals is normally higher in an uncontaminated environment probably due to the absence of metal binding proteins. Adjusting metal uptake and depuration are the primary means of maintaining metal homeostasis that occurs naturally for essential elements. This control is low in lead and cadmium that show lower excretory rate even in the uncontaminated environment. Regulation of sewage input or temporary stoppage of sewage effluents in aquaculture ponds can reap positive results to reduce metal related health hazards in human being.

Conclusion

From the public health point of view, it has been observed that lead and cadmium level in fish muscle is more and above the set international standard specified by WHO (Table 21.2). Thus regular consumption of sewage fed fish may cause lead and cadmium toxicity to its consumers.

Aquatic environment is fragile, dynamic and in most water bodies metals are present only in mixture or are associated with other compounds so the effect of any one metal becomes difficult to predict. Among the fish species studied, *C. punctatus* shows highest accumulation of metals in its tissues (Maiti and Banerjee, 2011). It feeds on a variety of benthic invertebrates, including cladocerans, nematodes, copepods, dipteran larvae as well as on plants and algae that remain in contact with excess metals in the sediment. This dietary uptake possibly allows for biomagnification of metals in it (Van den Broek *et al.*, 2002; Asaolu and Olaofe, 2005). Being great accumulator, it loses its suitability as a food fish. Burger *et al.* (2002) suggest that fish of higher trophic level generally accumulate more metals. Besides, exotic omnivore *O. nilotica* being ubiquitous in this sewage fed wetlands shows positive correlation of metal accumulation with body weight. It was observed that copper in liver and intestine while zinc in muscle and liver generally rise as the fish grows in size in the polluted system (Figures 21.6 and 21.7). Lead in muscle and liver however showed modest rise with chronic exposure (Figure 21.8) although the metal accumulates substantially in the younger fish. Thus this species may not be a good choice for daily consumption.

Summary

Polyculture of fish in the East Kolkata Wetlands employs extensive use of raw sewage. Although fish culture in this ecosystem requires minimum capital input but the huge amount of contaminants present along the waste is potentially harmful for the biota. The concentration of various contaminants in the wetlands increase beyond their natural level due to regular dumping of agricultural, domestic and industrial effluents. Of all the effective pollutants, the metals have drastic environmental impact on all organisms owing to their persistence and bioaccumulative property.

Metals form complexes with inorganic and organic compounds and settle at the sediments or remain in ionic form in the overlying waters. Consequently, aquatic organisms may acquire metals directly from the water, food or through gills and skin. These are concentrated by aquatic organisms and move along the food chain to their consumers, at a level that might affect their physiological state. In view

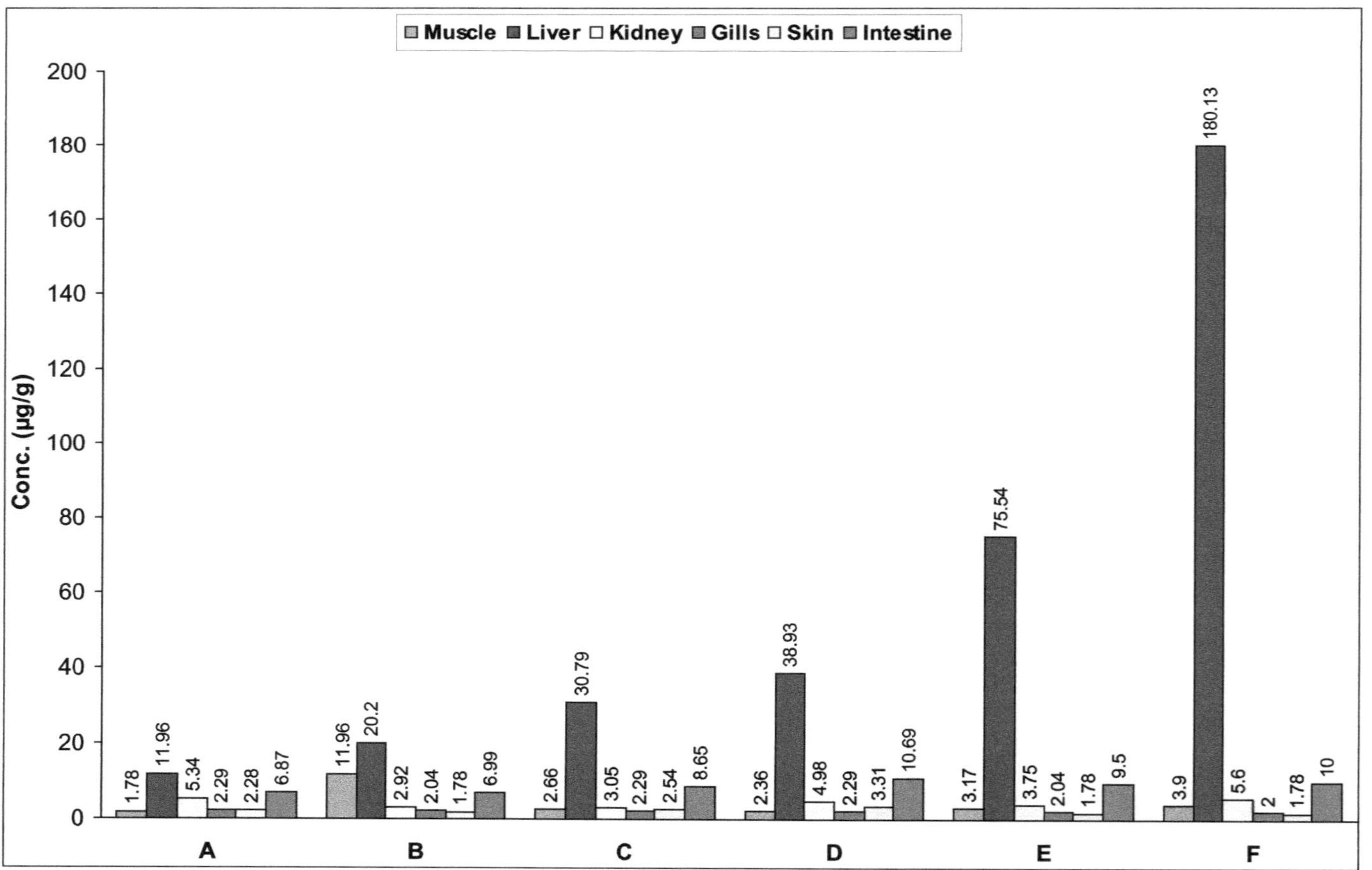

Figure 21.6: Accumulation of Copper in different Tissues of *Oreochromis nilotica* of Various Body Weights (A = 25 g; B = 50 g; C= 100 g; D = 200 g; E = 400 g; F = 1000 g) (Maity and Banerjee, 2010).

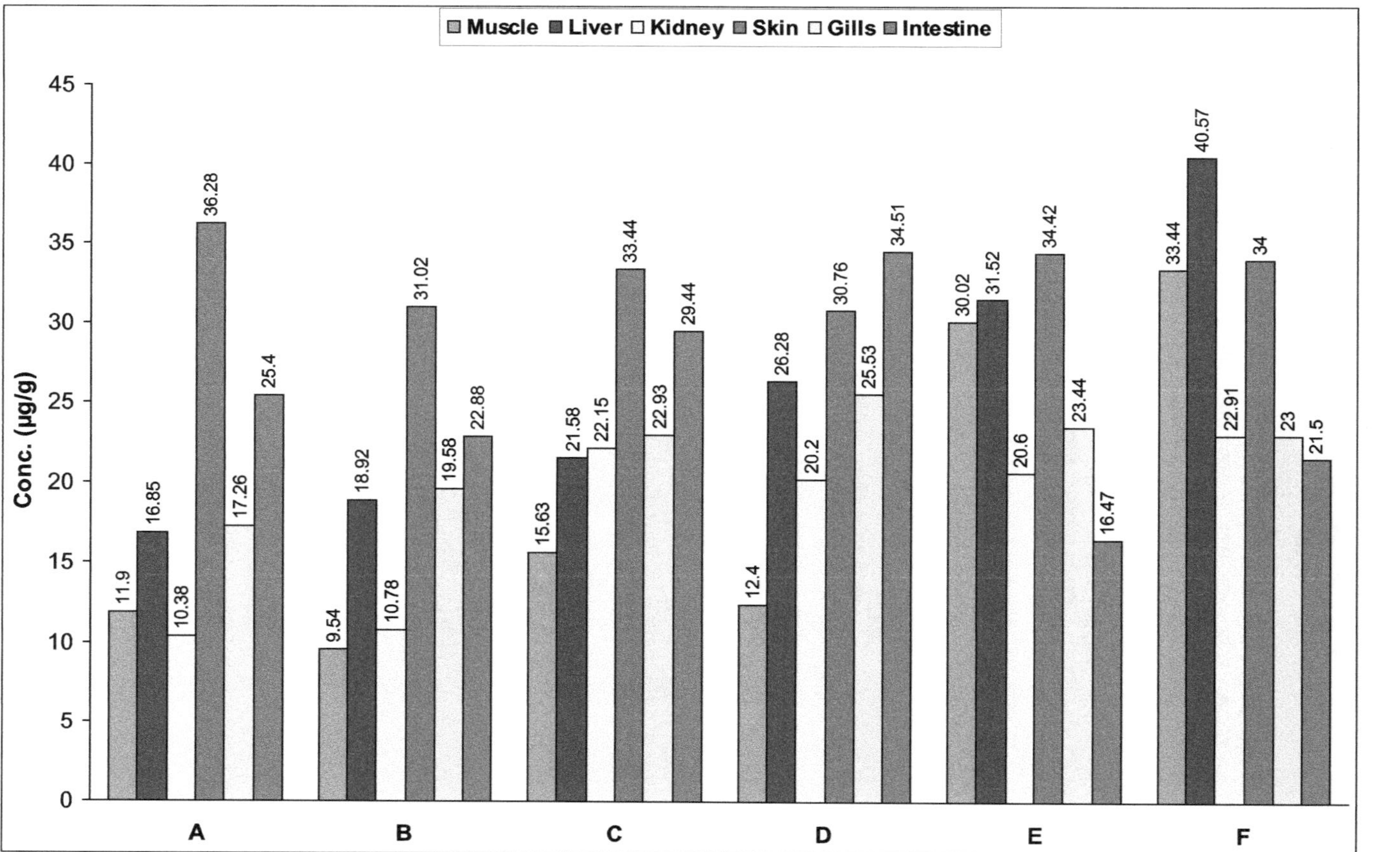

Figure 21.7. Accumulation of Zinc in different Tissues of *Oreochromis nilotica* of Various Body Weights (A=25g; B=50g; C=100g; D=200g; E=400g; F=1000g) (Maity and Banerjee, 2010).

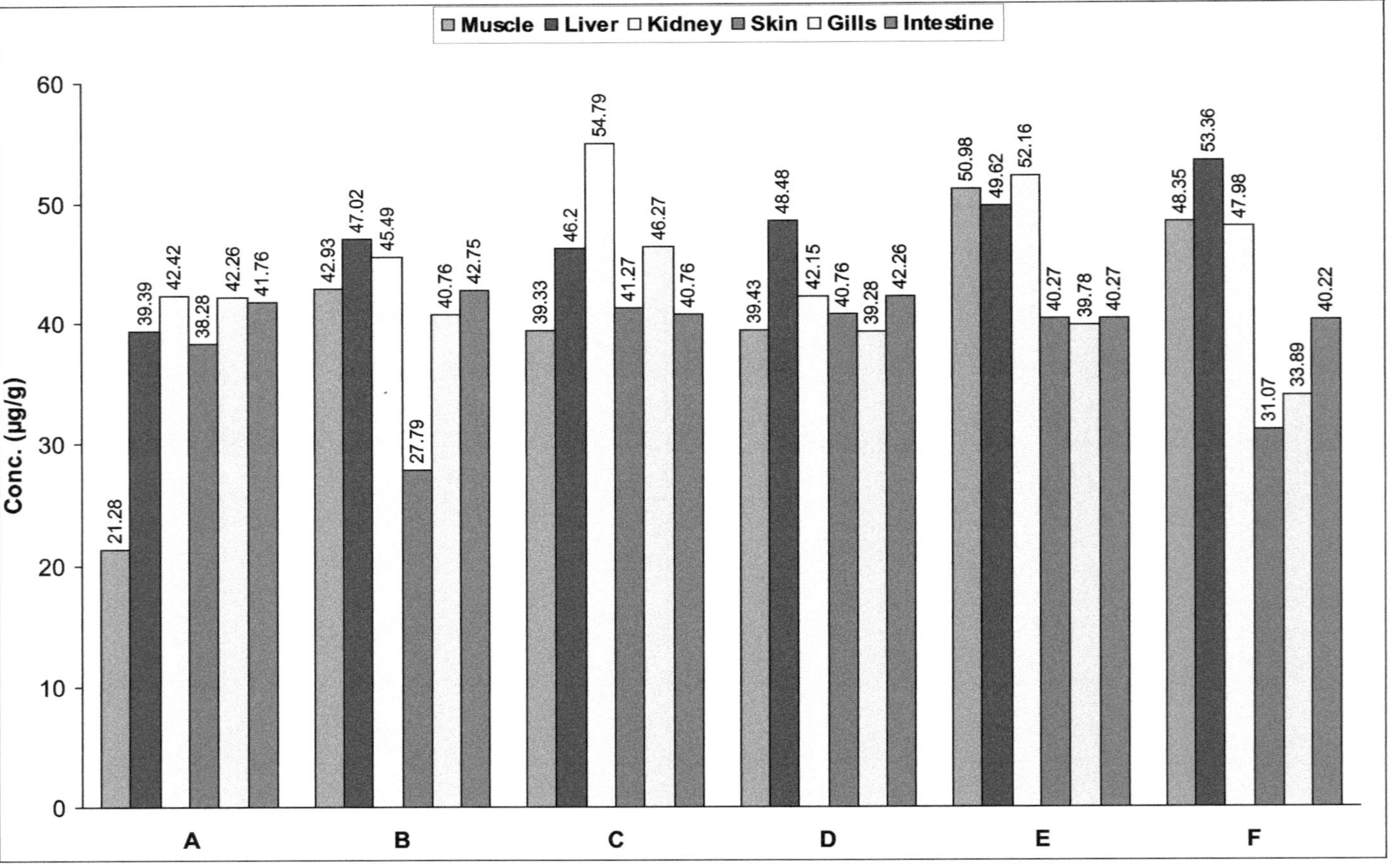

Figure 21.8: Accumulation of Lead in different Tissues of *Oreochromis nilotica* of Various Body Weights (A=25g; B=50g; C=100g; D=200g; E=400g; F=1000g) (Maity and Banerjee, 2010).

of public health, metal accumulation pattern in five commercially important fish species of various size and body weights has been extensively studied. Fish grown in the East Kolkata Wetlands are chosen to examine relationships among water, sediment, and tissue metal concentrations.

Higher concentration of copper, zinc and cadmium are observed in the soil while lead, is more in the overlying waters. Various fish species showed significant differences in metal accumulation in their tissues which reflected their food habit and trophic occupation. However, all the species showed high accumulation of copper in liver that increased with their body weight. The metal level remains low and invariant in other tissues. The metal is exceptionally high in the kidney of predatory *C. punctatus* and *C. mrigala* a benthic fish and omnivover *O. nilotica* also accumulates the metal considerably.

Zinc accumulates more in the gills of surface feeder *C. catla* and kidney of *C. punctatus*. It increased moderately with body weight especially in liver, gills and kidney of most of the species analyzed but *C. mrigala* shows gradual elimination of zinc from its tissues with age. Liver, scale and kidney tend to retain lead in most of the species while *C. punctatus* shows maximum accumulation of the metal in muscle and liver. In other species, lead and cadmium concentration remains high in the muscular tissue that exceeded the permissible limits set by WHO. Cadmium level however, remains closely invariant in all the species of all the age groups excepting in *C. punctatus*. Generally, predatory, benthivore and omnivorous fish and those higher in the food chain show increased concentration of metals in their tissues.

In another study, sewage input in fish ponds was stopped for six months. Metal body burden was measured before and after this period. It was observed that metal concentration in water decreased considerably but those bound to the soil fraction did not show any appreciable change. In fish tissues copper showed complete clearance excepting in liver and intestine. Zinc remained stored in gills especially in *C. catla*. Excretion rate of lead and cadmium from fish tissues seemed to be much lower which indicates chronic toxicity associated with these metals.

Acknowledgement

Thanks are due to CSIR New Delhi for their financial support and the Principal, Lady Brabourne College, Kolkata.

References

American Public Health Association. 2005. Standard methods for the examination of water and wastewater. 21th ed. APHA, AWWA, WPCP Publications. Washington DC, USA.

Asaolu SS and Olaofe O. 2005. Biomagnifications of some heavy and essential metals in sediments, fishes and crayfish from Ondo State coastal region, Nigeria. *Pakistan Journal of Scientific and Industrial Research*, 48 (2) : 96-102.

Authman MMN. 2008. *Oreochromis nilotica* as a biomonitoring of heavy metal pollution with emphasis on potential risk and relation to some biological aspects. *Global Veterinaria*, 2(3): 104-109.

Begum A, Harikrishna S and Khan I. 2009. Analysis of heavy metals, sediments and fish samples of Madivala Lakes of Bangalore, Karnataka. *Int. J. Chem. Tech. Res.*, 1(2): 245-249.

Bervoets L, Blust R and Verheyen R. 2001. Accumulation of metals in the tissues of three spined Stickleback (*Gasterosteus aculeatus*) from natural freshwaters. *Ecotoxicology and Environmental Safety*, 48: 117-127.

Bohn A and McElroy RO. 1976. Trace metals (As, Cd, Cu, Fe and Zn) in Arctic cod, *Boreogadus saida* and selected zooplankton from Strath cona sound, Northern Baftin Island. *J. Fish Res. Bd. Can.*, 33: 2836-2840.

Burger J, Gaines KF and Borin CS. *et al.*, 2002. Metal levels in fish from the Savannah River: potential hazards to fish and other receptors. *Environmental Research*, 89: 85-97.

Chernoff B. 1975. A method for wet digestion of fish tissue for heavy metal analyses. *Trans. Am. Fish. Soc.*, 4 : 803-804.

Ekeanyanwu CR, Ouguiya CA and Etienajirhevwe OF. 2011. Trace metal distribution in fish tissues, bottom sediments and water from Okumeshi River in Delta state, Nigeria. Fish and Periwinkles of Lagos Lagoon. *American-Eurasian J. Agric. and Environ. Sci.*, 5 (5): 609-617.

Gomez KA and Gomez AA. 1984. *Statistical Procedures for Agricultural Research*, 2nd ed. Willey Interscience, New York.

Heath AG. 1995. *Water Pollution and Fish Physiology*. CRC Press, NC. Florida, USA.

Hughes GM and Floss R. 1978. Zinc treatment of the gills of rainbow trout (*S. gairdneri*) after treatment with zinc solutions under normoxic and hypoxic conditions. *J. Fish. Biol.*, 13: 717-728.

Kotze PH, Preez HH. and VanVuren JHJ. 1999. Bioaccumulation of copper and zinc in *Oreochromis mossambicus* and *Clarias garicpinus*, from the Olifants River, Mpumalanga, South Africa. Water SA, 25: 99–110.

Kumar P. and Singh A. 2010. Cadmium toxicity in fish: An overview. *GERF Bulletin of Biosciences*. 1(1): 41-47

Kundu N, Pal M and Saha S. 2008. East Kolkata Wetlands: A resource recovery system through productive activities. In. *Proceedings of Taal* 2007. The 12th World Lake Conference. pp. 868-881.

Linnik PM and Zubenko IB. 2000. Role of bottom sediments in the secondary pollution of aquatic environments by heavy metal compounds. *Lakes and Reservoirs: Research and Management*, 5: 11-21.

Lock WH. (1994) Research Report (Urban Water Research Association of Australia) No.79: Heavy metals and organics in domestic wastewater, Urban Water Research Association of Australia.

Maiti P and Banerjee S. 2001. Hepatic accumulation of Cu in some sewage-fed fish species. *Ind. J. Environ. Ecoplan.*, 3(2) : 265-269.

Maiti P and Banerjee S. 2002. Size dependent accumulation of Zn in a wastewater raised fish species, *Oreochromis nilotica. Eco. Env. and Cons.*, 8(2) : 161-165.

Maiti P and Banerjee S. 2005. Depuration of metals in a sewage-fed fish during post harvest conditioning in uncontaminated water. *Poll.Res.*, 24(4): 753-761.

Maiti P and Banerjee S. 2007. Strategy of metal accumulation in fish with various food preferences. *Zoological Research in Human Welfare*. 39: 381-388.

Maiti P and Banerjee S. 2010. Size gradient accumulation of metals in a fish species (*Oreochromis nilotica*) exposed in wastewater ecosystem. *Science and Culture,* 17(1-2) : 40-45.

Maiti P. and Banerjee, S. 2011.Accumulation of Cu, Pb, Zn and Cd in three benthic fish species from the Kolkata wetlands. *J. Curr. Sci.* 16(1): 79-86

Maiti P and Banerjee S. 2012. Dynamics of metal concentration in relation to body size and feeding habit of sewage-fed carps. *World Journal of Fish and Marine Sciences*, 4 (4): 407-417.

Mastan SA. 2014. Heavy metals concentration in various tissues of two freshwater fishes, *Labeo rohita* and *Channa striatus*. *African Journal of Environmental Science and Technology*, 8(2): 166-170.

More TG, Rajput RA and Bandela NN. 2003. Impact of heavy metals on DNA content in the whole body of freshwater bivalve, *Lamellidens marginalis*. *Environmental Science and Pollution Research*, 22: 605-616.

Moriarty F, Hanson HM and Freestone P. 1984. Limitation of body burden as an index of environmental contamination of heavy metal in fish *Cottus gobio* L. from the river Ecclesbourne, Derbyshire. *Environ. Pollut. Ser.* A., 34: 297-320.

Murphy BR, Atchison GJ, McIntosh AW and Kolar DJ. 1978. Cadmium and zinc content of fish from an industrially contaminated lake. *J. Fish. Biol.*, 13: 327-335.

Nafde AS, Kondawar VK and Hasan MZ. 1998. Precision and accuracy control in the determination of heavy metals in sediment and water by atomic absorption spectrophotometry. *JIEAM*, 25: 83-91.

Paulson AJ, Richard AF, Herbert CC and James GF. 1984. Behavior of Fe, Mn, Cu, and Cd in the Duwamish river estuary downstream of a sewage treatment plant. *Water Res.*,18 (5): 633-641.

Pourang N. 1995. Heavy metal accumulation in different tissues of two fish species with regards to their feeding habits and tropic level. *Environmental Monitoring and Assessment*, 35: 207-219.

Schulz BM. 1974. Lead uptake from sea water and food and lead loss in common mussel, *Mytitus edulis*. *Mar. Biol.*, 25: 177-193.

Schmitt CJ and Finger SE. 1987. The effect of sample preparation on measured concentrations of eight elements in edible tissues of fish from streams contaminated by lead mining. *Arch. Environ. Contam. Toxicol.*, 16: 185-207.

Shrivastava AK. 2009. A review on Copper pollution and its removal from water bodies by pollution control technologies. *IJEP*, 29 (6): 552-560

Sivaperumal P, Sankar TVand Nair PGV. 2007. Heavy metal concentrations in fish, shellfish and fish products from internal markets of India vis-à-vis international standards. *Food Chemistry,* 102: 612-620. doi: org/10.1016/j. foodchem.2006.05.041

Taweel A, Shuhaimi-Othman M and Ahmad AK. 2011. Heavy metal concentration in different organs of tilapia fish (*Oreochromis niloticus*) from selected areas of Bangi, Selangor, Malaysia. *Afr. J. Biotechnol.,* 10(55): 11562-11566.

Van den Broek JL, Gledhill KS and Morgan DG. 2002. Heavy metal concentrations in the mosquito fish, *Gambusia holbrooki,* in the Manly Lagoon Catchment. In: *UTS Freshwater Ecology Report,* Department of Environmental Sciences, University of Technology.

WHO. 1992. Environmental Health Criteria, No. 134, Environmental Aspects, Geneva WHO.

Chapter 22

Evaluating Wetland Services: Ecology to Economy

Moumit Roy Goswami[1] and Aniruddha Mukhopadhayay[2]

[1]Department of Environmental Science,
Netaji Nagar College for Women, Kolkata
[2]Department of Environmental Science,
University of Calcutta, Kolkata

Introduction

Wetlands are considered as one of the most productive ecosystems of the world. It is estimated that wetlands support extensive food webs and rich biodiversity (e.g., in one square meter of coral reef there can be up to 3000 species) and are termed as "biological supermarkets." They are also described as "the kidneys of the landscape", because of the functions they perform in the hydrological and chemical cycles. The significance of wetlands and their protection measures gained momentum in the 1960s, mainly because of their importance as habitat for migratory species. A series of conferences and technical meetings culminated in the 'Convention on Wetlands of International Importance especially as Waterfowl Habitat' (well known as the Ramsar Convention) which came into force in 1975. Therefore, wetlands became the only single group of ecosystems to have their own international convention. Up to November, 1999, 116 countries are Ramsar Contracting Parties; with 1005 wetland sites included in the Ramsar List of Wetlands of International Importance. These sites cover about 71.7 million hectares, which correspond to about 0.5 per cent of the world's land surface (Turner *et al.*, 2000).

The Ecosystems of the world provide a myriad of services that are vital for human well-being, health, livelihoods, and survival (Costanza *et al.*, 1997; Millennium Ecosystem Assessment (MEA), 2005; TEEB, 2010). The monetary valuation of the ecosystem services started with various researches from 1960s onwards, however with the publication of Costanza *et al.* (1997) on "The value of the world's ecosystem services and natural capital" the field received widespread attention. Following this, a sound development in the number of articles and reports on the monetary valuation of natural resources, ecosystem services and biodiversity were seen. The values of ecosystem services are important to the human society in ecological, socio-cultural and economic aspects. Assessing the value of ecosystem services in monetary units is a significant tool to raise awareness, providing guidance and conveying the role and relative importance of ecosystems services to the policymakers. The monetary valuation allows more efficient use of funds for the protection and restoration of ecosystems at lowermost cost (Crossman and Bryan, 2009; Crossman *et al.*, 2011). Also the degree to compensation that should be paid for the loss of ecosystem services in liability regimes are assisted by monetary valuation (Payne and Sand, 2011).Unlike the services of various ecosystems of the world, the ecosystem services of wetlands are diverse (Barbier *et al.*,1997; Russi *et al.*, 2013). Historically, wetlands were viewed as a waste of valuable land that could only be 'improved' through drainage and destruction (Mitsch and Gosselink, 1986). However today, there is widespread recognition that wetlands provide valuable ecological services. It has been reported that wetlands provide around 40 per cent of global ecosystem services disproportionately (Zedler and Kercher, 2005). The ecosystem services of wetlands includes climate regulation that occurs as wetlands capture carbon dioxide, methane and nitrous oxide; nursery provision for fish population; flood protection to inland settlements; regulating groundwater recharge; filtering nutrients and pollutants from surface water; and providing opportunities for recreational and educational activities (Dise, 2009; Pramod *et al.*, 2011; Sarkar and Upadhyay, 2000). Also wetlands provide food and non-food products that contribute to income and food security. The deterioration of the state of wetlands in terms of both area and ecological superiority place the ecosystem services of wetlands at the verge of risk. It has been reported that in total, about 50 per cent of the world's original wetland area has been lost, ranging from relatively minor losses in boreal countries to extreme losses of >90 per cent in parts of Europe (Mitsch and Gosselink, 1986) due to agriculture, urban expansion, and other developmental activities. It has long been recognized by researchers that losing wetland ecosystems disrupts human welfare (Westman, 1977; Daily, 1997). The cause behind the overexploitation of natural resources in general and wetlands in particular, is that there is a lack in integrated methodologies in the decision making process regarding the full set of supports or benefits provided by these resources. As a result the economic incentive for their conservation is too scarce in comparison to the economic benefits provided by them. The ecosystem services framework as proposed by the Millennium Ecosystem Assessment (MEA, 2005) and the economics of ecosystems and biodiversity (TEEB, 2008, 2010) provide necessary information and show the path to the policymakers about the benefits of wetlands. The valuation of wetlands' ecological services is fairly a recent phenomenon that can be done with the help of both primary and secondary data. The estimation of wetland services and values can only be obtained through non-market valuation techniques as these services normally do not have a market price. Many wetland valuation studies have been conducted and the range of the estimates is noteworthy

(Barbieret *al.*, 1997; Bardecki, 1998; Brouwer *et al.*, 1999; Kazmierczak, 2001; Stanley, 2001; Woodward and Wui, 2001; Smith and Pattanayak, 2002). However, in assessing the value of wetlands, the differentiation between the systems' ecological functions from the associated services that are directly valued by humans is highly important (Costanza *et al.*, 1997).

Economic Valuation of Wetlands

The steps involved in ecological characterization to economic valuation provide a vital link between wetland ecology or functioning and wetland economics and values. The TEEB (2013) study translated the values of ecosystem services in terms of dollar. For instance, the economic value of inland wetland ecosystem services was estimated at up to US$ 44,000 per hectare per year.The most common structure for understanding the full economic value of environmental resources such as wetlands is the Total Economic Value (TEV) framework. The TEV structure identifies not only the value of financial or commercial outputs, but also non-consumptive values that may be environmental or social in nature. The economic value of the wetland ecosystem services can be categorized into direct value (DV), indirect value (IV), option value (OV) and existence value (EV) as shown in Figure 22.1. The direct value denotes the physical use of resources such as wild fish capture, timber, firewood, etc. An indirect value denotes to the ecosystem services such as watershed protection, carbon sequestration, water quality attenuation and supply. The option values denote to future economic options such as industrial, pharmaceutical, recreational applications. Lastly existence values denote the intrinsic worth, regardless of use such as biodiversity, landscape, aesthetic, heritage, bequest and culture (Kyophilavong, 2011).Therefore the total economic value of wetlands can be represented by the equation: Total Economic Value (TEV) = DV + IV+ OV + EV.

The classifications of total economic value for wetlands as represented in Table 22.1, reflect the direct uses of wetlands involving both commercial (marketed value) and non-commercial activities whereas indirect use values are unmarketed, and are connected indirectly to economic activities. A special category of value is option value, which arises because an individual may be uncertain about his or her future demand for a resource and/or its availability in the wetland in the future. The quasi-option value is the value that society would place on wetlands, if the complex functions of wetlands are known. The, non-use values are often referred as intrinsic or existence values that are difficult to measure. They involve subjective valuations by individuals distinct to either their own or others' use, whether current or future generations.

Wetlands Valuation Methodologies

A wide range of methods are available for estimation of economic values of wetlands (Table 22.2). This estimation methodologies solely depends on the type of value (direct use values, indirect use values and non-use values) being estimated. Usually, direct use values of a wetland are estimated by using market price approach method and indirect use values and non-use values of wetland are estimated mostly by Contingent Valuation Method (CVM). The Benefit Transfer

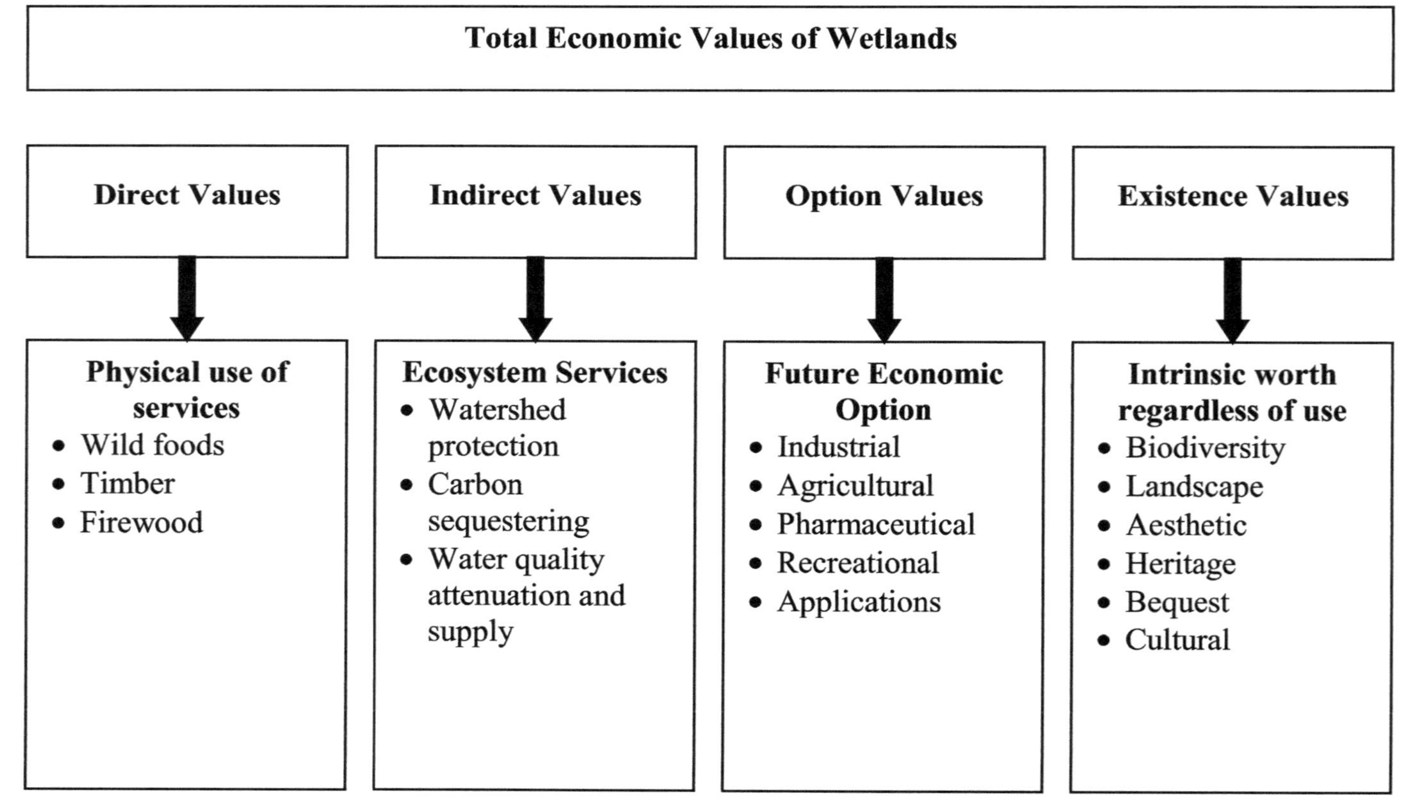

Figure 22.1: The Total Economic Value of a Wetland (Source: IUCN, 2006).

Table 22.1: Classification of Economic Value for Wetlands

Use Value/Benefits		*Non-Use Value/Benefits*	
Direct Use Benefits	*Indirect Use Benefits*	*Option and Quasi-Option Benefits*	*Existence Benefits*
Recreation	– Nutrient retention	– Potential future uses (as per direct and indirect uses)	– Biodiversity
– Boating	– Flood control	– Future value of information, *e.g.*, pharmaceuticals, education	– Culture
– Fauna (birds, etc.)	– Storm protection		– Heritage
– Wildlife	– Groundwater recharge		– Bequest
– Viewing	– External ecosystem support		
– Walking	– Micro-climatic stabilisation		
– Fishing	– Shorelines tabilisation, etc.		
Commercial harvest	– Water filtration		
– Fish	– Erosion control		
– Fuelwood			
– Transport			
– Nuts			
– Berries			
– Grains			
– Peat/energy			
Forestry			
– Wildlife harvesting			
– Agriculture			

Source: Barbier *et al.*, 1997.

Method (BTM) is another popular method normally practiced when time and cost are constraints. It uses the results from earlier studies of wetland values at some other sites for estimating the value of wetland services in the current study site. The market price approach is the simplest and most straight forward way of finding out the value of wetland goods because one can find out directly the consumption and sale value of wetland goods such as fish, animals, and other wetland "goods". It uses questionnaires to collect data regarding market price of buying and selling wetland goods. As mentioned above, CVM is one of the popular non-market valuation methodology. It requires construction of a hypothetical market that contains a description of the proposed policy or development that will have an effect on the wetland resource. The choosing of an appropriate methodology for wetland valuation therefore should be based on the following three factors:

1. Time and cost for study
2. Capacity, experience and expertise of those carrying out the study
3. Information related to characteristics of wetland

Table 22.2. Different Methodologies for Estimating Economic Value of Wetlands

Valuation Method	Description	Direct Use Values	Indirect Use Values	Nonuse Values
Market Analysis	Where market prices of outputs (and inputs) are available. Marginal productivity net of human effort/cost. Could approximate with market price of close substitute. Requires shadow pricing.	√	√	
(Productivity Losses)	Change in net return from marketed goods: a form of (does-response) market analysis.	√	√	
(Production Function)	Wetlands treated as one input into the production of other goods: based on ecological linkages and market analysis.		√	
(Public Pricing)	Public investment, for instance via land purchase or monetary incentives, as a surrogate for market transaction.	√	√	√
Hedonic Price Method (HPM)	Derive an implicit price for an environmental goods from analysis of goods for which markets exist and which incorporate particular environmental characteristics.	√	√	
Travel Cost Method (TCM)	Costs incurred in reaching a recreation site as a proxy for the value of recreation. Expenses differ between sites (or for the same site over time) with different environmental attributes.	√	√	
Contingent Valuation (CVM)	Construction of a hypothetical market by direct surveying of a sample of individuals and aggregation to encompass the relevant population. Problems of potential biases.	√	√	√
Damage Costs Avoided	The costs that would be incurred if the wetland function were not present; eg flood prevention.		√	
Defensive Expenditures	Costs incurred in mitigating the effects of reduced environmental quality. Represent a minimum value for the environmental function.		√	
(Relocation Costs)	Expenditures involved in relocation of affected agents or facilities:: a particular form of defensive expenditure.		√	
Replacement/ Substitute Costs	Potential expenditures incurred in replacing the function that is lost; for instance by the use of substitute facilities or 'shadow projects'.	√	√	√
Restoration Costs	Costs of returning the degraded wetland to its original state. A total value approach; improtant ecological, temporal and cultural dimensions.	√	√	√

Source: Turner *et al.*, 1997.

Stepped Approach to Wetland Valuation

The stepped approach as represented in Figure 2 in wetland valuation includes the following seven steps as proposed by Kyophilavong (2011).

Step 1 Defining the scope of the wetland to be valued - to define the type and boundary of wetland and to be valued in context of their complex nature.

Step 2 Identifying the primary wetland benefits/functions/services - to identify the major/primary benefits/functions/services provided by the wetland in order to determine the inclusion of these benefits/functions/services provided by the respective wetland for valuation. The wetland may provide significant direct values/benefits, indirect values/benefits or both. Therefore, identifying the wetland's benefits/functions is crucial. However, it is tough to assess all the value of all functions/benefits of wetland. Therefore, it is vital to identify the most important functions first and then moving to functions of lower rank.

Step 3 Identifying the wetland beneficiaries and stakeholders - to identify target group of beneficiaries and stakeholders for each of the wetland benefits/functions/services to be valued as the stakeholders and beneficiaries of wetlands varies subject to the type of wetland and their functions or benefits being valued.

Step 4 Identifying the constraints under which the valuation will be carried out - the time, budget and capacity of the person(s) involved in collection of set of data with respect to wetland to be valued are important parameters to be considered. The criteria to be chosen in expressing the constraints in time, budget, staff capacity and basic information of wetland are identified as 'highly constraints', 'medium constraints', and 'small constraints'. Steps should be undertaken to identify the most appropriate method of valuation based on the constraints on time budget and staff capacity. Based on the constraints perception, the options available for selecting an appropriate valuation method are BTM for 'Highly constraints', MPA for 'Medium constraints' and CVM for 'Low constraints'.

Step 5 Choosing a valuation method - in respect to the valuation methodologies for the perceived constraints, the detailed methodology as specified for a particular option can be followed. However, combination of the four options together based on constraints can also be followed. As for example, the BTM for valuing the flood control function of a wetland, the MPA for estimating direct benefits of a wetland such as wild fish capture and the CVM for estimating the biodiversity value of the wetland.

Step 6 Following steps of selected method the detailed methodology of the selected method is followed to generate data regarding the values of the particular wetland and their functions.

Step 7 Calculating the Total Economic Value in the final step, all the values estimated for each of the wetland benefits/functions/services are added to estimate a Total Economic Value (TEV) for the wetlands.

Benefit Transfer Method (BTM)

It is a common method when time and cost are constraints. In this method, results of previous study from a different site are used to estimate the value of the current study site (Smith *et al.*, 2002). BTM methods can be divided into four categories:

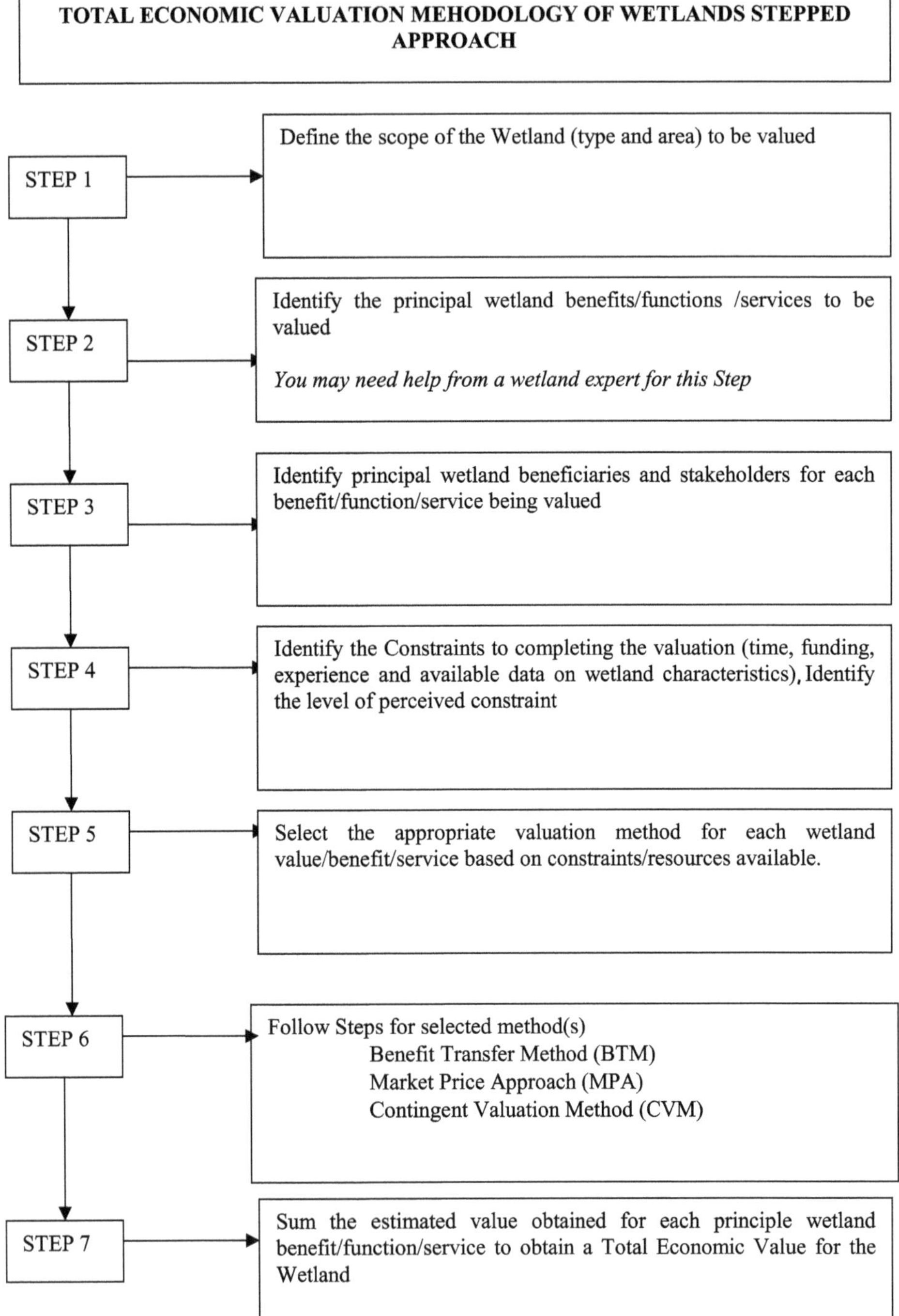

Figure 22.2: Total Economic Wetland Valuation Methodology (*Source*: Kyophilavong, 2011).

- Unit Benefit Transfer - involves estimating the value of an ecosystem service at a policy site by multiplying a mean unit value estimated at a study site by the quantity of that ecosystem service at the policy site. It is either expressed as values per household or as values per unit of area.
- Adjusted Unit Benefit Transfer - involves making simple adjustments to the transferred unit values to reflect differences in site characteristics.
- Value Function Transfer - use functions estimated through valuation applications (travel cost, hedonic pricing, contingent valuation, or choice modeling) for a study site together with information on parameter values for the policy site to transfer values.
- Meta-analytic Function Transfer - use a value function estimated from multiple study results together with information on parameter values for the policy site to estimate values.

The stages involved in benefit transfer method are:

- Review of literature on different wetland studies - the results of relevant studies from literature is reviewed.
- Selection of relevant papers and studies to conduct BTM- after review of literature on different studies relevant papers are selected to conduct the BTM. The selection of the previous study is based on sample size. The data obtained can be subjected to meta-regression analysis for estimating wetland value (Woodward and Wui, 2001; Enjolras and Boisson, 2008; Ghermandi *et al.*, 2010; Brander *et al.*, 2011)
- Estimation of economic value of wetlands - based on result of meta-regression analysis values for each benefit/function/services of wetlands have been calculated and these values, the "indirect" value and direct values are calculated in terms of US$/ha/year in a rank wise manner.
- Adjustment of wetland value - the adjustment of economic value of wetland for a particular year can be estimated from economic value for wetland services from a base year data following adjustment. This adjustment is based on increasing the value considering inflation.

Benefit transfer is typically less costly than conducting an original valuation study. However, benefit transfer may not be accurate, except for making gross estimates of values, unless the sites share all of the site, location, and user specific characteristics.

Market Price Approach (MPA)

Market price approach is the simplest method where the value of wetland goods can directly be calculated from the income generated from the sale of wetland goods. The method uses a questionnaire to collect data about the market price of buying and selling wetland goods. For example, catching fish from the wetland for sale in local markets. The approaches of direct market valuation are divided into (a) market price-based approaches, (b) cost-based approaches, and (c) approaches based on production functions. Market price-based approaches are most often used to obtain the value of provisioning services, since the commodities produced by provisioning services are often sold on, *e.g.*, agricultural markets. Cost-based approaches are based on estimations of the costs that would be incurred if ecosystem service benefits need to be recreated through artificial means (Garrod and Willis, 1999). Production function-based approaches (PF) estimate how much a given ecosystem service (*e.g.*, regulating service) contributes to the delivery of another service or commodity which is traded on an existing market. The PF approach is based on the contribution of ecosystem services to the enhancement of income or productivity (Mäler, 1994; Patanayak and Kramer, 2001).There is six stages for conducting market price approach method (Kyophilavong, 2008, 2011).

Stages for Conducting Market Price Approach

1. Setting the scope of wetland valuation - The scope depends on the area of wetland to be studied, the wetland services to be valued and the time and budget accessible for the survey. The following three factors are considered for setting the scope of study are as follows.
   - ☆ Determination of wetland area.
   - ☆ Determination of number of users/beneficiaries/villages to be included in the survey to choose the most relevant village for the survey.
   - ☆ Determination of the most relevant direct benefits to be valued by consultation with experts.
2. Designing of questionnaire for survey is done considering the following.
   - ☆ Respondent ability - based on capacity of answering the questions and memory.
   - ☆ Complexity of the questionnaire - based on local context and education status of village.
   - ☆ Time and cost.
3. Decision on the sample size and composition - The three steps involved in an appropriate sampling size and compositions are as follows: -
   - ☆ Dividing into urban and rural households on the basis of location of wetland in urban and rural areas and its characteristic features.
   - ☆ Dividing sample into rich, medium and poor household after dividing household into urban and rural.
   - ☆ Deciding the sample size - Stratified random sampling should be applied for selection of sample size of a population. The following equation is then used to calculate the sample size as follows.

     $n= N/(1+N^*e^2)$

     Where, n: Sample size, N: Population, e: error
4. Preparation for the survey - The steps for preparation of a survey includes training of the interviewer, decision on survey methodology (keeping in mind the educational status of villagers and cost and time needed for

the survey), budgeting and preparation of checklist incorporating all the arrangements including transportation before the survey.

5. Data analysis - Data analysis is a very important step of wetland valuation after survey.The following steps must be undertaken during data analysis (Kyophilavong, 2005)
   - ☆ Making coding table
   - ☆ Input data into Excel spread sheet
   - ☆ Checking the validity of data
6. Estimation of the direct value of a wetland - In the final step the total economic value of the direct benefits of a wetland is calculated using the following equation (Kyophilavong, 2008)with some modifications -
   (1) D total= D1 + D2 + D3....Dn (Where, D total: Total of direct benefit; D1: Type of direct benefit 1; D2: Type of direct benefit 2; D3: Type of direct benefit 3

   (2) D1or D2 or D3 = Mean1 or 2 or 3* Population1 or 2 or 3

   Where, Mean1or 2 or 3: mean of direct benefit type (1or 2 or 3) from all the questionnaires in the samplePopulation1or 2 or 3: Population use or collect in type (1 or 2 or 3) direct value

   (3) Mean1or 2 or 3 =Sum (D1 or D2 or D3)/n1or n2 or n3

   Where, Sum (D1/D2/D3): sum of direct benefit type (1/2/3) from all the questionnaires in a sample; n1/n2/n3: sample size.

Direct market valuation approaches are dependent on production or cost data, which are easily available than kinds of data required to establish demand for ecosystem services (Ellis and Fisher, 1987). However, in case of ecosystem service valuation, direct market valuation approaches have limitations as ecosystem services do not having markets or markets being distorted.

Contingent Valuation Method (CVM) or Willingness to Pay (WTP)

One of the most common methods used by different researchers to value wetland services are stated preference methods. Among the stated preference method, the most common being the contingent valuation method (Bateman *et al.*, 1992; Stevens *et al.*, 1995; Oglethorpe and Miliadou, 2000; Wattage and Mardle, 2008). This method is based on surveys and aims to assess how individuals would hypothetically react to changes in environmental quality and cast their vote for or against the various costs to their household (Haab and McConnell, 2002). Here an open-end questionnaire is used to question people directly how much they are willing to pay to conserve a particular benefit/function/service of a wetland (Whitehead and Blomquist, 2006). It is widely accepted that the CVM is one of the methods capable of estimating the full array of use, non-use, and existence values provided by environmental assets (Arrow *et al.*, 1993). Studies have shown that hypothetical values are not statistically different from real values when respondents are certain of their responses (Champ *et al.*, 1997).

The stages for conducting Contingent Valuation Method (CVM) are as follows.

1. Selection of the wetland benefit (in particular biodiversity, cultural heritage etc. having an intrinsic value) one wish to value that do not have a market price/value.
2. Selection of relevant stakeholders/beneficiaries benefitting from wetland at local, national and international regime to create a representative sample of the most important stakeholders/beneficiaries.
3. Designing of questionnaire on the basis of capacity of respondents, the available time and budget. The three steps involved in to designing a questionnaire, similar to market price approach are discussing with key informants, doing preparatory test and revision based on feedbacks. The following items are important to be considered in the questionnaire *i.e.*, WTP to preserve biodiversity; socio-economic features of residents; opinion on common environmental problems; opinion on endangered species and the impact of declining biodiversity and endangered species. The willingness to pay (WTP) part of the survey should include the following four components described or explained depending on the hypothetical scenario or the benefit the person want to value:
   - ✰ The significance of conservation of wetland.
   - ✰ Details of a theoretical project to conserve endangered wetland.
   - ✰ The impact of this theoretical project on biodiversity and their benefits.
   - ✰ Pictorial diagrams/photograph of current wetland attached in the questionnaire (to be shown to interviewees during interviews)
4. Decision on sampling size and composition - Here also like the market price approach stratified random sampling is to be implemented to ensure that all variations in WTP due to location, social class, income etc. is included in the sample. The steps involved in deciding sampling size and composition are defining the relevant "population" which benefits from the wetland benefits being valued; dividing into urban (inside-city) and rural (outside-city) populations as urban and rural dwellers may have different perspective on the importance of the wetland benefit being valued and therefore would have a different "willingness to pay. Therefore, it is important to divide the population into urban (inside-city) and rural (outside-city) dwellers.
   - ✰ Dividing in-city and outside-city into blocks.
   - ✰ Dividing block to rich, medium and poor household.
   - ✰ Deciding number in sample.
5. Preparation for the survey - The steps involved in preparation of survey include a) training of the interviewer, b) preparation of the survey manual and guidelines for those conducting the survey, c) decision on a survey method (based on educational status of respondents, time, place and cost involved during the survey), d) realistic budgets, transportation and

preparation of other relevant documents, e) preparation of checklist and defined questionnaires.

6. Data entry - Input of data involves organization of data, assigning of definite code and inputting of data in the excel sheet. The following steps are involved in this exercise such as making coding table, inputting data into excel spread sheet and above all checking the validity of data along with correction of typological errors (Kyophilavong, 2005).
7. Estimating willingness to Pay (WTP) - In the last step, the total willingness to pay for different communities is calculated by the following equation to obtain total WTP (Kyophilavong, 2011)

   (1) TWtp= Wtp1 + Wtp2 + Wtp3......Wtpn

   Where, TWtp : Total of willingness to pay (WTP)

   Wtp1 or 2 or 3: WTP for community1or 2 or 3

   (2) Wtp1 or 2 or 3 = Mean1 or 2 or 3 * Population1 or 2 or 3

   Where, Mean1or 2 or 3: mean of sample WTP in community (1 or 2 or 3)

   Population 1 or 2 or 3: Number of population in community (1or 2 or 3)

   (3) Mean1 or 2 or 3 =Sum (Wtp1or 2or 3)/n1 or 2 or 3

   Where, Mean1 or 2 or 3: mean of sample WTP in community 1

   Sum (Wtp1): sum of WTP in sample community (1 or 2 or 3)

   n1or 2 or 3: number of sample from community (1 or 2 or 3)

   Although the contingent valuation method (CVM) is the only method that can measure option and existence values and provide a true measure of total economic value but it suffers from the results related to numerous sources of bias in survey design and implementation (Barbier *et al.*, 1997).

Conclusion

In conclusion, it can be said that the services of ecological systems including wetlands are important to the functioning of the Earth's life-support system. These services contribute both directly and indirectly to human welfare. Wetlands being one of most important and productive ecosystem of the world provide a myriad of ecosystem services including habitat for species, stocking of biodiversity, protection against floods, water purification, climate regulation, amenities and recreational opportunities. However, many wetlands of the world have been lost and the remaining are under high pressure due to nutrient enrichment, invasive plants and animals, and encroachment from urban and agricultural development and policy intervention failures. This has caused the loss of their beneficial ecosystem services also. Many of the ecosystem services from different ecosystems in general and wetlands in particular remain unnoticed as they don't have a market price inspite of their crucial role. Therefore a multidisciplinary approach involving economic valuation of all the values and services from wetlands through various methodologies are highly required for policy/decision making process regarding framing of proper strategies for conservation of wetland and their sustainable use.

Summary

Wetlands are intermediate zone between terrestrial and aquatic ecosystems, found throughout the world in different climatic conditions. They differ widely in character due to regional and local differences in climate, soils, topography, hydrology, water chemistry, vegetation, and other factors. They are broadly classified into five categories *i.e.*, estuarine, marine, riverine, palustrine and lacustrine. However, recent classification also includes man-made wetlands under wetland classes. Wetlands are considered as one of the most productive ecosystems of the world having an array of components, functions and attributes. Wetlands provide important and diverse benefits to people around the world, contributing provisioning, regulating habitat and cultural services. The ecosystem services of wetlands includes climate regulation that occurs as wetlands capture carbon dioxide, methane and nitrous oxide; nursery provision for fish populations; flood protection to inland settlements; regulating groundwater recharge; filtering nutrients and pollutants from surface water; and providing opportunities for recreational and educational activities. However, about half of global wetland areas have been lost, and the condition of remaining wetlands is declining because of nutrient enrichment, invasive plants and animals, and encroachment from urban and agricultural development and policy intervention failures due to a lack of consistency among different policies in different areas involving economics, environment, nature protection, physical planning, etc. As wetland is complex multifunctional system and services provided by ecosystem possesses significant economic value. The economic value of the wetland ecosystem services can be categorized as direct, indirect, option and existence values. Economic valuation of wetland represents an important area for a clear understanding about the role of natural systems in economic development. A wide range of methods are available for estimation of economic values of wetlands. Valuation methods that follow the TEV approach can be divided into three main categories, direct market approaches, revealed preference and stated preference techniques, the latter of which is being increasingly combined with deliberative methods from political science to develop formal procedures for deliberative group valuation of ecosystem values.This estimation methodologies solely depends on the type of value (direct use values, indirect use values and non-use values) being estimated. Usually, direct use values of a wetland are estimated by using market price approach and indirect use values and non-use values of wetland are estimated by Contingent Valuation Method (CVM). The choice of the methods depends solely on preferences. The Benefit Transfer Method (BTM) is another popular method normally practiced when time and cost are constraints. The present article provides a comprehensive overview of the wetland valuation literature and has attempted to identify some of the wetland valuation methodologies in detail for estimation of total economic value (TEV) of wetland services. The total economic value (TEV) approach encompasses all the components of ecosystem services and expresses them in monetary or any market-based unit of measurement that helps not only in identifying the importance of the benefits derived from the services, but also allows their incorporation in policy/decision making process for conservation of wetland and their sustainable use.

References

Arrow K, Solow R, Portney PR. *et al.*1993. Report of the NOAA panel on contingent valuation. *Fed Reg.*, 58: 4601-4614.

Barbier EB, Acreman M and Knowler D. 1997. *Economic Valuation of Wetlands: A Guide for Policy Makers and Planners*, Ramsar Convention Bureau. Gland,Switzerland.

Bardecki MJ. 1998. *Wetlands and Economics: An Annotated Review of the Literature,* 1988–1998. Environment Canada,Ontario.

Bateman IJ, Willis KG, Garrod GD.*et al.*1992. *Recreation and environmental preservation value of the Norfolk broads: a contingent valuation study.* Technical report, Environmental Appraisal Group, University of East Anglia, UK.

Brander LM, Brauer I, Gerdes H. *et al.*, 2011. Using meta-analysis and GIS for value transfer and scaling up: valuing climate change induced losses of European wetlands. *Environmental and Resource Economics,* 52(3): 395-413.

Brouwer R, Langford IH, Bateman IJ. *et al.*1999. 'A Meta- Analysis of Wetland Contingent Valuation Studies', *Regional Environmental Change,* 1: 47-57.

Champ PA, Bishop RC, Brown TC and McCollum DW. 1997. Using donation mechanisms to value non use benefits of public goods. *J. Environ. Econ. Manag.*, 33: 151–162.

Costanza R, d'Arg R, de Groot R. *et al.*1997. The value of the world's ecosystem services and natural capital. *Nature,* 387: 253-260.

Crossman ND and Bryan BA. 2009. Identifying cost-effective hotspots for restoring natural capital and enhancing landscape multi-functionality. *Ecological Economics,* 68: 654-668.

Crossman ND, Bryan BA and Summers DM. 2011.Carbon payments and low-cost conservation. *Conservation Biology,* 25: 835-845.

Daily G. 1997. *Nature's Services. Societal Dependence on Natural Ecosystems.* Island Press, Washington DC, USA.

Dise NB. 2009. Peatland response to global change. *Science,* 326: 810-811.

Ellis GM and Fisher AC. 1987. Valuing environment as input. *Jr.Environmental Management,* 25: 149-156.

Enjolras G and Boisson J. 2008. Valuing Lagoons Usinga Meta-analytical Approach: Methodological and Practical Issues. Working Papers DRno.2008-05. LAMETA Universite Montpellier1, Montpelleir.

Garrod G and Willis KG. 1999. *Economic Valuation of the Environment.* Edward Elgar, Cheltenham.

Ghermandi A, vanden Bergh JCJM, Brander LM. *et al.*, 2010.The values of natural and human-made wetlands: a meta-analysis. *Water Resources Research,*46: W 12516.

Haab T and McConnell K. 2002. Valuing environmental and natural resources: the econometrics of non-market valuation. Edward Elgar Publishing Limited, Cheltenham.

IUCN 2006. Can Lao PDR afford not to invest in conserving its biodiversity? Exploring the need for innovative financial mechanisms, Vientiane, Laos: The World Conservation Union (IUCN).

Kazmierczak RF. 2001. *Economic Linkages between Coastal Wetlands and Hunting and Fishing: A Review of Value Estimates Reported in the Published Literature.* Baton Rouge: Louisiana State University Agricultural Center, Staff Paper 2001- 03.

Kyophilavong P. 2005. *Applied Econometrics,* Faculty of Economics and Business Management, National University of Laos, Vientiane.

Kyophilavong P. 2008. Impact of Smallholder Irrigation on the Aquatic Water Resources - Case Study: That Luang Marsh, Vientiane, Laos., EEPSEA Research Report, the Economy and Environment Program for Southeast Asia.

Kyophilavong P. 2011. Economic evaluation of wetland ecosystems in Lao PDR – case study of BKN, Champasak, Laos, UNDP.

Mäler K, Gren I and Folke C. 1994. Multiple use of environmental resources: a household production function approach to valuing natural capital. In. *Investing in natural capital* (Jansson A, Hammar M, Folke C and Costanza R. eds.). Island Press, Washington DC. pp. 234-249.

Millennium Ecosystem Assessment (MEA) 2005. *Ecosystems and Human Well-Being: Synthesis.* Island Press, Washington DC, USA.

Mitsch WJ and Gosselink JG. 1986. *Wetlands.* Van Nostrand Reinhold, New York.

Mitsch WJ and Gosselink JG. 2000. *Wetlands,* 4th edn. John Wiley, New York.

Oglethorpe DR and Miliadou D. 2000. Economic valuation of the non-use attributes of a wetland: a case-study for Lake Kerkani. *Environ. Plan. Manag.,* 43(6): 755-767.

Pattanayak SK and Kramer RA. 2001. Worth of watersheds: a producer surplus approach for valuing drought mitigation in Eastern Indonesia. *Environment and Development Economics,* 6(01): 123-146.

Payne CR and Sand P. 2011. Environmental liability: Gulf War Reparations and the UN Compensation Commission. Oxford University Press, London, New York.

Pramod A, Kumara V and Gowda R. 2011. A study on Physico-Chemical Characteristics of Water in Wetlands of Hebbe Range in Bhadra Wildlife Sanctuary, Mid-Western Ghat Region, *Indian Journal of Experimental Sciences,* 2(10): 9-15.

Russi D, ten Brink P, Farmer A. *et al.,* 2013. *The Economics of Ecosystems and Biodiversity for Water and Wetlands;* Institute for European Environmental Policy (IEEP), London, UK. and Ramsar Secretariat, Brussels, Belgium.

Sarkar A and Upadhyay B. 2013. Assessment of the Variations in Physico-Chemical Characteristics of Water Quality of the Wetlands in District Manipuri (UP),

India. *International Journal of Geology, Earth and Environmental Sciences*, 3 (1): 95-103.

Smith VK and Pattanayak K. 2002. 'Is Meta-Analysis A Noah's Ark for Non-MarketValuation?' *Environmental and Resource Economics*, 22: 271-296.

Smith VK, van Houtven G and Pattanayak SK. 2002. Benefit transfer via preference calibration: 'Prudential algebra' for policy. *Land Econ.*, 78: 132-152.

Stanley TD. 2001. 'Wheat from Chaff: Meta-Analysis as Quantitative Literature Review'. *Journal of Economic Perspectives*, 15: 131-150.

Stevens TH, Benin S, Larson JS. 1995. Public attitudes and economic values for wetland preservation in New England. *Wetlands*, 15: 181-194.

TEEB. 2008. *The Economics of Ecosystems and Biodiversity: An Interim Report*. European Communities. Brussels, Belgium.

TEEB. 2010.*The Economics of Ecosystems and Biodiversity: Ecological and Economic Foundations* (Kumar P. ed.). Earthscan, London, UK.

TEEB. 2013. *The Economics of Ecosystems and Biodiversity for water and wetlands.* Institute for European Environmental Policy (IEEP) and Ramsar Secretariat. London and Brussels. pp. 78.

Turner RK, van de Bergh JCJM and Barendregt A. 1997. Ecological-economic analysis of wetlands: Science and Social Science Integration, Global Wetland Economic Network (GWEN).

Turner RK, van den Bergh JCJM, Soderqvist T. *et al.*, 2000. 'Ecological-Economic Analysis of Wetlands: Scientific Integration for Management and Policy'. *Ecological Economics*, 35: 7-23.

Wattage P and Mardle S. 2008. Total economic value of wetland conservation in Sri Lanka: Identifying use and non-use values. *Wetlands Ecol. Manage.*, 16: 359-369.

Westman W. 1977. How much are nature's services worth? *Science*,197: 960-964.

Whitehead JC and Blomquist CG. 2006. Contingent valuation and benefit-cost analysis. In. *Handbook on contingent valuation* (Alberini A and Kahn JR eds.). Edward Elgar Publishing, Cheltenham, UK. pp. 66-91.

Woodward RT and Wui YS. 2001. 'The Economic Value of Wetland Services: A Meta-Analysis'. *Ecological Economics*, 37: 257-270.

Zedler JB and Kercher S. 2005. Wetland resources: status, trends, ecosystem services and restorability. *Annual Review of Environment and Resources*, 30: 39-74.

Part III

Wetlands: Challenges and Conservation Initiatives

Chapter 23

Wetland Conservation: Indian Perspective

Asish Kumar Ghosh

Centre for Environment and Development, Kolkata

Introduction

Wetlands have been defined under different system of classification from time to time and cannot be defined by a single ecologically sound and universally acceptable definition. U.S. Fish and Wildlife Service categorized them as Marine, Estuarine, Riverine, Lacustrine and Palustrine (Cowardene *et al.*,1979). Trisal and Zutshi (1985), proposed an Indian system and divide them as natural and man-made wetlands. However, Himalayan wetlands, Peninsular wetlands, both Coastal and Inland, have since attracted attention of researchers over the years. Wetlands in India have been variably calculated from 7.6 million hectares to 58.3 million hectares. Wetlands mapping have also been carried out at 1: 2,50,000 to 1: 50,000 scale between 1993 to 2011. The area under wetlands in India has been calculated with a wide range of variations. In 1993, it was calculated at 58.3 mha., by WWF-India and AWB, but inclusion of paddy field covering 71 per cent of such area, caused a confusion, while, MoEF (1996), estimated it at 4.1 mha., excluding mangroves. Although these wetlands provide fishes and other protein food, help carbon sequestration, abatement of pollution, conserve biodiversity and also in flood control, rapid urbanization, expansion of agriculture, increasing industrial pollution are making the wetlands as one of the most endangered and vulnerable natural resource. The Man and Biosphere Programme (MAB-India) initiated in 1970's, included studies on effects of human activities on lakes, marshes, river-deltas, estuaries and coastal zone, from domestic sewage, agricultural practices, industrial wastes, recreational activities and exotic biota. Wetlands services, besides the goods derived from the ecosystem have now been well defined. However, a trend of conversion of wetlands seems to be on the rise, especially in the peri-urban region of India, to meet ever increasing demand of land. Urban wetlands have undoubtedly been the worst victims of wetland conservation in a country of ever increasing population. Studies over last 25 years would reveal both serious research outputs for better understanding of this precious ecosystem and also increasing public concern in protecting urban wetlands. An analysis of extant policy and laws show the inadequacy of the system of governance. Suggestions have been given as how an effective national law may be implemented to protect the wetlands.

Wetland Mapping

The first wetland mapping by Space Application Centre (SAC) was carried out in 1993 at 1: 250,000 scale, using IRS 1A LISS-I II data of 1992-93. This was followed up in 2004-05. In 2007, a project on National Wetland Inventory and Assessment was again formulated by SAC, which was approved and funded by Ministry of Environment and Forests. The main objectives of the projects were:

- To map the wetlands on 1: 50000 scale using two-date (pre and post-monsoon) IRS LISS III digital data following a standard wetland classification system.
- Integration of ancillary theme layers (road, rail, settlements, drainage, administrative boundaries)
- Creation of a seamless database of the states and country in GIS environment.
- Preparation of State-wise wetland atlases.

All states of India, at that time a total of 28 were covered under this programme by 2011 and in addition, separate atlases have been published for Delhi, Andaman and Nicobar Islands, Lakshadweep, North East India and Union Territories of India (Figure 23.1). Space Application Centre (Garg *et al.*, 1998; Garg, 2013) put it at 7.6 mha, while, National Wetland Atlas (2011) mapped a total of 201503 wetland at 1: 500000 scale, besides 555,557 small wetlands (<2.25ha) covering a total area of 15.3 mha; this includes 69 per cent inland wetlands, 27 per cent coastal wetlands and 4 per cent other wetlands (Bassi *et al.*, 2014).

Wetland Services

The Millennium Assessment Report used four typologies to focus on ecosystem services.

- **Provisioning services**: Wetland resources for example, food (fishes, molluscs, crustaceans, edible plants), raw material (reed and pith), genetic resources (bacteria, algae, fungi, aquatic flora and fauna), medicinal resources (biota of medicinal value) ornamental resources (flowers like lotus and molluscs and shells) as reported by Alfred and Nandi (2000, 2002a, 2002b).
- **Regulating services**: Supporting essential ecological process like controlling flood, recharging ground water, controlling microclimate, help in the treatment of waste water at primary level etc.

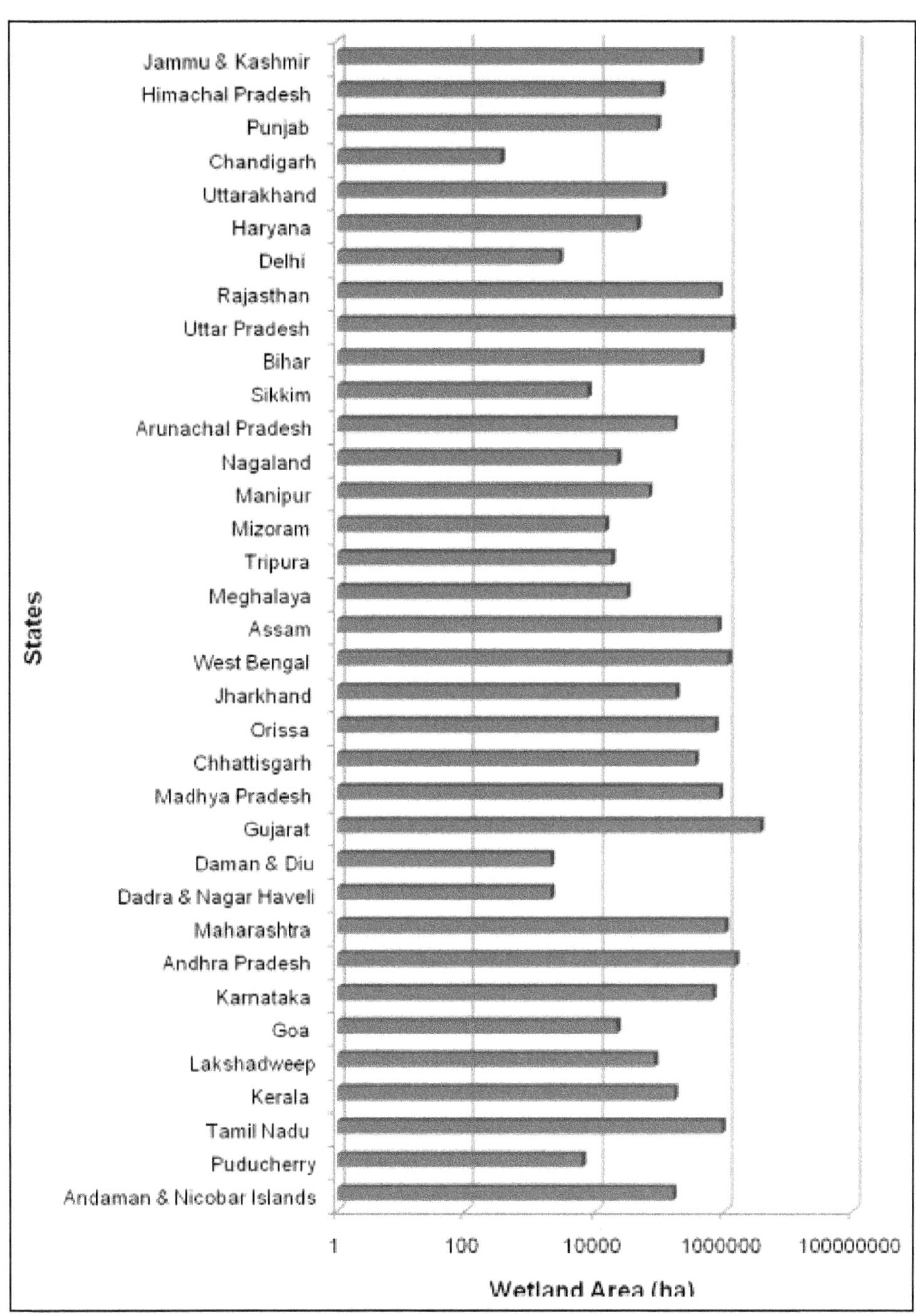

Figure 23.1: States and UTs with Wetland Area.
(*Source*: SAC/ISRO, 2011).

- **Cultural and amenity services**: Wetlands for recreational (boating, swimming, water sports), and cultural purposes (religious rituals).
- **Supporting services**: Wetlands acting as habitat for diverse living organisms, from bacteria to mammals, from algae to macrophytes etc. and also offering habitat for seasonal migratory birds. It is interesting to note that an estimate of wetland biota from the freshwater wetlands of India has been put at a total of 8000 only for the faunal resources, approximating nearly 8 per cent of the Indian fauna. Wetland ecosystem including lakes can be evaluated on the basis of each of the criteria. The Assessment Report recorded that 7 per cent of global space and 45 per cent of global productivity, depend on the wetlands in the world.

Vanishing Wetlands

The process of urbanization and increasing human population, has been putting tremendous pressure, searching for new land area. The worst victim of such a process is undoubtedly, the wetlands in the urban and peri-urban areas in India. The following table will reveal the trend of declining number of wetlands in major urban centre of India (Table 23.1).

Table 23.1: Vanishing Wetlands of India

Name of City	*No. of Wetlands*	
	Year (1962)	*Year (2012)*
Ahmadabad[1]	137	72
Bangalore	262	10
Delhi	44	21
Kolkata (Ward No. 2)[2]	32	4

1 CSE, 2013. Briefing paper: Legal institutional and technical framework for lake/wetland protection. 14pp. June, 2003.

2 ENDEV, 2005. A report on the status of wetlands in Kolkata Municipal Corporation. Ward No.2.KMC Project. Society for Environment & Development.

The process of urbanization is perhaps the single most triggering factor for conversion of wetlands. In a study carried out by the Institute of Town Planners, Kharagpur Regional Centre (1990), entitled "Development pressure on water bodies in Suburban area of Calcutta with special reference to Chandernagore, distribution of water bodies in 1970-71 vis-à-vis 1990 were plotted in maps. Out of 22 Wards, (where 6, 21,790 sq. km of wetlands existed) an area of 51,322 sq.km was filled up between 1957-1984, the proportion of filling up varied between 0.50 per cent to 42.94 per cent. A more recent study could have revealed the current situation. In the report under reference it was recommended that no water body above 500 sq.km., should be converted by land fill up. Needless to say that such an advice does not find support.

Very close to Kolkata city, A survey undertaken by ENDEV-Society for Environment and Development for Kolkata Municipal Corporation (KMC), within KMC area, recorded 1177 wetlands, while 46 listed wetlands could not be traced in Ward 101-125. KMC had provided a list of only 603 wetlands in 1999, which means more than 600 wetlands were never recorded by KMC. At least 41 of these wetlands were maintained by Fishery Cooperatives, 105 by Housing Societies, 51 by Corporations and other Government Departments, 795 under Private ownership etc., in Ward 126-141, 1263 wetlands were listed by KMC but 167 wetlands were recorded additionally. During the survey, land filling and encroachment were recorded in some wetlands and owners prohibited photography with the intention of land filling. A survey in 2015 may provide current situation.

Again in peri-urban region, onslaught on wetlands continues especially in East Kolkata area. Across the river Hooghly, protest against wetland conversion led to murder of prominent political leader in Howrah; one can remember, similar incidents in South Dumdum municipality in 1990's, resulting in the murder of the Chairman of the Municipality. More recently, during 2013-14, conversion of 1300 acres of wetlands, commonly called *Dankuni Jala*, in open daylight, attracted attention of the civil society and the media; however, media report failed to stop the massive process of draining out and filling up and it was only through Public Interest Litigation case in the Division Bench of Kolkata High Court, that the conversion was stopped. The district administration was directed to recover the cost of filling up of the drained wetland from the offending company. The most blatant violation of East Kolkata Wetland Act was witnessed in 2013, when a large *Bheri*, measuring 43 bighas (1 bigha=14,440 sq ft) was openly drained out in broad daylight and a 3 m brick wall was constructed around the drained out wetland in Ward No. 108, under Kolkata Municipal Corporation. This was a part of the Ramsar Site and public outcry and media support, finally compelled the authorities to demolish the wall and allow the wetland to naturally regenerate. It is to be noted that no offender was traced or penalized for the blatant violation of law.

While the filling up of wetlands in and around the capital city of Kolkata, attract public attention which in turn is often reflected in the media, but rampant conversion of the wetlands in the districts of the state, remain mostly unreported. Unless a state level Act recognizing such offences as cognizable and non-bailable comes into effect, it seems that the current Inland Fisheries (Amendment) Act, 1994, remains ineffective. A mechanism of fast action by Police Department when reported through the concerned Environment/Fisheries Department is urgently needed.

Wetland Restoration

The largest coastal lagoon in Asia, Lake Chilika in Odisha, provides one of the best examples of wetland restoration in the country. Lake Chilika, covering nearly 1000 sq km, spread along three districts of Odisha, offer an excellent example to understand the contribution of both wetland goods and services. While fish resources of Lake Chilika provide livelihood to 132 villages around the lake, the annual tourist traffic during the winter months who witnessed the largest flock of migratory birds, benefit the local market, travel agencies, boat operators, hospitality services, by way of facilitating revenue earning from ecosystem services. Chilika also is well known for harbouring a biota of more than 800 species, which was first surveyed in the second decade of the 20th century and resurveyed after 60 years. The lake underwent drastic changes due to heavy silt load brought by freshwater streams of Eastern Ghats, resulting in infestation of weed (*Potamegaton*); the change in the lease policy of Lake Chilika allowing prawn culture within the lake by using zero-mesh netting which led to disturbance in hydrological flow and helped in the quick spread of the weed. The mouth of Chilika was closed with sand bars, preventing inflow of the marine water and so was the only other connectivity with Bay of Bengal, namely the Palur canal in Ganjam district, which was blocked by the prawn cultivators. Due to drastic fall in economic income, more than 10,000 fishermen from the area joined a protest in the capital city of Bhubaneshwar and police firing led to death of four fishermen. Chilika was the most prominent Ramsar Site in India in the 1990's and was put in the Red List by the Ramsar Bureau because of its highly degraded condition. A World Bank sponsored Odisha Water Resource Consolidation Project (ORWCP) in late 1990's, provided an opportunity to prepare an Integrated Management Plan for restoration of Lake Chilika. The plan was subsequently implemented by the Chilika Development Authority, following every component of the action suggested; a new mouth was opened to allow marine water intrusion and new plantations came up in the catchment area to prevent soil erosion; an Interpretation Centre has been set up for public awareness and education; regular post-project monitoring by competent agency was made; community participation was ensured at every step and Chilika was awarded the best Ramsar Site award in 2003 (Ghosh *et al.*, 2006).

Policy and Laws

The conservation of lakes, wetlands, rivers have been concern of environmental policies, announced in 1992 and 2006 but supporting laws for protection of wetlands, still demand serious attention. An analysis carried out by the present author for WWF-India revealed that there is no effective and a stringent law to protect numerous water bodies across the country. Ramchandra and Ahalya (2001) has pointed out that "wetlands do face tragedy of commons" and stated that prevailing laws related to wetlands (a total of 11) are ineffective and India badly needs a National Wetland Conservation Act. It may be noted that in 2010 "Wetland Conservation and Management Rules" were formed under Environment (Protection) Act of 1986. The Rules are focused on Ramsar Site, wetlands in ecologically sensitive areas, wetlands in UNESCO Heritage Sites, wetlands in high altitude and wetlands located below 2500 m, but more than 500 ha in area. The Rules do not mention about a penalty clause for the violation. However, as has been pointed out by Raju (2012) that three lists- Union, State and Concurrent do not clearly mention as to who will legislate for matters relating to wetlands; this subject is being neglected when India is losing this valuable resource at a fast pace since, the process of globalization. It is further suggested that Union of India cannot enact an umbrella law unless the subject of water and agriculture, is shifted from List II to List III of the Seventh Schedule of the Constitution. Nevertheless, the Government of India has the power to protect all 25 of the Ramsar Sites by using article 253 of the Constitution, which empowers the government to implement "any treaty, agreement or convention with any other country or countries or any decision made at any international conference". By virtue of the fact that India is a signatory of the Ramsar Convention of 1971, the use of above Article can perhaps be strongly argued. It is argued that since water is a state subject, "states have exclusive jurisdiction past legislation upon wetlands in the federal system of India."

Lakes and wetlands by physical and hydrological characteristics cannot be separated; while all lakes are wetlands, all wetlands cannot be designated as lakes. The Government of India through programmes of Ministry of Environment, Forests and Climate Change (MoEFCC), have been undertaking implementation of Action Plan under two centrally sponsored schemes *viz.*, National Wetland Conservation Programme (NWCP) started in 1987 and National Lake Conservation Programme (NLCP) started in 2001. As for Annual Report of MoEF (2013-2014), a total of 61 lakes in 14 states have come under NLCP, at a cost of Rs. 1031.18 crores, while Rs. 133 crores were allotted for conservation of identified wetlands. However, the Government of India has, in February 2013, approved merger of these two schemes into "National Plan for Conservation of Aquatic Ecosystem" (NPCA). During 2014-2017, the merged scheme will have a total outlay of Rs. 525 crores, with a cost sharing basis of 70: 30 between Central and State governments but in case of North-East India, state's share is fixed at 10 per cent. Nationally, a total of 115 wetlands have been included under this programme for conservation and management. The need for a regulatory authority, long failed, has also been endorsed during 2013-2014, under Rule 5 of Environment (Protection) Act of 1986, and a Central Wetland Regulatory Authority (CWRA) has been formed with the Secretary of MoEFCC in the Chair, whose functions are as follows:

- ✰ "Appraise proposals for identification of new wetlands;
- ✰ To enforce the provisions contained under these rules along with other laws in force;
- ✰ Grant clearances or identify the areas for the grant of clearance for regulated activities in the wetlands under jurisdictions;
- ✰ Issue whatever directions, from time to time necessary for the conservation, preservation and wise use of wetlands to the State Governments;
- ✰ Review the list of wetlands and the details of prohibited and regulated activities under the rules and the mode and methodology for execution" (MoEF, 2013-14).

It is interesting to note that the Government of India had, while initially allocated Rs. 70.50 crores for implementation of the programme, has drastically reduced it by more than 20 per cent in the revised budget to Rs. 55.85 crores.

Conclusion

The forgoing account clearly indicate a state of crisis that the wetland resources of India continue to face. While wetlands in rural India are always considered an wealth of common public interest, the process of urbanization specially after 1991, in a country with limited land area, is posing greatest threat towards vast numbers of wetlands across the country. Unless the value of wetland can be appreciated by the system of governance and unless a stringent mechanism of putting appropriate laws into place and an effective process of implementation through strict enforcement and exemplary punishment for violation, is witnessed and recorded, it appears that the country will be losing its resources at a fastest pace, in a drive to achieve a higher percentage of GDP in near future. The trend of environmental governance in India is fast changing, and that too changing for the worst. Environmental objections are often termed as 'Green blockade' and such a trend does not speak well about the future. Wetlands will continue to be a case of tragedy of commons unless actions are taken now.

References

Alfred JRB and Nandi NC. 2000. Fauna diversity in Indian wetlands. *ENVIS Newsletter*, Zoological Survey of India, Kolkata, 6(2): 1-3.

Alfred JRB and Nandi NC. 2002a. Wetlands: Freshwater. In. *Ecosystems of India.* ENVIS, Zoological Survey of India. pp.165-193,

Alfred JRB and Nandi N.C. 2002b. Wetlands: Brackish Water. In. *Ecosystems of India.* ENVIS, Zoological Survey of India. pp.195-214.

Bassi N, Dinesh Kumar M, Sharma A. and Pardha-Saradhi P. 2014. Status of wetlands in India: A review of extant, ecosystem benefits, threats and management strategy. *J. Hydrol.,* 2: 1-19.

Cowardene LM, Carter V, Golet FC and LaRoe ET. 1979. Classification of wetlands and deep water habitats in the United States. US Fish and Wildlife Services, Washington DC. pp.103.

CSE. 2013. Briefing paper: *Legal institutional and technical framework for lake/wetland protection.* pp.14.

ENDEV. 2005. *A report on the status of wetlands in Kolkata Municipal Corporation. Ward No.2.* KMC Project. Society for Environment and Development.

Garg JK, Singh TS and Murthi TVR. 1998. Wetlands in India, Project Report. Space Application Centre, Ahmadabad.

Garg JK. 2013. Wetland assessment, monitoring and management in India using geospatial technique. *J.Env.Mng.,* 148: 112-123.

Ghosh AK, Pattnaik A and Ballatore TJ. 2006. Chilika lagoon: restoring ecological balance and livelihoods through re-salinization. *Lakes and Reservoirs: Research and Management,* 11: 239-255.

MoEF. 1996. Annual Report, Ministry of Environment and Forests, Government of India.

MoEF. 2013-14. Annual Report, Ministry of Environment and Forests, Government of India.

Ramachandra TV and Ahalya N. 2001. Wetland restoration and conservation. www. wbgis.ces.ernet.in.

Raju K. 2012. The wetland jurisprudence in India. India Environmental Portal.org.

SAC/ISRO. 2011. National Wetland Atlas. Space Application Centre/Indian Space Research Organisation, Ahmedabad.

Trisal CL and Zutshi DP. 1985. 'Ecology and Management of Wetland Ecosystems in India', *Proc. National MAB Committee of Central and South Asian Countries,* New Delhi. pp.195-214.

Chapter 24

Wetlands in Urban Landscapes: Problems and Potentials

Subhendu Mazumdar[1] and Goutam Kumar Saha[2]

[1]Shibpur Dinobundhoo Institution (College), Howrah
[2]Entomology and Wildlife Biology Research Laboratory, Department of Zoology, University of Calcutta, Kolkata

Introduction

Wetlands are one of the most productive, as well as, fragile ecosystems (Mulamotti *et al.*,1996) and are defined as 'lands transitional between terrestrial and aquatic ecosystems where the water table is usually at or near the surface or the land is covered by shallow water (Mitsch and Gosselink, 2000). These globally important, ecologically sensitive and adaptive ecosystems (Zhao *et al.*, 2005) are present in different climatic zones of the world in almost every continent occupying about 6.4 percent of the earth surface (IUCN, 1990). Most of the earliest civilization flourished along the rivers and riverine floodplains. Despite such long association with human beings, wetlands were neglected over the years and were often considered as 'waste lands'. Fortunately this perception has now largely changed and now-a-days wetlands are valued as important ecosystems throughout the globe for their multifarious contribution towards a healthy environment. But the values of wetlands vary among different communities (Mulamotti *et al.*, 1996). Harvesting of fish, prawn and other aquaculture products, as well as, collection of various wetland flora for consumptive use were traditionally important, and also play pivotal role in sustenance of many local communities presently living around these landscapes (Raburu *et al.*, 2012). In addition to the socioeconomic values of these floral and faunal resources, wetlands are also valuable from ecological perspective. Ecological benefits of wetlands include reduction in soil erosion, shoreline protection, flood attenuation, drought control, groundwater recharge, retention of nutrients and pollutants, cleaning of water, modification of microclimates and carbon sequestration (Prasad *et al.*, 2002; Sheoran and Sheoran, 2006). But as compared to the direct consumptive values, such non-consumptive values received lesser priority in the management plans of wetlands over the years. The underlying reason is possibly that these values though ecologically and aesthetically significant, yet are difficult to quantify (Das *et al.*, 2000). Nevertheless such values are increasingly gaining importance in different parts of the world since recent past. According to Turner and Jones (1991) such benefits and values of wetlands extend well beyond its boundaries. In addition to various ecological services, wetland habitats are also rich in biodiversity (Reid *et al.*, 2005). Despite being present over only around 6 percent of Earth's surface, wetlands support one fifth of the global biodiversity, provide suitable habitat for a wide array of flora and fauna (Van der Valk, 2006) and some wetlands even serve as last refuge for many threatened species. At least 20 percent of the threatened bird species inhabit various wetlands of Asia (Kumar *et al.*, 2005). Thus wetlands undoubtedly play vital role in maintaining populations of wild flora and fauna (Zhao *et al.*, 2005) and, therefore, continued existence of these wetlands is vital for wildlife conservation of a region.

Concept of Urban Wetlands

Natural and man-made wetlands of varied shape and size exist in many metropolitan areas (Mackintosh and Davis, 2013). In this chapter, we considered wetlands (natural or man-made) situated in urban areas as 'urban wetlands', which are highly influenced by human activities and form an integral part of the cityscapes. Such urban wetlands are found all over the country and have served as lifeline of many cities in India through ages. A total of 67429 wetlands are present in India covering an area of about 4.1 million hectares, most of which (96.77 per cent, n=67429) are man-made (MoEF, 1990). Likewise, in West Bengal, different types of wetlands account for about 12.5 per cent geographical area of the state and out of a total of 8670 inland wetlands, 4995 (57.6 per cent) are man-made (Anon, 2011). Despite of the fact that many wetlands are found to be present in metropolis, cities and towns, yet researches on urban wetlands in this country are still in its nascent stage. It is to be noted here that population explosion is a global phenomenon (Brown *et al.*, 1998) and is often associated with an increasing urban footprint (McDonald *et al.*, 2009). As a consequence, urban sprawl is noticed throughout the globe (UNDP, 2005). About half of the world's populations presently live in cities, which are projected to rise to 60 per cent by 2030 (Mackintosh and Davis, 2013). Rate of urbanization in India is also not an exception. About 25.72 per cent of country's population lived in cities and towns in 1991, which subsequently increased to 31.16 per cent in 2011 (Bhagat, 2011). Consequently, many natural landscapes along with many rural areas are rapidly been converted to urban areas (O'Meara, 1999; Antrop, 2004). Therefore, prioritizing research on urban wetlands is of paramount importance to assess their socioeconomic values, to evaluate their ecological services and also to perceive the threats to these important ecosystems.

Crisis to Urban Wetlands

Over half of the wetlands in the world have been lost in the last century and the remaining wetlands have been degraded to different degrees because of the adverse influence of human activities (Fraser and Keddy, 2005). Many wetlands have been drained and transformed by various anthropogenic activities, *viz.* unplanned urban and agricultural development, industrial sites, road construction, impoundment, over extraction of resources, etc. (Ramachandra *et al.*, 2002). Wetlands in urban areas are also continually subjected to various anthropogenic disturbances. The threats to urban wetlands may broadly be grouped under five major categories, *viz.* loss of wetland area, changes to water regime, changes to water quality, overexploitation of wetland products and introduction of exotic or invasive species (Ramachandra, 2001).

The land values in urban and peri-urban areas have increased manifolds and the real estate sector has also flourished incredibly in recent times. Despite of high land values, suitable upland for development of residential, commercial and industrial facilities have been exhausted due to increasing population in metropolitan areas and economic growth of people in urban areas (Zedler and Leach, 1998). Any undeveloped stretch of land in urban areas is, therefore, a valuable commodity. This dearth of land, subsequently, exerts immense pressure on urban wetlands (Ramachandra *et al.*, 2002). Landowners are often enticed with large financial incentives to drain privately-owned wetlands. Besides, comprehensive policy measures to protect inland wetlands are also inadequate, seldom properly executed and mostly non-existent in many urban areas. The scenario becomes even bleak as the present legal perception of wetland is often acreage based and, unfortunately, not based on their function (Ramachandra *et al.*, 2002). Exploiting these lacunae on legal provisions, wetlands in many urban areas are frequently being targeted for construction of residential and commercial apartments, infrastructural facilities, establishment of road, *etc.* (Boyer and Polasky, 2004). Though it apparently seems beneficial for landowners to fill wetlands and make profit from private development, but if the values of various resources and ecological services rendered by these urban wetlands are taken into consideration, such short-term financial profits are indeed negligible and are often unsustainable. However, the larger ecological services rendered by these wetlands are never taken into consideration since it is often difficult to place a monetary value to such services (Mackintosh and Davis, 2013). Thus, draining and infilling of wetlands and removal of its fringe vegetations for construction of housing and urban infrastructure has become a major threat in the existence of many urban wetlands (Zedler and Leach, 1998).

Another serious and recent concern to urban wetlands is the landfills. Many cities, with rising population and limited area, also face serious problem in disposing the increased volume of waste. Therefore, some wetland areas are often designated as landfill sites, where waste from different parts of the city is dumped. Construction of landfill may seriously alter the hydrology, alarmingly degrade the water quality and might even trigger rapid invasion by certain aggressive, highly-tolerant and exotic species. Exotic and invasive plants encroach and adversely alter the composition of indigenous wetland plant communities (Mackintosh and Davis, 2013).

Besides, most urban wetlands are surrounded by altered landscapes and dense human habitation in its immediate vicinity (Windham *et al.*, 2004). They receive contaminants from roads, gutters, storm drains, and occasional wastewater spills (Zedler and Leach, 1998). With roads, concrete structures and other build-up areas, there is very little scope for rainwater to percolate and reach the underground water table. Such increase in impervious area also increase the volume of urban runoff, which carry many sediments, contaminants, pollutants, mud, pesticides, fertilizers, organic matter, animal wastes, heavy metals, hydrocarbons, road salts and debris from adjacent uplands and accumulate in those urban wetlands (Ehrenfeld, 2000; Boyer and Polasky, 2004). Temperature of water rise as it moves over warmed impermeable surfaces and subsequently, the dissolved oxygen content of the runoff water decrease and the siltation rate increase. Urban wetlands often lack broad upland buffers that help filter materials before the urban run-off reaches these wetlands. Thus, the entire physico-chemical parameters of water are altered making it unsuitable for human consumption, as well as, threaten the existence of other species inhabiting these urban water bodies.

Although accurate data on wetland loss in India are unavailable, yet it would not be an exaggeration to state that many urban wetlands of this country are also facing similar threats in various degrees. A survey conducted by the Wildlife Institute of India indicated that 70-80 percent of fresh water marshes and lakes in the Gangetic flood plains have disappeared in the last five decades and presently are being lost at a rate of 2-3 percent every year (Ramachandra *et al.*, 2002). In India, smaller wetlands in and around urban areas are often used for dumping of solid and liquid wastes. Eutrophication also results in the untimely death of many smaller lakes in this country.

Urban Wetlands: A Valued Landscape

Despite being mired with various pressures, urban wetlands are intrinsically valuable for urban landscapes as they perform multiple ecosystem services including improving water quality, pollutant removal, providing hydrologic services and sustaining regional biodiversity (Zedler and Leach, 1998; Grayson *et al.*, 1999; MEA, 2005; Moore and Hunt, 2012). As the runoff passes through these wetlands, many pollutants are removed, trapped or utilized by various flora and fauna (Ramachandra *et al.*, 2002). Many large metropolitan areas utilize wetland areas for water recycling and flood control (Lawrenz *et al.*, 2012). Urban wetlands also provide intangible benefits that contribute to the quality of urban life (Bouland and Hunhammar, 1999), of which recreational, educational and aesthetic values are particularly important (West, 1995). For this reason, citizens in metropolitan areas of many developed nations value urban wetlands, which provide them with the opportunities of enjoyment and recreation through interactions with nature (Bouland and Hunhammar, 1999) that are otherwise scarce in those urban landscapes (Grayson *et al.*, 1999; Ehrenfeld, 2000). It has been found that in some cities the property values were higher in areas adjoining urban wetlands (Tapsuwan *et al.*, 2009). Emerging evidences further indicate that access to green spaces and wetlands in urban environment are important for physical health, psychological well being, cognitive ability, social cohesion and general well being of the citizens (Dallimer *et al.*, 2012; Keniger *et al.*, 2013) as shown in Table 24.1. Keniger *et al.* (2013) are of the opinion that interactions with nature during childhood influence the attitudes towards nature later in life, as well as, positively impact self esteem and mental well-being.

Table 24.1: The Positive Impacts on Well-being of Urban Citizens Provided Increased Access to Natural Environments within Urbanscapes (Adapted from Mackintosh and Davis, 2013)

Benefit	*Description*	*Examples*
Psychological	Positive effect on mental processes	☆ Increased self esteem ☆ Improved mood ☆ Reduced anger/frustration ☆ Reduced anxiety ☆ Improved behaviour
Cognitive	Positive effect on cognitive mental processes	☆ Reduced mental fatigue ☆ Improved academic performance ☆ Education/learning opportunities ☆ Improved ability to perform tasks ☆ Improved cognitive function in children ☆ Improved productivity
Physiological	Positive effect on physical function and/or physical health	☆ Stress reduction ☆ Reduced blood pressure ☆ Reduced headaches ☆ Reduced mortality rates from circulatory disease ☆ Faster healing ☆ Addiction recovery ☆ Perceived health/well being ☆ Reduced cardiovascular, respiratory disease and long-term illness ☆ Reduced occurrence of illness
Social	Positive social effect at an individual, community or national scale	☆ Facilitated social interaction ☆ Enables social empowerment ☆ Reduced crime rate ☆ Reduced violence ☆ Enables interracial interaction ☆ Social cohesion ☆ Social support
Spiritual	Positive effects on individual religious pursuits or spiritual well being	☆ Increased inspiration ☆ Increased spiritual well being
Tangible	Material goods that an individual can accrue for wealth or possession	☆ Food supply ☆ Money

Urban Wetlands: Safe haven for Wildlife

Wetlands situated in urban landscapes are important refuge for many wild fauna and some of those even serve as the only habitat supporting wildlife within the area (Hamer *et al.*, 2012). Isolated wetlands in nonurban areas may have lower species richness and be underutilized by wildlife, but similar habitats in an urban settings may be used by a wide variety of wild species (Ehrenfeld, 2000). Hence, urban wetlands are rich in biodiversity, often serve as important nursery areas for many commercial fish stalks and harbour many endemic, as well as, endangered species. According to Kiviat and MacDonald (2004), the floral and faunal diversity of urban wetlands are even higher than the adjoining terrestrial ecosystems. Such biodiversity support functions of urban wetlands are likely to become increasingly important under the 'climate change scenario'. Mackintosh and Davis (2013) opined that during dry seasons and in times of drought these wetlands will provide critical refuge for many aquatic species.

They also serve as 'stop-over sites' for many migratory birds, as well as, essential nesting and foraging sites for a variety of resident avifauna (Ma *et al.*, 2004; Benassi *et al.*, 2007; Okes *et al.*, 2008). In India, 310 species are dependent on different fresh and salt water wetlands of which 107 are winter migrants (Kumar *et al.*, 2005). Wetlands, even the ephemeral water bodies, along with other remnant natural habitats provide necessary resources such as food and shelter that facilitates the presence of many species of resident, resident migratory and migratory birds in urban areas (Murgui, 2009; Caula *et al.*, 2010). Some urban wetlands attract a wide variety of surface foragers (like teals, ducks etc.) and waders (such as sandpipers, stilts etc.). In addition to being aesthetically pleasing, these wetlands also help mitigate urbanization effects on bird species richness by providing enhanced habitat and resources (Mitsch and Gosselink, 2000) and play a crucial role in the population dynamics of many avian species, as well (Kushlan, 1986). Few of these water birds may occur only seasonally, yet these wetlands remain as vital support system for the entire population (Kushlan, 1986).

Birds are potential 'indicator species' in wetland management as they are extremely sensitive to minute and varied hydrological fluctuations. But, precise understanding about the parameters and processes that significantly influence the abundance and diversity of water birds is still lacking. The outcomes of researches done so far widely fluctuate and even come up with completely different propositions. For instance, some authors found that the area of open water (Patterson, 1976), size of wetland (Gonzalez-Gajardo *et al.*, 2009) and the size of the catchment area (Herrmaan, 2012) as some of the important factors in determining bird congregation and diversity. Conversely, McKinney *et al.* (2011) established that bird associations were independent of wetland size, suggesting that even small wetlands may provide valuable resources and act as potential bird habitat. Some other authors have opined that water depth (Colwell and Taft, 2000; Darnell and Smith, 2004) and habitat heterogeneity (Zarate-Ovando *et al.*, 2008; Gonzalez-Gazardo *et al.*, 2009) significantly influence richness, abundance and diversity of wetland birds. It has also been stated that distribution and diversity of avian species is influenced and regulated by richness and diversity of food resource (Johnson and Sherry, 2001), as well as, the diversity and cover of submerged and emergent vegetation present in wetland habitat (Mackintosh and Davis, 2013). Some authors even advocate the importance of floating vegetation in the breeding success of some wetland birds (Froneman *et al.*, 2001; Sanchez-Zapata *et al.*, 2005). On the other hand, invasive weeds pose serious threat to the water birds as they tend to reduce the foraging area by covering the water surface (Baral and Inskipp, 2004). Such lack of precise understanding on why and how the urban wetlands serve as avian abode is often a major lacuna for implementing appropriate management strategies to optimize avian diversity in those wetlands. Long-term monitoring of avian population and their habitat use in urban wetlands can thus serve as an important tool in setting appropriate management goals for these sites (Kushlan, 1986).

Migratory Waterfowls in Urban Wetlands: A Case Study

Mazumdar *et al.* (2007, 2008) conducted avifaunal surveys in six wetlands of southern West Bengal (*viz.* Nalban Bheri, Santragachi Jheel, Saheb Badh, Bakreshwar Barrage, Tilpara Barrage and in the wetlands inside Ballavpur Wildlife Sanctuary), to assess the composition, diversity and dominance of migratory and resident avifauna. Nalban Bheri (NB) is a part of the East Kolkata Wetlands (a Ramsar site), consisting of 152 ha of commercial sewage-fed fish farms owned by the Department of Fisheries, Government of West Bengal. Santragachi Jheel (SJ) and Saheb Badh (SB) are two large wetlands under constant anthropogenic pressure as both these wetlands are surrounded by roads and human habitations. The Bakreshwar (BB) and Tilpara (TB) barrages are two man-made wetlands. TB is a reservoir on Mayurakshi river and BB is constructed on the Bakreshwar river. As compared to other waterbodies surveyed, both BB and TB are more deep (average depth ~20m), and are almost devoid of any emergent vegetation and/or floating macrophytes (Mazumdar *et al.*, 2007). Excepting wetlands inside Ballavpur Wildlife Sanctuary, remaining five wetlands were located in urban/peri-urban areas. Hence, wetlands of Ballavpur Wildlife Sanctuary were kept out of the following discussion on urban wetlands.

In southern West Bengal, the population of migratory water birds reaches the peak during last week of December and the first week of January (Mazumdar *et al.*, 2005). Hence, 'total count' of water birds was carried out during January 2006 when supposedly maximum numbers of birds, including the migratory, resident migrant and resident ones, were present in those wetlands. During the survey, water birds, wetland dependent and wetland associated birds were collectively considered as 'wetland birds'.

Number and composition of migratory, resident migrant and resident species were found to be different across the sites (Figure 24.1), the dominance and diversity indices varied as well (Figure 24.2). TB supported maximum diversity of wetland birds (33 species) followed by SB (26 species), NB (25 species), BB (21 species) and SJ (16 species). Despite being lowest in diversity of wetland birds, maximum number of water birds and wetland-associated species was recorded in SJ (9,869), followed by SB (3,563), BB (1,887), TB (1,458) and NB (1,278). Highest numbers of trans-Himalayan migratory species was recorded from TB (10 spp.) and number of resident-migrant species was maximum at NB (12 spp.). The survey emphasized the importance of these urban/peri-urban wetlands as wintering grounds of migratory waterfowls, as well as, an abode for many wetland-associated avian species (Table 24.2).

Recently, some other workers also studied the avian diversity in some of these wetlands. Bhattacharyya *et al.* (2008) reported higher number of avian species (*i.e.* 39 species) from NB. This is possibly because they had compiled secondary data on water birds of a longer timescale spanning over four decades (1964-69,

Table 24.2: Composition of Water Birds and Wetland-Associated Species Recorded at different Wetlands in Southern West Bengal in January 2006 (abridged from Mazumdar *et al.*, 2008)

Water Birds	*Status*	*NB*	*SJ*	*SB*	*BB*	*TB*
Tachybaptus ruficollis	R	0	0	✓	✓	0
Podiceps cristatus	M	0	0	0	✓	✓
Phalacrocorax niger	RM	✓	✓	✓	✓	✓
Phalacrocorax fuscicollis	RM	✓	✓	✓	0	0
Phalacrocorax carbo	RM	✓	0	✓	✓	✓
Anhinga melanogaster	RM	0	0	0	0	0
Egretta garzetta	R	0	0	0	✓	✓
Ardea cinerea	RM	✓	0	0	0	0
Casmerodius albus	R	0	0	0	0	✓
Bubulcus ibis	R	✓	✓	✓	✓	✓
Ardeola grayii	R	✓	✓	✓	0	✓
Ixobrychus sinensis	RM	✓	0	0	0	0
Dendrocygna bicolor	RM	0	✓	✓	0	0
Dendrocygna javanica	R	0	✓	✓	0	✓
Anser anser	M	0	0	0	✓	✓
Anser indicus	RM	0	0	0	✓	✓
Tadorna ferruginea	RM	0	0	0	✓	✓
Nettapus coromandelianus	R	0		✓	0	✓
Anas strepera	M	✓	✓	✓	✓	✓
Anas falcata	V	0	0	0	✓	0
Anas penelope	M	0	0	0	✓	✓
Anas clypeata	M	✓	✓	✓	✓	✓
Anas acuta	M	0	✓	✓	✓	✓
Anas querquedula	M	✓	0	✓	0	0
Anas crecca	M	0	0	✓	0	0
Rhodonessa rufina	M	0	0	✓	✓	✓
Aythya ferina	M	0	0	✓	0	0
Aythya nyroca	RM	0	✓	✓	0	✓
Aythya fuligula	M	0	0	✓	✓	✓
Amaurornis phoenicurus	R	✓	0	✓	0	0
Porphyrio porphyrio	R	0	0	✓	0	✓
Gallinula chloropus	R	✓	✓	✓	0	✓
Fulica atra	RM	0	0	✓	✓	0
Hydrophasianus chirurgus	R	0	0	0	0	✓
Metopidius indicus	R	0	✓	✓	0	✓

Contd...

Table 24.2–*Contd...*

Water Birds	*Status*	*NB*	*SJ*	*SB*	*BB*	*TB*
Charadrius dubius	RM	✓	0	0	0	0
Vanellus indicus	R	0	0	0	0	✓
Vanellus cinereus	M	0	0	0	0	✓
Gallinago stenura	M	✓	✓	0	0	0
Gallinago megala	M	✓	✓	0	0	0
Tringa tetanus	RM	0	0	0	✓	✓
Tringa ochropus	M	✓	0	0	0	✓
Tringa glareola	M	✓	0	0	0	0
Actitis hypoleucos	RM	✓	0	0	0	✓
	Wetland-associated species					
Anastomus oscitans	R	0	0	0	✓	✓
Ciconia episcopus	RM	0	0	0	0	0
Circus aeruginosus	M	0	0	0	0	0
Circus melanoleucos	RM	0	0	0	0	✓
Aquila clanga	RM	0	0	0	0	0
Halcyon capensis	R	0	0	0	0	0
Halcyon smyrnensis	R	✓	✓	✓	0	0
Alcedo atthis	RM	✓	0	✓	0	0
Ceryle rudius	R	✓	0	0	0	0
Hirundo rustica	RM	✓	0	0	✓	✓
Motacilla maderaspatensis	R	0	0	0	✓	0
Motacilla alba	RM	✓	0	✓	0	✓
Motacilla citreola	RM	✓	0	0	0	0
Motacilla flava	RM	✓	0	0	0	✓

R: Resident; RM: Resident-migrant; M: Trans-Himalayan Migrant.

1978-83 and 1984-97). However, they had opined that the avian diversity in this site has dwindled from the past mainly due to reclamation of Salt Lake and other changes in land use pattern. Sinha *et al.* (2012) found a maximum of 40 species in BB, which is fairly high as compared to only 21 species observed by Mazumdar *et al.* (2007, 2008) in the same site. Sinha *et al.* (2012) mentioned that construction of Bakreswar Reservoir (or BB) was completed in 1999 and the water birds began to arrive in large numbers from 2001 onwards. Expansion of suitable wetland habitat and elevated amount of protection extended by the concerned authorities are possibly the important underlying cause beneath such consistent increase in the abundance and diversity of water birds at BB in the recent years. In addition to these management and conservation initiatives, it is also to be noted that wetland vegetation and faunal composition positively influence water bird abundance and diversity (Bellrose, 1980; Helmers, 1992). Thus, congregation of large numbers and rich variety of water birds in any newly constructed man-made wetland always

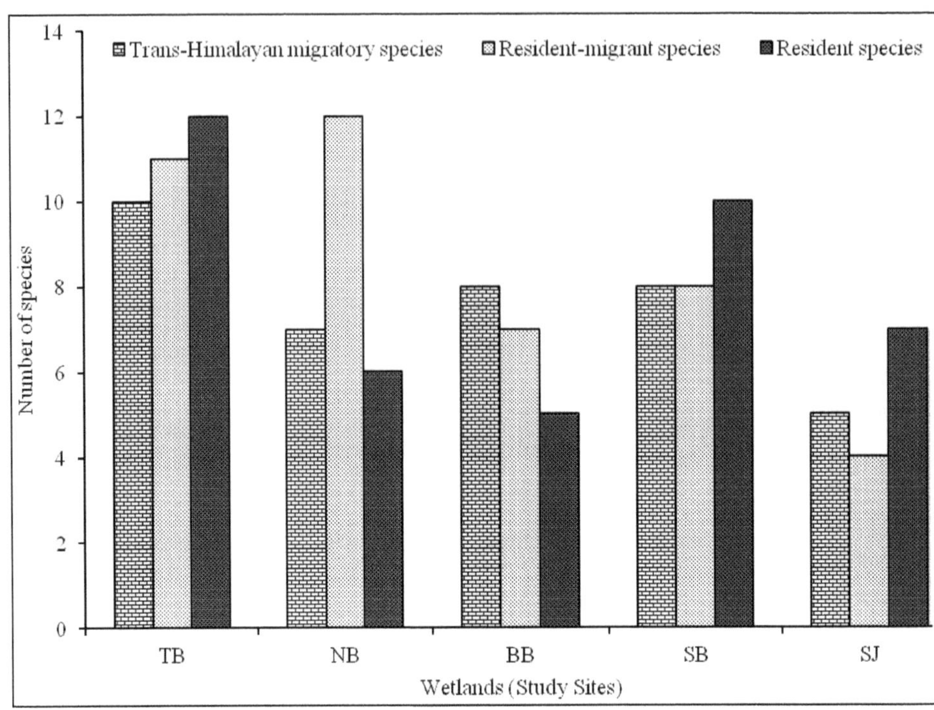

Figure 24.1: Number of Migratory, Resident-Migrant and Resident Species of Wetland Avifauna in the Study Sites (Abridged from Mazumdar *et al.*, 2008).

begins after a certain time span during which the wetland get enriched in various floral and faunal resources and gradually become a suitable and favoured habitat for the water birds. In the case of BB, all these factors seem to be important causes for lower water bird diversity in this wetland during the earlier surveys carried out by Mazumdar *et al.* (2007, 2008) and subsequent rise in the number of species as reported by Sinha *et al.* (2012). From both these surveys it has become evident that BB is coming up as a very important abode for both migratory and resident wetland birds in southern West Bengal. On the other hand, Khan (2010) reported a declining trend of the diversity of water birds in SJ due to loss of surface water area caused by rapid expansion of water hyacinth cover over the wetland.

Conclusion

Finally it is to be admitted that wetlands are integral part of many urbanized landscapes with potential of multiple ecosystem services, *viz.* supply of freshwater, flood control, nutrient retention, pollution amelioration, microclimate alteration, carbon sequestration, as well as, acting as abode of wildlife and biodiversity (Mackintosh and Davis, 2013). Urban wetlands also significantly contribute towards healthy environment and well-being of the city dwellers. In India, existence of healthy wetlands are absolutely indispensable to ensure the availability of water for humans and livestock, for sustainable agriculture, as well as, for the continued existence of its diverse floral and faunal populations, particularly those which are

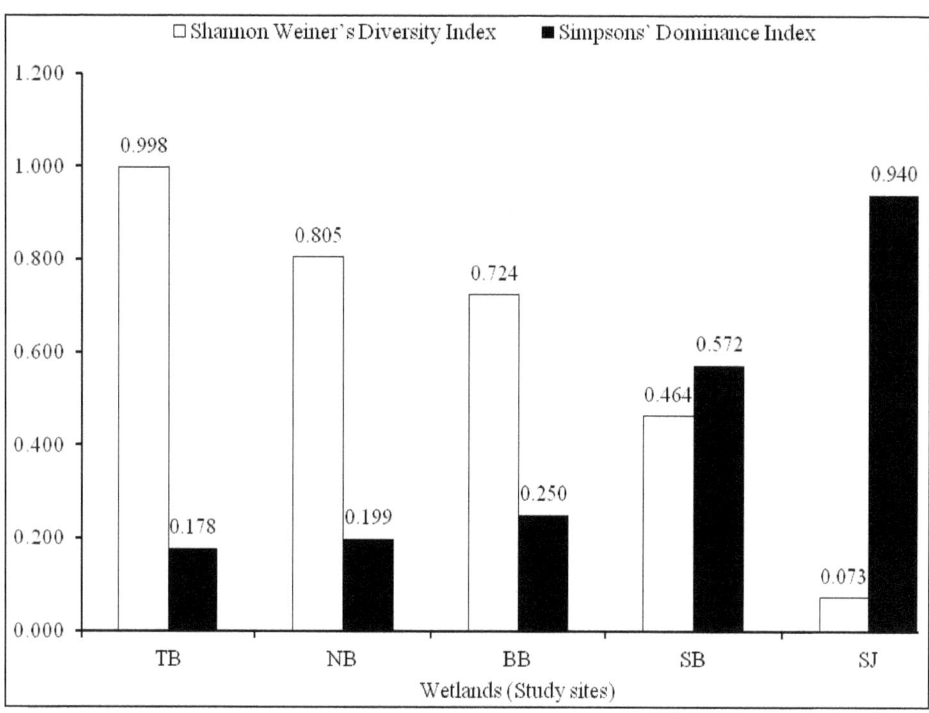

Figure 24.2: Comparative Account of the Dominance and Diversity Indices in the Study Sites (Abridged from Mazumdar *et al.*, 2008).

wetland dependent. Despite of providing various important functions and services, many wetlands of this country, particularly those situated in urban landscapes, are facing serious threats from rapid urbanization and burgeoning population. Conservation and management of these urban wetlands need to be based on scientific knowledge, but studies appraising the value of wetlands within urban areas of this country are almost nonexistent and awareness regarding the ecological services rendered by these water bodies is also very poor among the citizens in many urban areas of this country. Conservation of urban wetlands is thus undoubtedly challenging.

Many of these wetlands are safe haven for a number of wetland birds, which are sensitive to ecosystem changes. Hence, along with inventory and in depth socioeconomic and scientific researches on various ecosystem functions of urban wetlands, studies on diversity and habitat use of these avian species also need to be prioritized. Over and above, strong public support for protection of wetlands is indispensable to combat the persistent pressure to modify wetlands. To gain societal support wetlands managers need to reconnect local people with these ecosystems and garner public appreciation by providing opportunities for passive recreation and low-impact activities (such as bird-watching), as far as practicable. Also the local communities need to be empowered so that they develop a sense of ownership with the wetland. While ensuring the livelihood of the communities depending on these wetlands, these people also need to be made aware of the significance of the wetland not only in their own sustenance but also in the survival of other species as well. Such augmented ecological knowledge can reduce barriers to environmental activism (Taylor, 1989). In conclusion, collaborative scientific and public efforts, political motivation, mass awareness and activism are decisive in conserving these productive and precious ecosystems amidst urbanized landscapes.

Summary

Wetlands are highly adaptive, ecologically sensitive and fragile ecosystems. In addition to the socioeconomic values of wetland resources, wetlands are also important for their multifarious ecological benefits like reduction in erosion, shoreline protection, flood attenuation and drought control, groundwater recharge, retention of nutrients and pollutants, cleaning of water, modification of microclimates, carbon sequestration and biodiversity support. Due to rapid urbanization, many natural landscapes along with many rural areas are rapidly been converted to urban areas throughout the globe. Wetlands are often an integral part and serve as lifeline of many such urban landscapes. But, the importance of wetlands in urban areas was hardly realized by the citizens and policy makers, even in the recent past. Unfortunately, many such wetlands were drained and filled up for construction of residential and commercial apartments and other developmental activities. Many others have been degraded by inflicting changes to water regime, water quality, as well as, by overexploitation of wetland products and introduction of invasive species. Despite of such anthropogenic pressures, urban wetlands play vital role in improving water quality, pollutant removal, groundwater recharge and sustaining regional biodiversity. Besides, they are essentially precious and often aesthetically pleasing for urban landscapes. Floral and faunal diversity of urban wetlands are often high as compared to the adjoining terrestrial ecosystems. They also attract a wide variety of migratory and resident birds. These wetland birds are extremely sensitive to hydrological fluctuations. Hence, monitoring status and diversity of avian population and their habitat use may be of importance to assess the condition of the wetland. But, such avifaunal studies on urban wetlands are still very few in this country. Lack of understanding of the processes governing water bird abundance and diversity is also a major lacuna for implementing appropriate management strategies. Ornithological researches on urban wetlands, thus, need to be prioritized. Over and above, awareness and activism from the citizens, fortitude of the wetland-managers and strong political will is essential to protect these pristine and valuable urban ecosystems.

References

Anon. 2011. *National Wetland Atlas* (SAC/EPSA/ABHG/NWIA/ATLAS/34/2011). Space Applications Centre (ISRO), Ahmedabad, India. pp. 310.

Antrop M. 2004. Landscape change and the urbanization process in Europe. *Landscape and Urban Planning*, 67: 9-26.

Baral HS and Inskipp C. 2004. *The State of Nepal's Birds 2004.* Department of National Parks and Wildlife Conservation, Bird Conservation Nepal and IUCN Nepal, Kathmandu. pp. 76.

Bellrose, FC. 1980. *Ducks, Geese and Swans of North America*, Stackpole Books, PA, USA. pp. 544.

Benassi G, Battisti C and Luiselli L. 2007. Area effect on bird species richness of an arcipelago of wetland fragments of Central Italy. *Community Ecology*, 8: 229-237.

Bhagat RB. 2011. Emerging patterns of urbanization in India. *Economic and Political Weekly*, 66 (34): 10-12.

Bhattacharyya A, Sen S, Roy PK and Mazumdar A. 2008. A critical study on status of East Kolkata Wetlands with special emphasis on water birds as bio-indicator. In. *Proc. of Taal 2007: The 12th World Lake Conference* (Sengupta M and Dalwani R. eds.),Ministry of Environment and Forests, Govt of India. pp. 1561-1570.

Bouland P and Hunhammar S. 1999. Ecosystem services in urban areas. *Ecological Economics*, 29: 293-301.

Boyer T and Polasky S. 2004. Valuing urban wetlands: A review of non-market valuation studies. *Wetlands*, 24: 744-755.

Brown L, Gardner G and Halweil B. 1998. *Beyond Malthus: 16 dimensions of the population problem.* Worldwatch Institute, Washington DC. pp. 89.

Caula SA, Sirami C, Marty P and Martin JL. 2010. Value of an urban habitat for the native Mediterranean avifauna. *Urban Ecosystem*, 13: 73-89.

Colwell MA and Taft OW. 2000. Water bird communities in managed wetlands of varying water depth. *Water birds*, 23(1): 45-55.

Dallimer M, Irvine KN, Skinner AMJ. *et al.*, 2012. Biodiversity and the feel-good factor: Understanding associations between self-reported human well-being and species richness. *Bioscience*, 62: 47-55.

Darnell T and Smith EH. 2004. Avian use of natural and created salt marsh in Texas, USA. *Water birds*, 27(3): 355-361.

Das TK, Moitra B, Raichaudhuri A, and Jash T. 2000. Degradation of water bodies and wetlands in West Bengal: Interaction with economic development, *EERC Working Paper Series: WB-3 (Theme: Wetlands and Biodiversity)*, Jadavpur University, Kolkata. pp.102.

Ehrenfeld JG. 2000. Evaluating wetlands within an urban context. *Ecological Engineering*, 15(3): 253-265.

Fraser LH and Keddy PA. 2005. *The World's Largest Wetlands: Ecology and Conservation.* Cambridge University Press, Cambridge. pp. 498.

Froneman A, Mangnall MJ, Little RM and Crowe TM. 2001. Waterbird assemblages and associated habitat characteristics of farm ponds in the Western Cape, South Africa. *Biodiversity and Conservation*, 10(2): 251-270.

Gonzalez-Gajardo A, Sepulveda PV and Schlatter R. 2009. Waterbird assemblages and habitat characteristics in wetlands: influence of temporal variability on species-habitat relationships. *Water birds*, 32(2): 225-233.

Grayson JE, Chapman MG and Underwood AJ. 1999. The assessment of restoration of habitat in urban wetlands. *Landscape and Urban Planning,* 43: 227-236.

Hamer AJ, Smith PJ and McDonnell MJ. 2012. The importance of habitat design and aquatic connectivity in amphibian use of urban storm water retention ponds. *Urban Ecosystems,* 15: 451-471.

Helmers DL. 1992. *Shorebird Management Manual,* Western Hemisphere, Shorebird Reserve Network, Manomet, MA, USA. pp. 58.

Herrmann J. 2012. Chemical and biological benefits in a storm water wetland in Kalmar, SE Sweden. *Limnologia,* 42: 299-309.

IUCN. 1990. *Directory of Wetlands of International Importance.* Ramsar Convention Bureau, Gland Switzerland. pp. 796.

Johnson MD and Sherry TW. 2001. Effects of food availability on the distribution of migratory warblers among habitats in Jamaica. *Journal of Animal Ecology,* 70: 546-560.

Keniger LE, Gaston KJ, Irvine KN and Fuller RA. 2013. What are the benefits of interacting with Nature? *International Journal of Environmental Research and Public Health,* 10: 913-935.

Khan TN. 2010. Temporal changes to the abundance and community structure of migratory water birds in Santragachhi Lake, West Bengal, and their relationship with water hyacinth cover. *Curr. Sci.,* 99(11): 1570-1577.

Kiviat E and MacDonald K. 2004. Biodiversity pattern and conservation in the Hacensack Meadow lands, New Jersey. *Urban Habitat,* 2(1): 28-61.

Kumar A, Sati JP, Tak PC and Alfred JRB. 2005. *Handbook on Indian Wetland Birds and their Conservation.* Zoological Survey of India, Kolkata, India. pp. 468.

Kushlan JA. 1986. The Management of Wetlands for Aquatic Birds. *Colonial Water birds,* 9(2): 246-248.

Lawrenz D, Mallonee C and Simmons K. 2012. Urban Wetlands. *Earth Science Student Presentations (ES 767 Wetland Environments),* Emporia State University.

Ma Z, Li B, Zhao B. *et al.,* 2004. Are artificial wetlands good alternatives to natural wetlands for water birds? – a case study on Chongming Island, China. *Biodiversity Conservation,* 13: 333–350.

Mackintosh T and Davis J. 2013.*The Importance of Urban Wetlands.* In: *Workbook for Managing Urban Wetlands in Australia* (Paul S. ed.). Sydney Olympic Authority, Australia, pp. 1-17.

Mazumdar S, Ghosh P and Saha GK. 2005. Diversity and behaviour of waterfowl in Santragachi Jheel, West Bengal, India, during winter season. *Indian Birds,* 1(3): 68-69.

Mazumdar S, Mookherjee K and Saha GK. 2007. Migratory water birds of wetlands of southern West Bengal, India. *Indian Birds,* 3(2): 42-45.

Mazumdar S, Mookherjee K and Saha GK. 2008. Diversity and dominance of migratory water birds in six wetlands of southern West Bengal, India. In. *Zoological Research in Human Welfare* (Ramakrishna and Chatterjee RN. eds.). Zoological Survey of India, Kolkata, India. pp. 161-166.

McDonald RI, Foreman RTT, Kareiva P. *et al.*, 2009. Urban effects, distance, and protected areas in an urbanizing world. *Landscape and Urban Planning*, 93: 63-75.

McKinney RA, Raposa KB and Cournoyer RM. 2011. Wetlands as habitat in urbanizing landscapes: Patterns of bird abundance and occupancy. *Landscape and Urban Planning*,100: 144–152

Millennium Ecosystem Assessment (MEA). 2005. *Ecosystem and Human Well being*. Retrieved from: http://www.millenniumassessment.org/en/index.html

Mitsch WJ and Gosselink JG. 2000. *Wetlands* (3rd edn.). John Wiley and Sons, New York, pp. 920.

MoEF. 1990. *Wetlands of India - A directory*. Ministry of Environment and Forests, Govt. of India, New Delhi.

Moore TLC and Hunt WF. 2012. Ecosystem service provision by stormwater wetlands ponds – A means or evolution? *Water Research*,46: 6811-6823.

Mulamotti G, Warner GB and McBean EA (eds.). 1996. *Wetlands: Environmental gradients, boundaries and buffers*. CRC Press, Boca Raton, FIC, pp. 320.

Murgui E. 2009. Influence of urban landscape structure on bird fauna: a case study across seasons in the city of Valencia (Spain). *Urban Ecosystem*. 12: 249–263.

Nature, Environment and Wildlife Society (NEWS). 1998. *Wetlands and Water birds of West Bengal*. Vision Publications, Kolkata, pp. 119.

O'Meara M. 1999. *Reinventing cities for people and planet (Worldwatch Paper)*. Worldwatch Institute, Washington DC. pp.147.

Okes NC, Hockey PAR and Cumming GS. 2008. Habitat use and life history as predictors of bird responses to habitat change. *Conservation Biology*, 22: 151-162.

Patterson JH. 1976. The role of environmental heterogeneity in the regulation of duck populations. *Journal of Wildlife Management*, 40: 22-32.

Prasad SN, Ramachandra TV, Ahalya N. *et al.*, 2002. Conservation of wetlands of India – a review. *Tropical Ecology*, 43(1): 173-186.

Raburu PO, Onyango FO and Obiero KO. 2012. The value of consumptive goods and services of Nyando Wetlands. In: *Community based approach to the management of Nyando Wetlands, Lake Victoria Basin, Kenya* (Raburu PO, Okeyo-Owuor JB, Kwena F. eds.) UNDP, Nairobi, Kenya, pp. 53-67.

Ramachandra TV. 2001. Restoration and management strategies of wetlands in developing countries. *Electronic Green Journal*, 15: 1-13. Retrieved from: http://wgbis.ces.iisc.ernet.in/energy/water/paper/wetland_tvr.htm

Ramchandra TV, Kiran A, Ahylaya N and Deepa RS. 2002. *Status of wetlands of Banglore (Technical Report 86)*. Indian Institute of Science, Bangalore. Retrieved from: http://wgbis.ces.iisc.ernet.in/energy/TR86/intro.html

Reid WV, Mooney HA, Cropper A. *et al.* (eds.). 2005. *Ecosystems and Human Well-being: Synthesis.* Island Press, Washington DC, pp. 155.

Sánchez-Zapata JA, Anadón JD, Carrete M. *et al.*, 2005. Breeding water birds in relation to artificial pond attributes: implications for the design of irrigation facilities. *Biodiversity and Conservation,* 14(7): 1627–1639.

Sheoran AS and Sheoran V. 2006. Heavy metal removal mechanism and acid mine drainage in wetlands: a critical review. *Minerals Engineering,* 19: 105-116.

Sinha A, Hazra P and Khan TN. 2012. Emergence of a wetland with the potential for an avian abode of global significance in South Bengal, India. *Current Science,* 102(4): 613-616.

Tapsuwan S, Ingram G, Burton M and Brennan D. 2009. Capitalized amenity value of urban wetlands: a hedonic property price approach to urban wetlands in Perth, Western Australia. *Australian Journal of Agricultural and Resource Economics,* 53 (4): 527-545.

Taylor D. 1989. Blacks and the environment: Toward an explanation of the concern and action gap between blacks and whites. *Environment and Behavior,* 21: 175-205.

Turner K and Jones T. 1991. *Wetlands: Market Interventions Failures.* Earth Scan Publications, London, pp.202.

UNDP. 2005. *World urbanization prospects: the 2005 revision.* United Nations Population Division, New York, pp.196.

Van der Valk AG. 2006. *The biology of freshwater wetlands.* Oxford University Press, Oxford, pp. 173.

West RJ. 1995. Rehabilitation of seagrass and mangrove sites: success and failures in NSW. *Wetlands Australia,* 14: 13-19.

Windham L, Laska MS and Wollenberg J. 2004. Evaluating Urban Wetland Restorations: Case Studies for Assessing Connectivity and Function. *Urban Habitats,* 2 (1): 130-146.

Zárate-Ovando B, Palacios E and Reyes-Bonilla H. 2008. Community structure and association of water birds with spatial heterogeneity in the Bahia Magdalena-Almejas wetland complex, Baja California Sur, Mexico. *Revista de Biologia Tropical,* 56(1): 371–389.

Zedler JB and Leach MK. 1998. Managing urban wetlands for multiple use: research, restoration, and recreation. *Urban Ecosystems,* 2: 189–204.

Zhao BT, Li BC, Zhong Y. *et al.*, 2005. Estimation of ecological service values of wetlands in Shanghai, China. *Chinese Geographical Science,* 15(2): 151-156.

Chapter 25

Population Trends of Waterbirds in an Urban Wetland in Response to Habitat Degradation

Anirban Sinha, Prantik Hazra and Tarak Nath Khan

Ecology and Wildlife Biology Laboratory, Department of Zoology, Maulana Azad College, Kolkata

Introduction

Wetlands are among the most productive biomes in the world (Alongi, 1998) and provide many important ecosystem services (Woodward and Wui, 2001). They also serve as important habitats for a wide variety of waterbirds. For instance, almost all of the 655 prioritised Indian wetlands (by MOEF, 2005) are listed for their importance as preferred habitat for waterbirds. However, due to increasing anthropogenic activities, they have become increasingly threatened, such that their value as waterbird habitat is being eroded (Khan *et al.*, 2005). Habitat degradation, conversion, overexploitation and pollution have been the primary threats to the wetlands and all these threats are particularly important in heavily populated areas, where anthropogenic demands on water bodies are highest. During the last 50 years, many wetlands have been filled, while the others have been degraded. Santragachi Lake, located within an industrial belt of district Howrah, West Bengal provides an example of how the waterbird communities have been changing over time due to human impact. In 1990s it occasionally supported internationally significant populations of Asian openbill *Anastomus oscitans*, lesser whistling duck *Dendrocygna javanica* and cotton pygmy goose *Nettapus coromandelianus* (Wetlands International, 2009). Besides, invasive species, including water hyacinth (*Eichhornia crassipes*) poses serious threats to wetland biodiversity (Mack *et al.*, 2000) specially to waterbirds since it covers most surface areas of water and thus reducing their feeding areas (Baral and Inskipp, 2004). It removes nutrients (Aoi and Hayashi, 1996; Tiwari *et al.*, 2007; Zimmels *et al.*, 2007) and decreases phytoplankton productivity leading to dissolved oxygen depletion, which can have negative consequences on waterbird prey populations including fish (Miranda and Hodges, 2000; Meerhoff *et al.*, 2003; Toft *et al.*, 2003; Kateregga and Sterner, 2009). Conversely, water hyacinth promotes zooplankton, macro invertebrate (Haag *et al.*, 1987; Sharitz and Batzer, 1999; Masifwa *et al.*, 2001) and fish populations (Brendonck *et al.*, 2003; Toft *et al.*, 2003). It also provides habitat structure and refuge from predators for many waterbirds (Bartodziej and Weymouth, 1995; Svingen and Anderson, 1998; Brendonck *et al.*, 2003). Thus, we might expect higher waterbird abundance and diversity in water bodies invaded by water hyacinth. However, based on its negative effects there must have a threshold at which waterbirds become negatively affected by increasing water hyacinth cover. Khan (2010) found a threshold of 20 per cent beyond which water hyacinth cover may impact negatively on waterbird numbers in India. The mechanisms for the negative effect are unclear but may be related to dense mats physically preventing waterbirds to move and, hence, affect access to prey or to nutrient depletion and dissolved oxygen depletion causing negative effects on their prey populations. In fact, little attention has been paid to the effects of water hyacinth on waterbirds (Gibbons *et al.*, 1994) and data are required to understand the mechanisms by which invasive aquatic species, such as water hyacinth, can impact on their population. Water hyacinth causes not just a physical barrier but can also have chemical effects on its surroundings. Water hyacinth is known to accumulate chromium (Dixit and Tiwari, 2007). This heavy metal has negative impacts on aquatic plant growth (Hörcsik and Balogh, 2002), phytoplankton productivity (Kaladharan *et al.*, 1990), zooplankton (Eisler, 1986), molluscs (Satyaparameshwar *et al.*, 2006) and fish (Vinodhini and Narayanan, 2008). Chromium may well affect the waterbirds directly but there is little published works on its toxicity to birds.

This paper reports on a 16-years study on the diversity and population dynamics of migratory waterbirds in Santragachi Lake between 1998 and 2013 with an attempt to understand the trends in the abundance and community composition of these waterbirds and their relationships with water hyacinth and other anthropogenic impacts (*i.e.* BOD and chromium concentration). Biochemical Oxygen Demand (BOD) and chromium content were included in this study to understand whether the population trends were resulted from the effects of water hyacinth alone or other anthropogenic factors. The key questions being addressed by this paper are:

1. Are there significant temporal trends in waterbird abundance?
2. Whether water hyacinth is the main factor responsible for the observed decline in migratory waterbirds?
3. If so, what was the mechanism through which it impacted on waterbirds (*i.e.* by reducing the available habitat or by altering the prey bases of waterbirds)?
4. Did all migratory waterbirds decline at the same rate or were there changes in the assemblage?

Materials and Methods

Study Site

The study was conducted at Santragachi Lake, situated on the western bank of the river Hooghly about five km west of Kolkata city (22.580° N; 88.283° E). It extends over an area of 24 hectares, of which 18 hectares constitute a lake that provides suitable habitat for waterbirds (Figure 25.1). Being located inside a densely populated industrial area near Kolkata, the lake has become subjected to degradation due to anthropogenic activities. A sizeable portion of the waterbody remains clogged with water hyacinth throughout the year. The lake supports a wide variety of zooplankton, molluscs and fish that are consumed by various species of waterbirds (Khan, 2010).

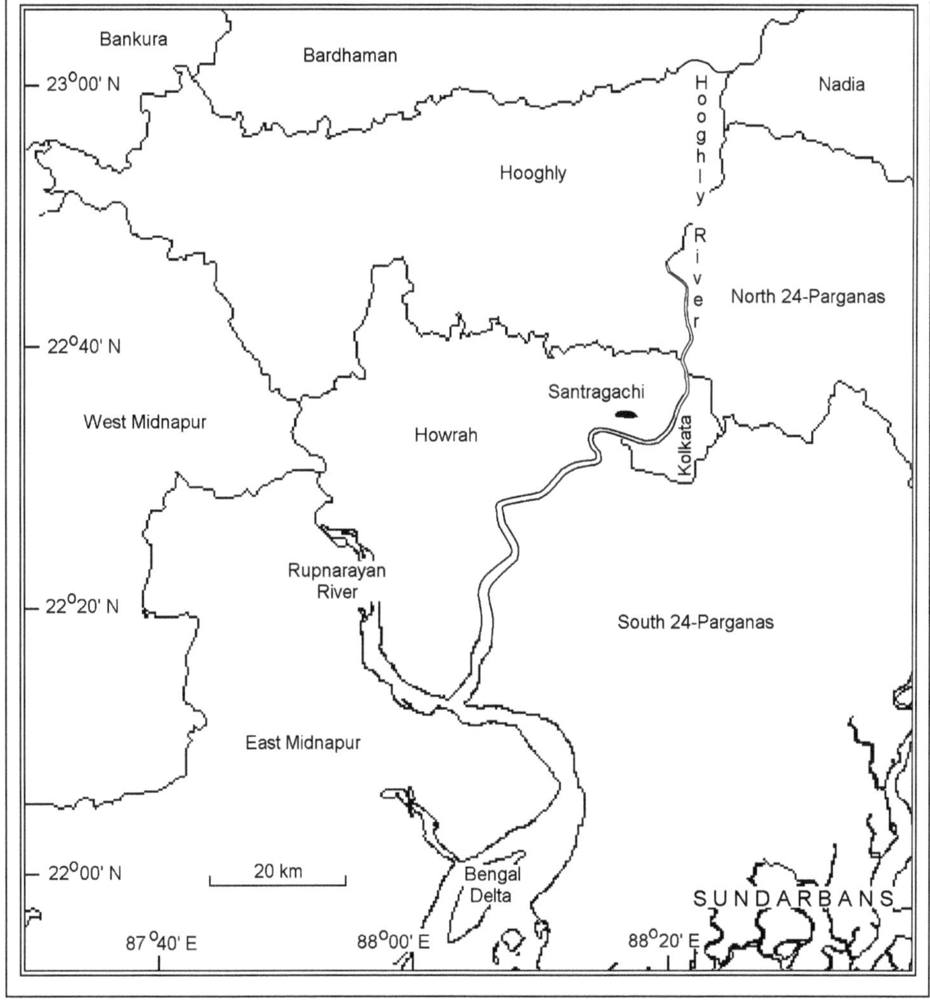

Figure 25.1: A Portion of South Bengal Showing the Districts along with the Location of Santragachi Lake.

Waterbird Census

Total counts of all waterbirds were performed annually between 1998 and 2013. These counts were made between January 1st and 30th. The census period was chosen to coincide with large-scale Asian Waterfowl Census Programme coordinated by Wetlands International. During that period, three censuses were made (between January 1-10; 11-20; 21-30) to have a comprehensive idea of the waterbird abundance (3 censuses per year for 16 years; n = 3 x 16 = 48). Waterbird counts were performed on foot by 5-8 observers trained in bird identification and census, as per guideline of Wetlands International (1997) and followed by Khan (2010). Observers walked around the wetland and counted birds using binoculars and telescopes. During each census, counts began at 08-00 hrs IST and continued till the total waterbird count was completed. From 2004 onwards aerial photographs of different segments of the wetland were also taken, which were later examined to have a crosscheck of the census data. These examinations provided a second set of census data which were compared with the data obtained from direct bird counts. These comparisons were necessary to make accurate estimates as far as possible for the year 2004 onwards because of the proliferation of water hyacinth, which created difficulties in detecting some waterbird species, especially lesser whistling duck (*Dendrocygna javanica*), fulvous whistling duck (*Dendrocygna bicolor*) and garganey (*Anas querquedula*) as they often prefer to rest on this vegetation. This crosscheck helped mitigating census errors including observer bias (Khan, 2010).

Changes in Waterbird Diversity and Population Trends

Comparisons of the abundance of each waterbird species between 1998 and 2013 were performed between the values for the counts performed in 1998, and in 2013 using the following formula:

Percent change between 1998 and 2013 = $\{(M_{2013} - M_{1998})/M_{1998}\} \times 100$

[where M_{2013} is the mean value for the year 2013 and M_{1998} is the mean value obtained in 1998]

Changes in the waterbird diversity during the study period was studied using three most widely used diversity estimates:

1. Dominance (D) Simpson (1949), calculated as $1-\Sigma(n/N)^2$, where n is the number of individuals of one particular species and N is the total number of individuals recorded.
2. Shannon-Weiner diversity (Shannon, 1948), calculated as $-\Sigma(_{pi} \ln _{pi})$, where $_{pi}$ is the proportion of the ^{ith} species.
3. Pielou's Evenness (J2) (Pielou, 1966), calculated as $H2/H2_{max}$ where H2 is the Shannon-Weiner diversity and $H2_{max}$ is the Shannon-Weiner maximum diversity given by ln (S).

Population trends were calculated using loglinear Poisson-based modelling framework using the programme Tends and Indices for Monitoring Data (TRIM, Version 3.54; Pannekoek and van Strien, 2001). TRIM analyses time series of counts and produces yearly indices of abundance and trends using Poisson regression.

From the modelling options in the TRIM, 'Time Effect Model' was selected for the estimation of the imputed yearly population indices. The population trend of each species was described by this TRIM Imputed index. As summary statistics, overall trends in yearly indices were computed, taking into account the uncertainty of the indices. These trends were expressed as multiplicative slopes, *i.e.* as yearly multiplication factors (1 = stable) and were classified into the following categories according to statistical significance and magnitude (Pannekoek and van Strien, 2001):

a. *Strong increase* – increase significantly > 5 per cent per year and thus the lower limit of the confidence interval of the slope estimate is >1.05;
b. *Moderate increase* – significant increase, but not significantly > 5 per cent per year, and thus the lower limit of the confidence interval is >1.0 but <1.05;
c. *Stable* – no significant increase or decline, and it is certain that trends are less than 5 per cent per year; thus the confidence interval encloses 1.00 but the lower limit is >0.95 and the upper limit is <1.05;
d. *Moderate decline* – significant decline, but not significantly > 5 per cent per year; thus the upper limit of the confidence interval is >0.95 but <1.00;
e. *Steep decline* – decline significantly > 5 per cent per year; thus the upper limit of the confidence interval is <0.95.

Habitat Survey

The condition of the lake, with respect to building of new landmasses inside it and the area clogged with water hyacinth was assessed during all the census years. The percent cover of the water hyacinth was estimated, with the help of professional surveyors, mainly from aerial photographs and Global Positioning system (GPS) (Garmin; e Trex®) data on Map Maker Pro Package, Version 3.5 [Map Maker Ltd, The Pier, Carradale, Kintyre, PA28 6SQ, UK (*www.mapmaker.com)]*. Other factors posing threats to the wetland and the waterbirds inhabiting the lake were also recorded. The dissolved oxygen content (DO), BOD_5 and chromium concentration of the wetland were measured during all the census periods. BOD_5 was estimated using standard method with five-day incubation at 20°C. The oxygen concentration, both for DO and BOD, was estimated using Aquamerck® Oxygen Testing Kit (E Merck, Germany), while chromium concentrations in water was measured using standard methods (APHA, 2005). Attempts were also made to estimate the abundances of prey populations (*i.e.* macro-invertebrate and small fish) following the methods described by Sinha *et al.* (2011) and Sarkar *et al.* (2014). They were sampled once in January of each study year from 1998 to 2013. Zooplankton sampling and their density estimates were made following standard methods as suggested by Ramachandra and Solanki (2007). Fish and insect abundances were estimated using catch-effort method (Johnson, 1965) utilising locally available fine-meshed fishing nets. Benthic macro invertebrates were sampled using Ekman's Dredge. In each case, abundances were estimated from 30 random efforts.

Data Analysis

Relationship between Wetland Variables and Waterbird Abundances

A complete list of environmental variables considered in this study is presented in Table 25.1. The data does not include waterbird prey abundances in these analyses because the data were incomplete. Canonical Correspondence Analysis (CCA) using CANOCO 4.5 (ter Braak and Šmilauer, 2002) was used to determine the variables that best predicted the composition of waterbird communities. CCA is an eigenvector ordination technique for multivariate direct gradient analysis, which iteratively develops an ordination of species and sampling sites, combined with multiple regressions on a series of environmental gradients. The environmental variables are reduced to a few orthogonal axes as composite environmental gradients structuring species distribution patterns (ter Braak, 1986). The significance of the relationship of each wetland variable with the waterbird data was determined by the magnitude of the additional variation the variable explained ("conditional effects", λ_a). Stepwise forward selection was used to include significant variables ($P<0.05$) in the model. The significance of the first canonical axis and of all canonical axes together was tested by the distribution-free Monte Carlo simulation (998 permutations). Multicollinearity among the wetland variables was not excessive [variance inflation factor (VIF) <7; range 3.27-4.35], except for chromium content (8.49). However, we did not exclude this variable from CCA since its VIF did not exceed 10 [*i.e.* Rule of 10 (Belsley *et al.*, 1980; O'Brien, 2007)].

Table 25.1: Wetland Habitat Variables (mean±SD and range), Selected as Potential Predictors of Migratory Waterbird Abundances at Santragachi Lake.

Variables	*Description*	*Mean±SD*	*Range*
Water hyacinth cover	Percentage of water surface area	22.69±11.27	7.89-43.00 per cent
Dissolved oxygen content	Surface water DO estimation (mg/l)	3.78±1.24	2.2-6.3 mg/l
Biochemical oxygen demand	BOD_5; as estimated after 5 days of incubation at 20ºC mg/l)	1.41±0.49	0.81-2.38 (mg/l)
Chromium content	Estimated from surface water (mg/l)	133.60±3.04	127-138 (mg/l)

Although CCA provides information of the habitat association of individual species, its main goal is to determine relative effects of environmental variables on the species assemblage as a whole. Therefore, we used generalised linear models (GLMs) with Poisson distribution and logarithmic link (typical for count data, Bolker *et al.*, 2008) (SPSS; Version 21) to identify the habitat variables most important in determining the abundances of individual waterbird species at the wetland. We used a backward elimination procedure, starting with a full model including all the four environmental variables and we dropped variables one by one (depending on their contribution to the model) until the final model is achieved. All the waterbird data were log (x+1) transformed to improve normality before GLM analysis; the wetland variable data were not log transformed. We did not apply the Bonferroni correction when multiple tests on different response variables (waterbird species) addressed the same hypothesis (Moran, 2003).

Variations in the Waterbird Community Composition

To have an insight into the spatial and temporal variations in waterbird community composition, we adopted multi-response permutation procedure (MRPP) (PC-ORD 6.04; McCune and Mefford, 2006) on the census data, using study years as the grouping variable. MRPP compared intra-group distances (in these case, Bray-Curtis dissimilarities) with the average distances that resulted from all possible combinations in the data.

CCA community plot (MVSP 3.2; Kovach Computing Services (www.kovcomp.com) was also made to substantiate the temporal variations in the community composition of the waterbirds, as well as to identify any shift in the waterbird community during the study period.

Results

Waterbird Diversity and Abundance

A total of 253,131 migratory waterbirds, belonging to 13 species, were recorded during the course of this study (Table 25.2). Among them four species (lesser whistling duck *Dendrocygna javanica,* fulvous whistling duck *Dendrocygna bicolor,* knob-billed duck *Sarkidiornis melanotos* and cotton pygmy goose *Nettapus coromandelianus*) were local migrants while the others were long-distance winter visitors (Kazmierczak, 2003) mainly from the northern areas of Eurasia and western Asia (Clements, 2007). Three species (spot-billed duck *Anas poecilorhycha,* mallard

Table 25.2: Checklist of Migratory Waterbirds Recorded in Santragachi Lake during all the Study Years from 1998 to 2013 along with their Mean (*i.e.* 16-year mean) and Total (16- year total) Number of Birds

Waterbird	*Mean*	*Total*	*Resident Status*
Lesser Whistling Duck (*Dendrocygna javanica*)	4857	77715	Local Migrant
Fulvous Whistling Duck (*Dendrocygna bicolor*)	17	275	Local Migrant
Knob-billed Duck (*Sarkidiornis melanotos*)	3	54	Resident
Northern Pintail (*Anas acuta*)	84	1336	Long-distance migrant
Common Teal (*Anas crecca*)	45	716	Long-distance migrant
Gadwall (*Anas strepera*)	167	2665	Long-distance migrant
Garganey (*Anas querquedula*)	28	446	Long-distance migrant
Northern Shoveler(*Anas clypeata*)	23	371	Long-distance migrant
Cotton Pygmy Goose (*Nettapus coromandelianus*)	24	379	Local Migrant
Ferruginous Duck (*Aythya nyroca*)	6	98	Long-distance migrant
Tufted Duck (*Aythya fuligula*)	7	107	Long-distance migrant
Eurasian Coot (*Fulica atra*)	5	72	Local Migrant
Swinhoe's Snipe (*Gallinago megala*)	9	143	Long-distance migrant

Anas platyrhynchos and baikal teal *Anas formosa*) were vagrants and were not considered in the population trend or habitat relationship analysis

All the waterbirds exhibited significant declining trends during the period between 1998 and 2013 (Figure 25.2). The species which declined dramatically included ferruginous duck (*Aythya nyroca*) (77.78 per cent decline; $P < 0.001$), tufted duck (*Aythya fuligula*), Eurasian coot (*Fulica atra*) and swinhoe' snipe (*Gallinago megala*) (75.00 per cent decline in each case; $P < 0.001$). All the dabbling ducks declined by almost 50 per cent during that period, with common teal (*Anas crecca*) (73.08 per cent decline; $P < 0.001$) and northern pintail (*Anas acuta*) (71.21 per cent decline; $P < 0.001$) exhibiting drastic declines. The lesser (38.47 per cent decline; $P < 0.001$) and fulvous (25.00 per cent decline; $P < 0.001$) whistling ducks exhibited the lowest levels of declines (Table 25.4). Total number of waterbirds declined by 41.19 per cent ($P < 0.001$) during this period.

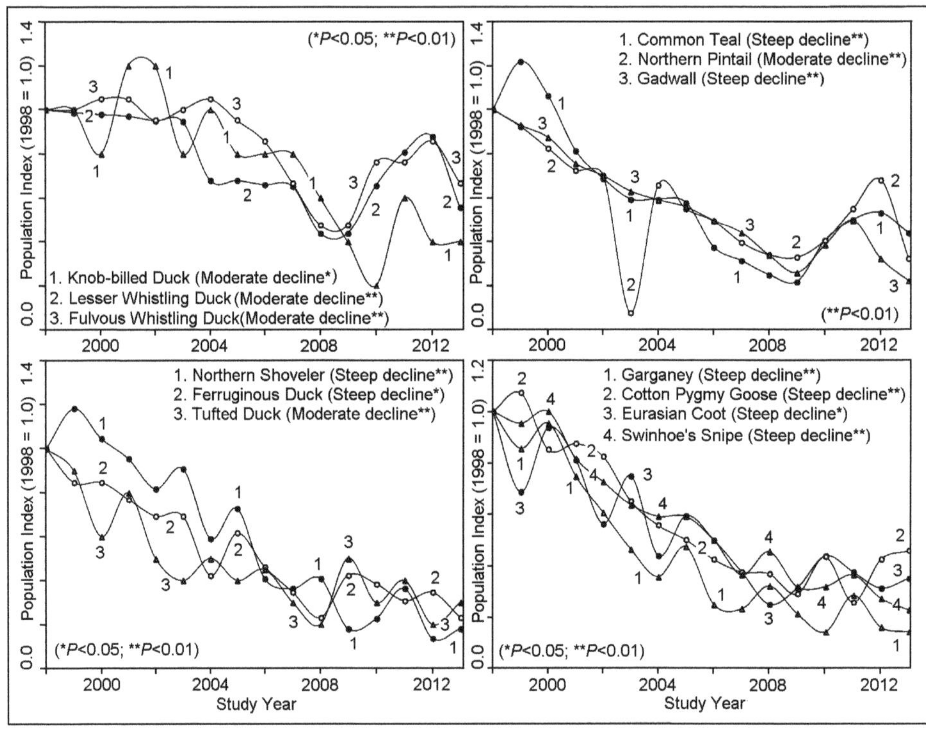

Figure 25.2: Waterbird Population Trends in Santragachi Lake during the Period between 1998 and 2013, Derived by the TRIM Imputed Index. Starting year was 1998 for all species (population index = 1.00). Population indices >1.00 indicate increases, while <1.00 denote declines (see 'Methods').

Habitat Deterioration and its Relationship with Migratory Waterbirds

Relationship between Environmental Variables and Waterbird Abundance

The four environmental variables included by the CCA forward selection as best predicting the waterbird abundances were water hyacinth cover, DO, BOD and chromium content. The species-environment correlations for axes 1 (r = 0.80), 2 (r = 0.67) and 3 (r = 0.78) (Figure 25.3) indicated a strong relationship (Monte-Carlo permutation P=0.001) between the waterbird species abundance and environmental variables used for CCA. The permutation test on the eigenvalues of the CCA axes also revealed that the variables included in the model explained a significant variation in the waterbird species data (P = 0.001). The total variance (*i.e.* inertia) in the species data was 0.003, of which the axis 1 explained 33.9 per cent and the axis 2 as 12.6 per cent. Of these variables water hyacinth cover was negatively correlated (at<0.05) with DO (r = -0.774), but positively with BOD and chromium content (r = 0.878 and 0.630 respectively). The DO was negatively related with BOD and chromium content (r = -0.870 and -0.800 respectively); the chromium content positively associated with BOD (r = 0.704).

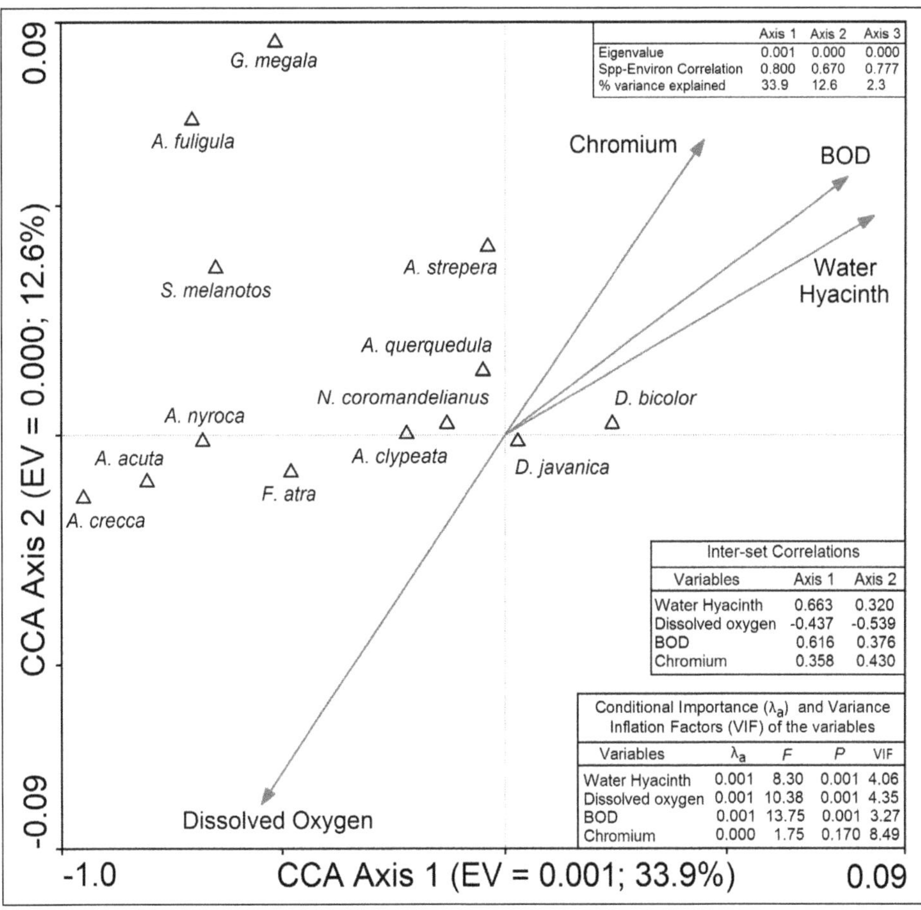

	Axis 1	Axis 2	Axis 3
Eigenvalue	0.001	0.000	0.000
Spp-Environ Correlation	0.800	0.670	0.777
% variance explained	33.9	12.6	2.3

Inter-set Correlations

Variables	Axis 1	Axis 2
Water Hyacinth	0.663	0.320
Dissolved oxygen	-0.437	-0.539
BOD	0.616	0.376
Chromium	0.358	0.430

Conditional Importance (λ_a) and Variance Inflation Factors (VIF) of the variables

Variables	λ_a	F	P	VIF
Water Hyacinth	0.001	8.30	0.001	4.06
Dissolved oxygen	0.001	10.38	0.001	4.35
BOD	0.001	13.75	0.001	3.27
Chromium	0.000	1.75	0.170	8.49

Figure 25.3: CCA Ordination Biplot of the Waterbird Species for the Period between 1998 and 2013. EV means the eigenvalue, while percentages shown on axes represent the amount of variance explained by that axis. Total variance (inertia) in the waterbird species data = 0.003.

The inter-set correlations indicated that CCA Axis 1 reflected trends across the water hyacinth cover and BOD, where as Axis 2 was largely defined by the gradient of DO and chromium contents (Figure 25.3).

Table 25.3: Correlations between the Wetland Variables Selected by CCA Forward Selection and Waterbirds Abundances (n = 48). [*Significant P values after Bonferroni corrections for as for 120 comparisons (for 16 years); comparison-wise error rate 0.0004].

	Water Hyacinth	*Dissolved Oxygen*	*BOD*	*Chromium Content*
		Wetland variables		
Water Hyacinth cover	1	–0.774*	0.878*	0.800*
Dissolved Oxygen	–0.774*	1	–0.870*	–0.630*
BOD	0.878*	–0.870*	1	0.795*
Chromium	0.800*	–0.630*	0.795*	1
		Waterbird species		
Lesser Whistling Duck	–0.489	0.886*	–0.912*	–0.484
Fulvous Whistling Duck	–0.677*	0.872*	–0.780*	–0.469
Knob-billed Duck	–0.694*	0.627*	–0.681*	–0.295
Northern Pintail	–0.913*	0.795*	–0.909*	–0.643*
Common Teal	–0.948*	0.764*	–0.885*	–0.674*
Gadwall	–0.892*	0.745*	–0.887*	–0.608*
Garganey	–0.877*	0.778*	–0.923*	–0.764*
Northern Shoveler	–0.852*	0.827*	–0.925*	–0.678*
Cotton Pygmy Goose	–0.862*	0.843*	–0.870*	–0.615*
Ferruginous Duck	–0.851*	0.748*	–0.868*	–0.606*
Tufted Duck	–0.665*	0.468	–0.640*	–0.335
Eurasian Coot	–0.888*	0.699*	–0.823*	–0.728*
Swinhoe's Snipe	–0.437	0.373	–0.491*	–0.107

Individual ordination scores (Table 25.3) for waterbird species indicated that their abundances were mostly influenced by water hyacinth cover, DO and BOD. Eleven of 13 waterbird species showed significant negative correlations with water hyacinth cover. Common teal (r = -0.948; P<0.001), northern pintail (r = -0.913; P<0.001), gadwall (r = -0.892; P<0.001), Eurasian coot (r = -0.888; P<0.001) and garganey (r = -0.877; P<0.001) abundances were most strongly correlated with water hyacinth cover. Conversely, 11 waterbirds exhibited strong positive correlations with DO; the lesser and fulvous whistling ducks (r = 0.886 and 0.872 respectively; P<0.001), cotton pygmy goose (r = 0.842; P<0.001), northern shoveler (r = 0.827; P<0.001) and northern pintail (r = 0.795; P<0.001) densities being the best correlated with DO and BOD most strongly affected northern shoveler (r = -0.925; P<0.001) and garganey (r = -0.923; P<0.001)). Many of these waterbird species exhibited significant negative correlations with chromium content of the water body. Garganey (r = -0.764; P<0.001), Eurasian coot (r = -0.728; P<0.001), northern shoveler (r = 0.678; P<0.001) and common teal (r = -0.674; P<0.001) exhibited stronger correlations with this variable than the other waterbirds. Although the most common species, lesser whistling duck, showed strong relationships with DO and BOD, secured its position almost at the origin of the ordination, suggesting that it was a generalist species.

Although GLMs failed to detect the associations of some of the waterbird species determined by CCA, they generally confirmed that abundance of individual species were typically influenced by water hyacinth cover and BOD (Table 25.4). Eleven and 13 waterbird species showed negative relationships with water hyacinth cover and BOD respectively. In contrast, only 6 waterbirds exhibited positive relationship with DO. However, chromium content failed to show any significant relationship with most of the waterbird included in this study. Total number of waterbirds was negatively correlated with both water hyacinth cover and BOD, while positively with DO (Table 25.4).

Relationship between Environmental Variables and Waterbird Community Composition

The CCA community plot (Figure 25.4) suggested clear shifts of the waterbird communities in correspondence with water hyacinth cover and other wetland

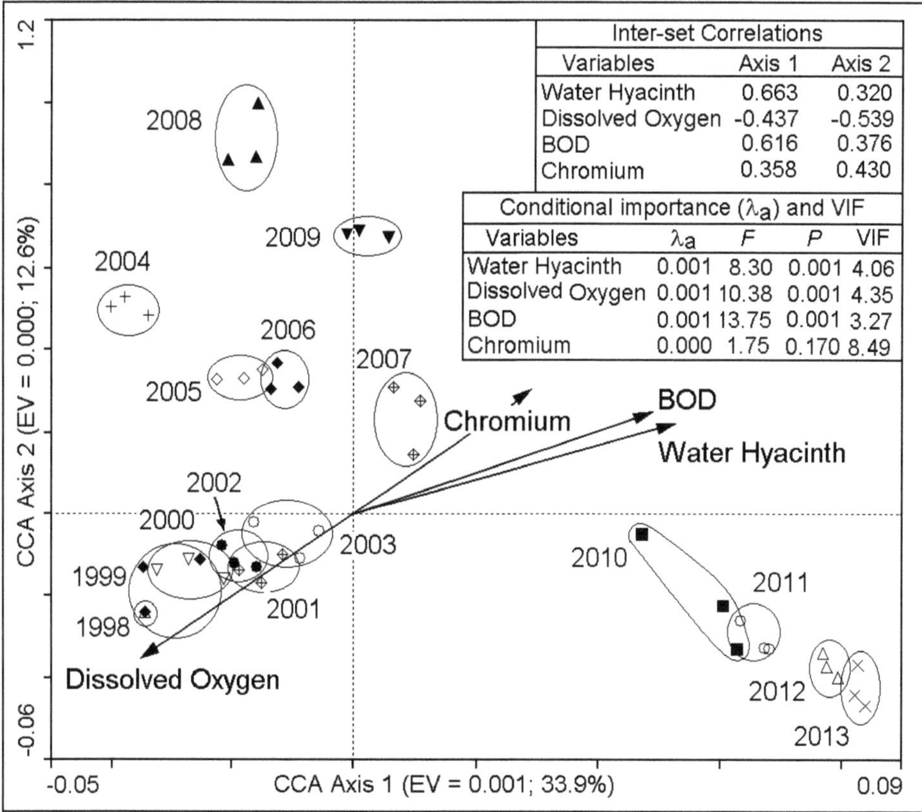

Inter-set Correlations		
Variables	Axis 1	Axis 2
Water Hyacinth	0.663	0.320
Dissolved Oxygen	-0.437	-0.539
BOD	0.616	0.376
Chromium	0.358	0.430

Conditional importance (λ_a) and VIF				
Variables	λa	*F*	*P*	VIF
Water Hyacinth	0.001	8.30	0.001	4.06
Dissolved Oxygen	0.001	10.38	0.001	4.35
BOD	0.001	13.75	0.001	3.27
Chromium	0.000	1.75	0.170	8.49

Figure 25.4: CCA Ordination Biplot of the Waterbird Community for the Period between 1998 and 2013. EV means the eigenvalues, while percentages shown on axes represent the amount of variance explained by that axis.

Table 25.4: Generalised Linear Models for the Relationship between the Wetland Variables and Waterbird Abundances (Number of birds for each species) in Santragachi Lake

Waterbird Species	*Per cent Decline*	*Effect*	*Estimate±SE*	*Wald* χ^2 *(df=1)*	*P*
Lesser Whistling Duck (*Dendrocygna javanica*)	38.47	DO	0.146±0.034	18.55	<0.001
		BOD	–0.123±0.046	7.23	0.007
Fulvous Whistling Duck (*Dendrocygna bicolor*)	25	DO	0.236±0.045	27.37	<0.001
		Cr	0.023±0.009	5.68	0.017
Knob-billed Duck (*Sarkidiornis melanotos*)	50	WH	–0.023±0.007	10.12	0.001
		BOD	–0.392±0.116	11.46	0.001
		Cr	0.121±0.024	26.17	<0.001
Northern Pintail (*Anas acuta*)	71.21	WH	–0.014±0.003	20.37	<0.001
		BOD	–0.456±0.056	66.95	<0.001
Common Teal (*Anas crecca*)	73.08	WH	–0.022±0.004	33.41	<0.001
		BOD	–0.344±0.066	26.95	<0.001
Gadwall (*Anas strepera*)	47.62	WH	–0.009±0.002	27.82	<0.001
		DO	–0.058±0.028	4.36	0.037
		BOD	–0.262±0.041	40.22	<0.001
Garganey (*Anas querquedula*)	61.9	WH	–0.013±0.004	7.68	0.006
		BOD	–0.260±0.039	43.83	<0.001
Northern Shoveler (*Anas clypeata*)	62.5	WH	–0.016±0.006	7.27	0.007
		BOD	–0.357±0.053	45.72	<0.001
Cotton Pygmy Goose (*Nettapus coromandelianus*)	50	WH	–0.011±0.005	5.23	0.022
		BOD	–0.455±0.085	28.69	<0.001
		DO	0.121±0.049	6.18	0.013

Contd...

Table 25.4–*Contd...*

Waterbird Species	*Per cent Decline*	*Effect*	*Estimate±SE*	*Wald* χ^2 *(df=1)*	*P*
Ferruginous Duck (*Aythya nyroca*)	77.78	WH	–0.011±0.005	5.23	0.022
		BOD	–0.455±0.085	28.69	<0.001
Tufted Duck (*Aythya fuligula*)	75	WH	–0.021±0.008	7.8	0.005
		DO	–0.314±0.121	6.73	0.009
		BOD	–0.651±0.185	12.37	<0.001
Eurasian Coot (*Fulica atra*)	75	WH	–0.015±0.004	11.92	0.001
		BOD	–0.192±0.077	6.26	0.012
Swinhoe's Snipe (*Gallinago megala*)	75	BOD	–0.785±0.220	12.74	<0.001
Total number of waterbirds	41.19	WH	–0.004±0.002	3.9	0.048
		DO	0.125±0.032	14.73	<0.001
		BOD	–0.156±0.047	11.06	0.001

For the sake of brevity, the reduced models resulting from stepwise backward elimination of insignificant terms are presented. Statistics and *P* value of significant terms were taken from the minimal models. The full model included 4 wetland variables: Water Hyacinth cover, Dissolved Oxygen content, BOD and Chromium content.

variables. There were clear trends in the community shifts of waterbirds exhibiting discernible changes in ordination scores with the progress of time and concurrent changes in the habitat variables from 1998 to 2013. The waterbird community has also changed and shifted in its composition along with an increase in the level of dominance and declines in the diversity and equitability (Figure 25.5). These changes were associated with the decline in the relative abundances of dabbling and diving waterbirds that depend primarily on macro-invertebrates and fish, while an increase in the relative abundances of the herbivorous species (Table 25.5).

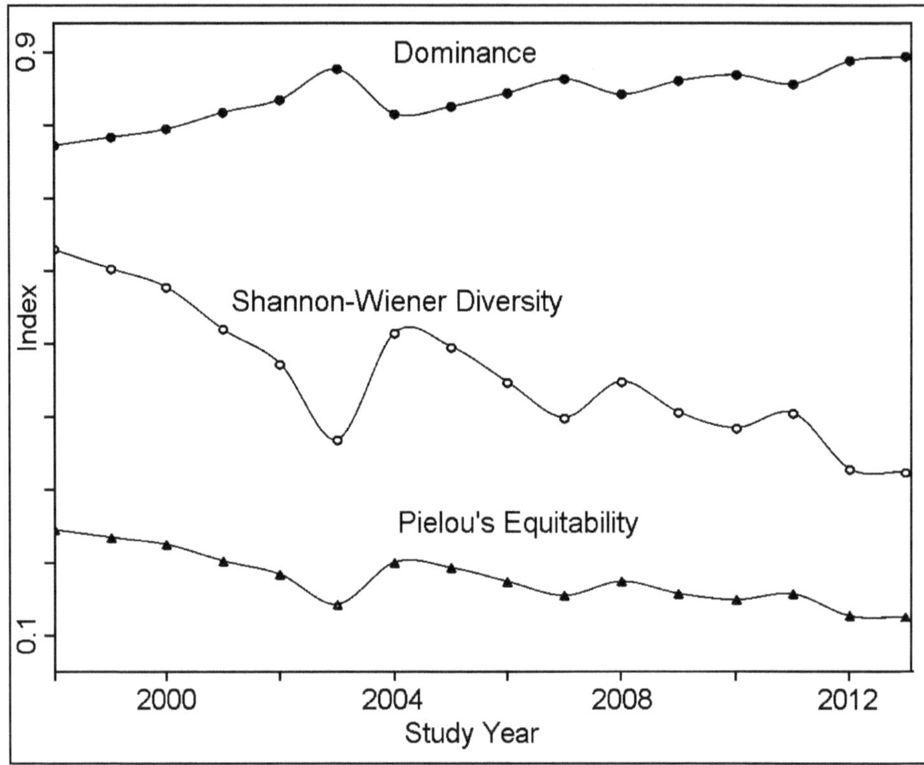

Figure 25.5: Trends in Waterbird Diversity in Santragachi Lake during the Period between 1998 and 2013.

The results of MRPP on the waterbirds species (T=-10.213, A=0.958, $P < 0.0001$) revealed high levels of between-year and almost negligible levels of within-years variations in their communities.

Discussion

Freshwater ecosystems are among the most significantly human-altered systems in the world (Millennium Ecosystem Assessment, 2005) and Santragachi Lake exemplifies this. The present study clearly shows that the migratory waterbirds have declined significantly (25.0 per cent for fulvous whistling duck to 77.8 per cent for ferruginous duck) during 1998–2013, but how far is this decline related to the water hyacinth cover or its associated environmental variables? Studies of

Table 25.5: Feeding Habits of Migratory Waterbird Species at Santragachhi Lake along with the Changes in their Relative Abundances. Feeding habits were mostly available from Bird Life International (2011).

Waterbird	*Nature*	*Trophic level*	*Main diet*	*Change (per cent)*
Lesser Whistling Duck	Dabbler	Herbivorous	Seeds and other parts of aquatic vegetation	4.13
Fulvous Whistling Duck	Dabbler	Herbivorous	Seeds and other parts of aquatic vegetation	24.61
Knob-belled Duck	Dabbler	Omnivorous	Aquatic plant parts, aquatic insect larvae, fish	–68.85
Northern Pintail	Dabbler	Omnivorous	Aquatic plants, invertebrates, amphibians, fish	–56.86
Common Teal	Dabbler	Herbivorous	Grasses, sedges and seeds of aquatic plants	–18.22
Gadwall	Dabbler	Herbivorous	Various parts of aquatic plants	–2.13
Garganey	Dabbler	Omnivorous	Aquatic plants, Macro-invertebrates, fish	–73.3
Northern Shoveler	Dabbler	Omnivorous	Macro-invertebrates, fish, aquatic plants	–88.34
Cotton Pygmy Goose	Dabbler	Omnivorous	Macro-invertebrates and aquatic plants	–36.05
Ferruginous Duck	Diver	Omnivorous	Macro-invertebrates, fish, aquatic plants	–56.86
Tufted Duck	Diver	Omnivorous	Macro-invertebrates, aquatic plants	–90.65
Eurasian Coot	Diver	Omnivorous	Macro-invertebrates, fish, aquatic plants	–53.27
Swinhoe' Snipe	Wader	Carnivorous	Macro-invertebrates	–57.52

Weller and Spatcher (1965) and Miranda and Hodges (2000) suggested maximum waterbird abundance at about 50 per cent vegetation coverage in accordance with maximum densities of their prey populations at this level. No such relationship was found in this study. Although the waterbird prey abundance data were incomplete, their prey organisms declined along with increasing water hyacinth cover as did the waterbirds. However, as indicated by GLMs, three other associated environmental variables played important roles in this decline. The CCA community plot suggests a clear shift in the waterbird community in correspondence with water hyacinth cover and other three environmental variables. The proximation of the waterbird communities for census years from 1998 to 2003 forming a distinct cluster might indicate that water hyacinth cover and its associated environmental variables failed to affect the waterbird communities during this period. Notable effects, as indicated by distinct shifts in the waterbird communities from this cluster, were observed in the years 2004 onwards. Since water hyacinth cover in 2004 was 32.0 per cent the results might suggest that a water hyacinth cover of less than 32 per cent may not be sufficient enough to affect waterbird species. Villamagna (2009) found a similar threshold level in Lake Chapala, Mexico.

On the other hand, variable rates of waterbird species decline and a shift in their communities suggested that water hyacinth impact was not restricted to the reduction in the availability of open water and foraging ground. Rather, water hyacinth cover might also have acted by altering their food sources, especially their prey assemblages. A detailed analysis of the invertebrate and fish communities is beyond the scope of this study. However, a functional analysis based on diet and feeding techniques indicates that carnivorous and omnivorous waterbirds deriving major food from invertebrates and fish were the most affected ones. Herbivores were the least affected functional group (Table 25.5).

Therefore, the shifts in waterbird assemblage might well be attributed to not only the imposition of a physical barrier, but also changes in the key food resources, such as plant materials, macro invertebrates and fish. Increasing water hyacinth cover along with decreasing DO might have reduced the abundance of these resources as it is true for many other wetlands (e.g. Miranda and Hodges, 2000; Meerhoff *et al.*, 2003; Toft *et al.*, 2003). Water hyacinth also affects phytoplankton and other aquatic vegetation through allelopathy. Presence of an alkaloid (18, 19-Seco-15 beta-yohimb) and four phthalate derivatives in water hyacinth are known to exhibit antimicrobial and antialgal activities (Sanaa *et al.*, 2010). In addition to the reduction in phytoplankton productivity and DO (Rommens *et al.*, 2003; Perna and Burrows, 2005) this negative impact affects the food sources of many invertebrates and fish resulting in decline in their populations. In fact, photosynthesis is limited beneath water hyacinth mats, and the plant itself does not release oxygen into the water unlike phytoplankton and other submerged vegetation (Meerhoff *et al.*, 2003), resulting in decreased DO. McVea and Boyd (1975) found reduced fish production at water hyacinth covers around 25 per cent due to decreases in phytoplankton production with increased cover of water hyacinth which might result in declining waterbird abundance. Increasing pollution load, as indicated by increases in BOD, might have augmented the process by lowering the oxygen level further. Therefore, along with the reduction in the availability of open water area and thereby in the foraging ground, food resources of the waterbird species appear to have shifted in distribution, resulting in concurrent changes in their abundance.

The observed temporal variation of waterbird species generally reflects the distribution of food resource availability, both in relation to the abundance of food items, and ease of access to these food materials by the waterbirds (Paton *et al.*, 2009). This is especially the case for the dabbling and diving waterbirds at Santragachi depending primarily on macro-invertebrates and fish such as northern shoveler, garganey, ferruginous duck, tufted duck and Eurasian coot. However, the abundance of lesser whistling duck was not so intimately linked to food resources, as indicated by its position almost at the origin of the ordination. This might indicate that it did not primarily depend on the food sources within Santragachi Lake. In fact, lesser whistling duck mainly used Santragachi Lake as a roosting ground, mostly for resting and rarely observed feeding here. Moreover, during the course of this study the species was found to leave the wetland at dusk and returning in the next dawn.

Chromium showed significant negative correlation with the numbers of almost all the waterbird species. There is a lack of detailed information on the effects of chromium on waterbirds. However, increased levels of chromium are known to negatively affect the growth and development of plants (Shanker *et al.*, 2005), including phytoplankton (Hörcsik and Balogh, 2002), thereby influencing the species composition of the plankton community (Walsh, 1975). It also affects aquatic animals (Ashraj, 2005; Vosyliene and Jankaite, 2006) including zooplankton (Eisler, 1986), molluscs (Satyaparameshwar *et al.*, 2006) and fish (Vinodhini and Narayanan, 2008). It appears, therefore, that chromium might also have indirectly affected the waterbirds through its negative effects on their food resources, although direct toxic effects cannot be over ruled, since this heavy metal has negative effects on herring gull *Larus argentatus*, chicks (Burger and Gochfeld, 1995) and Japanese quail *Coturnix coturnix japonica* (Butkauskas and Sruoga, 2004). Certain species of fish and aquatic invertebrates are sensitive to chromium, showing reduced survival or growth at concentration > 10 mg/l. Their elimination from environments contaminated with chromium may have detrimental effects on wild birds that depend on such organisms for food (Outridge and Scheuhammer, 1993). Since the chromium concentration at Santragachi wetland varied between 127 and 138 mg/l, it might have negative effects on carnivorous waterbirds.

Conclusion

This study shows that (1) migratory waterbirds at Santragachi Lake exhibited significant declining trends during the period between 1998 and 2013 with some of the species declined drastically; (2) Water hyacinth cover was the principle driving factor for temporal variations in abundance and community structure in Santragachi. Anthropogenic activities in the form of increased BOD and chromium concentration contribute to worsen the wetland, thereby boosting the process of this decline; (3) Water hyacinth cover affected the waterbirds through its negative impacts on their prey bases, as well as by causing reduction in the availability of open water area and thereby in the foraging ground; (4) migratory waterbirds declined at variable rates;

the carnivorous and omnivorous birds depending primarily on macro invertebrates and fish were most seriously affected; and (5) their assemblages exhibited clear shifts.

The processes governing fluctuations in waterbird abundance and community composition are not yet well understood. This lack of knowledge will certainly limit the options for implementing management strategies aiming at optimising biodiversity. Many factors may be involved to drive the changes in abundance and community composition of waterbirds. Future research will need to focus on the links between these important drivers, and the way in which they influence the waterbird abundance and community structure. At the same time monitoring programme should be undertaken on a long-term basis to update the persistence of population trends. These will certainly throw light into the strategy for wetland management along with their denizens.

Summary

Waterbirds face serious threats mainly due to increasing rates of habitat degradation and disturbance driven by human activities. Invasive species, especially water hyacinth, add to the problem of habitat degradation. Santragachi Lake, India, located within the industrial belt, provides an example of how the waterbird abundance and community composition have been changing over time due to proliferation of water hyacinth and human impact on wetlands. The study, carried out during 1998–2013, showed that migratory waterbirds in the Santragachi Lake have declined significantly over a period of sixteen years. The most affected waterbirds included the long-distance migrants like northern pintail, common teal, ferruginous duck, tufted duck and swinhoe' snipe. All of these waterbirds suffered declines by 70 per cent or more. The local migrant Eurasian coot also declined by 75 per cent. The local migrants fulvous whistling duck and the most common species lesser whistling duck exhibited the lowest levels of declines by 25 per cent and 38 per cent respectively. These changes were associated with the reduction of the surface water area due to proliferation of water hyacinth and the consequent changes in prey bases of waterbirds. Ten of 13 waterbirds considered in this study showed significant negative relationships with increasing water hyacinth cover. The waterbird community has also changed and shifted in its composition along with an increase in the level of dominance and declines in the diversity and equitability. These changes were associated with the decline in the relative abundances of dabbling and diving waterbirds that depend primarily on macro invertebrates and fish, and an increase in the relative abundances of the herbivorous species. Anthropogenic activities in the form of increased BOD and chromium concentration enhanced the process of this decline. Chromium was related to the decline in fulvous whistling duck, while BOD was involved in the decline of all the waterbirds excepting fulvous whistling duck. Decline in knob-billed duck was related to both the variables.

Acknowledgements

A part of this study was supported by funding from University Grants Commission and we gratefully acknowledge this support. We are thankful to the Higher Education Department, Government of West Bengal and Principal, Maulana

Azad College, for allowing us to carry out this work. Special thanks are due to the Post Graduate students at Maulana Azad College for their active participation in survey works. We also gratefully acknowledge the active cooperation of the people inhabiting around Santragachi wetland in carrying out this long-term study and providing important information about the wetland and its denizens. Anirban Sinha and Prantik Harza gratefully acknowledge CSIR and UGC respectively for their Fellowship (NET) grants.

References

Alongi DM. 1998. *Coastal ecosystem processes*. CRC Press, Boston.

Aoi T and Hayashi T. 1996. Nutrient removal by water lettuce (*Pisitia stratiotes*). *Water Science and Technology*, 34: 407-412.

APHA 2005. *Standard methods for the examination of water and wastewater*, 21st edn. Washington DC.

Ashraj W. 2005. Accumulation of heavy metals in kidney and heart tissues of *Epinephelus microdon* fish from the Arabian Gulf. *Environ. Monit. Assess.*, 101 (1-3): 311-316.

Baral HS and Inskipp C. 2004. *The state of Nepal's birds 2004*. Kathmandu: Department of National Parks and Wildlife Conservation, Bird Conservation Nepal and IUCN Nepal.

Bartodziej W and Weymouth G. 1995. Waterbird abundance and activity on water-hyacinth and *Egeria* in the St-Marks River, Florida. *Journal of Aquatic Plant Management*, 33: 19-22.

Belsley DA, Kuh E and Welsch RE. 1980. Regression diagnostics identifying influential data and sources of collinearity. John Wiley and Sons, New York.

Bird Life International 2011. IUCN Red List for birds. Downloaded from http://www.birdlife.org on 19/03/2014.

Bolker BM, Brooks ME, Clark CJ. *et al.*, 2008. Generalized linear mixed models: a practical guide for ecology and evolution. *Trends in Ecology and Evolution*, 24(3): 127-135.

Brendonck L, Maes J, Rommens W. *et al.*, 2003. The impact of water hyacinth (*Eichhornia crassipes*) in a eutrophic subtropical impoundment (Lake Chivero, Zimbabwe). II. Species diversity. *Archiv Fur Hydrobiologie*, 158: 389-405.

Burger J and Gochfeld M. 1995. Growth and behavioral effects of early postnatal chromium and manganese exposure in herring gull (*Larus argentatus*) chicks. *Pharmacology Biochemistry and Behavior.*, 50(4): 607-612.

Butkauskas D and Sruoga A. 2004. Effect of lead and chromium on reproductive success of Japanese quail. *Environmental Toxicology*, 19(4): 412-415.

Clements J. 2007. *The Clements Checklist of the Birds of the World*, Cornell University Press, Ithaca.

Dixit S and Tiwari S. 2007. Effective utilization of an aquatic weed in an eco-friendly treatment of polluted water bodies. *J. Appl. Sci. Environ. Manage.*, 11(3): 41-44.

Eisler R. 1986. Chromium hazards to fish, wildlife, and invertebrates: a synoptic review. *Biological Report 85(1.6)*, Patuxent Wildlife Research Center, U.S. Fish and Wildlife Service, Laurel, MD 20708.

Gibbons M, Gibbons H Jr and Systma M. 1994. A Citizen's Manual for Developing Integrated Aquatic Vegetation Management Plants. Water Environmental Services. http://www.ecy.wa.gov/programs/wa/plants/management/manual/index.html.

Haag KH, Joyce JC, Hetrick WM and Jordan JC. 1987. Predation on Water Hyacinth weevils and other aquatic insects by three wetland birds in Florida. *The Florida Entomologist*, 70: 457-471.

Hörcsik ZT and Balogh Á. 2002. Intracellular distribution of chromium and toxicity on growth *in Chlorella pyrenoidosa. Acta Biologica Szegediensis*, 46(3-4): 57-58.

Johnson MG. 1965. Estimates of fish populations in warm water streams by the removal method. *Transactions of the American Fisheries Society*, 94: 350-357.

Kaladharan P, Alavandi SV, Pillai VK and Balachandran VK. 1990. Inhibition of primary production as induced by heavy metal ions on phytoplankton population off Cochin. *Indian J. Fish*, 37 (1): 51-54.

Kateregga E and Sterner T. 2009. Lake Victoria fish stocks and the effects of water hyacinth. *The Journal of Environment and Development*, 18: 62-78.

Kazmierczak K. 2003. A *Field Guide to the Birds of India, Sri Lanka, Pakistan, Nepal, Bhutan, Bangladesh and the Maldives*, Om Book Service, New Delhi, India.

Khan TN. 2010. Temporal changes to the abundance and community structure of migratory waterbirds in Santragachi Lake, West Bengal, and their relationship with water hyacinth cover. *Current Science*, 99(11): 1570-1577.

Khan TN, Sinha A, Basu D and Chakraborty S. 2005. A preliminary observation of diversity and threats to the survival of migratory waterfowls in an avian abode near Kolkata. *Journal of Natural History*, 1(1): 56-67.

Mack RN, Simberloff D, Lonsdale WM. *et al.*, 2000. Biotic invasions: Causes, epidemiology, global consequences, and control. *Ecological Applications*, 10: 689-710.

Masifwa WF, Twongo T and Denny P. 2001. The impact of Water Hyacinth, *Eichhornia crassipes* (Mart) Solms on the abundance and diversity of aquatic macroinvertebrates along the shores of northern Lake Victoria, Uganda. *Hydrobiologia*, 452: 79-88.

McCune B and Mefford MJ. 2006. PC-ORD. Multivariate analysis of ecological data, Version 5. Gleneden Beach, OR: MjM Software.

McVea C and Boyd CE. 1975. Effects of water-hyacinth cover on water chemistry, phytoplankton, and fish in Ponds. *Journal of Environmental Quality*, 4: 375-378.

Meerhoff M, Mazzeo N, Moss B and Rodriguez-Gallego L. 2003. The structuring role of free-floating versus submerged plants in a subtropical shallow lake. *Aquatic Ecology*, 37: 377-391.

Millennium Ecosystem Assessment, Ecosystems and Human Well-being: Biodiversity Synthesis 2005. World Resources Institute, Washington, DC.

Miranda LE and Hodges KB. 2000. Role of aquatic vegetation cover on hypoxia and sunfish abundance in bays of a eutrophic reservoir. *Hydrobiologia,* 427: 51-57.

Moran MD. 2003. Arguments for rejecting the sequential Bonferroni in ecological studies. *Oikos,* 100: 403-405.

O'Brien RM. 2007. A Caution Regarding Rules of Thumb for Variance Inflation Factors. *Quality and Quantity,* 41: 673-690.

Outridge PM and Scheuhammer AM. 1993. Bioaccumulation and toxicology of chromium: implications for wildlife. *Rev. Environ. Contam. Toxicol.,* 130: 31-77.

Pannekoek J and van Strien A. 2001. TRIM (Trends and Indices for Monitoring Data). Vooburg: Centraal Bureau voor de Statistiek (CBS). (Research paper no. 1020).

Paton DC, Rogers DJB, Hill MC. *et al.,* 2009. Temporal changes to spatially stratified waterbird communities of the Coorong, South Australia: implications for the management of heterogeneous wetlands. *Anim. Conserv.,* 12: 408-417.

Perna C and Burrows D. 2005. Improved dissolved oxygen status following removal of exotic weed mats in important fish habitat lagoons of the tropical Burdekin River floodplain, Australia. *Marine Pollution Bulletin,* 51: 138-148.

Pielou EC. 1966. The measurement of diversity in different types of biological collections. *Journal of Theoretical Biology,* 13: 131-44.

Ramachandra TV and Solanki M. 2007. Ecological assessment of lentic water bodies of Bangalore. *ENVIS Technical Report* No. 25, pp. 103.

Rommens W, Maes J, Dekeza N. *et al.,* 2003. The impact of Water Hyacinth (*Eichhornia crassipes*) in a eutrophic subtropical impoundment (Lake Chivero, Zimbabwe). I. Water quality. *Archiv Fur Hydrobiologie,* 158: 373-388.

Sanaa M, Shanab M, Shalaby EA. *et al.,* 2010. Allelopathic Effects of Water Hyacinth [*Eichhornia crassipes*]. *PLoS One,* 5(10): e13200.

Sarkar B, Hazra P, Pawan Kumar S. *et al.,* 2014. Habitat Attributes and waterbird-use of four Wetlands in Manas National Park, Assam, India. *Proc. Zool. Soc.,* 67(2): 94-107.

Satyaparameshwar K, Ravinder Reddy T and Vijaya Kumar N. 2006. Effect of chromium on protein metabolism of fresh water mussel, *Lamellidens marginalis. J. Environ. Biol.,* 27(2): 401-403.

Shanker AK, Cervantes C, Loza-Tavera H and Avudainayagam, S. 2005. Chromium toxicity in plants. *Environment International,* 31: 739-753.

Shannon CE. 1948. A mathematical theory of communication. *Bell System Technical Journal,* 27: 379-423.

Sharitz RR and Batzer DP. 1999. An introduction to freshwater wetlands in North America and their invertebrates. In. *Invertebrates in freshwater wetlands of North America: ecology and management* (Batzer DP, Rader RB and Wissinger SA. Eds.) John Willey and Sons, New York. pp. 1-21.

Simpson EH. 1949. Measurement of Diversity. Nature, *163: 688.*

Sinha A, Hazra P and Khan TN. 2011. Population Trends and Spatio-temporal Changes to the Community Structure of Waterbirds in Birbhum District, West Bengal, India. *Proc. Zool. Soc.*, 64: 96 -108.

Svingen D and Anderson SH. 1998. Waterfowl management on grass-sage stock ponds. *Wetlands*, 18: 84-89.

ter Braak CJF.1986. Canonical correspondence analysis: A new eigenvector technique for multivariate direct gradient analysis. *Ecology*, 67: 1167-1179.

ter Braak CJF and Šmilauer P. 2002. CANOCO reference manual and CanoDraw for Windows User's guide: software for Canonical Community Ordination (version 4.5). Microcomputer Power, Ithaca.

Tiwari S, Dixit S and Verma N. 2007. An effective means of biofiltration of heavy metal contaminated water bodies using aquatic weed *Eichhornia crassipes*. *Environmental Monitoring and Assessment*, 129: 253-256.

Toft JD, Simenstad CA, Cordell JR and Grimaldo LF. 2003. The effects of introduced Water Hyacinth on habitat structure, invertebrate assemblages, and fish diets. *Estuaries*, 26: 746-758.

Villamagna A. 2009. *The ecological effects of Water Hyacinth (Eichhornia crassipes) on Lake Chapala, Mexico.* Ph.D. Thesis. Virginia Polytechnic Institute and State University, Blacksburg.

Vinodhini R and Narayanan M. 2008. Bioaccumulation of heavy metals in organs of fresh water fish *Cyprinus carpio* (Common carp). *Int. J. Environ. Sci. Tech.*, 5 (2): 179-182.

Vosyliene MZ and Jankaite A. 2006. Effect of heavy metal model mixture on rainbow trout biological parameters. *Ekologija*, 4: 12-17.

Walsh GE. 1975. Utilization of energy by primary producers in four ponds in northwestern Florida. In. *Proceedings Biostimulation and Nutrient Assessment.* U.S. Environmental Protection Agency. pp. 249-274.

Weller MW and Spatcher CS. 1965. Role of habitat in the distribution and abundance of marsh birds. Special Report 43. Agricultural and Home Economics Experiment Station. Iowa State College, Iowa, USA.

Wetlands International 1997. *The Asian Waterfowl Census.* Lopez, A. and Mundkur, T. (eds.).

Wetlands International 2009. *Status of waterbirds in Asia: Results of the Asian Waterbird census 1987-2007.* Wei *et al.* (eds.).

Woodward RT and Wui YS. 2001. The economic value of wetland services: a meta-analysis. *Ecological Economics*, 37: 257-270.

Zimmels Y, Kirzhner F and Malkovskaja A. 2007. Advanced extraction and lower bounds for removal of pollutants from wastewater by water plants. *Water Environment Research*, 79: 287-296.

Chapter 26

Ecotoxicological Threats of Agro and Geogenic Chemicals in the Freshwater Molluscs of India: A Review

Anindya Sundar Bhunia, Mitali Ray and Sajal Ray

Aquatic Toxicology Laboratory, Department of Zoology, University of Calcutta, Kolkata

Introduction

Mollusca is the second largest invertebrate Phylum and exhibits a wide range of diversity and with nearly 200,000 of living species (Ponder and Lindberg, 2008). Evolutionarily, molluscs are ancient group of animals which evolved around 450 million years ago and occupied a wide range of habitats including terrestrial, marine, estuarine and freshwater ecosystems. Molluscs exist due to their evolutionary and reproductive success which depends on various factors including highly evolved immune system to protect themselves from various pathogens and toxins in different environment. *Lamellidens marginalis*, *Pila globosa* and *Bellamya bengalensis* are aquatic molluscs which are widely distributed in the freshwater ponds, lakes and other water bodies of the different states of India including West Bengal. All of these molluscan species bear ecological, economical, medicinal (Prabhakar and Roy, 2009), nutritional (Baby *et al.*, 2010), industrial and biotechnological importance and considered as important biological resource.

Use of pesticide is a common method in the agricultural practice to reduce the loss of crop production due to pest attack. In recent years, various classes of pyrethroids and neem based pesticides are in use in the agricultural sector. The indiscriminate use of pesticides not only destroys pests but affects the non-target species adversely. The farmers with limited knowledge of rational use of pesticides plays a causative role in the destruction of non-target species and their ecology leading to reduction in population (Figure 26.1). Anwar (1997) reported that pesticides contaminate entire environment and is traced in majority of global biodiversity. The harmful effect of pesticide on human and environment was first reported and elaborately described in a constructive way by Rachel Carson in her

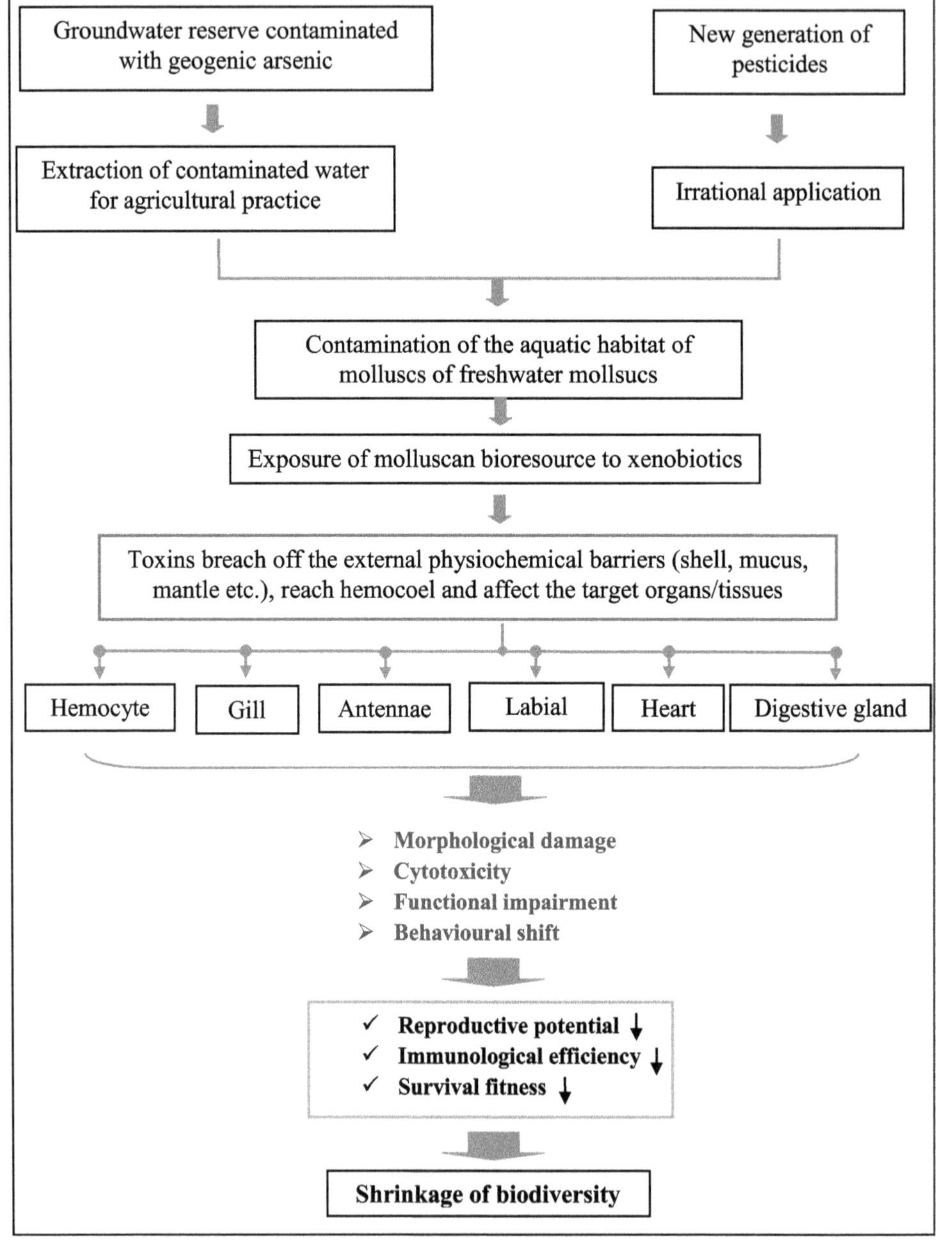

Figure 26.1: Mobility and Consequence of Geogenic and Agrotoxins in the Freshwater Ecosystem.

book Silent Spring (1962) in USA (Paull, 2007). This historical publication influenced the ban of DDT in 1972 in the United States.

In the last two decades, the use of diverse groups of insecticides has been increased greatly including organophosphate, organochlorine, carbamate and pyrethroids. Pyrethroids in broad spectrum insecticides and gained popularity for their low toxicity in birds and mammals. Pyrethroid group of pesticides affect adversely on the flying, crawling, chewing, and sucking insects of the orders Coleoptera, Diptera, Homoptera, Heteroptera, Hymenoptera, Lepidoptera, Orthoptera and Thysanoptera (Deborah Gnana Selvam *et al.*, 2013). Pyrethroids are synthetic analogues of the naturally occuring pyrethrins. *Chrysanthemum* genus of plants is the chief source of highly concentrated natural pyrethrins. Synthetic pyrethroid insecticides are usually derived from pyrethrin and produce several types of insecticides including tetramethrin, resmethrin, fenvalerate, permethrin, cypermethrin, λ-cyhalothrin, and deltamethrin (Indira Rani and Kumaraguru, 2014). Cypermethrin and fenvalerate belong to the second generation synthetic pyrethroids and widely used in the agricultural sectors of world including India.

Inorganic sodium arsenite is a geogenic xenobiotic which is reported as a serious immunotoxin of important aquatic mollusc (Chakraborty and Ray, 2009; Ray *et al.*, 2011). Contamination of surface water by arsenic appeared to be a new ecotoxicological threat to filter feeding bivalves and crabs (Saha *et al.*, 2008). However, report of ecotoxicity of arsenic in freshwater invertebrates is relatively inadequate. Common contaminants of freshwater ponds and lakes include cypermethrin, fenvalerate, azadiractin, sodium arsenite etc.

Aquatoxins and Invertebrate Blood Cells

Hemocytes are the chief immune effector free circulating blood cells of many invertebrate Phyla, which play various physiological functions like shell repair, nutrient digestion, metabolite transport, transport of calcium, excretion, secretion of exoskeleton, protein regeneration and wound repair. Hemocytes attribute to the defence mechanism by internalization of nonself material, detoxification of xenobiotics, disease resistance, chemotaxis, lectin mediated pathogen recognition, encapsulation and production of antimicrobial peptides.

Hemocytes of molluscs under the exposure to pyrethroids present the symptoms of substantial damages in morphology as observed by bright field (Figure 26.2) and phase contrast microscope (Ray *et al.*, 2013a). They reported morphological damages of hemocytes including hypervacuolation, hypergranulation, nuclear disintegration, rounding up of cell, alteration of cellular shape, membrane disintegration and membrane blebbing under the exposure of pyrethroid pesticides in two species of molluscs, *B. bengalensis* and *L. marginalis*. The morphological damages of hemocytes induced by cypermethrin and fenvalerate were indicative of possible impairment of immunological and physiological functions of the blood cells of invertebrates.

Total hemocyte count and differential hemocyte count are reported as the significant immunological parameters (Mello *et al.*, 2010) of invertebrates and vertebrates. Alteration of hemocyte density is the early indication of disturbance of cellular homeostasis. Fluctuation of hemocyte density under different environmental

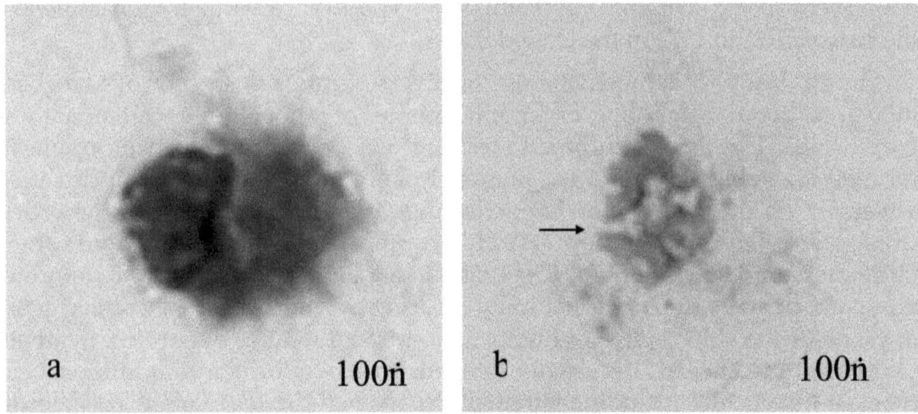

Figure 26.2: Cypermethrin Induced Morphological Damage of the hemocytes of *P. globosa*.
[a: Control, b: Treated (1.5ppm/96 hour)].

toxins had been reported by many authors. Russo and Madec (2007) reported that the exposure to fomesafen pesticide elevated total number of circulating hemocytes in snail, *Lymnaea stagnalis*. The total cell density reduced in the hemolymph of *Villorita cyprinoides* under the treatment of copper (Suresh and Mohandas, 1990). Russo and Lagadic (2000, 2004) reported an elevation in hemocyte number on exposures to atrazine and hexachlorobenzene in molluscs (*Lymnaea stagnalis* and *L. palustris*). A significant increase in THC was observed on exposure to phenanthrene in the temperate scallop *Pecten maximus* (Hannam *et al.*, 2010). Earlier, Mukherjee *et al.* (2006) reported the shift in the total hemocyte count in *L. marginalis* under the treatment of different sublethal concentrations of azadirachtin, a neem based biopesticide, commonly used in the agricultural fields of India. The authors proposed that the toxin induced shift in the cell density may lead to a steady declination of this species in its natural habitat. Arsenic, an environmental pollutant, reduced the total count of hemocytes of the freshwater bivalve, *L. marginalis* (Chakraborty *et al.*, 2008). In the recent past, Ray *et al.* (2013b) reported a significant elevation in density of hemocytes by the pyrethroids in gastropods and bivalves. The shift in the hemocyte density under the effect of pyrethroids appeared to be non-linear and dose independent.

Molluscan hemocytes exhibit variable morphologies with different sizes, shapes and characterestics. Ray *et al.* (2013a) proposed a general scheme of classification of hemocyte morphotypes employing flow cytometric cell sorting and light microscopy on the basis of relative size and relative granularity of cells in the freshwater molluscs. Authors reported agranulocytes, semigranulocytes and granulocytes as three morphotypes of hemocytes. Each morphotype had been further divided into various sub populations. Blast like cells, round and spindle hyalinocytes were the subpopulations of agranulocytes. Blast like cells (Chang *et al.*, 2005; Chakraborty *et al.*, 2008) were smaller in size, had centrally placed large nucleus with a thin rim of scanty cytoplasm without granules and exhibit high nucleus to cytoplasmic ratio. Round hyalinocytes contained central or eccentric nuclei with clear cytoplasm, around while few hyalinocytes exhibited spindle and slender shape without granules in the cytoplasm. Semigranulocytes included two different sub-populations as semigranular asterocytes and round semigranulocytes. Round semigranulocytes were with central or eccentric nucleus and the star shaped semigranular asterocytes were with numerous angular cytoplasmic projections and were spreading in nature. Granulocytes consisted of three subpopulations including round granulocytes, spindle granulocytes and granular asterocytes. All the sub-populations exhibited high density of granules in their cytoplasm. Round granulocytes were with central or eccentric nucleus, spindle shaped granulocytes were few in numbers and granular asterocytes were star shaped with spreading appearance including angular cytoplasmic projections. Different morphotypes of hemocyte performed different physiological functions. Semigranulocytes of *P. globosa*, agranulocytes of *B. bengalensis* and granulocytes of *L. marginalis* exhibited maximum efficiency of phagocytosis against yeast challenge. The authors noticed a highest generation of superoxide anion in semigranulocytes of *P. globosa* and agranulocytes of *B. bengalensis*. Beside this, semigranulocytes of *P. globosa*, granulocytes of *B. bengalensis*, granulocytes of *L. marginalis* generated the nitric oxide in the morphotypes of hemocytes. On the other hand, only agranulocytes of *P. globosa* exhibited the highest activity of phenoloxidase. Variations in percent populations of hemocyte sub-populations under different toxin exposure altered the various physiological functions related to environmental stress. Mukherjee *et al.* (2008) reported the pesticide induced modulation in percent population of different hemocytes of *L. marginalis*. They recorded a decline in the percent density of granulocytes, agranulocytes, hyalinocytes and asterocytes in *L. marginalis* under the sublethal exposure of azadirachtin, a neem based pesticide. The authors concluded that the alteration of hemocyte populations was indicative to possible impairment of cellular functions of organism distributed in the contaminated habitat. In the oysters *Crassostrea virginica* and *C. gigas* collected from polluted sites exhibited increased numbers of granulocytes when compared to the control (Ruddell and Rains, 1975). Cadmium exposed *C. virginica* showed an increased percentage of hyalinocytes (Cheng, 1988a, 1988b). Sami *et al.* (1992) stated that small hemocyte populations was elevated under the treatment of polycyclic aromatic hydrocarbon-contaminated sediments and oppositely reduced the populations of the larger hemocytes. Hyalinocytes population was increased in hexachlorobenzene and atrazine exposed *Lymnaea palustris* (Russo and Lagadic, 2000).

Exposure of different sublethal concentrations of cypermethrin altered the relative percentage of hemocyte subpopulation of *L. marginalis* (Das *et al.*, 2012). Percent population of agranulocytes and asterocytes were declined but blast like cells and granulocytes were elevated with respect to control hemocyte population of *L. marginalis* under the cypermethrin exposure. Ray *et al.* (2013) reported that the relative density of blast like cells and round hyalinocytes were elevated significantly under 0.5 ppm of cypermethrin for 96 hours of exposure and a significant decline of granular asterocyte population was also observed in *B. bengalensis* when compared to control. On the other hand, relative percentage of round hyalinocyte was decreased under the treatment of 3 ppm of fenvalerate in *L. marginalis* with respect to control.

The cypermethrin and fenvalerate influenced shift in the percent populations of hemocyte may lead to impairment of cellular homeostasis in the freshwater molluscs distributed in pyrethroid contaminated environment.

Phagocytosis and Environmental Toxins

Phagocytosis is regarded as a pivotal component of classical innate immune behaviour in most of the animal Phyla. Phagocytosis (Figure 26.3) is an immune response which plays a significant role in the defense system of a cell to engulf the foreign particulates or pathogens. Hemocyte mediated phagocytosis of nonself particles provides natural immunity in mollusc (Lopez *et al.*, 1994) and it is reported as biomarker of aquatic pollution (Oliver and Fisher, 1999). The phagocytic response was inhibited by arsenic in the hemocytes of juvenile mud crab, *Scylla serrata* of Sunderbans Biosphere Reserve of West Bengal under the challenge of yeast (Ray *et al.*, 2011). Significant shift in the phagocytic response by arsenic may reduce the propagation potential and survival fitness of mud crab population by increasing its susceptibility to pathogen invasion and various diseases. Cooper and Roch (2003) reported that the phagocytic efficiency was decreased under carbaryl exposure in the earthworm, *Eisenia fetida*. Similarly, glyphosate negatively affected the phagocytic activity in *Cyprinus carpio* (Terek-Majewsca *et al.*, 2004). Hemocyte phagocytic activity was reported to be decreased by phenol (Fries and Tripp, 1980) and butyltin compounds (Bouchard *et al.*, 1999) in clam *Mercenaria mercenaria*. Exposure to sodium arsenite impaired the ability of phagocytosis in hemocytes of freshwater bivalve *L. marginalis* (Chakraborty *et al.*, 2010). Mukherjee *et al.* (2011) reported that the phagocytic efficacy was significantly reduced in the hemocytes of *L. marginalis* under the treatment of different sublethal exposure to azadirachtin against the challenge of charcoal particulate. Righi and Palermo-Neto (2005) reported that the cyhalothrin, a synthetic type II pyrethroid, suppressed the phagocytic response and spreading behaviour of macrophage. Mikula *et al.* (1992) reported supercypermethrine induced decline of phagocytic index in sheep. The effect of various pyrethroids like cypermethrin, supercypermethrin and deltamethrin suppressed the immune response of humoral and cell-mediated immunity in mice, rats and goats (Stelzer

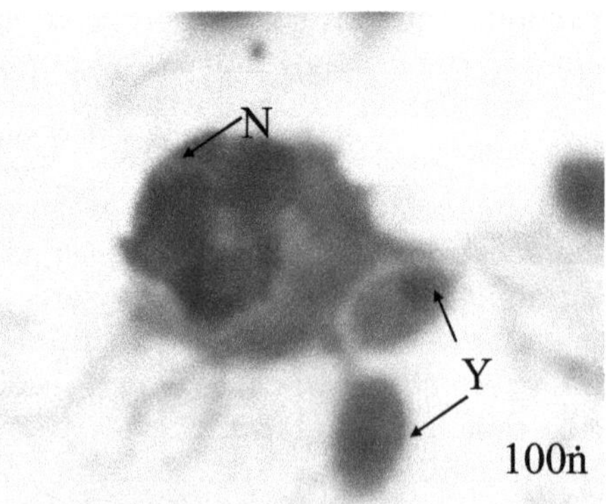

Figure 26.3: Phagocytosis of Yeast (Y) Particles by the Hemocytes of *P. globosa* Exposed to Cypermethrin (1.5 ppm/96 hours), (N-nucleus).

and Michel, 1984; Tamag *et al.*, 1988; Tulinska *et al.*, 1995). The effect of pyrethroids and other pesticides reduced the phagocytic response which indicated the state of immune suppression in the organism distributed in the contaminated habitat.

Cytotoxicity and Environmental Toxins

Cytotoxicity of a cell is represented by the generation of various cytotoxic molecules to destroy and inactivation of pathogens. Superoxide anion, nitric oxide and phenoloxidase are the major cytotoxic molecules which are reported in various mollusc including *L. marginalis, B. bengalensis* and *P. globosa*. The oxygen dependent powerful defense mechanism consists of the generation of reactive oxygen intermediates. Superoxide anion is a free radical which is generated from molecular oxygen (O_2). Xanthine oxidase produces it from oxidation of xanthine. Likewise NAD(P)H oxidase and nitric oxide synthase also produce superoxide anion. In invertebrates, this microbicidal system has been first demonstrated in gastropods (Dikkeboom *et al.*, 1988). The production of reactive oxygen species by phagocytic hemocytes of bivalves is believed to accompany with the release of degradative enzymes for the destruction of foreign material (Pipe, 1992; Noλl *et al.*, 1993). Arumugam *et al.* (2000) reported that phorbol myristate acetate, laminarin and lipopolysaccharide of *Vibrio cholera* or *Pseudomonas aeruginosa* are nonself molecules which stimulates the generation of superoxide anions in the hemocytes of mollusc, *Mytilus galloprovincialis*. Generation of active oxygen species has been reported in several bivalves, such as *Patinopecten yessoensis* (Nakamura *et al.*, 1985), *Crassostrea virginica* (Larson *et al.*, 1989), *Pecten maximus* (Le Gall *et al.*, 1991), *Crassostrea gigas, Ostrea edulis* (Bachère *et al.*, 1991), *Mytillus edulis* (Pipe, 1992) and *M. galloprovincialis* (Dailianis, 2009). Superoxide anion is also generated by giant clam, *Tridacna crocea* (Nakayama and Maruyama, 1998). The levels of superoxide anion generation was altered by toxins exposure have been well documented in invertebrates (Adema *et al.*, 1991; Coles *et al.*, 1994, 1995; Pipe and Coles 1995; Cajaraville *et al.*, 1996; Arumugam *et al.*, 2000; Wootton and Pipe, 2003). Mukherjee *et al.* (2012) reported that superoxide anion generation was increased in a dose dependent manner in the hemocytes of freshwater edible bivalve, *L. marginalis* under the exposure of azadirachtin, a neem based pesticide.

Nitric oxide is a short-lived electrically neutral, very reactive, free radical. It readily diffuses through membranes. Nitric oxide synthase generates nitric oxide by the oxidation of guanadino nitrogen of L-arginine and produces equal amount of L-citrulline and nitric oxide. Nitric oxide is important for its diverse physiological functions. Nitric oxide conjugates with reactive $O_2^{\bullet-}$ to form peroxynitrite which is a strong bactericidal agent. Nitric oxide is significantly suppressed by the exposure of tributyltin in *Mytilus edulis* (Smith *et al.*, 2000). The intrahemocytic generation of nitric oxide was elevated under the exposure of arsenic in estuarine mud crab, *Scylla serrata* (Saha *et al.*, 2008). Authors proposed that significant elevation of nitric oxide generation may be considered as a biomarker of toxin contamination in polluted environment. Chakraborty *et al.* (2010) reported that the generation of nitric oxide was suppressed under the exposure of arsenic in *L. marginalis* under prolonged exposure and the authors suggested this studied parameter as possible biomarker of arsenic toxicity in aquatic environment.

Phenoloxidase system is another form of cytotoxic agent in the hemocytes of invertebrate. Prophenoloxidase is an inactive form of phenoloxidase which is present in the hemolymph of arthropod. Phenoloxidase is a combination of three copper proteins including tyrosinase, catecholase and laccase. Phenoloxidase had been detected and identified in different tissues like hemolymph, digestive gland, gills and mantle of many molluscan species. Alteration of the phenoloxidase activity by environmental factors has been studied in molluscs (Gagnaire *et al.*, 2004, Bado-Nilles *et al.*, 2008, Luna-Acosta *et al.*, 2011). Endosulfan pesticide significantly reduced the phenoloxidase activity and haemocyte numbers (Campero *et al.*, 2008) in damsel fly (*Coenagrion puella*) larvae after a short term starvation. Phenoloxidase activity was decreased significantly in the larvae of cotton leafworm *Spodoptera littoralis* treated with buprofezin and pyriproxyfen. The activity of phenoloxidase in the gill and digestive gland of freshwater bivalve, *L. marginalis* was suppressed under the sublethal exposures of sodium arsenite (Chakraborty *et al.*, 2010, 2013). According to these authors, arsenic induced modulation in the cytotoxic status in the hemocytes and vital organs suggested a possible state of immunosuppression in the animal.

Apoptosis and Environmental Toxins

Apoptosis is a form of cell death in which a programmed sequence of events often leads to the removal of cells without releasing harmful substances around the dying cell. Apoptosis plays a significant role in embryonic development, cellular and tissue homeostasis, maintaining the health of the body and immune defense by eliminating the old cells, unnecessary cells, unhealthy and infected cells in multicellular organisms. Morphologically, apoptotic cells exhibit some important characteristic features including chromatin condensation, DNA fragmentation, cell shrinkage, membrane blebbing which are described in some of molluscan species (Sunila and LaBanca, 2003; Sokolova *et al.*, 2004; Buckland-Nicks and Tompkins, 2005). Hughes *et al.* (2010) reported that apoptosis was a host defense strategy. Environmental factors like high temperatures, salinity alterations, and pollutants exposure exhibited modulatory effects on apoptosis in molluscan cells. Environmental toxins like heavy metals, tributyltin, and polycyclic aromatic hydrocarbon increased apoptosis in molluscan hemocytes. The intensity of modulation of apoptosis depended on the dose and duration of exposure of toxins and the mode of exposure (Barsiene *et al.*, 2008). Russo and Madec (2007) reported fomesafen, a herbicide, induced hemocyte apoptosis which is an established immune response of aquatic mollusc *Lymnaea stagnalis*. Apoptosis was affected under fomesafen exposure in a dose independent manner. Casco *et al.* (2006) reported that apoptosis was elevated under cypermethrin exposure in tadpole *Bufo arenarum*. Ray *et al.* (2013b) reported that the pyrethroid induced apoptosis and necrosis were quantified in hemocytes of *B. bengalensis* (Figures 26.4 and 26.5) and *L. marginalis* under the various experimental concentrations of cypermethrin and fenvalerate respectively. Apoptosis and necrosis of hemocyte morphotypes of *B. bengalensis* and *L. marginalis* were declined under the treatment of cypermethrin and fenvalerate with respect to control. Fenvalerate exposure of hemocytes of *L. marginalis* marked a dose dependent suppression of apoptosis and necrosis.

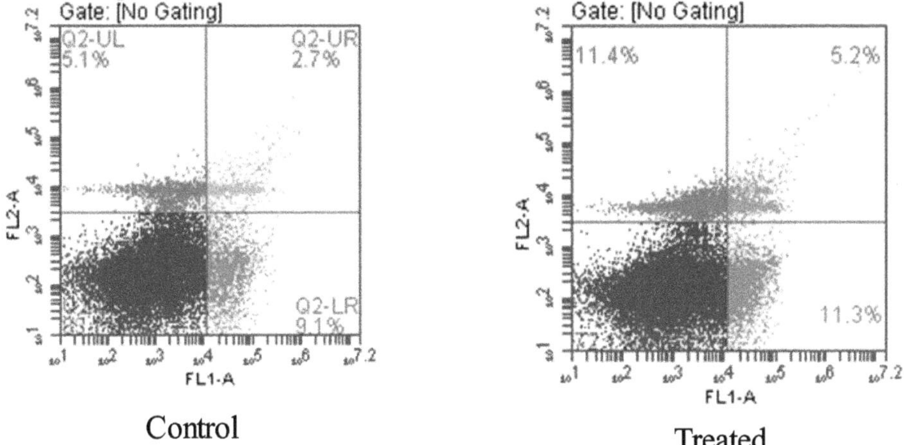

Figure 26.4: Dot Plot Presentation of FACS Analyses of Hemocyte Apoptosis of *B. bengalensis* Exposed to Fenvalerate (3.0 ppm/15 days).

Figure 26.5: Immunofluorescent Detection of Apoptosis in the Hemocytes of *B. bengalensis* by Annexin-V-FITC and Propidium Iodide Staining (Magnification: 100×). Apoptotic hemocytes generate a relatively intense green (FITC) fluorescence (a) and low red (PI) fluorescence (b) which appears to be yellow upon colocalization (c).

Histopathology

Histopathology refers to the microscopic examination of tissue in order to study the manifestations of symptoms of disease and damages. Pyrethroids and arsenic induced histopathological changes of different vital organs like labial palp, gills and digestive gland were observed in different organisms including mollusc. Labial palps of bivalve *L. marginalis* act as an important site for ingestion of food particulates and capable of effecting particle selection. Das *et al.* (2015) reported that the structural damage in the labial palp under the exposure of cypermethrin may affect the food particle selection and transport mechanism (Figures 26.6a and 26.b). The labial palp tissue exposed to cypermethrin exhibited a severe morphological damage in the oral groove, palp crest and ridged oral epithelium. Increase in gap between ciliated columnar epithelium in the oral groove region was recorded along with hypervacuolization in the degenerated muscle fiber region between inner and outer surface layer of labial palp. Scattered infiltrated hemocytes were

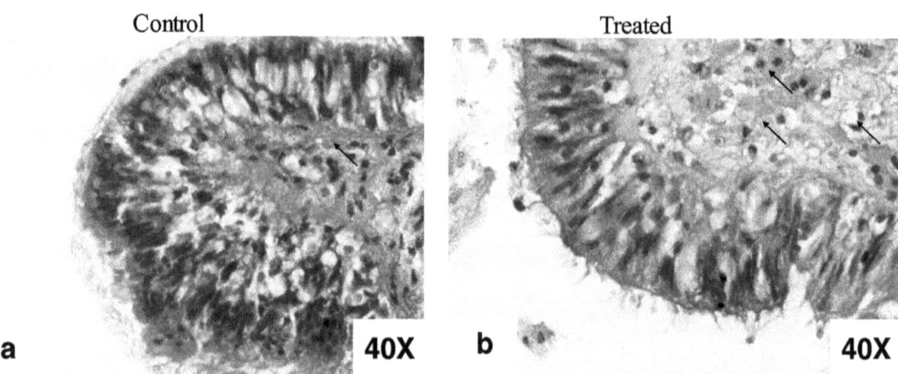

Figure 26.6: Histopathology of Labial Palps of *L. marginalis* Exposed to Cypermethrin (0.09 ppm/30days); a: Control, b: Treated exhibiting hypervacuolation, degeneration of muscle fibres.

recorded in the palp tissue. Densely packed ciliated epithelium in the dorsal fold region apparently lost their cellular integrity and structural integrity. The authors suggested that severe damage on labial palp of *L. marginalis* under cypermethrin exposure provided a clear view on impairment and destruction of the functional status of the organ.

Antenna is an important sensory organ which is involved in different physiological activities like collection of food, movement and reproduction of *B. bengalensis*. Mandal *et al.* (2012) reported the fenvarelate induced structural alterations like vacuolation and substantial disintegration of cells in the treated antenna in *B. bengalensis* (Figures 26.7a–26.7d). Authors suggested that fenvarelate induced damage of antennae indicate possible impairment of food collection and mate recognition of the organism in its natural habitat.

Pyrethroid induced histological alteration in gill tissue of *L. marginalis* (Figures 26.8a and 26.8b) and *B. bengalensis* (Figures 26.8c–26.8f) was studied by Das *et al.* (2012) and Mandal *et al.* (2012). Under sublethal concentration of cypermethrin for 30 days, connective tissue and epithelial cells of gill of *L. marginalis* lost their normal histoarchitecture (Das *et al.*, 2012). Reduction of space between water channel and interlamellar junctions were recorded. Epithelial lining at the tip of gill filaments with swelled epithelium were found to be disintegrated in comparison to untreated gill tissue. Hyperchomatic anaplastic cells infiltrated in the affected gill tissue and cilia get clumped (Figures 26.8b and 26.8d). Ruptured and swelled mucus secretory cells were studied with their fused form in gill filaments. Reduction in thickness is followed by detachment of epithelial cells from the muscular layer. Severe damage in the tip regions of lamellar epithelium was noticed under pyrethroid exposure. Dense fibrosis in the matrix of gill filaments was also noticed. Mandal *et al.* (2012) studied the fenvalerate induced histological alterations (Figures 26.8d and 26.8f) like membrane damage, tissue clumping within the matrix of gill filaments of *B. bengalensis.* Water channels of gill filaments were found to be clogged in the cypermethrin and fenvalerate treated gill filaments of *L. marginalis* (Das *et al.*, 2012) and *B. bengalensis* (Mandal *et al.*, 2012) respectively. The gill of the cypermtherin

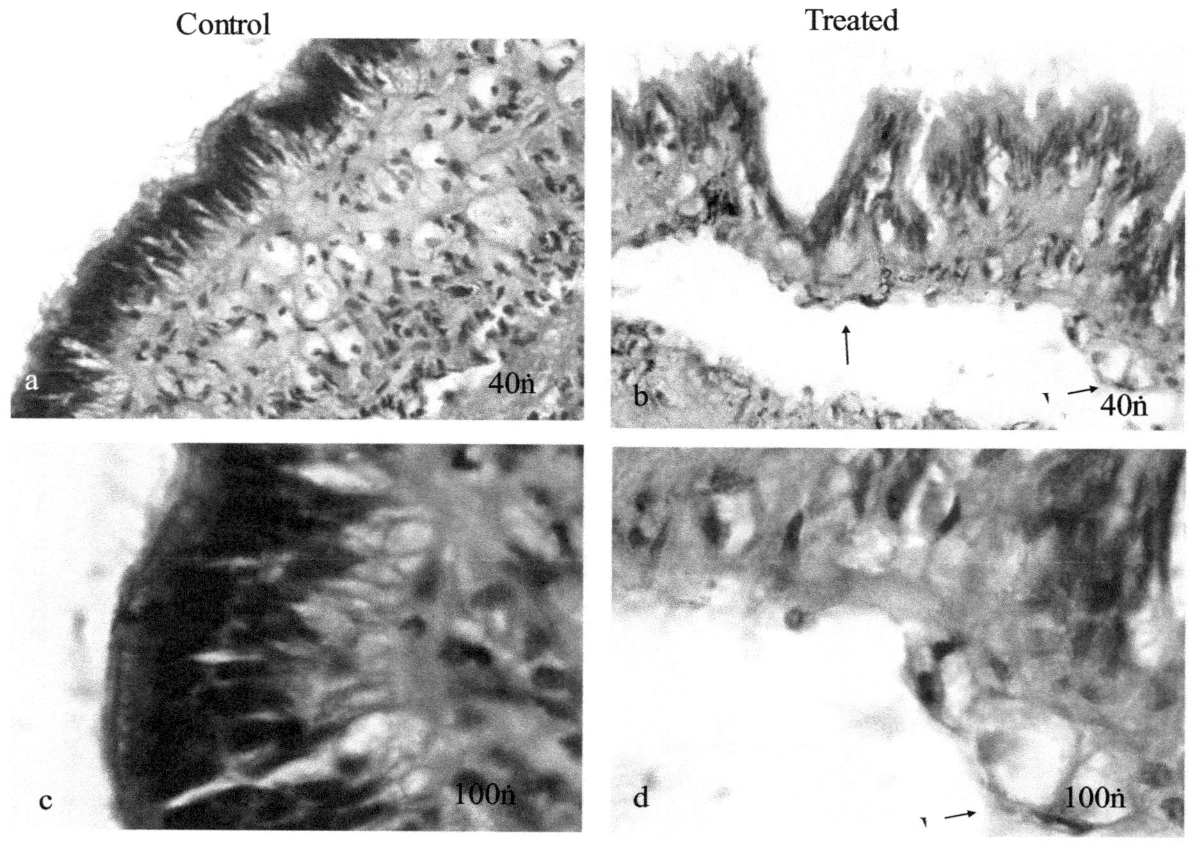

Figure 26.7: Histopathology of Antenne of *B. bengalensis* Exposed to Fenvalerate (3.0 ppm/96 hours); a, c: Control; b, d: Treated exhibiting vacuole formation (v), tissue disruption.

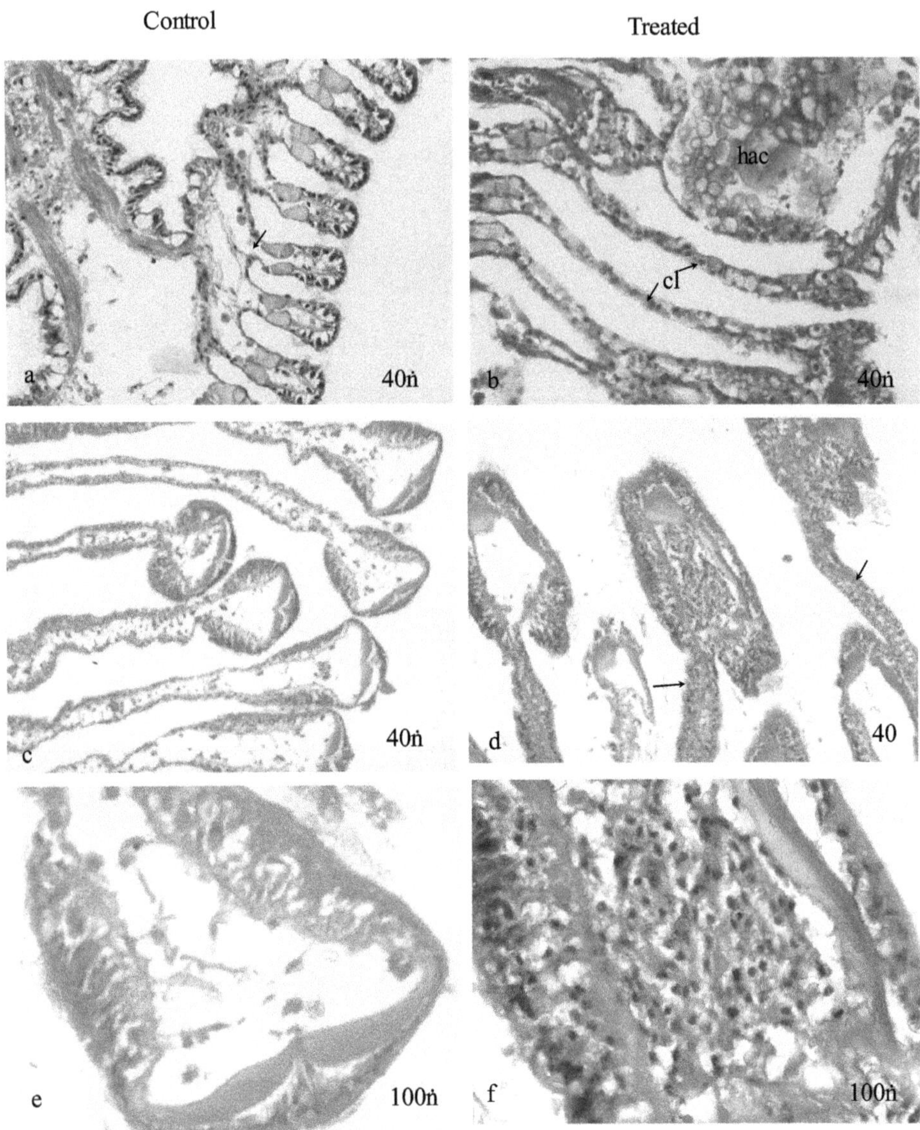

Figure 26.8: Transverse Sections of Gill of *L .marginalis* (a: Control, b: Treated with 0.09 ppm cypermethrin for 30days) and *B. bengalensis* (c, e: Control and d, f: Treated with 3.0 ppm fenvalerate for 96 hours). Pyrethroid exposure resulted in clogging of water channel (cl), fibrosis (fr) and appearance of hyperchromatic anaplastic (hac) cells.

exposed specimens exhibited a number of cell damaging features. Cypermethrin induced osmotic imbalance in the ruptured mucus secretory cells reduced the food transport system. Disintegrated gill lamellae reduced the filtration and ingestion rate (Payne *et al.*, 1995; Mehedinti *et al.*, 1999). Degeneration of ciliary surface area on gill epithelial cells reduces the rate of refreshing water penetration phase into lamellar network. Gill lesions affected gas exchange and food transport efficiency (Sunila, 1987). The occurrence of swollen membrane folds in epithelial cells of gill lamellar reflected functional changes linked to increased pinocytic activity or intercellular transport. Presence of infiltrated hemocytes in the hemocelic areas of *L. marginalis* suggested functional immunological significance.

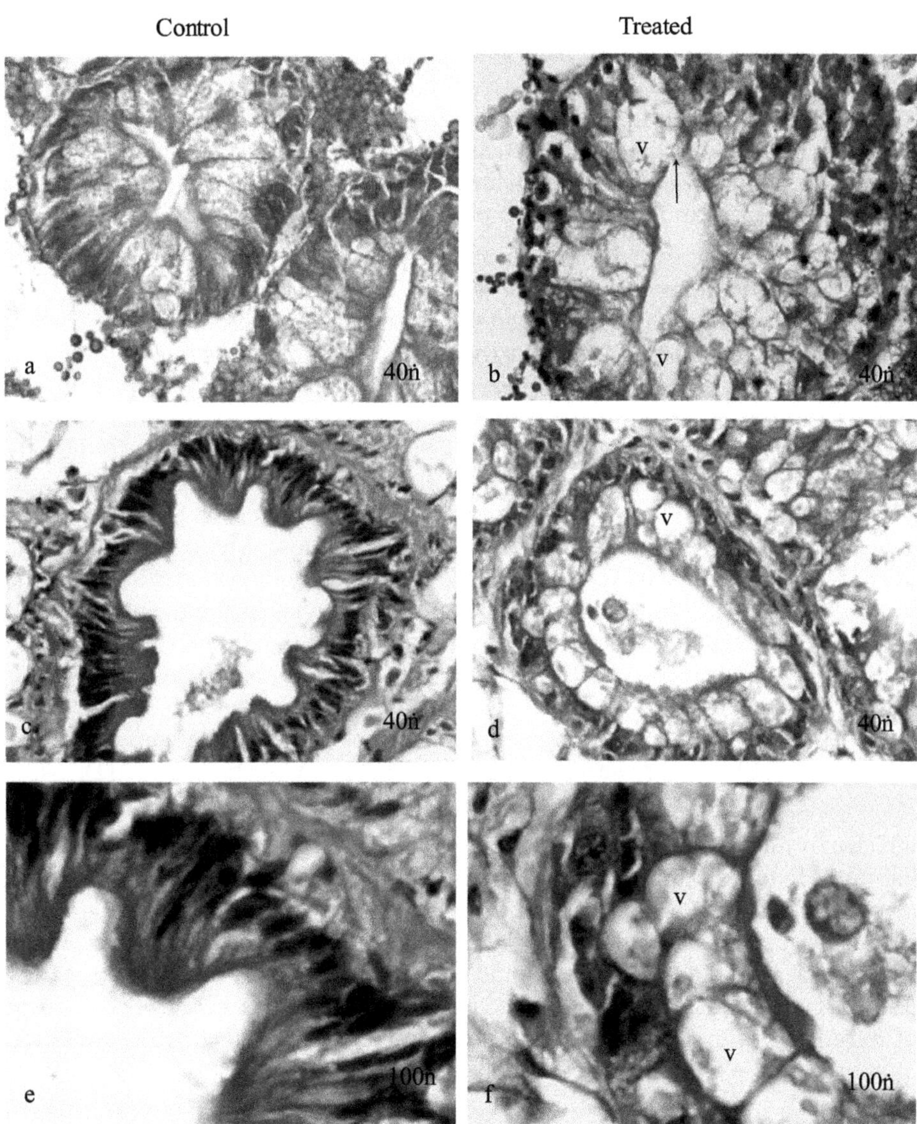

Figure 26.9: Transverse Sections of Digestive Gland of *L. marginalis* (a: Control, b: Treated with 0.09 ppm cypermethrin for 30days) and *B. bengalensis* (c, e: Control; d, f: Treated with 3.0 ppm fenvalerate for 96 hours). Pyrethroid exposure resulted in vacuolation (v) and necrotic tissue.

Microscopic study shows that digestive gland is composed of numerous hepatic lobules. Each lobule is lined by columnar epithelial cells resting on the basement membrane (Figure 26.9a). Hepatic lobules in its core region exhibit narrow lumen. Interlobular space is filled with connective tissue layer. The histopathological details of digestive glands of *L. marginalis* was studied by various workers on bivalves (Usheva *et al.*, 2006; Agwuocha *et al.*, 2011; Kumar *et al.*, 2011).

Das *et al.* (2012) observed that the digestive gland upon exposure to cypermethrin showed atrophy, edema and lesion in epithelial cells of the organ (Figure 26.7b). In addition to vacuolization (Mandal *et al.*, 2012) in the monolayer epithelium of digestive tubules, many of the digestive tubules were found to be fused with each other under cypermethrin (Figure 26.9 b) and fenvarelate (Figures 26.9d and 26.9f) exposure. Some epithelial cells lost their attachment with the basement membrane and hemocytic infiltrations in the disintegrated connective tissue were noticed in both cypermethrin and fenvarelate exposed *L. marginalis* and *B. bengalensis*.

Digestive glands of cypermethrin and fenvarelate exposed specimens exhibited a number of alterations in the hepatic lobules, lumen and connective tissue (Das *et al.*, 2012; Mandal *et al.*, 2012). Reduced thickness of epithelial cells in cypermethrin treated digestive glands may be due to entire or partial release of calcium during the detoxification process (Marigomez, 1998). Sublethal dose of cypermethrin leads to increase in tubular lumen diameter (Andhale *et al.*, 2011), irregular shape of digestive cells and presence of vacuoles. Histopathological changes in the digestive glands suggest the reduced activity in food absorption, digestion and detoxification (Agwuocha *et al.*, 2011). Increases in basophilic cells are closely associated with degenerative changes in digestive cells of marine mollusc, *Mytilus edulis, Cardiurn edule* and *Littorina littorea* (Thompson *et al.*, 1974; Cajaraville *et al.*, 1990; Marigomez *et al.*, 1990). Increase in number of infiltrated hemocytes in the connective tissue region was correlated with environmental pollution and detoxification. Reduction in digestive gland epithelium thickness may be due to the loss of apical cytoplasm during the detoxification (Johnson *et al.*, 1996; Snyman *et al.*, 2005). Such adverse damages in digestive epithelium are considered as generalized stress response of bivalves under toxin exposure (Usheva *et al.*, 2006; Agwuocha *et al.*, 2011).

Degenerated digestive cells are subjected to the state of insufficient nutrition which may affect the survival fitness of *L. marginalis* distributed in the contaminated habitat. Histopathological analysis confirmed the alteration in the histology of gills and digestive gland consequent to cypermethrin toxicity. Pyrethroid induced structural damage of gill and digestive gland were indicative to possible impairment of respiratory, digestive and detoxification efficacy of *L. marginalis* and *B. bengalensis* distributed in contaminated habitat. Chronic and uncontrolled exposure of pyrethroid may lead to develop a detrimental toxic effect in the molluscan bioresource of West Bengal to gradual decline in biodiversity.

Summary

Freshwater wetlands of West Bengal harbours a rich assemblage of molluscan bioresource. Many of these less researched organisms are of economical, nutritional, ecological, ethno -medicinal and industrial importance. Major environmental threats of this bioresource have been identified as destruction and contamination of their natural habitat. Habitat destruction is intimately associated with the problem of rapid urbanization and industrialization. Habitat contamination, on the other hand is associated with the alarming ecotoxicity caused by geogenic toxins like arsenic and newer generation of pesticides like pyrethroids and neem based azadiractin. Present article reviews the recent trend of cell biological and toxicological research carried out on selective molluscs of economical importance. Acute, subacute and chronic toxicity of geogenic and agricultural toxins had been reported to create a precarious level of toxicity in molluscs at their tissue, cellular and subcellular levels. Many of these studied parameters have recently been studied as effective immunological markers of environmental contamination. Environmental toxins generated from anthropogenic activities often contaminate the freshwater habitat of molluscs. Arsenic is a metalloid toxin bearing the potentiality of carcinogenesis. This geogenic xenobiotic is a reported contaminant of ground water. Drinking of arsenic for longer period of time may initiate carcinogenesis in human. Many of the districts of the state of West Bengal have been reported as affected by arsenic pollution. Extraction and contaminated ground water for the purpose of agriculture poses a serious environmental threat to human and other organisms. Contamination of the natural habitat of freshwater molluscs by arsenic is assumed to be a causative factor of toxicity in many invertebrates. Report of toxicity of arsenic in mollusc is scanty in the current scientific literature. Arsenic exposure has been reported to affect adversely the specific cell biological and immunological parameters of molluscs. Hemocytes, the chief immune effector cells of molluscs perform diverse physiological functions. Arsenic induced stress had been reported to affect the blood cell homeostasis, cytotoxic and apoptotic potential of hemocytes. Arsenic treatment is also reported to create histological damage in multiple target organs like gill, labial palp, mantle and heart respectively in undesirable shift in their functional attributes. Newer generation of agrotoxins had been reported as another group of contaminants of the natural habitat of molluscs. Agricultural runoff containing pyrethroids and neem based azadirachtin pesticides often pollute the freshwater ponds and lakes during monsoon. Thus, several aquatic invertebrates are being exposed to these agrotoxins for a long period of time. Agrotoxin exposures, like geogenic arsenic may result in a shift in the functional parameters of several target organs like gill, mantle, labial palp and digestive gland and hemocytes. These parameters include relative density, apoptotic response, morphological status, and cytotoxic potential of blood cells. Moreover, the agrotoxin induced morphological damage of gill, digestive gland, mantle and labial palp are in report. Unrestricted and continuous contamination of the natural habitat of molluscan species is thus apprehended to cause a gradual and alarming reduction of the existing aquatic biodiversity.

Acknowledgements

Authors thankfully acknowledge the University Grant Commission for departmental award of SAP DRS-II and RFSMS to Anindya Sundar Bhunia. DST WOS-A award to Dr. Mitali Ray is acknowledged too. We extend our thanks to the personnels of Center for Research in Nanoscience and Nanotechnology of the University of Calcutta for their cooperation in FACS analyses. Additionally, research contributions of Suman Mukherjee, Sudipta Chakraborty, Chiranjib Mandal, Krishnendu Das and Soumalya Mukherjee are also acknowledged.

References

Adema CM, Knaap WPW and Sminia T. 1991. Molluscan haemocyte mediated cytotoxicity. The role of reactive oxygen intermediates. *Reviews in Aquatic Sciences*, 4: 201-223.

Agwuocha S, Kulkarni BG and Pandey AK. 2011. Histopathological alterations in hepatopancreas of *Gafrarium divaricatum* exposed to xylene, benzene and gear oil-Wsf. *Journal of Environmental Biology*, 32(1): 35-38.

Andhale AV, Bhosale PA and Zambare SP. 2011. Histopathological study of nickel induced alterations in fresh water bivalve, *Lamellidens marginalis*. *Journal of Experimental Sciences*, 2(4): 1-3.

Anwar WA. 1997. Biomarkers of human exposure to pesticides. *Environmental Health Perspectives*. 4: 801-806.

Arumugam M, Romestand B and Torreilles J. 2000. Nitrite released in haemocytes from *Mytilus galloprovincialis, Crassostrea gigas* and *Ruditapes decussatus* upon stimulation with phorbol myristate acetate. *Aquatic Living Resources*, 13: 173-177.

Baby RL, Hasan I, Kabir KA and Naser MN. 2010. Nutrient analysis of some commercially important molluscs of Bangladesh. *Journal of Scientometric Research*, 2 (2): 390-396.

Bachère E, Hervio D and Mialhe E. 1991. Luminol dependent chemiluminescence by hemocytes of two marine bivalves *Ostrea edulis* and *Crassostrea gigas*. *Diseases of Aquatic Organisms*, 11: 173-180.

Bado-Nilles A, Gagnaire B, Thomas-Guyon H. *et al.*, 2008. Effects of 16 pure hydrocarbons and two oils on haemocyte and haemolymphatic parameters in the Pacific oyster, *Crassostrea gigas* (Thunberg). *Toxicology in Vitro*, 22: 1610-1617.

Barðienλ J, Andreikλnaitλ L, Garnaga G and Rybakovas A. 2008. Genotoxic and cytotoxic effects in bivalve molluscs *Macoma balthica* and *Mytilus edulis* from the Baltic Sea. *Ekologija*, 54(1): 44-50.

Bouchard N, Pelletier E and Founier M. 1999. Effects of butyltin compounds on phagocytic activity of haemocytes from marine bivalves. *Environmental Toxicology and Chemistry*, 18: 519-522.

Buckland-Nicks J and Tompkins G. 2005. Paraspermatogenesis in *Ceratostoma foliatum* (Neogastropoda): Confirmation of programmed nuclear death. *Journal of Experimental* Zoology, 303A: 723-741.

Cajaraville MP, Diez G, Marigomez IA and Angulo E. 1990. Responses of basophilic cells of the digestive gland of mussels to petroleum hydrocarbon exposure. *Diseases of Aquatic Organisms*, 9: 221-228.

Cajaraville MP, Olabarrieta I and Marigomez I. 1996. In *Vitro* Activities in Mussel Hemocytes as Biomarkers of Environmental Quality: A case study in the Abra estuary (Biscay Bay). *Ecotoxicology and Environmental Safety*, 35: 253-260.

Campero M, De Block M, Ollevier F and Stoks R. 2008. Metamorphosis offsets the link between larval stress, adult asymmetry and individual quality. *Functional Ecology*, 22: 271-277.

Casco VH, Izaguirre MF, Vergara MN. *et al.*, 2006. Apoptotic cell death in the central nervous system of *Bufo arenarum* tadpoles induced by cypermethrin. *Cell Biology and Toxicology*, 22: 199-211.

Chakraborty S, Ray M and Ray S. 2008. Sodium arsenite induced alteration of hemocyte density of *Lamellidens marginalis* -an edible mollusc from India. *Clean soil air water*, 36(2): 195-200.

Chakraborty S, Ray M and Ray S. 2010. Toxicity of sodium arsenite in the gill of an economically important mollusc of India. *Fish and Shellfish Immunology*, 29: 136-148.

Chakraborty S, Ray M and Ray S. 2013. Cell to organ: physiological, immunotoxic and oxidative stress responses of *Lamellidens marginalis* to inorganic arsenite. *Ecotoxicology and Environmental Safety*, 94: 153-163.

Chakraborty S and Ray S. 2009. Nuclear morphology and lysosomal stability of molluscan hemocyte as possible biomarkers of arsenic toxicity. *Clean soil air water*, 37(10): 769-775.

Chang SJ, Tseng SM and Chou HY. 2005. Morphological characterization via light and electron microscopy of the hemocytes of two cultured bivalves: a comparison study between the hard clam (*Meretrix lusoria*) and pacific oyster (*Crassostrea gigas*). *Zoological Studies*, 44(1): 144-153.

Cheng TC. 1976. Aspects of substrate utilization and energy requirement during molluscan phagocytosis. *Journal of Invertebrate Pathology*, 27: 263-268.

Cheng TC.1988a. In vivo effects of heavy metals on cellular defense mechanisms of *Crassostrea virginica*: total and different cell counts. *Journal of Invertebrate Pathology*, 51: 207-214.

Cheng TC. 1988b. In vivo effects of heavy metals on cellular defense mechanisms of *Crassostrea virginica*: phagocytic and endocytosis indices. *Journal of Invertebrate Pathology*, 51: 215-220.

Coles JA, Farley SR and Pipe RK. 1994. Effects of fluoranthene on the immunocompetence of the common marine mussel, *Mytilus edulis*. *Aquatic Toxicology*, 30: 367-379.

Coles JA, Farley SR and Pipe RK. 1995. Alteration of the immune response of the common marine mussel *Mytilus edulis* resulting from exposure to cadmium. *Diseases of Aquatic Organisms*, 22: 59-65.

Cooper EL and Roch P. 2003. Earthworm immunity: a model of immune competence. *Pedobiologia*, 47(5-6): 1-13.

Dailianis S. 2009. Production of superoxides and nitric oxide generation in haemocytes of mussel *Mytilus galloprovincialis* (Lmk.) after exposure to cadmium: A possible involvement of Na^+/H^+ exchanger in the induction of cadmium toxic effects. *Fish and Shellfish Immunology*, 27: 446-453.

Das K, Ray M and Ray S. 2012. Cypermethrin induced dynamics of hemocyte density of Indian mollusc *Lamellidens marginalis*. *Animal Biology Journal*, 3(1): 1-11.

Das K, Ray M and Ray S. 2015. Shift in foot protrusion frequency, cholinergic impairment in pedal ganglia and histopathology of labial palp in freshwater mussel exposed to cypermethrin. In. *Perspectives in Animal Ecology and Reproduction* (Gupta VK ed.). Daya Publishing House, New Delhi. pp. 123-134.

Deborah Gnana Selvam A, Thatheyus AJ and Vidhya R. 2013. Biodegradation of the Synthetic Pyrethroid, Fenvalerate by *Pseudomonas viridiflava*. *American Journal of Microbiological Research*, 1 (2): 32-38.

Dikkeboom R, Tijnagel JMGH and van der Knaap WPW. 1988. Monoclonal antibody recognized hemocyte subpopulations in juvenile and adult *Lymnaea stagnalis*: functional characteristics and lectin binding. *Developmental and Comparative Immunology*, 12: 17-32.

Fries CR and Tripp MR. 1980. Depression of phagocytosis in *Mercenaria mercenaria* following chemical stress. *Developmental and Comparative Immunology*, 4: 233-244.

Gagnairea B, Thomas-Guyonb H and Renaulta T. 2004. *In vitro* effects of cadmium and mercury on Pacific Oyster, *Crassostrea gigas* (Thunberg), haemocytes. *Fish and Shellfish Immunology*, 16: 501-512.

Hannam ML, Bamber SD, Galloway TS. *et al.*, 2010. Effects of the model PAH phenanthrene on immune function and oxidative stress in the haemolymph of the temperate scallop *Pecten maximus*. *Chemosphere*, 78(7): 779-84.

Hughes FM, Foster B, Grewal S and Sokolova IM. 2010. Apoptosis as a host defense mechanism in *Crassostrea virginica* and its modulation by *Perkinsus marinus*. *Fish and Shellfish Immunology*, 29: 247-257.

Indira Rani G and Kumaraguru AK. 2014. Behavioural responses and acute toxicity of *Clarias batrachus* to synthetic pyrethroid insecticide, λ- cyhalothrin. *Journal of Environmental and Applied Bioresearch*, 2 (1): 19-24.

Johnson MA, Paulet YM, Donval A and Pennee ML.1996. Histology, histo-chemistry, and enzyme biochemistry in the digestive system of the endosymbiont bearing bivalve (*Loripes lucinalis* Lamarck). *Journal of Experimental Marine Biology and Ecology*, 197(1): 15-38.

Kumar S, Pandey RK, Das S and Das VK. 2011. Pathological changes in hepatopancreas of freshwater mussel (*Lamellidens marginalis* Lamarck) exposed to sub-lethal concentration of dimethoate. *GERF Bulletin of Biosciences*, 2(2): 18-23.

Larson KG, Roberson BS and Hetrick FM. 1989. Effect of environmental pollutants on the chemiluminescence of hemocytes from the American oyster *Crassostrea virginica*. *Diseases of Aquatic Organisms*, 6: 131-136.

Le Gall G, Bachère E and Mialhe E. 1991. Chemiluminescence analysis of the activity of *Pecten maximus* hemocytes stimulated with zymosan and host species Rickettsiales like organisms. *Diseases of Aquatic Organisms*, 11: 181-186.

Lopez C, Villalba A and Bachère E. 1994. Absence of generation of active oxygen radicals coupled with phagocytosis by hemocytes of the clam *Ruditapes decussatus* (Mollusca: Bivalvia). *Journal of Invertebrate Pathology*, 64: 188-192.

Luna-Acosta A, Saulnier D, Pommier M. *et al.*, 2011. First evidence of a potential antibacterial activity involving a laccase-type enzyme of the phenoloxidase system in Pacific oyster *Crassostrea gigas* haemocytes. *Fish and Shellfish Immunology*, 31: 795-780.

Mandal C, Ray M and Ray S. 2012. Histopathology of gill, digestive gland and antenna of *Bellamya bengalensis* exposed to fenvalerate. *Animal Biology Journal*, 3(2): 59-66.

Marigomez IA, Kortabitarte M and Dussart GBJ. 1998. Tissue-level biomarkers in sentinel stages as cost effective tools to assess metal pollution in soils. *Archives of Environmental Contamination and Toxicology*, 34: 167-176.

Marigomez IA, Cajaraville MP and Angulo E. 1990. Histopathology of the digestive gland-gonad complex of the marine prosobranch *Littorina littorea* exposed to cadmium. *Diseases of Aquatic Organisms*, 9: 229-238.

Mehedinti R, Hancu M, Coman M and Mehedinti T. 1999. Histological and histochemical alteration in certain black sea bivalves submitted to accidental pollution. *Morphol. Embryol.*, XLV: 143-152.

Mello DF, Proenca LA and Barracco MA. 2010. Comparative study of various immune parameters in three bivalve species during a natural bloom of *Dinophysis acuminata* in Santa Catarina Island, Brazil. *Toxins*, 2: 1166-1178.

Mikula I, Pistl J and Kacmar P. 1992. Immune response of sheep at subchronic intoxication by pyrethroid insecticide supercypermethrine. *Acta Veterinaria Brno.*, 61: 57-60.

Mukherjee S, Ray M and Ray S. 2006. Azadirachtin induced modulation of total count of hemocytes of an edible bivalve *Lamellidens marginalis*. *Proceedings of Zoological Society of Kolkata*, 59(2): 203-207.

Mukherjee S, Ray M and Ray S. 2008. Dynamics of hemocyte subpopulation of *Lamellidens marginalis* exposed to a neem based pesticide. In. Zoological Research in Human Welfare. Zoological Survey of India (Ramakrishna and Chatterjee RN eds.), Kolkata. pp. 389-394.

Mukherjee S, Ray M and Ray S. 2012. Oxidative stress in freshwater bivalve of India exposed to azadirachtin based pesticide. *Romanian journal of Biology- Zoology*, 57(1): 79-88.

Mukherjee S, Ray M and Ray S. 2015. Immunotoxicity of washing soda in a freshwater sponge of India. *Ecotoxicology and Environmental Safety*, 113: 112-123.

Mukherjee S, Ray M and Ray S. 2011. Phagocytosis of charcoal particulate as immunological marker of azadirachtin exposure. In. *Perspective of Animal Ecology and Reproduction* (Gupta VK ed.). Daya Publishing House, New Delhi, India. pp. 55-67.

Nakamura M, Mori K, Inooka S and Nomura T. 1985. *In vitro* production of hydrogen peroxide by the amoebocytes of the scallop *Patinopecten yessoensis* (Jay). *Developmental and Comparative Immunology*, 9: 407-417.

Nakayama K and Maruyama T. 1998. Differential production of active oxygen species in photo-symbiotic and non-symbiotic bivalves. *Developmental and Comparative Immunology*, 22(2): 151-159.

Noλl D, Bache're E and Mialhe E. 1993. Phagocytosis associated chemiluminescence of hemocytes in *Mytilus edulis* (Bivalvia). *Developmental and Comparative Immunology*, 17: 483-493.

Oliver LM and Fisher WS. 1999. Appraisal of prospective bivalve immunomarkers. *Biomarkers*, 4: 510-530.

Paull J. 2007. Rachel Carson, a voice for organics-the first hundred years. *Journal of Bio-dynamics Tasmania*, 86: 37-41.

Payne BS, Lel J, Miller AC and Hubertz ED. 1995. Adaptive variation in palps and gill size of the zebra mussel (*Dreissena polymorpha*) and Asian clam (*Corbicua fluminea*). *Canadian Journal of Fisheries and Aquatic Sciences*, 52: 1130-1134.

Pipe RK. 1992. Generation of reactive oxygen metabolites by the haemocytes of the mussel *Mytilus edulis*. *Developmental and Comparative Immunology*, 16: 111-122.

Pipe RK and Coles JA. 1995. Environmental contaminants influencing immune function in marine bivalve molluscs. *Fish and Shellfish Immunology*, 5: 581-595.

Ponder WF and Lindberg DR. 2008. Phylogeny and Evolution of the Mollusca. University of California Press.

Prabhakar AK and Roy SP. 2009. Ethno-medicinal Uses of Some Shell Fishes by People of Kosi River Basin of North-Bihar, India. *Journal of Ethnobiology and Ethnomedicine*, 3(1): 1-4.

Ray M, Bhunia NS, Bhunia AS and Ray S. 2013a. A comparative analyses of morphological variations, phagocytosis and generation of cytotoxic agents in flow cytometrically isolated hemocytes of Indian molluscs. *Fish and Shellfish Immunology*, 34: 244-253.

Ray M, Bhunia AS, Bhunia NS and Ray S. 2013b. Density shift, morphological damage, lysosomal fragility and apoptosis of hemocytes of Indian molluscs exposed to pyrethroid pesticides. *Fish and Shellfish Immunology*, 35: 499-512.

Ray S, Ray M, Chakraborty S and Mukherjee S. 2011. Immunotoxicity of environmental chemicals in the pearl forming mussel of India-a review. In. *Mussels: Anatomy, Habitat and Environmental Impact* (McGevin LE. ed.). Nova Science Publishers Inc. NY, USA. pp. 429-440.

Righi DA and Palermo-Neto J. 2005. Effects of type II pyrethroids cyhalothrin on peritoneal macrophage activity in rats. *Toxicology*, 212: 98-106.

Ruddell CL and Rains DW. 1975. The relationship between zinc, copper and the basophils of two crassostreid oysters, C. *gigas* and C. *virginica*. *Comperative Biochemical Physiology*, 51A: 585-591.

Russo J and Lagadic L. 2000. Effects of parasitism and pesticide exposure on characteristics and functions of hemocyte populations in the freshwater snail *Lymnaea palustris* (Gastropoda, Pulmonata). *Cell Biology and Toxicology*, 16: 15-30.

Russo J and Lagadic L. 2004. Effects of environmental concentrations of atrazine on hemocyte density and phagocytic activity in the pond snail *Lymnaea stagnalis* (Gastropoda, Pulmonata). *Environmental Pollution*, 127: 303-311.

Russo J and Madec L. 2007. Hemocyte apoptosis as a general cellular immune response of the snail, *Lymnaea stagnalis*, to a toxicant. *Cell and Tissue Research*, 328(2): 431-441.

Saha S, Ray M and Ray S. 2008. Nitric oxide generation by immunocytes of mud crab exposed to sodium arsenite. In. *Zoological Research in Human Welfare* (Ramakrishna and Chatterjee RN eds.). Zoological Survey of India, Kolkata. pp. 425-428.

Sami S, Faisal M and Huggett RJ. 1992. Alterations in cytometric characteristics of hemocytes from the American oyster *Crassostrea virginica* exposed to a polycyclic aromatic hydrocarbon (PAH) contaminated environment. *Marine Biology*, 113: 247-252.

Smith KL, Galloway TS and Depledge MH. 2000. Neuro-endocrine biomarkers of pollution-induced stress in marine invertebrates. *Science of The Total Environment*, 262(1-2): 185-190.

Snyman RG, Reinecke AJ and Reinecke SA. 2005. Quantitative changes in the digestive gland cells of the snail *Helix aspersa* after exposure to the fungicide copper oxychloride. *Ecotoxicology and Environmental Safety*, 60: 47-52.

Sokolova IM, Evans S and Hughes FM. 2004. Cadmium-induced apoptosis in oyster hemocytes involves disturbance of cellular energy balance but no mitochondrial permeability transition. *Journal of Experimental Biology*, 207: 3369-3380.

Stelzer KJ and Michel AG.1984. Effect of pyrethroid on lymphocyte mitogenic responsiveness. *Research communications in chemical pathology and pharmacology*, 46: 137-150.

Sunila I and LaBanca J. 2003. Apoptosis in the pathogenesis of infectious diseases of the eastern oyster *Crassostrea virginica*. *Diseases of Aquatic Organisms*, 56: 163-170.

Sunila I. 1987. Histopathology of mussels (*Mytilus edulis* L.) from the Tvarminne area, the Gulf of Finland (Baltic Sea). *Annales Zoologici Fennici*, 24: 55-69.

Suresh K and Mohandas A. 1990. Number and types of hemocytes in *Sunetta scripta* and *Villorita cyprinoids* var. *cochinensis* (Bivalvia) and leucocytosis subsequent to bacterial challenge. *Journal of Invertebrate Pathology*, 55: 312-318.

Tamag RK, Jha GJ, Chanhan HUS and Tiwari BK. 1988. In *vivo* immunosuppression by synthetic pyrethroid (cypermethrin) pesticide in mice and goats. *Veterinary Immunology and Immunopathology*, 19: 299-305.

Terek-Majewsca E, Siwicki AK and Szweda W. 2004. Modulative influence of lysosyme dimmer on defence mechanisms in the carp (*Cyprinus carpio*) and

European sheatfish (*Silurus glanis*) after suppression induced by herbicide Roundup. *Polish Journal of Veterinary Sciences*, 7: 123-128.

Thompson RJ, Ratcliffe NA and Bayne BL. 1974. Effects on starvation on structure and function in the digestive gland of the mussel (*Mytilus edulis* L.). *Journal of the Marine Biological Association*, UK., 54: 699-712.

Tulinska J, Kubova J, Janota S and Nyulassy S. 1995. Investigation of immunotoxicity of supercypermethrin forte in the Wistar rat. *Human and Experimental Toxicology*, 14: 399-403.

Usheva LN, Vaschenko MA and Durkina VB. 2006. Histopathology of the digestive gland of bivalve mollusk *Crenomytilus grayanus* (Dunker, 1853) from South Western Peter, The Great Bay, Sea of Japan. *Russ. Journal of Marine Biology*, 32(3): 166-172.

Wootton EC and Pipe RK. 2003. Structural and functional characterisation of the blood cells of the bivalve mollusc, *Scrobicularia plana. Fish and Shellfish Immunology*, 15: 249-262.

Chapter 27

High Altitude Lakes of Sikkim: Conservation Initiatives

Lak Tsheden Theengh, Priyadarshinee Shrestha and Partha Sarathi Ghose

WWF-India, Khangchendzonga Landscape Programme, Gangtok, Sikkim

Introduction

Sikkim is the 22nd state of Indian union and lies between 27°04′46′′ N to 28°07′48′′N latitudes and 88°00′58E′′ to 88°55′25′′E longitudes and occupies an area of 0.22 per cent of the total geographical area of India. Sikkim shares its border with Bhutan to the east, Nepal towards west, Tibetan plateau in the north and neighbouring state of West Bengal to the south. Sikkim, with population of 6.11 lakhs is one of the least populated states of India and has the fourth lowest population density in the country. The state forms an integral part of the Eastern Himalayan region which is included among the 34 Global Biodiversity Hotspots and Global 200 Ecoregions (Olson and Dinerstein, 1998; Myers *et al.*, 2000). With a geographical area of 7096 km^2 Sikkim demonstrates a great variation in altitudinal gradient ranging from 300 m to 8586 m, which is responsible for the great variation in eco-climatic conditions of the state ranging from sub-tropical in the south to the tundra type conditions in the north. The range of eco-climatic conditions of Sikkim support great diversity of flora and fauna in the state. Sikkim has two main drainage basins namely Teesta and Rangeet and river Teesta is described as the "lifeline of Sikkim". Almost 60 per cent of Sikkim's geographical area lies above 3000 m that are home to innumerable high altitude lakes (Tambe *et al.*, 2008).

High Altitude Wetlands of Sikkim

Wetlands are vital natural resources and are defined as areas of land that are either temporarily or permanently covered by water. Wetlands can be of diverse types and may vary considerably in characteristics on account of their geographical location, water regime, chemical and biotic composition (Sharma *et al.*, 2010). Natural resource conservation is impaired seriously due to lack of knowledge about its distribution, extent and threats it faces (Sharma *et al.*, 2010) and proper information on the wetlands of Sikkim were limited and has been a subject of speculation for a long time. Total surface area covered by these wetlands is 7477 ha, which is 1.05 per cent of total land area of the state (Sharma *et al.*, 2010). Approximately 55 per cent of this area is accounted for by the streams and rivers while the lakes account for the remaining 45 per cent (Sharma *et al.*, 2010). Recent remote sensing assessments, however, show that about 541 lakes occur across Sikkim (Sharma *et al.*, 2010). Two hundred and sixty of these wetlands have an area greater than 2.5 ha in size and there are about 315 lakes situated above 3000 m and majority of them are of glacial origin (Tambe *et al.*, 2008). These lakes together cover more than 3000 ha of Sikkim's geographical area (Sharma *et al.*, 2010). There are 85 glaciers in Sikkim, which cover an area of about 576 km^2. Khangchen Tso (Tso meaning Lake) that originates from Teesta Khangtse Glacier, the main sources of river Teesta, with an area of 164 ha is the largest lake of Sikkim (Tambe *et al.*, 2008). Gyamtsona, situated at an altitude of 4990 m, is the only brackish water lake of the state and is regarded as the vestige of the Tethys Sea. Most lakes in Sikkim are considered holy, of which the most important ones are Tsho Lhamo, Khecheopalri, Gurudongmar and Tsomgo (Tambe *et al.*, 2008).

Biodiversity of Sikkim's High Altitudes

Alpine meadows in the catchment areas of the high altitude lakes have high biodiversity values and are home to unique species of alpine flora and fauna. The alpine meadows and pastures (>3500 m - 5000 m) are predominated by alpine scrubs and grasslands and part of the extensive *krummholz* zone (stunted forests) and the key species found in this association include *Rhododendron wightii*, *R. fulgens*, *R. thomsonii*, *R. hodgsonii*, *R. lanatum*, *R. campanulatum*, *R. anthopogon* and *R. setosum* (Rawat and Tambe, 2011). The continuity of the scrublands are interrupted intermittently by pastures where species like *Aconitum ferox*, *Eriophyton wallichii*, *Pedicularis siphonantha*, *P. longiflora*, *Allium wallichi*, *Anaphalis margaritacea*, *Aster himalaicus*, *Bistorta macrophylla*, *Gentiana phyllocalyx*, *Meconopsis* spp., *Iris* spp., *Rheum nobile* and *Saussurea simpsonianan* among others grow profusely. Areas above 5000 m are predominated by Greater and Trans-Himalayan region (especially in North Sikkim), and is characterized by steep mountainous slopes, valleys and rocky outcrops (Chanchani *et al.*, 2010). Flora is dominated by *Stipa*, *Elymus*, *Carex*, *Ephedra*, *Artemisia*, *Androsace* and *Arenaria* and marsh meadows (*Kobresia*, *Pedicularis*) and scrub species (*Lonicera*, *Potentilla*) as reported by Chanchani *et al.* (2010).

The region is home to wide diversity of wildlife species. The high alpine areas support populations of six ungulate species, like Himalayan musk deer *Moschus chrysogaster*, Himalayan tahr *Hemitragus jemlahicus*, Tibetan argali *Ovis ammon hodgsoni*, Tibetan gazelle *Procapra picticaudata*, southern kiang *Equus kiang polyodon* and blue sheep *Pseudois nayaur* (Tambe *et al.*, 2008; WWF-India, unpublished report). The globally endangered snow leopard *Panthera uncia* is the flagship species of alpine habitats. Other interesting carnivores found in this region include the Tibetan wolf *Canis lupus chanco*, Tibetan sand fox *Vulpes ferrilata*, red fox *Vulpes vulpes* and martens *Mustella* spp. The Himalayan marmot *Marmota himalayana* and plateau pika *Ochotona curzoniae* are the predominant species found on the alpine plains. The Sikkim snow toad *Scutiger sikkimensis*, which is adapted to complete its life cycle in the high altitude wetlands, is found in all the water bodies dotted across the alpine areas of Sikkim (Tambe *et al.*, 2008; WWF-India, unpublished data). The region is also home to a diverse avian species; at least 30 water birds have been recorded from various high altitude wetlands of Sikkim. Among them the globally vulnerable black-necked crane *Grus nigricollis* is of high conservation significance, which has been recorded in the North Sikkim regularly.

Challenges and Issues of High Altitude Wetlands

Being situated at a higher altitude and remote areas with adverse climatic conditions, high security issues and lack of proper motorable roads had made most of the high altitude wetlands inaccessible in the past. However, situations have changed noticeably and in recent times, communication by road, to some of these areas, has improved considerably. Sikkim annually receives about 6.3 lac tourists, which is higher than the overall population of the state. Almost half of them visit high altitude wetlands like Tsomgo and Gurudongmar. The tourist volume is increasing annually and this has lead to steady increase in the vehicular traffic in those areas. Increasing vehicular traffic is a constant source of disturbance to the lake environment and hampers the sanctity of these high altitude lakes. The increasing tourism has also contributed to the increase in waste burden in these otherwise pristine high altitude wetlands. In recent times, to improve accessibility large-scale road development work has been initiated in areas passing Tsomgo and Gurudongmar Lakes and their catchment areas. This has added to the level of disturbance of the wetland environment and at times causing deterioration of the wetland quality. These areas of Sikkim having international borders are inhabited by defence forces, thereby adding to the stress on the environment. Moreover, study of the wetland ecosystems of the state has been very limited, and is an area that requires much focus, and even more so in the face of climate change.

Considering the vulnerable status and cultural and religious significance 11 high altitude lakes have been prioritized for conservation purpose including Omechho (Omai-stho, West Sikkim), Sungmoteng Chho (Tsho, West Sikkim), Lamchho (Lham-tso or Lham Pokhri, West Sikkim), Tolechho (Dhole-tso, West Sikkim), Kabur Lamchho (Gabur Lah-tso, West Sikkim), Khachoedpalri Pemachen Tsho (Khechoepalri lake, West Sikkim), Kathogtsho at Yuksam (Kathog Lake, West Sikkim), Tsho-mGo lake (Tsomgo Lake, East Sikkim), Guru Dongmar Lake (North Sikkim), Tsho-lhamo (North Sikkim) and Mulathingkhai-tsho (Green lake at Zema Glacier, North Sikkim) (Home Department; Notification No. 70/HOME/2001 Dated 20/09/2001. WWF-India have been working closely with the communities, Department of Forests, Environment and Wildlife Management (Government of

Sikkim) and other Government Departments and Non-Government agencies for conservation of high altitude wetlands.

Conservation Initiatives

Sikkim's Lake Conservation Guideline

The lakes of Sikkim are of environmental, cultural and spiritual importance and some of them being important tourist destinations, offering important livelihood opportunities to the local communities. However, impacts of various human related activities pose a serious threat to the long term ecological security of these lakes. For conservation of the lakes, the State Forest Department drafted a Lake Conservation Guideline for Sikkim in 2005. This was done through stakeholder consultations and with involvement of NGOs like WWF- India and The Mountain Institute. Lake Tsomgo was selected as a model to begin the lake conservation initiative in Sikkim and a participatory framework for lake conservation was developed. The framework was drafted and after modification, following suggestions from experts from various government and private sectors, was approved by the Government of Sikkim as "Guidelines for lake conservation in partnership with *Gram Panchayats* and *Pokhri Sanrakshan Samities* in Sikkim" (notification no. 355/F dated 31 July 2006). The lake conservation guideline outlines the following broad features-

- ✰ Formation of the *Pokhri Sanrakshan Samiti* (PSS) by the gram sabha with ward panchayat as the President.
- ✰ Generation of revenue through collection of conservation fee from visitors to the wetland (Rs. 10 from each visitor).
- ✰ Formulation of annual lake conservation plan with the involvement of all community stakeholders.
- ✰ Regular capacity building for the lake management group or the PSS.
- ✰ Formation of the State Lake Conservation Federation, the *Rajya Pokhri Sanrakshan Sangh*, with the PSS Presidents to review practices of the individual PSSs and formulate future action plans.

Case Study I: Tsomgo Lake

Tsomgo Lake ('Tso' = lake, 'Mgo' = head) is situated at an altitude of 3780 m in East District and derives its name from the Bhutia language and literally means 'source of the lake'. Seasonal snow and rain water feed this lake, while the main inlet flows in from the eastern side. The lake drains into a small stream from its western end falling steeply for a distance, and then gently flows through the forests of Menla and Kyongnosla. The stream then forms one of the branches of Rangpo Chu upon meeting the Lungze Chu flowing down from the Nathula catchment. The water of the lake is frozen during November and March. It is situated at a distance of 38 km away from the capital Gangtok on the Gangtok - Nathula highway. It is an oval-shaped lake with an area of 24.47 ha. Tsomgo, being situated within 2-3 hours drive from Gangtok, is a popular tourist destination of the state that receives approximately three lac tourists annually. The increased tourist traffic and sprawling tourism amenities was a key source of disturbance to the lake and its catchment.

Considering the tourism related impacts that this wetland was facing, the first PSS was constituted for the Tsomgo Lake. *Tsomgo Pokhri Sanrakshan Samiti* (TPSS), was formed at the Office of the Divisional Forest Officer, East Territorial Division in 2007 (Registration Sl. No. - 01/ED/T, p- 01), in accordance with rule 1(B)(4) of Government notification no. 335/F of Forest 01/ED/T/07 dated 03/12/2007. The TPSS was constituted by the Gram Sabha on 24 May 2006, in presence of the Divisional Forest Officer (Territorial Division, East) and the Conservator of Forest (Land Use). The Executive Body of the Samiti comprises of eight members – the President (the local Ward Panchayat), the Member Secretary (Range Officer, Territorial Division) and six additional executive members. The main objectives of PSS is to work in collaboration with the Department of Forests, Environment and Wildlife Management (Government of Sikkim) staff to regulate disturbance caused to the lake environment and its catchment and biodiversity that includes the migratory birds and breeding wildlife species. Since its formation, the Tsomgo PSS started collecting conservation fee of Rs. 10 per head from tourists from April, 2008 and as per the rule 50 per cent of the collected fee is deposited with the State Conservation Agency. Till now the Tsomgo PSS has collected approximately Rs. 85 lakhs and the fund has been utilized systematically for the following initiatives.

Conservation and Environmental Management

Garbage management of the lake area is one of the main activities of the PSS. A resource recovery center has been set up near the lake premises for segregation of garbage and recovery of recyclables. PSS members also conduct plantation programmes with its members. Regular water quality assessment for the Tsomgo lake is also conducted by PSS with technical support from WWF- India. *Pokhri Rakshaks* have also been employed under PSS from the community who patrol the lake area during peak tourist season to keep a check on irregular activities.

Environment Education and Awareness

Wetland conservation initiatives at Tsomgo are strengthened by regular awareness programmes with different stakeholders comprising of students, villagers, shopkeepers, and tourism stakeholders, which constitutes tour operators, drivers and tourists. These programmes are aimed to educate stakeholders on the importance of the lake, its cultural values and how responsible actions could help to maintain the pristine status of the lake and its catchment. To strengthen the waste management initiative at Tsomgo, the taxi drivers are regularly sensitized on their responsibility towards regulating waste generation by the tourists and are distributed with waste bags to keep in their vehicle for use by tourists travelling to the lake. Additionally, signages are set-up at regular intervals along the road leading to Tsomgo to educate tourists about their responsibility to maintain sanctity of the lake during the visit.

Asset Creation for Communities and Tourists

Tsomgo PSS also utilizes its funds for creating and maintenance of assets for the Tsomgo Lake community. These are mostly need based demands that are placed by the community during the consultation for the annual lake plan formulation process. Some of the assets created are drinking water supply, stairs for tourists at the shopping complex, booth for police personnel etc.

Case Study II: Lake Gurudongmar and its Catchment

Gurudongmar Lake, located at an altitude of 5183 metres in North Sikkim's cold desert area is regarded as one of the largest glacial wetland of Sikkim. It is a complex of 3 large water bodies, categorized as Gurudongmar A, B and C and fed by the Gurudongmar Glacier. Glacial studies have shown all three wetlands to have increased in size due to glacial melt (Kumar and Prabhu, 2013). The outlet of the river flows down to meet the streams from the Tso lamo Lake and Zemu glacier to form the Lachen Chu which combines with the Lachung Chu to form river Teesta at Chungthang.

From a religious perspective the Gurudongmar Lake holds great value since it is considered to have been blessed by Guru Padmasambhava himself. It is also believed that the portion of the Lake that the Guru touched while blessing does not freeze even when at extremely low temperatures during winter. Some people also collect water from the lake since it is supposed to have curative properties. Gurudongmar is gaining popularity as a tourist destination among visitors and receive up to 20 to 30 thousand tourists annually.

Lachen says No to packaged water

Empty plastic water bottles formed bulk of the garbage collected during a cleanliness drive organised by LTDC and Lachen Dzumsa with WWF support at Gurudongmar and Lachen Village in 2011. These plastic bottles seemed to be a big polluting factor for the lake area as well as the village. The community of Lachen realized this was a serious impact from tourism, as villagers were not using packaged water but water from the streams and springs for drinking.

Subsequently, with consent of the local communities and tourism stakeholders, Lachen Dzumsa (the village governing council of Lachen) passed the decision to ban packaged water in the village, making Lachen the first village in Sikkim to implement such a ban.

Water filters were distributed to the hotels and shops and regular water monitoring is conducted to ensure clean drinking water to the visitors.

Wetland conservation work at Gurudongmar was initiated in 2010, taking experiences from Tsomgo with the community at Lachen village, which is the last village towards the cold desert area. The Lachen Dzumsa and Lachen Tourism Development Committee are the main partners for the conservation initiative of WWF – India. Like in Tsomgo Lake, garbage accumulation in the high altitude areas is one of the main impacts from tourism. Initiatives for proper waste management system within the village and at Gurudongmar have been undertaken. Lachen community is encouraged to practice household segregation which will ensure that recyclable items from the waste is recovered, and minimizes the volume of waste to be ultimately discarded. Apart from awareness and sensitization for community, regular engagements with the defence personnel are also organized to sensitize them on the biodiversity of the area and its conservation.

Conclusion

The high altitude areas dotted with numerous wetlands are one of the most unique features of Sikkim's landscape that covers more than 50 per cent of the geographical areas of the state. While most of these were inaccessible in the past, that is rapidly changing with increasing tourism prospects and improved connectivity through road networks, which will have irreversible consequences on the pristine nature of these areas. Coupled with this, the risk of climate change and its impacts in these wetland ecosystems is emerging as one of the main threats.

At such times of change, study of these high altitude areas and wetland ecosystems must be taken up on a priority basis to enhance understanding on the biodiversity values, risk factors as well as the ecosystem services being provided. At the same time, the state Government's initiative for wetland conservation through a participatory framework with involvement of communities that has been tried and tested in Tsomgo Lake also needs to be expanded to cover other wetland complexes of the state.

Acknowledgements

We would like to thank our partners Forests, Environment and Wildlife Management Department (Government of Sikkim), Rural Management and Development Department (Government of Sikkim), Police Check Post (Gangtok), *Tsomgo Pokhri Sanrakshan Samiti*, Lachen Dzumsa, Lachen Tourism Development Committee and our other partner NGOs for their constant support towards the lake conservation initiative in Sikkim. We would also like to acknowledge Sri Sangay Sherpa and Sri Chewang Lachenpa for their tireless efforts in the field. We are grateful to Dr. Dipankar Ghose, Director, Landscape Programme, and colleagues at WWF-India, Secretariat, for their constant encouragement.

References

Chanchani P, Rawat GS and Goyal SP. 2010. Unveiling a wildlife haven: status and distribution of four Trans-Himalayan ungulates in Sikkim, India. *Oryx*, 44(3): 366 - 375.

Kumar B and Prabhu M. 2012. Impacts of climate change: glacial lake outburst floods (GLOFS). In. *Climate Change in Sikkim Patterns, Impacts and Initiatives* (Arrawatia ML and Tambe S. eds.). Information and Public Relations Department, Government of Sikkim, Gangtok, pp. 81-102.

Myers N, Mittermeier RA, Mittermeier CG. *et al.*, 2000. Biodiversity hotspots for conservation priorities. *Nature*, 403: 853-858.

Olson DM and Dinerstein E. 1998. The Global 2000: Priority ecoregions for global conservation. *Annals of the Missouri Botanical Gardens*, 89: 199-224.

Rawat GS and Tambe S. 2011. Sikkim Himalaya: Unique features of biogeography and ecology. In. *Biodiversity of Sikkim: Exploring and Conserving a Global Hotspot* (Arrawatia ML and Tambe S. eds.). Information and Public Relations Department, Government of Sikkim, Gangtok. pp.1-13.

Sharma N, Pradhan S, Arrawatia ML and Shrestha DG. 2010. Study on type and distribution of wetlands of Sikkim Himalayas using satellite imagery with remote sensing and GIS technique. *Proc. Lake 2010: Wetlands, Biodiversity and Climate Change*. pp.1-10.

Tambe S, Ghose D and Arrawatia ML. 2008. Designing participatory policy framework for the conservation of lakes in the Sikkim Himalaya. In. *Proceedings of Taal 2007: The 12th Lake Conference* (Sengupta M and Dalwani R. eds.), pp. 2056-2060.

Index

D

E

F

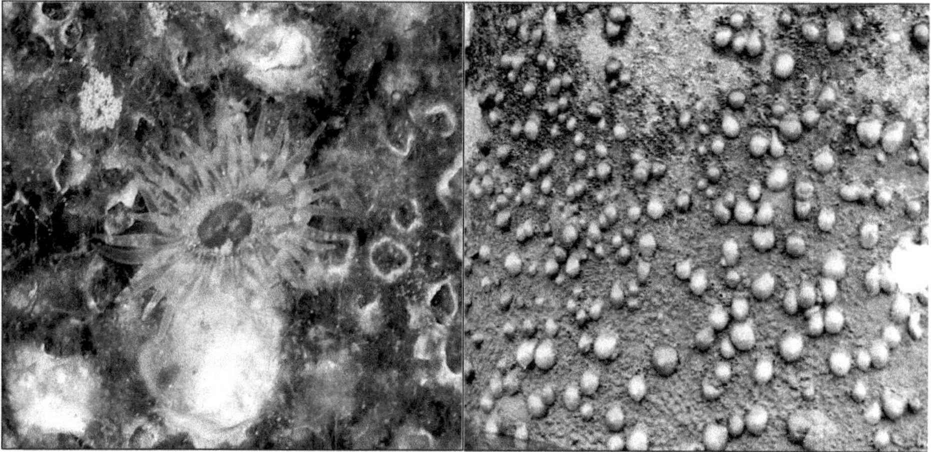

1. ***Diadumene schilleriana*** **with open tentacle in High tide**

2. **A colony of *Diadumene schilleriana* with closing their tentacle in low tide**

3. ***Paracondylactis sinensis*** **opening its tentacles at low tide at sandy habitat**

4. ***Edwardsia jonesii***

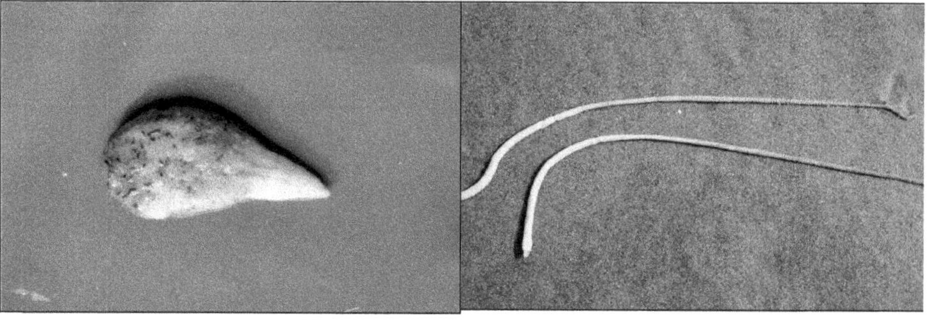

5. ***Cavernularia*** **sp.**

6. ***Virgularia*** **sp.**

Figure 5.1: Some Cnidarians from wetland ecosystems of West Bengal. (p.71)

Some Commercially Important Indigenous Ornamental Fishes of West Bengal

Amblypharyngodon mola

Notopterus notopterus

Badis badis

Danio rerio

Colisa fasciata

Colisa lalia

Chanda nama

Chanda ranga

(p.88)

Some Commercially Important Indigenous Ornamental Fishes of West Bengal

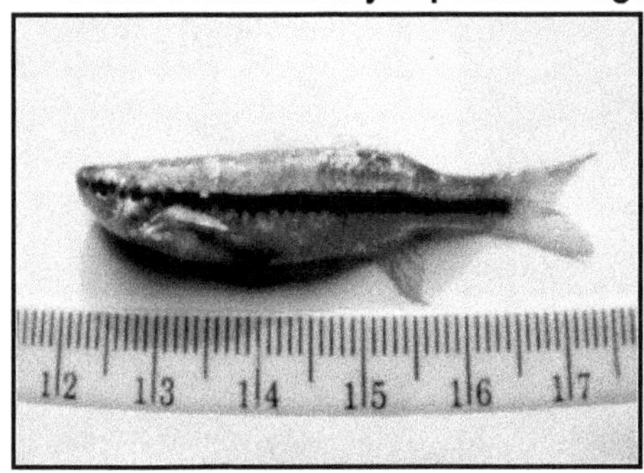

Esomus danricus

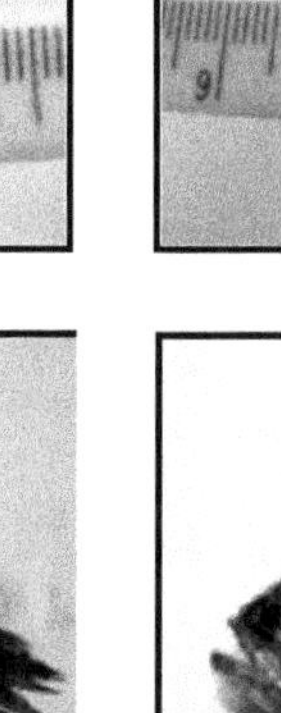

Channa marulius

Channa striata

Nandus nandus

Lepidocephalichthys guntea

Mystus vittatus

Macrognathus pancalus

(p.89)

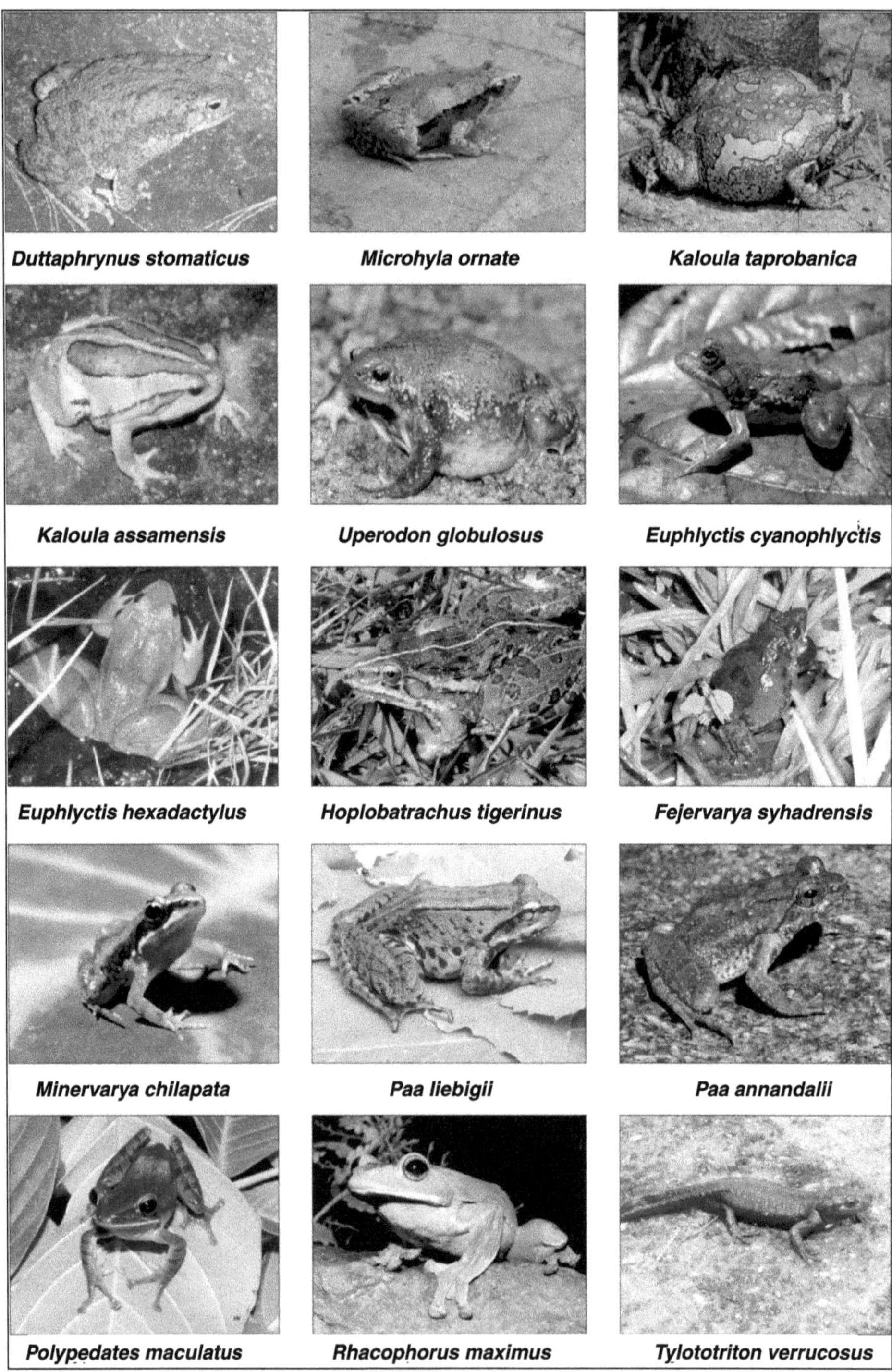

Figure 8.1: Some Common Amphibians of Wetland Ecosystems of West Bengal. (p. 119)

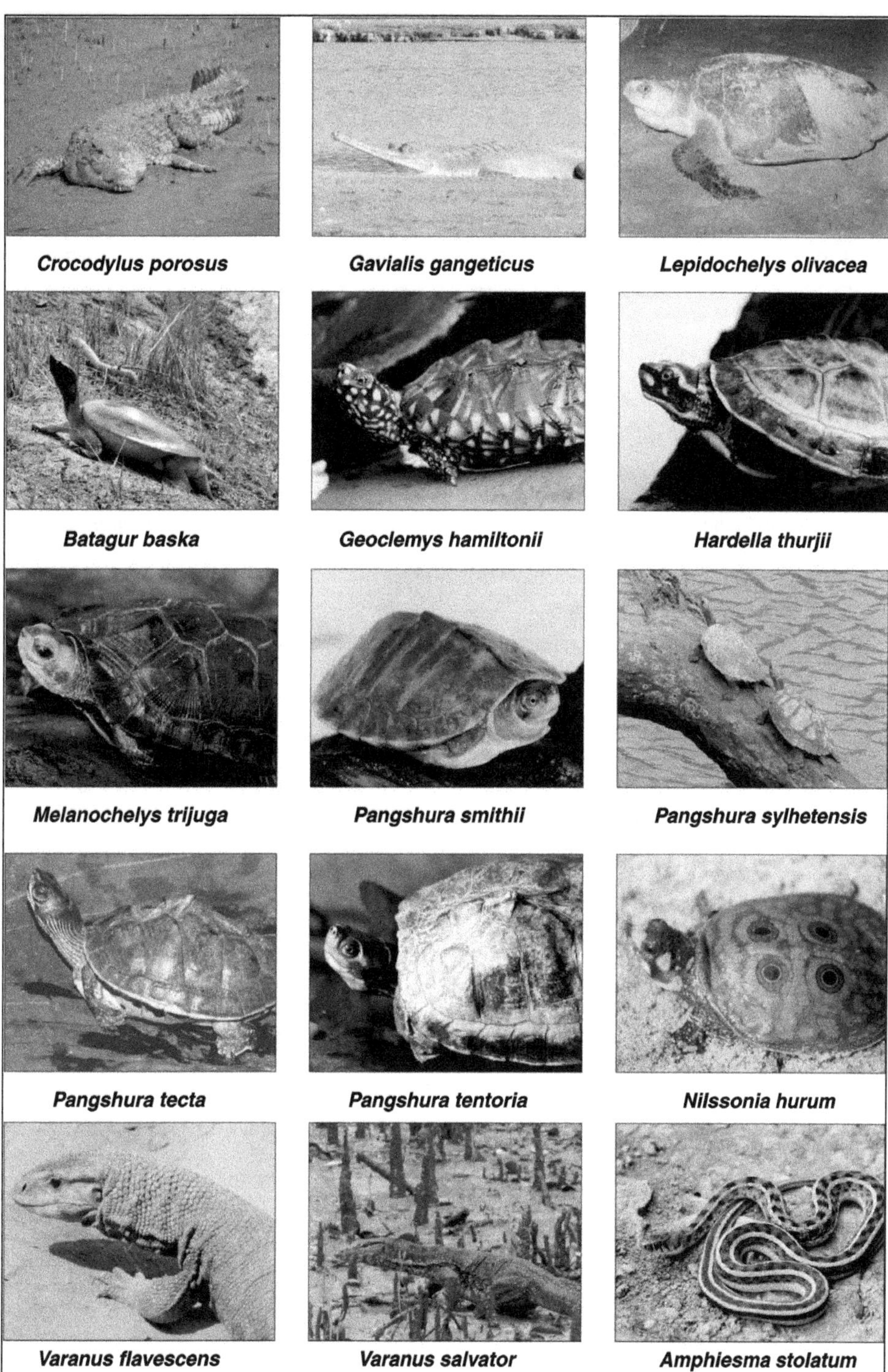

Crocodylus porosus | *Gavialis gangeticus* | *Lepidochelys olivacea*

Batagur baska | *Geoclemys hamiltonii* | *Hardella thurjii*

Melanochelys trijuga | *Pangshura smithii* | *Pangshura sylhetensis*

Pangshura tecta | *Pangshura tentoria* | *Nilssonia hurum*

Varanus flavescens | *Varanus salvator* | *Amphiesma stolatum*

Figure 9.1: Some Common Reptilian Fauna Associated with Wetland Ecosystems of West Bengal. (p. 133)

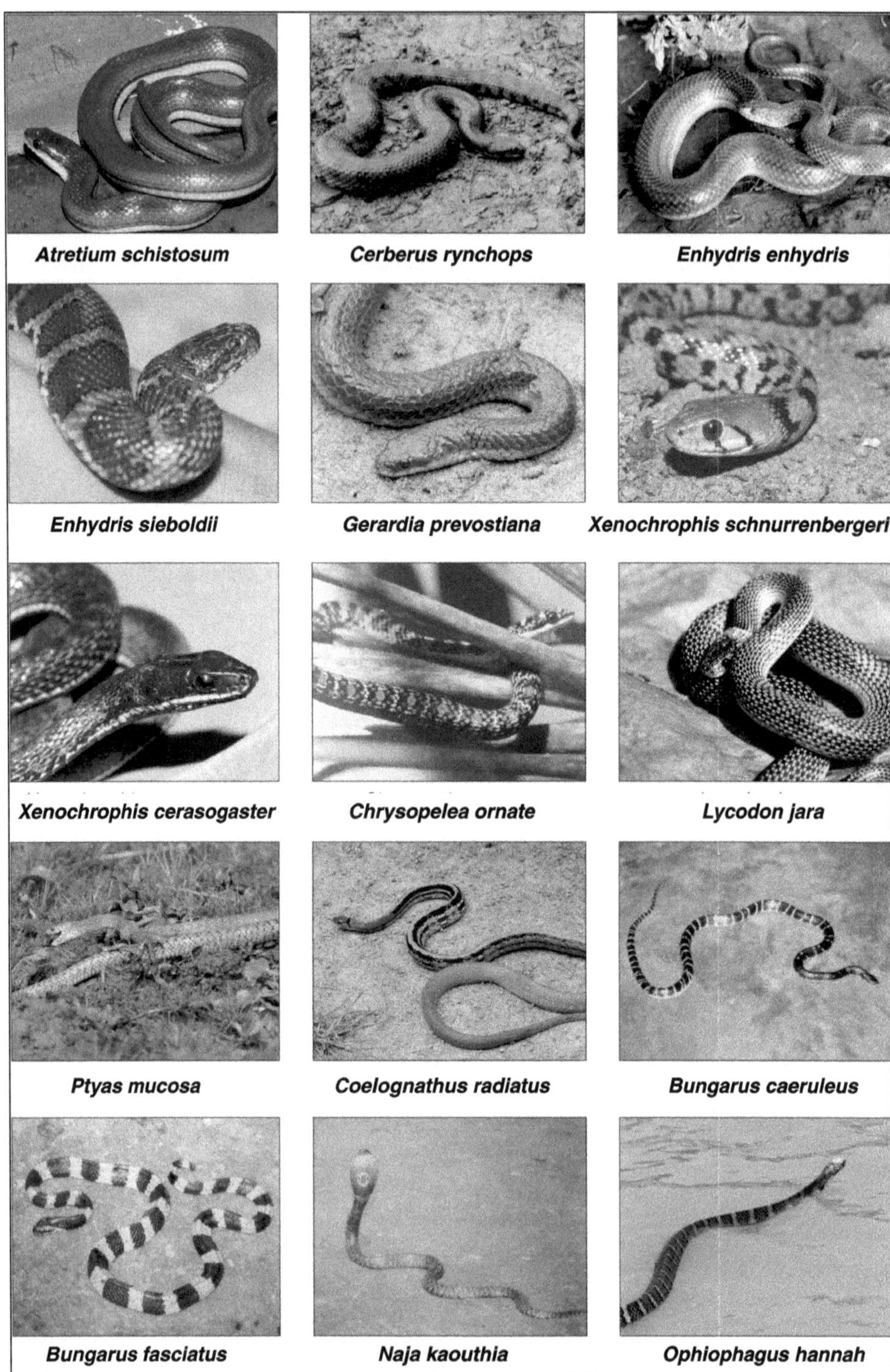

Figure 9.2: Some Common Reptilian Fauna Associated with Wetland Ecosystems of West Bengal. **(p. 134)**

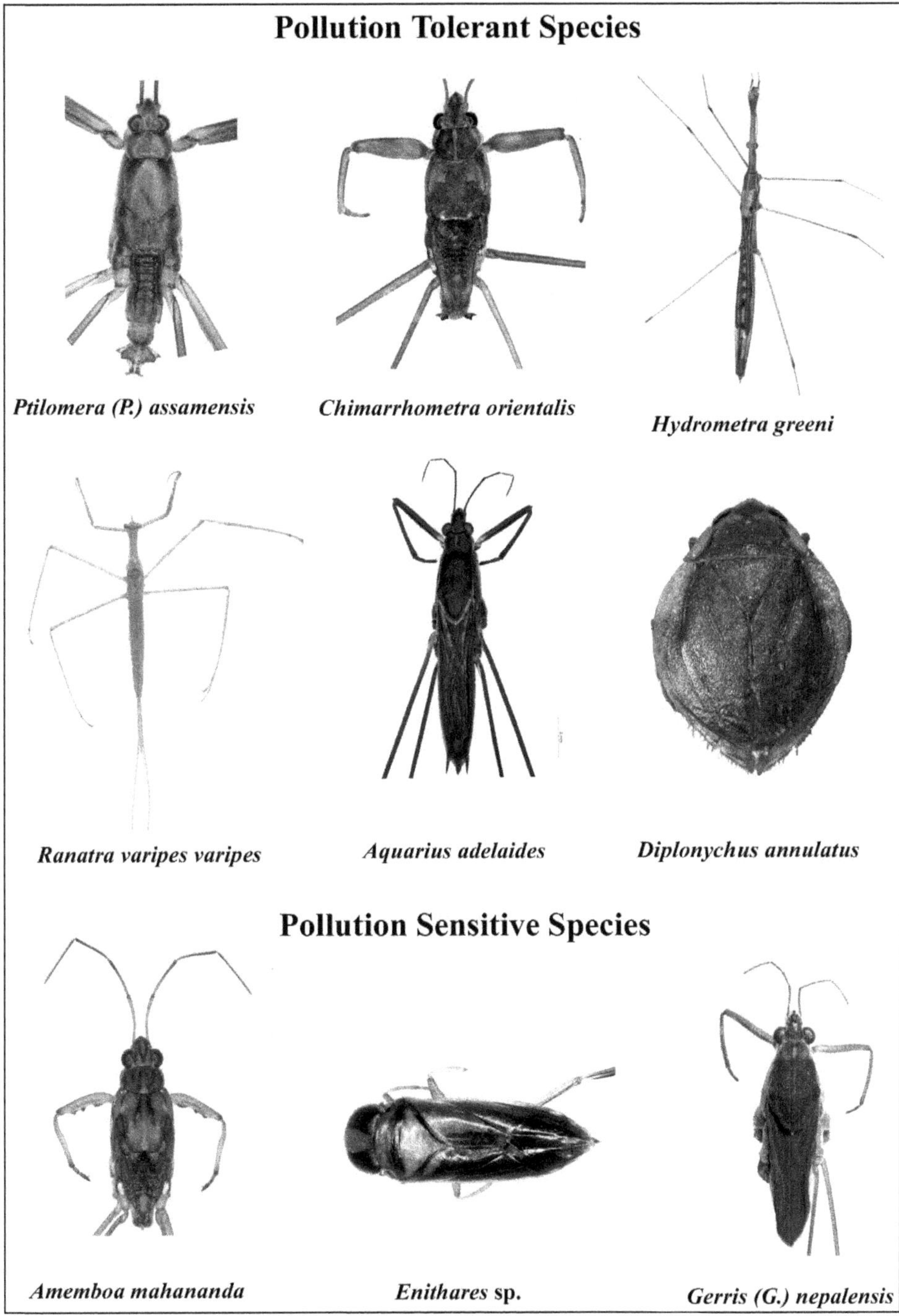

Figure 13.15 (p. 206)

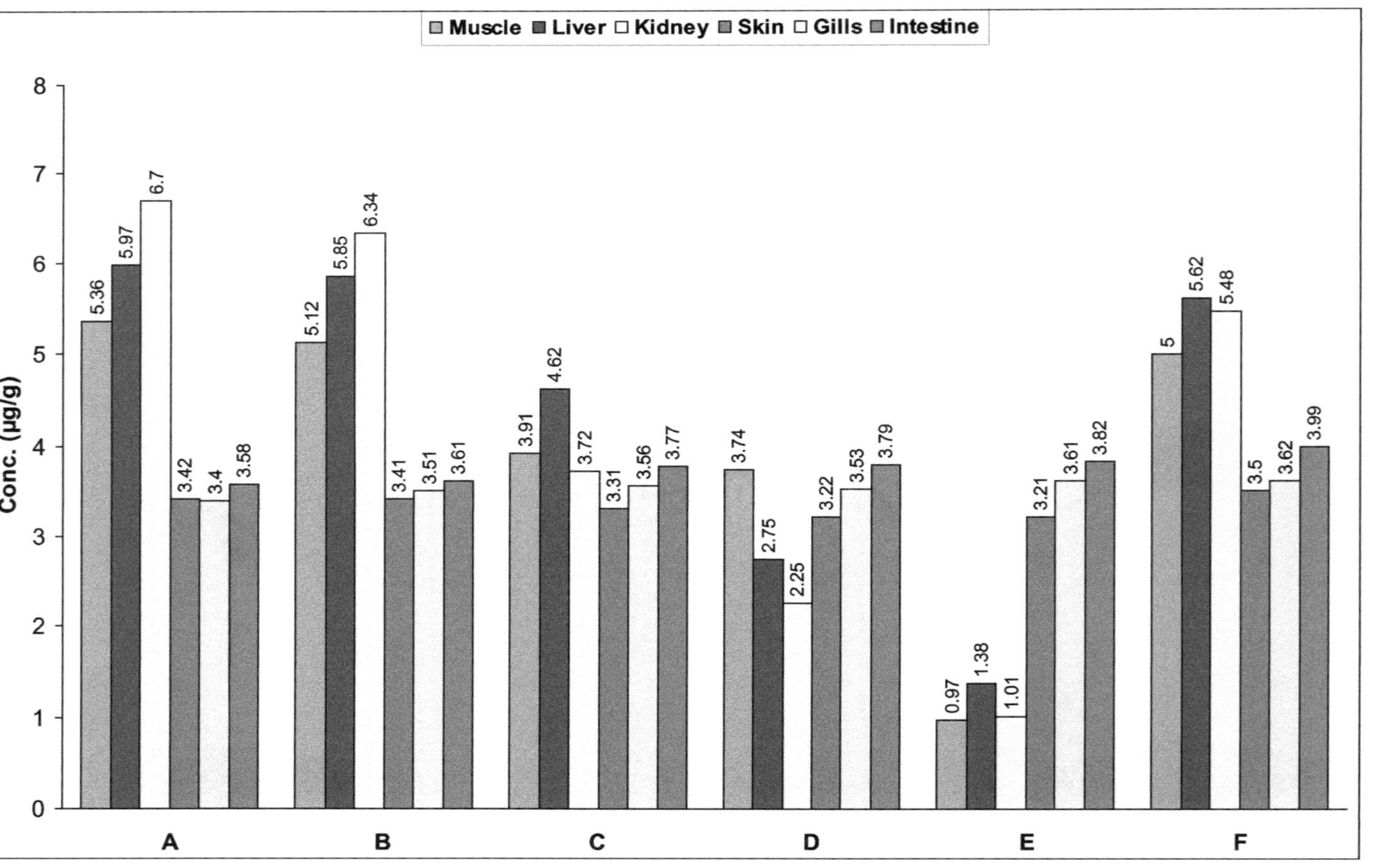

Figure 21.5: Accumulation of Cadmium in different Tissues of *Oreochromis nilotica* of Various Body Weights (A = 25 g; B = 50 g; C= 100 g; D = 200 g; E = 400 g; F = 1000 g) (Maity and Banerjee, 2010). (p. 345)

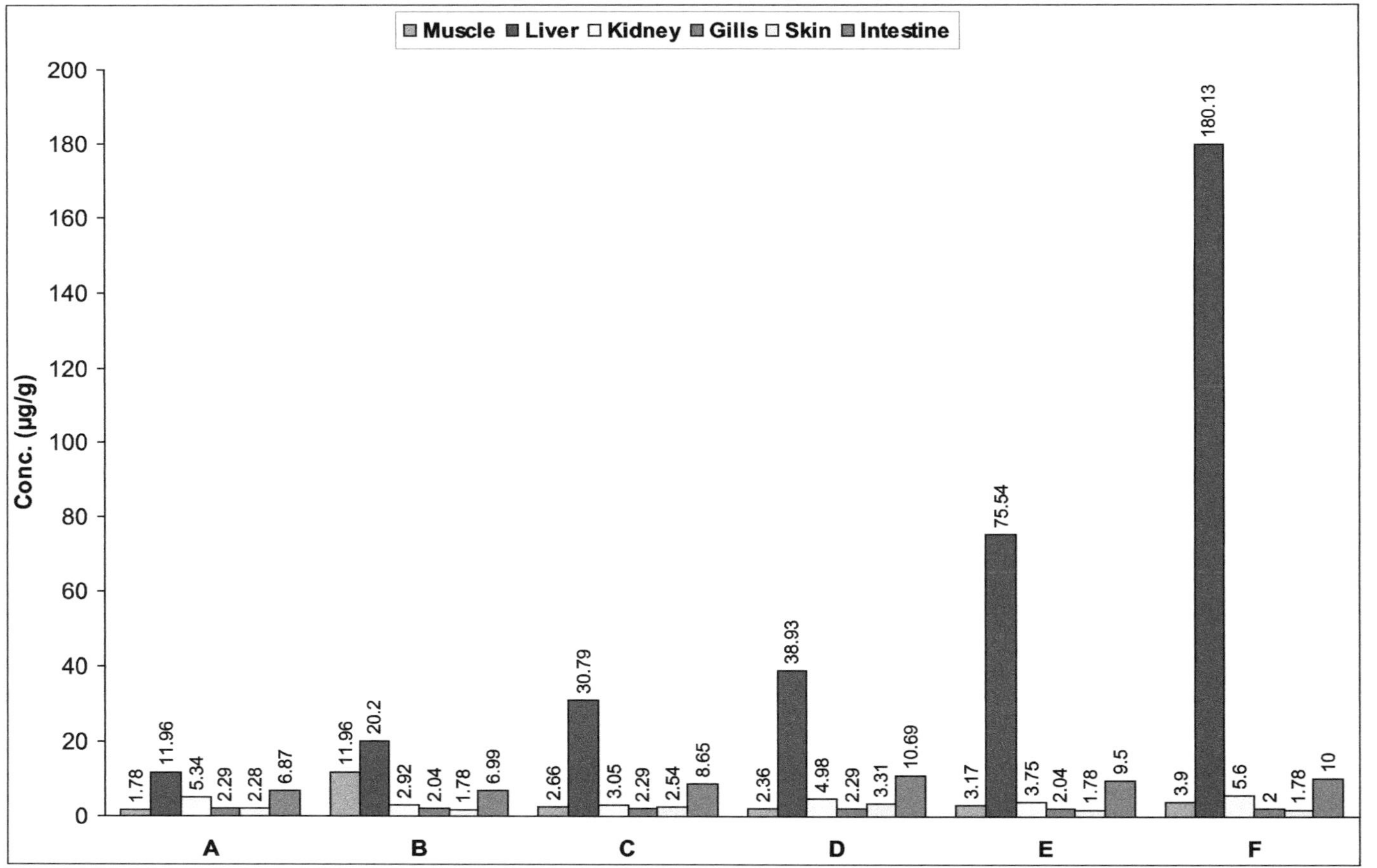

Figure 21.6: Accumulation of Copper in different Tissues of *Oreochromis nilotica* of Various Body Weights (A = 25 g; B = 50 g; C= 100 g; D = 200 g; E = 400 g; F = 1000 g) (Maity and Banerjee, 2010). (p. 348)

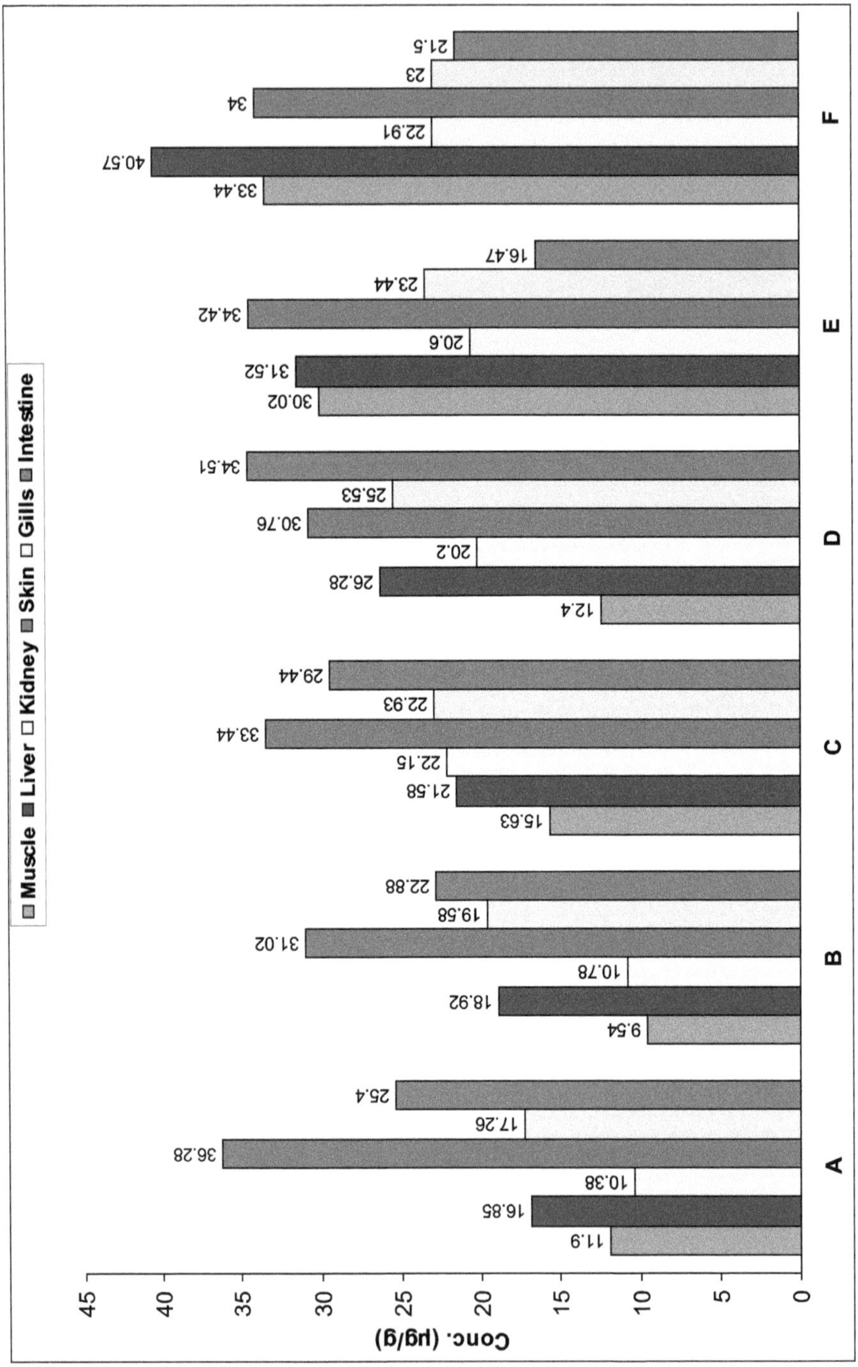

Figure 21.7. Accumulation of Zinc in different Tissues of *Oreochromis nilotica* of Various Body Weights (A=25g; B=50g; C=100g; D=200g; E=400g; F=1000g) (Maity and Banerjee, 2010). (p. 349)

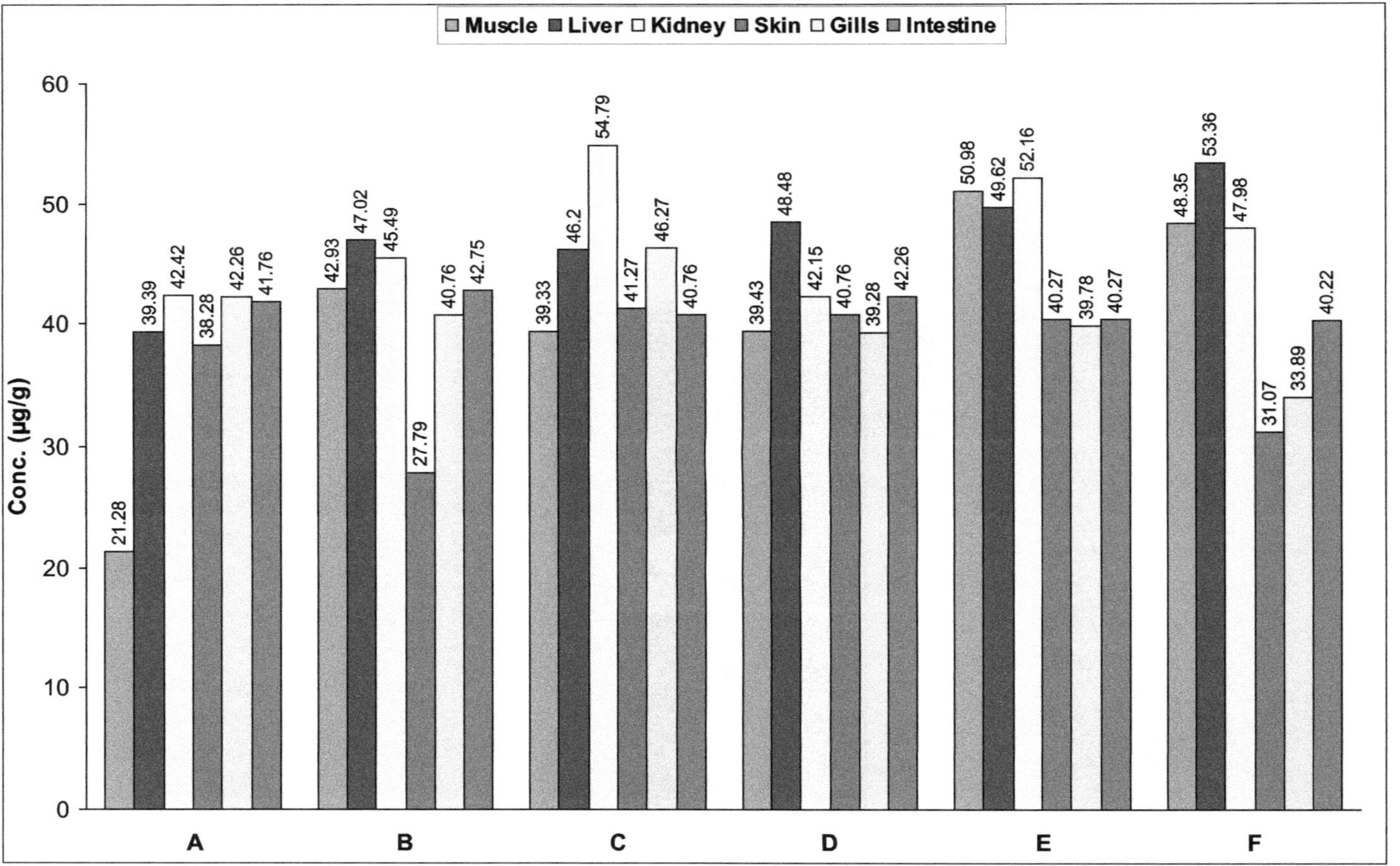

Figure 21.8: Accumulation of Lead in different Tissues of *Oreochromis nilotica* of Various Body Weights (A=25g; B=50g; C=100g; D=200g; E=400g; F=1000g) (Maity and Banerjee, 2010). (p. 350)

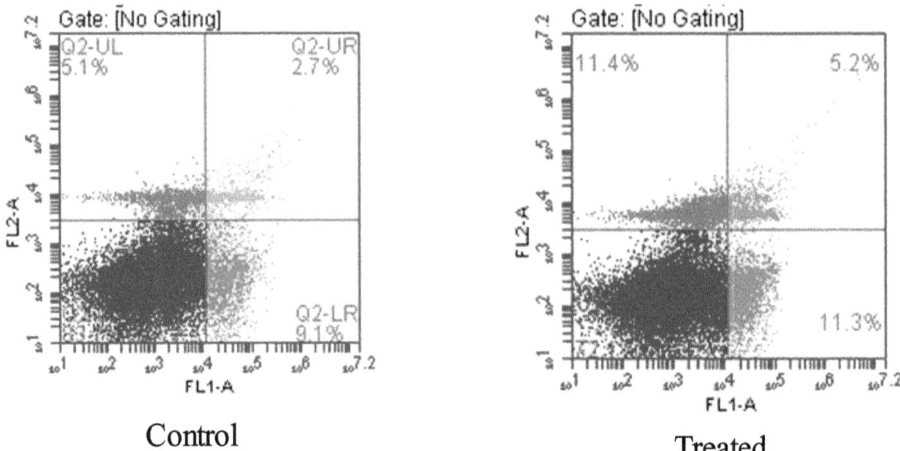

Figure 26.4: Dot Plot Presentation of FACS Analyses of Hemocyte Apoptosis of *B. bengalensis* Exposed to Fenvalerate (3.0 ppm/15 days).

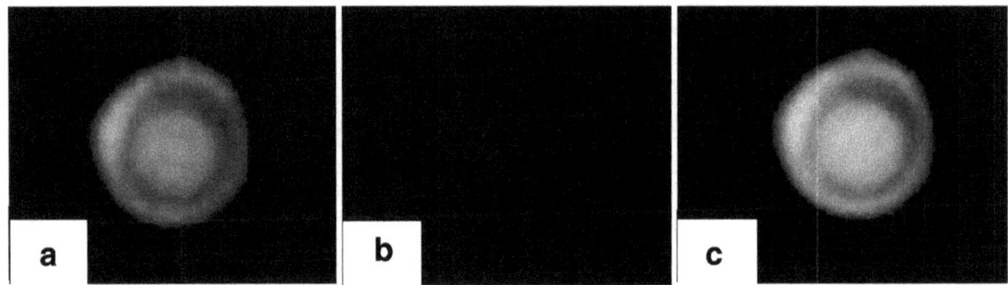

Figure 26.5: Immunofluorescent Detection of Apoptosis in the Hemocytes of *B. bengalensis* by Annexin-V-FITC and Propidium Iodide Staining (Magnification: 100×). Apoptotic hemocytes generate a relatively intense green (FITC) fluorescence (a) and low red (PI) fluorescence (b) which appears to be yellow upon colocalization (c). (p. 431)

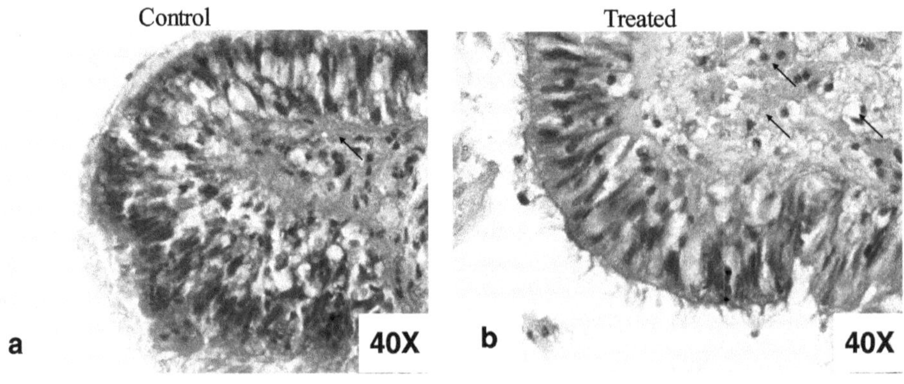

Figure 26.6: Histopathology of Labial Palps of *L. marginalis* Exposed to Cypermethrin (0.09 ppm/30days); a: Control, b: Treated exhibiting hypervacuolation, degeneration of muscle fibres. (p. 432)

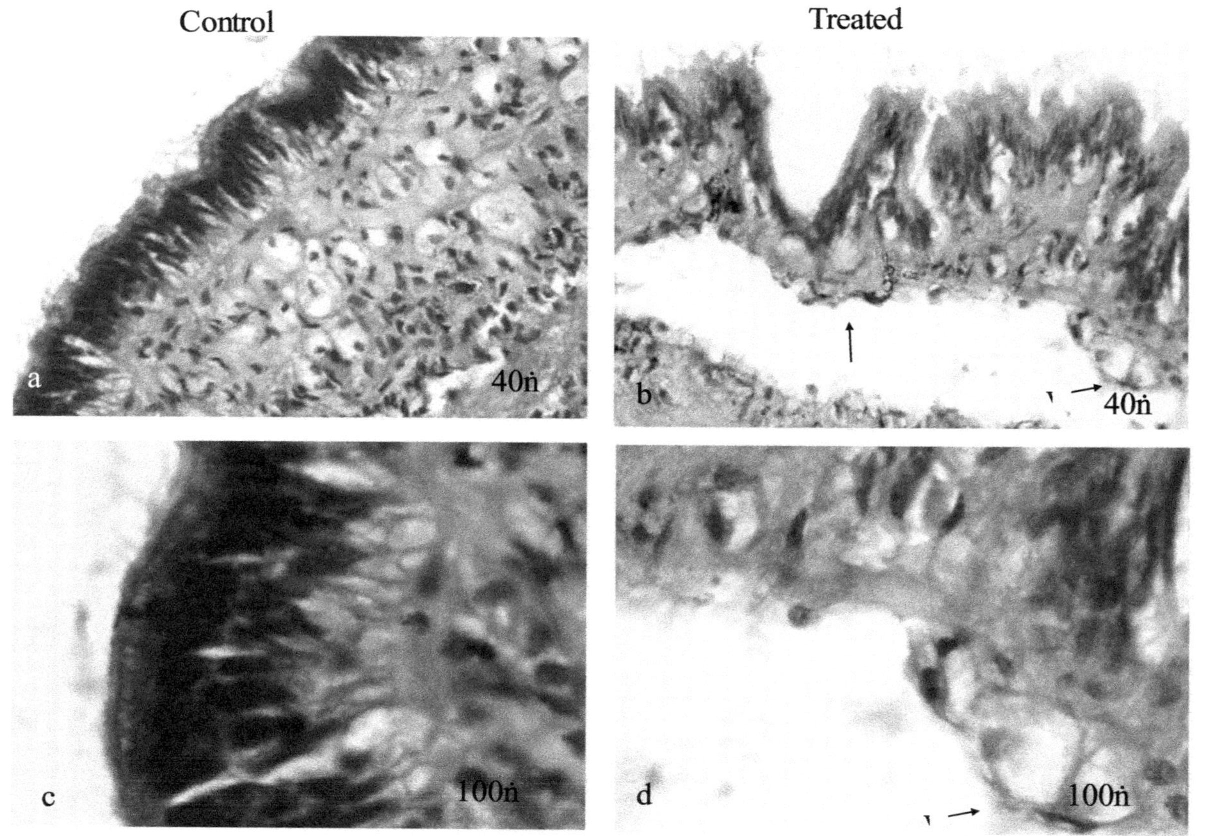

Figure 26.7: Histopathology of Antenne of *B. bengalensis* Exposed to Fenvalerate (3.0 ppm/96 hours); a, c: Control; b, d: Treated exhibiting vacuole formation (v), tissue disruption. (p. 433)

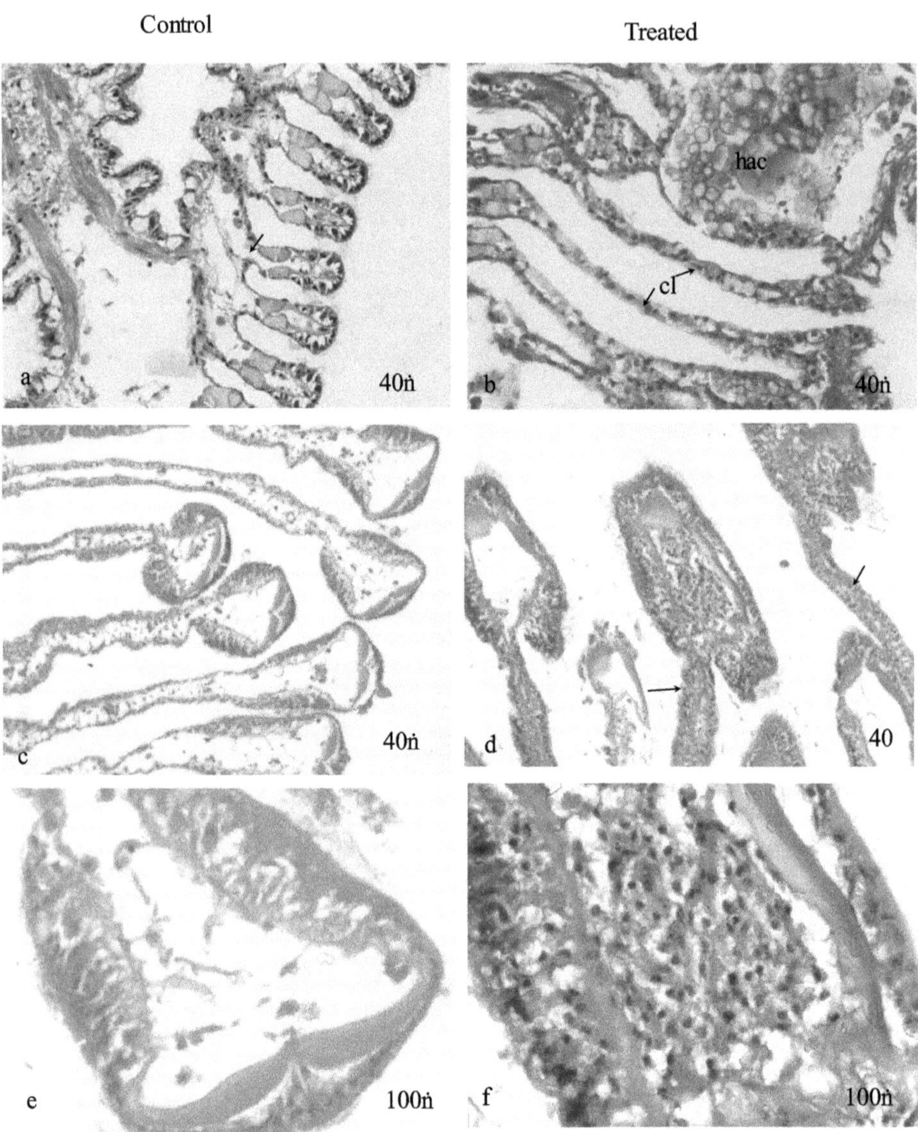

Figure 26.8: Transverse Sections of Gill of *L .marginalis* (a: Control, b: Treated with 0.09 ppm cypermethrin for 30days) and *B. bengalensis* (c, e: Control and d, f: Treated with 3.0 ppm fenvalerate for 96 hours). Pyrethroid exposure resulted in clogging of water channel (cl), fibrosis (fr) and appearance of hyperchromatic anaplastic (hac) cells. (p. 434)

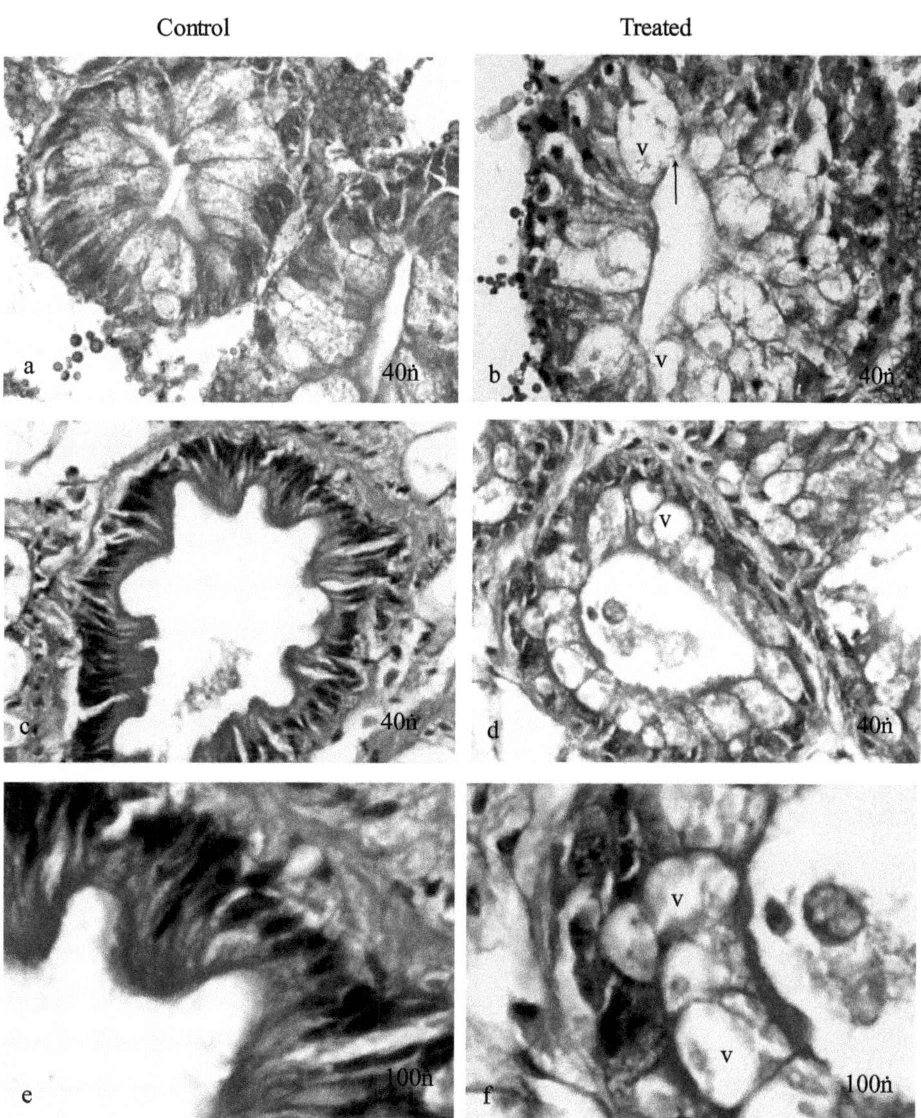

Figure 26.9: Transverse Sections of Digestive Gland of *L. marginalis* (a: Control, b: Treated with 0.09 ppm cypermethrin for 30days) and *B. bengalensis* (c, e: Control; d, f: Treated with 3.0 ppm fenvalerate for 96 hours). Pyrethroid exposure resulted in vacuolation (v) and necrotic tissue. (p. 435)

www.ingramcontent.com/pod-product-compliance
Ingram Content Group UK Ltd.
Pitfield, Milton Keynes, MK11 3LW, UK
UKHW021438280726
14060UKWH00001BA/134